Venomous Animals and Their Venoms
VOLUME II
Venomous Vertebrates

Contributors to This Volume

Moacyr de F. Amorim

Carlos A. Bonilla

D. K. Chaudhuri

Harold G. Cogger

John W. Daly

P. J. Deoras

Venancio Deulofeu

Anima Devi

B. N. Ghosh

Gerhard Habermehl

Bruce W. Halstead

Alphonse Richard Hoge

Norman Horner

E. Kaiser

Laurence M. Klauber

D. de Klobusitzky

Horst Linde

Wolfgang Luther

Bertha Lutz

S. R. Maitra

Kuno Meyer

H. Michl

Sylvia Alma R. W. D. L. Romano

G. Rosenfeld

Edmundo A. Rúveda

Wayne Seifert

Charles E. Shaw

Ernest R. Tinkham

E. R. Trethewie

Bernhard Witkop

VENOMOUS ANIMALS AND THEIR VENOMS

Edited by

WOLFGANG BÜCHERL
INSTITUTO BUTANTAN
SÃO PAULO, BRAZIL

ELEANOR E. BUCKLEY
WYETH LABORATORIES
PHILADELPHIA, PENNSYLVANIA

VOLUME II *Venomous Vertebrates*

ACADEMIC PRESS   New York · London   1971

ACADEMIC PRESS, INC.
111 Fifth Avenue, New York, New York 10003

United Kingdom Edition published by
ACADEMIC PRESS, INC. (LONDON) LTD.
Berkeley Square House, London W1X 6BA

LIBRARY OF CONGRESS CATALOG CARD NUMBER: 66 - 14892

PRINTED IN THE UNITED STATES OF AMERICA

Contents

VENOMOUS SNAKES

Chapter 21. Pharmacology and Toxicology of the Venoms of Asiatic Snakes
D. K. CHAUDHURI, S. R. MAITRA, AND B. N. GHOSH

Chapter 22. The Story of Some Indian Poisonous Snakes
P. J. DEORAS

Chapter 23. The Venomous Snakes of Australia and Melanesia
HAROLD G. COGGER

Chapter 24. The Pharmacology and Toxicology of the Venoms of the Snakes of Australia and Oceania

E. R. TRETHEWIE

Chapter 25. The Pathology, Symptomatology, and Treatment of Snake Bite in Australia

E. R. TRETHEWIE

Chapter 26. Classification, Distribution, and Biology of the Venomous Snakes of Northern Mexico, the United States, and Canada: Crotalus and Sistrurus

LAURENCE M. KLAUBER

Chapter 27. The Coral Snakes, Genera Micrurus and Micruroides, of the United States and Northern Mexico

CHARLES E. SHAW

VENOMOUS SAURIANS AND BATRACHIANS

Chapter 37. Venomous Toads and Frogs

BERTHA LUTZ

Chapter 38. The Basic Constituents of Toad Venoms

VENANCIO DEULOFEU AND EDMUNDO A. RÚVEDA

Chapter 39. Chemistry and Pharmacology of Frog Venoms

JOHN W. DALY AND BERNHARD WITKOP

Chapter 40. Collection of Toad Venoms and Chemistry of the Toad Venom Steroids

KUNO MEYER AND HORST LINDE

List of Contributors

Numbers in parentheses indicate the pages on which the authors' contributions begin.

Moacyr de F. Amorim, Department of Pathological Anatomy and Physiology, Escola Paulista de Medicina and Section of Pathological Anatomy, Instituto Butantan, São Paulo, Brazil (319)

Carlos A. Bonilla, Department of Biology, Midwestern University, Wichita Falls, Texas (203)

D. K. Chaudhuri, Department of Biochemistry, University Colleges of Science and Technology, Calcutta, India (3)

Harold G. Cogger, Australian Museum, Sydney, Australia (35)

John W. Daly, National Institute of Arthritis and Metabolic Diseases, National Institutes of Health, Bethesda, Maryland (497)

P. J. Deoras, Haffkine Institute, Bombay, India (19)

Venancio Deulofeu, Departmento de Química Orgánica, Facultad de Ciencias Exactas y Naturales, and Cátedra de Fitoquímica, Facultad de Farmacia y Bioquimica, Universidad de Buenos Aires, Buenos Aires, Argentina (475)

Anima Devi, Department of Physiology, University of Calcutta, Calcutta, India (175)

B. N. Ghosh, Department of Pure Chemistry, University Colleges of Science and Technology, Calcutta, India (3)

Gerhard Habermehl, Institut für Organische Chemie der Technische Hochschule, Darmstadt, Germany (569)

Bruce W. Halstead, World Life Research Institute, Colton, California (587)

Alphonse Richard Hoge, Department of Herpetology, Instituto Butantan, São Paulo, Brazil (211)

Norman Horner, Department of Biology, Midwestern University, Wichita Falls, Texas (203)

E. Kaiser, Hochschule für Bodenkultur, Institut für Chemie, Vienna, Austria (307)

Laurence M. Klauber,* San Diego, California (115)

D. de Klobusitzky, State University of São Paulo, National Academy of Medicine of Brazil, São Paulo, Brazil (295)

Horst Linde, Institute of Pharmaceutical Chemistry, University of Basel, Basel, Switzerland (521)

Wolfgang Luther,† Department of Zoology, Technological University, Darmstadt, Germany (557)

Bertha Lutz, Museu Nacional, Rio de Janeiro, National Research Council of Brazil, and the Federal University of Rio de Janeiro, Brazil (423)

S. R. Maitra, Department of Physiology, University Colleges of Science and Technology, Calcutta, India (3)

Kuno Meyer, Institute of Pharmaceutical Chemistry, University of Basel, Basel, Switzerland (521)

H. Michl, Hochschule für Bodenkultur, Institut für Chemie, Vienna, Austria (307)

Sylvia Alma R. W. D. L. Romano, Department of Herpetology, Instituto Butantan, São Paulo, Brazil (211)

G. Rosenfeld, Hematology Laboratory, Instituto Butantan, São Paulo, Brazil (345)

Edmundo A. Rúveda, Departamento de Quimica Orgánica, Facultad de Ciencias Exactas y Naturales, and Cátedra de Fitoquimica, Facultad de Farmacia y Bioquimica, Universidad de Buenos Aires, Buenos Aires, Argentina (475)

Wayne Seifert, Department of Biology, Midwestern University, Wichita Falls, Texas (203)

Charles E. Shaw, Curator of Reptiles, San Diego Zoological Gardens, San Diego, California (157)

Ernest R. Tinkham, Desert Naturalist, Indio, California (387, 415)

E. R. Trethewie, Department of Physiology, University of Melbourne, and Royal Melbourne Hospital, Melbourne, Australia (79, 103)

Bernhard Witkop, National Institute of Arthritis and Metabolic Diseases, National Institutes of Health, Bethesda, Maryland (497)

* Deceased.
† Deceased.

Preface

The modern trend in the study of the wide field of venomous animals and their venoms is directed toward basic research that emphasizes zoological ecology, biochemistry, pharmacology, and immunobiology. The increasing importance of this development, stimulated also by the political and industrial expansion into the undeveloped areas of the tropics, is reflected by the great number of publications on venoms of animal origin. Every year about 10,000 papers are published on this subject, scattered in hundreds of journals in many languages, thus making it impossible for the individual scientist to keep abreast of new developments.

The present treatise is an attempt to offer, for the first time, a comprehensive presentation of the entire field of the venomous members of the animal kingdom, of the chemistry and biochemistry of the venoms, of their pharmacological actions and their antigenic properties. The medical aspects, both symptomatology and therapy, are included. The work is the result of close cooperation of seventy-two scientists from thirty-two countries on all continents. The authors are highly qualified specialists in their specific areas of research; their concerted efforts make this work one of unusual scope and depth.

Volume I of this three-volume work is devoted to venomous mammals and begins the extensive section on snakes. Volume II continues the discussion on snakes and includes the saurians, batrachians, and fishes. The venomous invertebrates, such as insects, centipedes, spiders, and scorpions, venomous molluscs, and marine animals, will be considered in Volume III.

The interdisciplinary aspects of the subject necessitated assigning several chapters to a single group of animals and offering separate sections covering the zoological, chemical, and biomedical points of view.

It is hoped that these volumes will be valuable reference works and stimulating guides for future research to all investigators in the field; they will also serve the needs of physicians and veterinarians seeking information on the injuries caused by venomous animals. The volumes should also facilitate the teaching of this important topic and should prove a welcome source of instruction to students and to the large group of laymen interested in this fascinating field of natural science.

The editors wish to thank the authors for their cooperation and for generously contributing the results of their work and experience.

Our thanks are also due to the staff of Academic Press for helpful advice, patience, and understanding.

We cannot conclude this preface without expressing our gratitude to Professor Dionysio de Klobusitzky who conceived the idea of this treatise and outlined its initial organization.

WOLFGANG BÜCHERL
ELEANOR E. BUCKLEY

Contents of Other Volumes

VENOMOUS COELENTERATES, ECHINODERMS, AND ANNELIDS

Introduction

The so-called venomous animals described in these volumes possess at least one or more venom glands and mechanisms for excretion or extrusion of the venom, as well as apparatus with which to inflict wounds or inject the venomous substances. The venom often may be injected at will. These animals have been characterized by several authors as being "actively venomous." The "passively venomous" species have venom glands and venom-excreting ducts, but lack adequate apparatus for inflicting wounds or injecting venom (toads, frogs, and salamanders).

In their struggle for life, all venomous animals seem to be rigorously extroverted. Their energies are directed against the other animal and vegetable organisms in their environment. All the venom glands of these animals are of the exocrine type. Their venoms produced by special epithelial cells and stored in the lumina of glands are always extruded to the outer world, generally by biting or stinging such as is the case with shrews, serpents, saurians, some fishes, stinging social insects, scolopendrids, spiders, scorpions, molluscs, some echinoderms, and worms. Other animals envenomate their victims by direct bodily contact such as is true with caterpillars, a few moths, with several representatives of Coleoptera (*Paederus*, etc.), certain hydroids, jellyfishes, sea anemones, urchins, cucumbers, starfishes, and a few marine worms. The venomous compounds of toads and salamanders act generally in direct contact with mucous membranes (eyes, lips, or throats). All venomous animals possess characteristics which distinguish them from other members of the animal kingdom. Often venomous animals are hunters, predators, solitaries, and also enemies of other members of the animal kingdom. There are exceptions, of course, such as the social Hymenoptera.

The wounding apparatus is located on the head, on the hind portion of the body, or over the entire exposed surface of the animal. In shrews, serpents, Gila monsters, and some molluscs the venom apparatus is inside the mouth. The venom glands are in fact salivary glands; the bite is inflicted by modified teeth equipped with venom canals. In scolopendrids and spiders, the venom system is situated outside the mouth, but is in close proximity to it, and is designed for protection and acquisition of food. A strange situation is present in the scorpion: the venomous mechanism is found in the last segment of the body, the "cauda venenum." In fact, the scorpion sting must be considered "peribuccal." The scorpion is able to move its tail sufficiently far in front of its head to kill prey before eating it.

In the venomous Hymenoptera, such as ants, bees, wasps, and hornets, the wounding apparatus and the venom glands are also situated in the last segments of the abdomen, far from the mouth. The stinging mechanism often may function primarily as an ovipositor, having no connection with the mouth, and its venom-injecting function may only be secondary (honeybee queen).

In some venomous fishes and bristleworms, the venomous organs may be distributed over certain exposed portions of the body or may cover more or less the entire body surface, as in caterpillars, some echinoderms. and coelenterates, with no relation to the mouth.

The location of the venom system and the transformation of certain organs into venom-conducting channels may lead us to theorize as to the significance of venom in the animal kingdom. Why do venomous animals exist? What is the primary function of venom? Are venoms present principally for digestion of food, and is the wounding apparatus intended for self-defense and even attack in the never-ending struggle for survival? Is the stinging designed mainly for oviposition, or for defense and attack, and is it combined with the mechanism for obtaining food and provision for offspring, as is true for all the solitary wasps? Thus, the role of venoms immediately appears very complex.

Shrews, serpents, scolopendrids, spiders, scorpions, solitary wasps, and some coelenterates are exclusively carnivorous, but they never feed on an animal that is already dead, They are predators and active hunters, and they capture and kill their prey. The social wasps, bees, hornets, and caterpillars are exclusively herbivorous; other venomous animals may be omnivorous, i.e., they will feed on creatures that have died of other causes.

The venom and wounding apparatus must also be considered in relation to sex, particularly in venomous adult insects such as bees, wasps, and hornets. Only the adult female Hymenoptera are poisonous, not the adult males. In all other venomous animals both sexes may be equally poisonous, or the males, which often are much smaller, may do less serious harm, as is true of most spider species.

Consideration of the localization of the venom apparatus, the mode with which these animals take their prey or their food, and the fact that often only one sex bears a venom-conducting apparatus may guide us to another very important question: For what purpose is the venom used?

Toads, venomous frogs, salamanders, and other "passively" venomous animals certainly may use their toxic products for self-defense. Often these animals may not rely entirely on their venomous power, but may prefer to use other protective methods such as mimicry, flight, and concealment. Caterpillars and other Lepidoptera larvae are also in this category. The latter procure food from plants, and desire peace from other animals.

One habit of several solitary wasps is rather curious: They use their stinging apparatus to paralyze spiders and other insects. Then they bring the prey to the nest, deposit an egg over the body of it, and close the orifice of the nest. The wasp larva, hatched a few weeks later, thus is provided with fresh food. These wasps possess a nerve- or muscle-paralyzing venom with long-lasting effect and they may attack in order to protect their offspring. The social Hymenoptera, such as the bee, wasp, and hornet, may use the venom apparatus primarily for defense against enemies, even against other groups of the same family. Also, they may attack and kill, e.g., the females of bees kill the male after fertilization of the new queen. A newly hatched queen bee kills all the other queens present in the hive. Thus the stinging apparatus and venom have both defensive and offensive functions. The venomous fishes, coelenterates, and echinoderms, as well as the bristleworms, use the wounding mechanism for self-defense.

It is curious that in all these animals—toads, salamanders, bees, wasps, hornets, caterpillars, some fishes, molluscs, sea cucumbers, urchins, starfishes, sea anemones, and jellyfishes—the venom and the biting or stinging system have nothing to do with the acquisition of food. Consequently, the venom apparatus will have nothing in common with the digestive or salivary organs.

In scolopendrids, spiders, scorpions, venomous snakes, Gila monsters, and venomous shrews, the venom apparatus and the wounding system are designed primarily for food acquisition, and not so much for the predigestion of food. This is especially true of scorpions. Their venoms are paralyzing, not digestive agents. They use the sting only when the prey is large and vigorous in defending itself, as spiders. Small animals are captured directly with the pedipalps, and immediately killed and eaten; the sting is not needed. Scolopendrids and spiders use the wounding apparatus in two ways: to hold the prey and introduce it into the mouth, or, when resistance is offered, to inject and kill the prey with the venom. The salivary function of venom in scolopendrids, spiders, and epecially in scorpions may be questionable.

The situation appears to differ with snakes, venomous saurians, and In-sectivora. Since the venom glands and the venom-injecting apparatus are found in the mouth, with phylogenetic transformation of a few teeth, and the glands may be true salivary glands, with or without digestive ferments and enzymes, one might think that the main purposes of the venom mechanism are the capture of prey and the partial breakdown of body tissues. On the other hand, it is also true that venomous snakes may be force-fed with rats, birds, and other small animals, which they do not envenomate but which they digest very well. Without the venom apparatus it may be very difficult or even impossible for them to obtain their food. Venom may also activate the digestive processes in some manner, but probably it is not necessary for this purpose. Scholopendrids, scorpions, spiders, snakes, venomous saurians, and shrews may be considered primarily of the offensive type, their venom apparatus being used for the capturing of food; secondarily, of course, they use such apparatus for self-defense.

Exact knowledge of the biological habits of venomous animals would provide more accurate answers as to the real purpose of venoms. Too little is known about this broad subject.

Another very important issue to be clarified concerns the intensity of action of the venoms of all species. For example, a venom of one species of snake may be several times more active in rats, mice, and birds than in other animals. Human beings are extremely sensitive to certain animal venoms. One-tenth of one milligram of Loxosceles venom may seriously endanger human life. It is conservatively estimated that 40,000 to 50,000 people throughout the world may be killed every year by accidental contact with venomous animals. Every scientific effort must be directed toward the prevention of this tragedy.

WOLFGANG BÜCHERL

VENOMOUS SNAKES

Chapter 21

Pharmacology and Toxicology of the Venoms of Asiatic Snakes

D. K. CHAUDHURI, S. R. MAITRA, AND B. N. GHOSH

UNIVERSITY COLLEGES OF SCIENCE AND TECHNOLOGY, CALCUTTA, INDIA

I. INTRODUCTION

Broadly speaking the asiatic snakes can be classified into five families: Colubridae, Hydrophidae, Elapidae, Viperidae, and Crotalidae. The genera of the Hydrophidae family abound in the Persian Gulf, the Indian Ocean, the Bay of Bengal, the Straits of Malacca, the China Sea, the Philippines, and the Malay Archipelago. The genera belonging to the Elapidae family are mostly found in India, Ceylon, and Burma as well as Southeast Asian countries of South Vietnam, Southern China, Indonesia, Borneo, Malay, Japan, and the Philippines. The Viperidae and Crotalidae families are found in the larger areas of Asia like Siberia, Caucasus, Persia, China, India, Ceylon, the Himalayas, Arabia, Beluchistan, Afganistan, Ghana, Japan, and Southeast Asian countries. Some of these snakes are extremely poisonous and deaths due to snake bite among the local population are very frequent. In order to obtain a clear picture of the effect of the different components of snake venoms of Asiatic origin on the various organs of the body, studies on the pharmacology as well as the toxicology of snake

3

venom have been undertaken by different workers; a detailed discussion of this work follows.

On analyzing patients suffering from poisonous snakebite it has been found that different toxic phenomena are associated with different species of snakes. When a quantity of venom introduced into the animal body by a reptile is sufficient to produce fatal results, the venom causes its toxic action in two ways. The first is local and affects only the surrounding area of the bite while the second is more generalized and affects the circulatory or nervous system. With bites from Viperidae local disorders are extremely pronounced but with Colubridae they are almost insignificant. Consequently, a differential diagnosis of the type of snake may be made easily.

II. DETERMINATION OF THE LETHAL DOSE OF VENOMS

Venomous snakebites are not always fatal. Even in India where snakes are found in large numbers, only 30–40 % of snakebites results in death. By experimentation on animals with known doses of venom, it is possible to determine for each venom the minimum lethal dose per kilogram of body weight of the animal. Assuming that the toxicity of each type of venom is the same as that determined on mammals, the approximate lethal dose for man can be obtained. Based on the findings of different workers, Calmette (1908) compiled data on the minimum lethal doses of the different snake venoms on guinea pig as follows:

Minimum lethal dose (MLD) in 24 hours for a guinea pig weighing 600 to 700 gm:

Venom	Grams
Elapidae	
Naja tripudians	0.0002
Bungarus caeruleus	0.0006
Naja naja	0.003
Viperidae	
Vipera berus	0.04
Vipera russelli (Daboia)	0.001
Crotalidae	
Bothrops lanceolatus	0.02
Lachesis mutus (Surucucu)	0.02
Bothrops neuwiedi (urutu)	0.02
Trimeresurus flavoviridis	0.007
Agkistrodon contortrix	0.015

Minimum lethal dose of cobra venom in 24 hours for different animals:

Dog	0.0008	grams per kilogram
Rabbit	0.0006	grams per kilogram
Guinea pig	0.0004	grams per kilogram
Rat	0.0001	grams per kilogram

Mouse	0.000003	grams per 150 grams
Frog	0.0003	grams per 30 grams
Pigeon	0.0002	grams per 120 grams

Minimum lethal dose of venom of *Bungarus caeruleus* in 24 hours:

Frog	0.0006 grams
Rat	0.001 grams
Rabbit (by subcutaneous injection)	0.00008 grams per kilogram
Rabbit (by intravenous injection)	0.00004 grams

Minimum lethal dose of venom of *Vipera russelli (Daboia)* in 24 hours:

Rabbit (by intravenous injection)	0.00005 grams per kilogram

Venom of *Trimeresurus gramineus* (Green Pit Viper, India) in 24 hours:

Rabbit (by intravenous injection)	0.002 grams per kilogram.

The figures for minimum lethal dose provide a general idea of the toxicity of venoms of different species of snakes inhabiting the Asiatic region and indicate the extent of respective sensitiveness of the dog, cat, rabbit, guinea pig, rat, mouse, and frog with regard to the same venom. It will also be noticed that the minimum lethal dose is in no way proportional to the weight of these animals; for instance, weight for weight the monkey is much more susceptible to snake venom than the dog, and the ass is extremely sensitive (0.010 gm of cobra venom is sufficient to kill it); while the horse is less so and the pig is by far the most resistant. Rogers and Megaw (1904) deduced (see Table I) the toxicity of the snake venoms on human beings calculated on the basis of the lethal dose found for experimental animals.

It has been observed that the lethal dose of the snake venom obtained from the same species for the same experimental animal varies and that these variations in toxicity might possibly be due to the physical condition of the individual snake. It is also true that differences in toxicity are found in venoms in different specimens of the same species of snake collected at different times of the year. One of the authors (S.R.M.) found while working with toad heart perfusion experiments that the snake venoms of the same species collected in different times of the year were effective in stopping heart at concentrations ranging from 1/200 to 1/500. Calmette (1908) reported that the *Naja* and the *Lachesis* types of snakes reared in his laboratory produced venoms whose toxicity varied according to the length of time during which the snake was without food and the proximity of the molting period. Venoms collected near the moulting period were more toxic, and on evaporation a greater quantity of dry extract remained. There is evidence that immediately after moulting and also after a prolonged fast, venom is ten times more active than after a plentiful meal or before the molt. Such variations are found in all species of snakes. Even the same species inhabiting the different regions of the earth show varying toxicities.

TABLE I

CALCULATED MINIMUM LETHAL DOSE OF DIFFERENT SNAKE VENOM FOR MAN

Species of snake	Approximate dose at a single bite in milligrams	MLD for rat in milligrams	MLD for monkey in milligrams	Calculated fatal dose for man in milligrams
Common cobra (Naja naja)	211.3	0.12	2.4	15.0
King cobra (Naja bungarus)	100.0	0.05	1.8	12.0
Common Krait (Bungarus candidus)	5.4	0.20	0.15	1.0
Banded Krait (Bungarus fasciatus)	42.9	0.10	1.5	10.0
Indian Daboia (Vipera russelli)	72.0	2.5	7.5	42.0
Phoorsa (Echis carinatus)	12.3	1.0	0.5	5.0
Green Pit Viper (Trimeresurus gramineus)	14.1	0.5	16.0	100.00

III. EFFECT OF VENOM ON VARIOUS TISSUES OF THE ANIMAL

Histological lesions produced by snake venom studied by several workers showed profound action on the different organs of the body. Nowak (as cited by Calmette, 1908, pp. 182 and 183) observed that lesions in the liver and kidney were very pronounced. N. K. Sarkar (1947) noted changes in the heart muscle by perfusing the toad's heart with 1/200 dilution of crude cobra venom. He found that the muscle fibers were shortened considerably and many intervening spaces and fine striations of muscle fiber could be seen under the microscope. The same thing occurred when the heart was perfused with digitonin. However, the action of digitonin on the heart was reversible while that of cobra venom was not. No explanation of this was put forward by him.

Several workers like Ewing and Bailey (1900) and Lamb (1904, 1905) studied the effects of venom on nervous system. Their studies indicate that the intensity of the lesions depended primarily on the length of time between introduction of the venom into the body of the experimental animal and its death and secondarily on the venom. Viperidae venom has little effect on the nerve cell. N. K. Sarkar (1947) and Maitra (1949) experimented on the sciatic-gastrocnemius nerve or muscle of toad by dipping the nerve and muscle separately in two chambers constructed on a myograph board and excited the nerve or muscle in separate chamber from a secondary coil. They observed that nerve remained unaffected by the cobra venom solution, but muscle lost the functions of contraction and excitation by cobra venom solution. Their conclusion is that muscle and nerve endings might be affected by cobra venom but not the nerve trunk.

IV. EFFECT OF VENOM ON RED BLOOD CELLS

Action of venom on blood was studied by several workers. It was reported that the Elapidae venoms *(Naja naja, Naja tripudians)* kept the blood in the fluid state even after death, while the Viperidae venoms, on the other hand, brought about coagulation. Contrary to Viperidae venoms the Elapidae venoms retarded the coagulation of blood *in vivo* and also *in vitro*. Studies of several workers in the past showed that venoms of the Elapidae exerted strong anticoagulant action. However, Devi *et al.* (1956) claimed that cobra venom *(Naja naja)* had no retarding effect on human blood coagulation which is contrary to the findings of earlier workers. These authors noticed that the blood of man, ox, and monkey was not affected in their clotting time by cobra venom, but the blood of rabbit, sheep, and guinea pig was affected by cobra venom and did not clot in presence of cobra venom. One to two milligrams were required to prevent clotting of rabbit's plasma, whereas 0.05 to 0.1 mg was required to prevent the clotting of sheep and guinea pig blood. It was postulated (Devi *et al.*) that possibly there might be a plasma factor or venom cofactor present in cobra venom which had an inhibitory effect on the coagulation of human plasma. These authors further argued that cobra venom and venom cofactor of plasma might prevent in some unknown way the conversion of prothrombin to thrombin which ultimately resulted in retarding the conversion of fibrinogen to fibrin clot. According to them, there are two divisions of blood—one of man, ox, horse, and monkey, and other of rabbit, sheep, and guinea pig. The blood of the former group was not affected in its coagulation time by cobra venom while that of the latter group was either delayed or prevented completely.

Hemolytic properties of venoms on red blood corpuscles were extensively studied some time ago and it was found that different venoms are all hemolytic but with considerably variation. De (1942, 1944), working on crystalline hemolysin isolated from *Naja tripudians,* took one unit as the amount of crude venom which hemolyzed the blood mixture in 2 hours. As it was difficult to determine the complete hemolysis in a given time, the amount of hemoglobin liberated from a partial hemolysis in a given time was estimated in a colorimeter. Testing on red blood cells of different species, it was discovered that the red blood cells of ox, goat, sheep, and rabbit are less sensitive, and those of man, guinea pig, and rat far less sensitive than those of the horse to the hemolytic action of venom.

De (1942, 1944) estimated the hemolytic activity of venoms of *Naja tripudians* (monocellate and binocellate) and *Bungarus fasciatus* using guinea pig blood. His data follows:

Venom	1 unit of hemolysis contained in:	1 mg of venom contains:
Naja tripudians (variety monocellate)	5.2 μg	192 units of hemolysin
Naja tripudians (variety binocellate)	4.3 μg	232 units of hemolysin
Bungarus fasciatus	15.0 μg	66 units of hemolysin

De (1942, 1944) was able to purify hemolysin 17.5 times from *Naja tripudians* (monocellate variety) with 3360 units of hemolysin, 14.3 times from cobra venom of binocellate variety containing 3318 units of hemolysin per gram of the purified materials. De (1942, 1944) also purified the hemolysin from *Bungarus fasciatus* 25 times with 1650 units of hemolysin per gram of the purified material.

V. EFFECT OF THE DIFFERENT ACTIVE PRINCIPLES OF SNAKE VENOM ON VARIOUS ORGANS OF THE ANIMAL

Snake venoms are composed of several principles such as neurotoxin, hemolysin, cardiotoxin, coagulating factor, and cholinesterase. The toxicity of crude venom is either dependent on any one of these active principles or on the cumulative effect of more than one of these. Slotta and Fraenkel-Conrat (1938a,b) crystallized the toxic principle of *Crotalus durisssus terrificus* venom and showed that the crystalline material possessed both the neurotoxic and hemolytic activity. Ghosh and De (1938a), on the other hand, showed subsequently that the neurotoxic and hemolytic principles of *Crotalus durisssus terrificus* venom could be partially separated from each other, which demonstrated the existence of two separate entities. Ghosh and De (1938b) also succeeded in separating and purifying the neurotoxic principle from hemolysin present in *Naja naja* and *Bungarus fasciatus* venoms.

Feldberg and Kellaway (1937) and Trethewie (1939) showed that histamine and a coagulable protein were liberated by perfused tissues when venoms were added in perfused fluid. Trethewie (1939) and Kellaway and Trethewie (1940) also found a strict parallel between hemolytic activity and its capacity to liberate histamine and protein. This phenomenon was interpreted by them by assuming that the liberation of histamine and protein in cell injury was mediated by the formation of lysolecithin. Kellaway and Trethewie (1940) assumed that adenyl compounds were present in the perfusate from the perfused liver, kidney, and rabbit, and cat heart after injection of cobra *(Naja naja)* venom. Gauntrelet (1939) suggested that the liberation of histamine, adenyl compounds, etc., might be a secondary effect following certain primary action of venom on the cells. He thought that these substances were not indicative of the initial injury of the cells of the organism by the venom but resulted from the physiological effects that followed its injection. He also suggested that the fall of blood pressure caused by cobra venom *(Naja bungarus)* was not from cardiac nor central circulatory disturbances but was peripheral in origin and analogous to the action of histamine and proteotoxic substances. The action was essentially on the capillaries. According to Amuchastegui (1940) the venom of *Naja tripudians* caused arterial hypertension and venous hypertension; he suggested that this was due to myocardial insufficiency. Macht (1941) was of the opinion that the hypertensive effect of cobra neurotoxin, which was considered to be secondary to the paralysis of respiratory center, was of central origin. Kellaway and Trethewie (1940) indicated that the action of cobra venom was a complicated one and it affected all the cells of body. Gottdenker and Wachstein (1940) observed vasoconstriction of the vessels of the perfused ear of the rabbit and dilatation of the coronary vessels, but Gauntrelet (1939) did not observe any change in the vessels of the ear of the rabbit with *Naja bungarus* venom.

Electrocardiographic studies on the effect of intravenous injection of cobra venom *(Naja naja)* were made by several workers. Kellaway and Trethewie (1940) observed the following in rabbit heart after intravenous injection of cobra venom: bradycardia, increased P-R interval, QRST deviation, and terminal heart block. In cat heart, tachycardia, inversion of T-wave, extra systole, and increased P-R interval (irregularly) effects resembling bundle branch block and ventricular fibrillation. Amuchastegin (1940) also reported similar results with the venom of *Naja tripudians*. Injecting the venom of *Naja bungarus* into a dog, Gauntrelet (1939) observed a few extra systoles. Gottdenker and Wachstein (1940) observed profound changes in the electrocardiogram of cats which were earlier noticed by Kellaway and Trethewie (1940) in cats and rabbits. These electrocardiographic results showed variations in some cases. These differences are not surprising when the several variable factors such as the species, dose of venom used, experimental animals, and experimental procedures are considered. These differences might also indicate the complexity of cobra venom.

Ghosh, De, and Chaudhuri (1941) isolated the neurotoxic principle of *Naja naja* in an almost pure state and found its toxicity to be 16 times stronger than that of the crude venom. De (1944), isolated hemolysin from *Naja tripudians* and *Bungarus fasciatus* while Chaudhuri (1944) purified cholines-terase from *Naja tripudians* and *Bungarus fasciatus*. Physiological and phar-macological actions of these purified active principles as well as the crude venoms were studied by B. B. Sarkar *et al.* (1942). Experiments were con-ducted to study the action on toad heart perfusion, guinea pig heart *in situ*, and on rabbit blood pressure and respiration. It was noticed that the crude venom acting on toad heart caused augmentation of contraction lasting for a short time followed by diminution of height of contraction to less than normal in all concentrations. With higher concentrations, the heart became irregular with ventricular block which disappeared on washing with Ringer's solution; at 1/1000 concentration, the block remained. On increasing the dose up to 1/500 after initial stimulation, the heart stopped in systole and could not be reactivated even after prolonged washing.

Pure neurotoxin, which was isolated from cobra venom *(Naja naja),* also produced similar effects like crude cobra venom on toad heart. The main differences were that the auriculoventricular block was less evident and the systolic contracture of the heart did not occur with neurotoxin. On perfusing with neurotoxin, there was an immediate augmentation of both auricular and ventricular contractions, especially the latter. This was followed by diminu-tion of contraction, but in most cases the heart functioned normally. The heart beat was irregular if the concentration of the neurotoxin was increased to 1/10,000 or more, but on washing with Ringer's solution the beat became regular. No stoppage of heart with systolic contracture was obtained even with concentrated solution of neurotoxin.

On perfusing with pure hemolysin, which was isolated from cobra venom, there was an augmentation of both auricular and ventricular contraction followed by marked diminution of contraction and irregular beating of the heart. This irregularity with auriculoventricular block was found in every case even with a dilution of 1/100,000 and could not be reversed in most cases even after prolonged washing.

When toad heart was perfused with strong solution of cholinesterase (1/1000 or more), it showed slight augmentation of contraction which dis-appeared quickly after washing; the force and frequency of the heart beat remained regular. Comparing the action of cobra venom, neurotoxin, hemolysin, and cholinesterase on toad heart, crude cobra venom increased the heart rate during the augmentary period and with stronger doses the rate became irregular, but with neurotoxin, no distinct change was observed in the frequency of the heart. There was a slight quickening during the augmentary period. With hemolysin the heart rate increased with the diminution of the force of contraction and decreased again on washing. The heart also became irregular by hemolysin with the ventricular block persisting for longer or shorter periods. With cholinesterase there was no change in the frequency of the heart. It may be mentioned here that all of these active principles of neurotoxin, hemolysin, and cholinesterase were prepared from the same crude cobra venom, the pharmacological action of which was studied along with these purified active principles. Repeatedly perfusing the heart with neurotoxin and hemolysin and after washing with Ringer-Lock solution after each perfusion, it was found that the heart responded to each perfusion in the usual way, but the ventricular block persisted in the case of hemolysin while it appeared much later in the case of neurotoxin.

Action on intact heart of guinea pig was studied by maintaining the animal with artificial respiration. A dose of 1.5 mg of neurotoxin per kg of body weight produced no effect on mammalian heart; but increasing the dose to 6 mg per kg produced pronounced and prolonged stimulation of contraction. There was, however, no change in the frequency, and the heart beat was regular. A second effective dose produced further augmentation. Similar augmentation was obtained with a stronger dose (8 mg of hemolysin per kg body weight) but the heart beat became irregular and ultimately stopped. The blood taken from the heart of the experimental animal was found to be partially hemolyzed. A dose 0.5 mg of crude cobra venom per kg body weight stopped the mammalian heart.

While carrying out investigations to compare the action of crude cobra venom, neurotoxin, hemolysin, and cholinesterase on animals, it was noticed that the action of crude cobra venom with a dose of 2 mg per kg body weight of a rabbit was found to cause a preliminary rise and then a sharp fall in blood pressure. Epstein (1930) obtained a rise in blood pressure in cat and rabbit with 2 mg of crude cobra venom. Elliot (1905), Chopra and Iswariah (1931), Venkatachalam and Ratnagiriswaran (1934), and Gottdenker and Wachstein (1940) using perfused cobra venom obtained a steady rise in blood pressure with doses of 6 mg to 9 mg per kg body weight. Using a dose of 0.25 mg to 0.5 mg per kg body weight of cat, Chopra and Iswariah (1931) and Feldberg and Kellaway (1931) obtained a steep fall in blood pressure resulting in death in a few minutes. Gauntrelet et al. (1930) obtained fall in pressure in the dog with a dose of 0.1 mg.

Neurotoxin produced a slight rise in blood pressure of rabbit maintained on artificial respiration with a dose ranging from 0.1 mg to 8.5 mg per kg body weight; but the most striking effect was noticed in its action on respiration. Injecting 0.2 mg of neurotoxin, respiration stopped in 9 minutes. Failure of respiration produced asphyxia, and caused the rise in blood pressure. Applying artificial respiration, the blood pressure came back to normal and remained there as long as artificial respiration was continued. With 1.2 mg neurotoxin per kg body weight, respiration stopped in 7 minutes; with 2.5 mg, in 4.5 minutes, and with 8.3 mg, in 2 minutes. A second smaller dose of neurotoxin given to a rabbit receiving artificial respiration whose breathing had stopped because of a previous dose of neurotoxin produced a slight increase in blood pressure, a larger amount of cobra venom neurotoxin as the second dose caused a sharp rise in blood pressure followed by a rapid fall and stoppage of heart action. Vagosympathetic effects on the heart and blood pressure were found to be negative after neurotoxin injection.

Injection of 0.2 mg to 6 mg of hemolysin per kg body weight in rabbit produced a slight rise in blood pressure similar to neurotoxin but had no effect on respiration. Blood was found to be slightly hemolyzed with a dose of 6 mg hemolysin per kg body weight, but the action of the vagus nerve was intact. With a dose of 10 mg per kg body weight in a rabbit both circulation and respiration failed in 1 minute, and unlike neurotoxin, artificial respiration could not maintain the heart and blood pressure. On examining the blood, it was found to be hemolyzed. After a dose of hemolysin (4 mg per kg body weight) that was insufficient to cause failure of circulation and respiration, injection of neurotoxin produced the usual effects and the respiration stopped. In this case also artificial respiration restored circulation. Cholinesterase, however, as high as 13 mg per kg body weight was found to have no effect on respiration but produced a slight rise in blood pressure. B. B. Sarkar *et al.* (1942) further observed that weak concentration of crude venom stimulated the heart, and with medium concentration the stimulation of heart was followed by depression with irregularity of heart; but with high concentration the heart stopped entirely in systolic contracture. Regarding the action of the various purified principles obtained from crude cobra venom, neurotoxin differed from the crude venom by not causing heart stoppage even with high concentrations of neurotoxin and also had no effect on blood pressure. Its effect was mainly on the respiratory system, therefore, Ghosh (1942) named it a respiratory toxin. On the other hand, a small dose of hemolysin had practically no effect on circulation and respiration; with high doses there was a failure of both. The circulatory failure was probably due to the hemolytic activity. It was, therefore, postulated by B. B. Sarkar, Maitra, and Ghosh (1942) that in addition to the active principle of neurotoxin, hemolysin, and cholinesterase crude cobra venom probably contained another active principle that might be responsible for heart stoppage; this was named cardiotoxin.

On the basis of these findings N. K. Sarkar (1947) started work on the isolation of this active principle (cardiotoxin) from crude cobra venom and was successful in separating it from the other constituents by a salting-out process. The active principle, isolated in this way, was 15 times more toxic than the crude cobra venom. Physiological and pharmacological activities of cardiotoxin were later studied by N. K. Sarkar (1947) and Maitra (1949). It was noted by these authors that the failure of circulation, as indicated by the fall of the carotid blood pressure to zero, was obtained with 2 mg of cobra venom and 0.77 mg of cardiotoxin per kilogram of body weight of the cat.

Taking into account the fall of the blood pressure to zero with simultaneous failure of respiration as the basis of measurement, the ratio of the amount of cobra venom with that of cardiotoxin and neurotoxin required to effect such changes may be given as:

Cobra venom	Cardiotoxin	Neurotoxin
2 mg	0.77 mg	0.1 mg

Experiments conducted by perfusion technique on mammalian heart (kitten) with cobra venom, respiratory toxin, and cardiotoxin isolated from the same cobra venom *(Naja tripudians)* showed that 1 cc of 1/15,000 dilution with cobra venom and 1/34,000 with cardiotoxin stopped the heart in systole. A comparatively strong solution of 1 cc of 1/15,000 dilution of respiratory toxin, hemolysin, and cholinesterase produced no effect on isolated heart. Table II indicates the results obtained.

A similar study on amphibian heart (perfusion on toad's heart) showed that cobra venom stopped the heart with 5 cc of 1/200 concentration of cobra venom. A dose 5 cc of 1/50 concentration of purified hemolysin (noncrystalline) also stopped it in systole, but 5 cc of 1/50 concentration of crystalline hemolysin had no effect. On the other hand, 5 cc of 1/450 concentration of cardiotoxin (purified) stopped the heart in systole.

Table III shows the comparative action of cobra venom, hemolysin, and cardiotoxin isolated from the cobra venom.

From the experimental results cited on blood pressure and respiration of a normal cat, on mammalian and amphibian heart, it was clear that both crude venom and cardiotoxin acted directly on heart and stopped its action on the myocardium. But the crude cobra venom also contains active principles other than cardiotoxin so crude venom action on blood pressure and respiration was quite different from that of cardiotoxin which only paralyzed the cardiac muscle. The effect of cardiotoxin was the gradual fall of blood pressure and then increase in amplitude of respiration with subsequent failure along with the sudden fall in blood pressure. This indicated that the increase in amplitude of respiration was probably due to asphyxia which was caused by first gradual and then sudden cardiac failure and this could not be changed by artificial respiration; but with cobra venom there was a cessation of respiration with first a rise of blood pressure and finally a fall in blood pressure. This could be prevented by artificial respiration which indicated that the toxic effect of the respiratory toxin was first manifested on the animals.

It was also noticed that the effect of cardiotoxin and cobra venom in stopping mammalian and amphibial heart in systole as observed by perfusion technique was irreversible. B. B. Sarkar *et al.* (1942) found no action on amphibian heart or guinea pig heart *in situ* with neurotoxin and cholinesterase, but found a ventricular blocking effect with unchangeable irregularity

TABLE II

PERFUSION OF KITTEN HEART WITH CRUDE COBRA VENOM *(Naja tripudians)*,
RESPIRATORY TOXIN, HEMOLYSIN, AND CHOLINESTERASE[a]

Concentration of the venom solution	Condition of heart	
	After perfusing with venom solution	After washing with normal Ringer's solution
1 cc in 150,000	Stimulated	Regular and normal
1 cc in 60,000	Slightly stimulated with amplitude decreased	Regular, but beat with decreased amplitude
1 cc in 30,000	Regular with some occasional irregular beats but always accompanied by very short amplitude	Regular but with marked shortening in amplitude
1 cc in 27,000	Irregular beat with decreased amplitude	Irregular beat
1 cc in 15,000	Stopped	Stopped in systole
1 cc in 1,500 (respiratory toxin)	Regular and normal	Regular and normal
1 cc in 1,500 (hemolysin)	Regular and normal	Regular and normal
1 cc in 1,500 (cholinesterase)	Regular and normal	Regular and normal
1 cc in 90,000 (cardiotoxin)	Rate and amplitude increased followed by a decrease in rate and amplitude	Regular
1 cc in 45,000 (cardiotoxin)	Regular accompanied with occasional irregular heart-beat with very short amplitude	Regular but beat with decreased amplitude
1 cc in 39,000 (cardiotoxin)	Stopped in systole	Revived but finally stopped
1 cc in 3,400	Stopped in systole	Stopped in systole

[a] Note 1 cc was used for each perfusion experiment.

with the purified hemolysin but N. K. Sarkar (1947) while extending the work of B. B. Sarkar *et al.* (1942) obtained no such effect with crystalline hemolysin. These authors, therefore, claimed that the respiratory toxin, cholinesterase, and hemolysin isolated from crude cobra venom *(Naja tripudians* or *Naja naja)* had no effect on heart. According to N. K. Sarkar (1947) the effect of purified hemolysin on heart as noticed by B. B. Sarkar *et al.* (1942) might be due to the presence of some impurities in their preparation.

Studying the effect of cardiotoxin and cobra venom on gastrocnemius muscle and sciatic nerve preparation, N. K. Sarkar (1947) and Maitra (1949) observed that cobra venom acted on both muscle and neuromuscular junction but cardiotoxin acted on muscle only. Both cobra venom and cardiotoxin had no action on sciatic nerve or on the nerve trunk in general. Cobra venom acted first on the neuromuscular junction when response by stimulation to sciatic nerve had been stopped. It was found that when stimulation was

TABLE III

COMPARATIVE ACTION OF COBRA VENOM, HEMOLYSIN, AND CARDIOTOXIN[a] ON HEART

Venom	Concentration used	Condition of heart after perfusion with venom solution	Condition of heart after washing with normal Ringer's solution
Cobra venom (*Naja tripudians*)	1 in 10,000	Stimulation of rate and amplitude	Regular
	1 in 1,000	Same as before but followed by a decrease in both	Regular but beat with shorter amplitude
	1 in 500	Decrease in amplitude was marked accompanied by occasional irregular beat	Beat with still shorter amplitude
	1 in 200	Stopped in systolic contraction	Remained stopped in systole
Hemolysin isolated from cobra venom (*Naja tripudians*)	1 in 500 (noncrystalline)	Irregular beat with ventricular block	Irregular beat with ventricular block
	1 in 100 (noncrystalline)	Irregular beat was marked and amplitude became very short	Above effect was marked
	1 in 50 , (noncrystalline)	Stopped in systole	Remained stopped in systole
	1 in 50 (crystalline)	No irregular beat but amplitude was slightly shortened	Neither any irregular beat nor any shortening of amplitude
Cardiotoxin isolated from cobra venom (*Naja tripudians*)	1 in 50,000	Stimulating action with increase of both amplitude and rate	Regular
	1 in 5,000	Initial increase in rate and amplitude followed quickly by decrease in both	Regular but amplitude became slightly shorter
	1 in 1,000	Shortening of amplitude was marked	Still shorter amplitude
	1 in 450	Stopped in systolic contracture	Systolic contracture was maintained

[a] Note: 5 cc of the solutions were used for the perfusion experiments.

given directly to the muscle, this phenomenon was not found with cardiotoxin. These authors also observed that the action of cobra venom or cardiotoxin could not be altered by changing Na^+, K^+, or Ca^{++} ions of Ringer's solution and even by adding acetylcholine which produced a dilatory effect on the heart. Sarkar (1947) compared the action of cobra venom and cardiotoxin with saponin and digitalis on heart and skeletal muscle but found no similarity.

Further Sarkar (1947) and Maitra (1949) observed that ultraviolet radiation destroyed the toxicity of cardiotoxin in both cobra venom and also purified cardiotoxin. By exposing 5 % solution of cobra venom or cardiotoxin in Ringer's solution of pH 7.4 to 7.6 to ultraviolet radiation at a distance of 20 cm from the source, the toxic effect of cardiotoxin was completely destroyed after 150 minutes while it took 200 minutes to destroy cardiotoxin present in cobra venom as revealed by toad's heart perfusion technique. The effect of ultraviolet light on cardiotoxin with time as determined on toad heart is given in the Table IV.

The effect of temperature on the toxicity of cardiotoxin present in cobra

TABLE IV

EFFECT OF ULTRAVIOLET LIGHT ON CARDIOTOXIN AND COBRA VENOM[a]

Material	Duration of exposure (minutes)	Minimum concentration producing systolic arrest of toad heart	Percentage of activity remained after exposure
Cardiotoxin	0	1 in 450	100
	30	1 in 420	90
	50	1 in 350	78
	70	1 in 290	64
	90	1 in 200	45
	120	1 in 90	20
	150	Arrest of heart could not be produced at any concentration	nil
Cobra venom	0	1 in 200	100
	30	1 in 185	92
	50	1 in 160	80
	80	1 in 125	62.5
	120	1 in 70	35
	150	1 in 20	10
	200	Arrest of the heart could not be produced at any concentration	nil

[a] Note: 5 cc of the solutions were used for each perfusion.

venom as well as of the purified cardiotoxin was also studied by Sarkar (1947) and it was noted that both of them lost their toxicity if they were kept at 90°C for 30 minutes. Table V depicts the results.

It was found that the loss of toxicity of cardiotoxin in cobra venom and also in purified cardiotoxin due to heating was irreversible even when it was kept at 4°C for 8 hours. From the study of the toxicity and physiological action of cobra venom and cardiotoxin it appears that the active principle isolated by Sarkar (1947) was responsible for stopping the cardiac movement

TABLE V

EFFECT OF TEMPERATURE ON CARDIOTOXIN AND COBRA VENOM[a]

Active principle used	Temperature at which the substances were heated for 30 minutes	Minimum concentration producing the systolic arrest of toad's heart	Percentage of activity retained after heating
Cardiotoxin	32°C	1 in 450	100
	60°C	1 in 450	100
	70°C	1 in 425	95
	75°C	1 in 375	85
	80°C	1 in 300	67
	82°C	1 in 225	50
	85°C	1 in 110	24
	90°C	Arrest of the heart could not be produced	0
Cobra venom	60°C	1 in 200	100
	70°C	1 in 200	100
	75°C	1 in 175	87.5
	80°C	1 in 145	72.5
	82.5°C	1 in 110	55
	85.5°C	1 in 50	25
	90°C	Arrest of heart could not be produced at any concentration	0

[a] Note: 5 cc of the solutions were used for each perfusion.

while the respiratory toxin acted predominantly on the respiratory center; hemolysin and cholinesterase had no toxic effect. The isolation of these active principles from cobra venom has led to significant contributions in our understanding of the toxic effects of the venom *(Naja naja* and *Naja tripudians)*. A systematic investigation on the isolation of the active principles of the venoms of other poisonous snakes and a study of their physiological properties would be of fundamental importance.

REFERENCES

Amuchastegui, S. R. (1940). *Compt. Rend. Soc. Biol.* **133**, 317.

Calmette, A. (1908). "Venoms," p. 173.

Chaudhuri, D. K. (1944). *Ann. Biochem. Expt. Med. (Calcutta)* **4**, 77.

Chopra, R. N., and Iswariah, V. (1931). *Indian J. Med. Res.* **18**, 1113.

De, S. S. (1942). *Ann. Biochem. Exptl. Med. (Calcutta)* **2**, 236.

De, S. S. (1944). *Ann. Biochem. Exptl. Med. (Calcutta)* **4**, 45.

Devi, A., Mitra, S. N., and Sarkar, N. K. (1956). In "Venoms," Publ. No. 44, p. 218. Am. Assoc. Advance. Sci., Washington, D.C.

Elliot, R. H. (1905). *Phil. Trans. Roy. Soc. London* **B197**, 361.

Epstein, D. (1930). *Quart. J. Exptl. Physiol.* **20**, 7.

Ewing, and Bailey, K. (1900). *Med. Record* p. 15.

Feldberg, W., and Kellaway, C. H. (1937). *J. Physiol. (London)* **90**, 257.

Gauntrelet, J. (1939). *Bull. Acad. Sc . Med. (Paris)* **122**, 412.

Gauntrelet, J., Halpern, N., and Corteggiani, E. (1930). *Arch. Intern. Physiol.* **38**, 293.

Ghosh, B. N. (1942). Unpublished observations.

Ghosh, B. N., and De, S. S. (1938a). *Nature* **143**, 380.

Ghosh, B. N., and De, S. S. (1938b). *Indian J. Med. Res.* **25**, 779.

Ghosh, B. N., De, S. S., and Chaudhuri, D. K. (1941). *Indian J. Med. Res.* **29**, 367.

Gottdenker, F., and Wachstein, M. (1940). *J. Pharmacol. Exptl. Therap.* **69**, 117.

Kellaway, C. H., and Trethewie, E. R. (1940). *Australian J. Exptl. Biol. Med. Sci.* **18**, 63.

Lamb, G. (1904). *Lancet* Jan. 2, Aug. 20, Oct. 22.

Lamb, G. (1905). *Lancet* Sept. 23.

Macht, D. I. (1941). *Med. Res. Council, Spec. Rept. Ser.* **153**, 369.

Maitra, S. R. (1949). D.Phil. Thesis, Calcutta University.

Rogers, L., and Megaw, S. W. D. (1904). *Phil. Trans. Roy. Soc. London* **B197**, 123.

Sarkar, B. B., Maitra, S. R., and Ghosh, B. N. (1942). *Indian J. Med. Res.* **30**, 453.

Sarkar, N. K. (1947). D.Sc. Thesis, Calcutta University.

Slotta, K. H., and Fraenkel-Conrat, H. L. (1938a). *Nature* **142**, 213.

Slotta, K. H., and Fraenkel-Conrat, H. L. (1938b). *Mem. Inst. Butantan (Sao Paulo)* **12**, 505.

Trethewie, E. R. (1939). *Australian J. Exptl. Biol. Med. Sc .* **17**, 145.

Venkatachalam, K., and Ratnagiriswaran, A. N. (1934). *Indian J. Med. Res.* **22**, 289.

Chapter 22

The Story of Some Indian Poisonous Snakes

P. J. DEORAS*

HAFFKINE INSTITUTE, BOMBAY, INDIA

I. COBRA *(Naja naja)*

The cobra is found in all areas of India, up to an altitude of 15,000 feet above sea level. The maximum length recorded was 147.32 cm.

In Portuguese, the word cobra means snake, but in English it is the name given to the snake having a hood with markings on it. Scientifically, this is *Naja naja* Nikol. The snake is called by various names in different dialects of India:

State	Local name
Uttar Pradesh	
Madhya Pradesh	
Bihar	Nag or Naga
Orrissa	
Maharashtra	
Gujarat	

* Present address: Sahitya Sahawas, Bombay, India.

19

State	Local name
West Bengal	Gokhura
Madras	
Andhra Pradesh	Nale pambo
Mysore	Nagara havu
Malabar	Moorkhan pambu

The markings on the dorsal side of the hood may be bionocoellate, monocoellate, or there may be no markings at all. The snakes with the monocoellate markings are seen in some parts of Bengal. Bionocoellate and ascellate snakes are found all over India. Fig. 1 illustrates the various marking patterns seen on the dorsal side of the hood of a cobra. On the underside of the hood are broad faint black stripes. Above these stripes are two dark spots extending to four scales. These may be surrounded by white borders. It is interesting to note that these markings are always present whether or not the snake has any markings on the dorsal side. King cobra, *Naja hannah* Cantor, has three black cross stripes on the dorsal side of the hood and has spots and two cross stripes on the ventral side of the hood (Fig. 2). The color of the snake may be brown, when it is normally called Gehuwa or "of wheat color" in Hindi. It may be black and then it is called Domi which means "dark"

Fig. 1. The different patterns of markings on the hood of cobras. Left to right: bionocoellate, monocoellate, ascellate. (Photo by author.)

FIG. 2. Hood and markings in (A) cobra *(Naja naja)* and (B) king cobra *(Naja hannah)*. (Photo by author.)

in Hindi. In certain regions of central India an albino cobra has been recorded with faint brown markings on the dorsal side of the hood. Similarly, during the period after hibernation, one meets with a rare color in a cobra, which is very near yellow. Such specimens have been recovered from the crevices in deep wells. They have good bionocoellate markings. But out in the open, they turn brown. Aside from these, one finds a specimen that has chevron-shaped dark brown cross-bars on a brown body, which has been encountered in the Bombay region.

The neck region in this snake is dilatable and the ribs in this area are elongated. When the snake is disturbed it dilates this region and forms the structure commonly known as the hood. The cat snake *(Boiga gokool* Grey*)* often tries to inflate the neck region and raise the head in imitation of a cobra, but this snake does not have a hood.

A. Scalation and Measurements

The distinctive features of the scalation of this snake consist of three important scale arrangements on the head. The third supralabial scale touches the nasal area and the eye (Fig. 3). There are also three small scales immediately behind the eye and a triangular dark scale between the fourth and fifth infralabials.

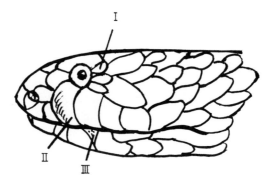

FIG. 3. Drawing of the distinctive scale features in a cobra head. I, three scales behind the eye; II, third supralabial; III, triangular wedge.

Other features are: intranasals, 3, supralabials 7–8, third and fourth touching the eye, temporals 2–3, infralabials 7–8, genials two pairs, frontals are usually longer than broad and intranasals as long as prefrontals. Maxillary bone extends forward beyond the palatine and is followed by 1–3 small teeth. The fangs are nonerectile and have a groove dorsally for the transmission of poison. Eyes are large and pupils are round.

The following are some of the scale measurements representing the average of 27 specimens ranging in length from 63.627 cm to 146.685 cm.

Length from tip of nose to anus 75.565 cm to 146.685 cm.
Length from anus to tip of tail is 13.97 cm to 28.75 cm.
Number of dorsal scales across body 19–35.
Caudals 17.
Ventrals 187–194.
Subcaudals 56–77.
Supralabials 7–8.
Infralabials 7–8.
Neck scales 27–31.
Dorsal bionocoellate markings on the head up to 10 rows of scales.
Total width of bionocoellate marking, 15 rows.
Length of each arm 4–6 scales.
Width of each arm 2 scales.
The dark stripes on the ventral side of head extend up to 10–17 of the ventral scale length.

Male and female snakes can be distinguished since the tail is longer and the head and hood are wider in males in comparison to females.

1. The ratio of body length to tail: males   1.0 : 0.18;
                                           females 1.0 : 0.17.
2. The ratio of length to width in the head: males   1.0 : 0.75;
                                               females 1.0 : 0.56.
3. The ratio of length to width of the hood: males   1.0 : 0.63;
                                               females 1.0 : 0.59.

The males are heavier than the females. A female of 1.47 m weighs 355 gm while a male of the same length is 560 gm. The maximum weight ever seen by us was 1247 gm for a male cobra 1.60 m long.

B. Habits

The cobra is very common in all areas of India and is even found near human habitations, particularly during the evening and early morning.

This snake, like other elapids, has only one lung, the other being aborted. The charming of a cobra with a gourd flute which is frequently seen in India (Fig. 4) is just showmanship since the cobra has no external ears, and the internal ear is quite reduced. In fact, the stapes bone is ankylosed to the quadrate bone of the cranium and the snake perceives sound as vibrations. The cobra is either disturbed by these vibrations on the basket or soothed by the sight of the charmer's swaying flute. The hood of the cobra sways, not because it is charmed by the music, but because the charmer is swaying his hand or flute and the snake is swaying in order to strike.

In the reproductive system the penis is paired and without spines. Mating takes place in a prone position during monsoon months from July onwards. At such times males may be seen twining erectly; these are fighting males. Experiments done by taking monthly cloacal smears have shown that the spermatozoa are found from February on, although copulation takes place in July–September. Eggs are laid in April. On our snake farm, 56 eggs were laid

FIG. 4. "Charming" the cobras by a flute made from a gourd. Village scene in India. (Photo by author.)

by one snake. The incubation period is about 56–59 days and the young snakes instinctively spread their hoods in a menacing way on the first day of life. The young shed their skin on the second, seventh and then from the twenty-first day onwards. Molting in the laboratory varies with the seasons. It is rare to see the complete skin cast by a snake. When a snake is unable to shed his skin properly or if the skin is pulled off, the injury to the snake may be fatal.

Snakes in captivity have a life span of only 5 years; in nature they may live for 25 years or more.

C. Venom

Venom can be removed by making a snake bite a wine glass over which a plastic cloth has been stretched. The liquid can be lyophilized and stored in vacuum-sealed ampules at −4°C. Fresh venom is a transparent liquid that becomes translucent when exposed to light and air. The venom obtained during the winter months has a slight yellowish tinge and is thicker than that obtained in summer.

Newly captive snakes give more venom; the output decreases in captivity. Black cobras give more venom than brown cobras of equal size. On the first day of arrival, venom from one black cobra weighed 1.50 gm after lyophilization. The output gradually decreased, and the snake died after 10 months in captivity. When snakes are kept outdoors, as on a snake farm, the output of venom is greater than when the snakes are caged. In output experiments, it was also noticed that the male cobras give more venom than the female. Venom yield records for snakes kept outdoors and in cages taken on the average for a number of years are 0.1688 gm per snake on the farm, and 0.1200 gm per snake in closed quarters. The records taken for male and female cobras kept in the farm and room separately have been seen to be 0.2027 gm and 0.1772 gm, 0.1789 gm and 0.1342 gm respectively. The output of venom showed periodicity during different months and there is a difference in the MLD of the venom given in different months. Venom is more toxic during summer months in comparison to winter. The MLD varied from 1/50,000 to 1/120,000 by the intravenous route.

The lethal dose of cobra venom for an average size man is 12 mg.

The venom is neurotoxic. The poison acts on the nervous system and chiefly affects the respiratory centre, the heart remains unaffected.

Cobra venom has been observed to play some role in relieving pain and venom solutions of different dilutions are used for this purpose. Of late enzymes such as neurotoxin, hemolysin, phosphodiesterase, and ophio-oxidase have been isolated from crude cobra venom and are being used in biochemical research.

D. Snake Bite and Symptoms of Poisoning

The snake normally attacks only when seriously disturbed. It raises its hood, makes intermittent hissing sounds, and then attacks. If a person keeps out of the radius made by the raised body and stands still, the snake will normally go away. But any movement will disturb the snake further and it will strike. However, the range of sustained run is very limited. Normally a person is bitten on the extremities, which if properly covered, will give protection against the bite. The fangs are small and will not penetrate the leather of ordinary hunting shoes. The mongoose *(Herpestes edwardsi)* and certain birds like the peacock *(Pavo cristatus),* owl *(Tyto albajanica),* and hawk *(Accipiter nisus)* have been known to attack the snake; consequently it stays away from open areas. It is a very delicate snake and extreme cold immobilizes it; similarly, extreme heat nearly kills it. The snake is gentle and if handled carefully and gently, it will not strike.

With cobra bite there are two deep marks with some redness accompanied by a number of small punctures. It should be remembered that there is very little pain at the site of the bite. The symptoms begin in about 8 minutes if a lethal dose of venom has been injected. The actual symptoms start with slight local pain at the site of the bite which radiates along the extremity and is followed by a slight paralysis. The swelling at the site is either absent or very small. The tissues around the puncture become slightly bluish or pinkish in color. There is then an onset of gradual weakness, and loss of power in the legs. Face color remains normal at first, but becomes quite flushed later. The skin is warm, the pulse is initially normal and regular. Breathing gradually becomes more and more labored; it becomes more rapid and there is frothing from the mouth and gasping toward the end. Paralysis sets in, in the extremities and travels toward the trunk; the neck muscles become weakened and the head droops. Swallowing becomes difficult; the lower lip hangs loosely and there is drooling. Articulation becomes increasingly difficult. Death is due to respiratory depression. These symptoms should be distinguished from symptoms that may be due to sheer fright which may occur even if the patient has not been bitten. In such cases obviously there is no paralysis and the skin is cold and clammy, the pulse is feeble, and breathing is shallow and unduly frequent; death may be due to cardiac depression.

First aid with cobra bite may be primarily of a psychological nature. Other immediate treatment consists of enlarging the wound, bleeding it by scarification, tourniquet, and later taking the victim for intravenous serum treatment. The only sure treatment is a specific antivenin, which should be given as soon as possible.

The mortality due to snake bite recorded in 1953 in India was 15,000 per year; considering the population of India this is about 0.0041 %. In fact, more people are killed by the bite of the cobra than by any other snake.

E. Legends and Beliefs

The cobra is considered sacred in many parts of India. In fact there are two days in the year, one in July and other in October, when it is venerated.

There are numbers of pictures showing the snake with Indian deities. Lord Krishna is depicted as a child jumping into the river Jumna to subjugate the naga (cobra) Kaliya. The Devas (gods) and Asuras (demons) churned an ocean using a serpent, Sheshnag, as a rope. Lord Vishnu is shown resting on a many-hooded cobra in an ocean of milk. Lord Shiva is depicted as wearing cobras around his neck. In one of the forms of Yoga, Kundalini Yoga, the supreme power, Kundalini, is depicted as a coiled cobra. In the *Ramayana,* the great epic, the naga (cobra) mother Surasa is mentioned, and in the *Mahabharata* the great warrior Arjuna was to marry the sister of the naga-king Vasuki. In Buddhist literature, Sanghamitra, the daughter of the emperor Asoka, is shown chasing away the magic power of nagas (cobras). In the *Vedas,* particularly the *Rigveda,* the king of the gods, Indra, is supposed to kill Ahi-Vritrasura, who is said to be a snake demon.

A number of temples have been dedicated to naga kings in Kangra valley in North India; these are Sheshnag, Basaknag, Takthanag, etc. Many of the deities worshipped in the temples are shown in human form with a number of hoods on the heads. In fact, throughout India, images of serpents with human heads are placed for worship beneath fig trees. At Ajanta in the cave temples, statues of kings are depicted with varying numbers of cobra hoods. In one of the stone bas reliefs at Sanchi (Fig. 5) a king is shown wearing head-gear with a number of cobra hoods; his queen is at his side wearing only one hood.

From the literature available in other countries, it can be surmised that the snake was probably the symbol or totem of sun worship in many parts of the world. In India this totem was in the form of the cobra snake hood and those kings who claimed their genealogy from the sun took it as a mark of honor. These kings lived in India before the Aryans were there, and when the older cultures and the Aryans intermingled the symbol of ancient kings, who were known as Naga-Rajas or "cobra-kings", was incorporated into a comprehensive symbol of culture and civilization. The symbol of the cobra hood which is shown in statues and shrines indicates the continuity of Indian civilization.

FIG. 5. Bas relief on the stone gate at Sanchi showing naga kings with cobra hoods as headgear. (Photo by author.)

II. THE COMMON KRAIT (*Bungarus caeruleus* Boulenger)

This snake is found throughout India up to elevations of 914.4 m. It reaches 1.2194 m in length.

It is commonly called by the following names in different states of India.

State	Name
Uttar Pradesh	
Bihar	
Orrissa	
Madhya Pradesh	Karayat
Rajastan	
Punjab	
Maharashtra	Maniyar or Dandekar
Gujarat	Konotaro
Madras	Yannabarian
Mysore	Gudinasar

The krait is normally steel blue in color with white cross circular bandlike markings over the entire body. These markings in snakes from certain areas are paired in the posterior parts of the body. In other areas there are snakes with a single band only or the bands may be merely a collection of white dots. The maximum number of bands is about 32. Some snakes are gray with only 2–3 white spots for the first few rows and circular markings after the fifth row. In another gray variety there is a white spot after every fifth bar and white cross bars after every fourth row middorsally. In dark-brown varieties there are continuous white cross circular bars nearer to the posterior region. A newborn snake is purple or blackish brown. Young snakes are purple and the bars are incomplete.

One distinctive characteristic of the krait is the chain of hexagonal scales on the middorsal side of the body. These are bigger than other scales and they are also marked by a dorsal vertebral ridge. Also the ventral scales beyond the vent are complete and not paired as with elapid snakes. There are two other snakes having color markings similar to the krait: *Lycodon* and *Oligodon* species. Both of these are deep brown in color with white cross stripes; however they do not have enlarged hexagonal dorsal scales and the ventral scales beyond the vent are paired. The eyes are large and the pupils are round.

Scalation:
Head scales: No loreal, supralabials 7, third and fourth touching the eye, postoculars 2, preocular 1, infralabials 4.
Body scales: Middorsal body scales 15–17, anal 11–13, caudal 11, total median hexagonal scales 245–255, ventrals 198–208, subcaudal 44–50.

Length from tip of the nose to anus 112 cm.
Length from anus to tip 14 cm.

The male snakes are longer and heavier than females. The average weight of ten males of 1.23 m length is 266 gm.

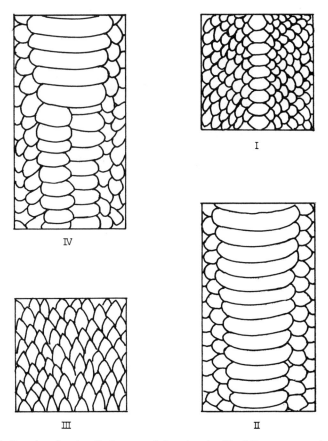

FIG. 6. Drawing showing the hexagonal dorsal scales (I) of *Bungarus caeruleus* and the unpaired scales beyond the vent (II). Dorsal scales in a viper (III) and paired scales beyond vent in a cobra (IV) are also shown.

Habits

This snake is very quiet and timid and lives in crevices between rocks, bricks, and even vegetation where it is seen to go inside the crevices of roots of many plants. It is nocturnal in habit and unless disturbed will not attack man. This snake does not hiss but it has been noticed that in the early hours of the morning it gives out a peculiar shrill noise similar to a faint noise

FIG. 7. *Bungarus caeruleus* seen in its habitat. (Photo by author.)

emitted by a kite. It travels late in the evening or early in the morning. The frequency of casting of the skin as seen in the laboratory was from 70 to 85 days in summer months and from 135 to 205 days in the rainy season and winter. This snake often feeds on other snakes. The most peculiar morphological feature of this snake is the reduction of the left lung to a mere stump. There are four anal glands which secrete dark purple liquid during a number of months in a year. Pairing takes place during July–September and eggs are laid in April. As many as 60 eggs are laid and these hatch in about 50 to 60 days. The newborn snakes are barely 15 cm in length and grow at the rate of

5–7 cm per year for the first 2 years. Snakes have been kept in captivity for 6 years.

The venom of this snake is neurotoxic. In experiments here, it has been noticed that the amount of venom of snakes kept outdoors is greater than the venom given by the snakes kept in cages indoors. Also male kraits secrete more venom than females. The average venom output after lyophilization over 5 years for both sexes is 0.0198 gm and 0.0134 gm for males and females, respectively. The output for male and female snakes kept outdoors is 0.0255 gm and 0.0112 gm; for those caged indoors 0.0228 gm and 0.0099 gm. The intravenous MLD is 1/250,000 and CLD is 0.004 mg. The lethal dose for an average-sized man is about 6 mg.

The symptoms of krait poisoning are similar but milder than with cobra envenomisation and there is little local swelling or pain. There is also no nausea or frothing at the mouth. The victims sleep and gradually die after the numbness in the extremities extends to the neck region. There is also albumin found in the urine.

III. THE SAW-SCALED VIPER (*Echis carinatus* Ingoldby)

This snake is also found in all parts of India on the plains and deserts; it attains a maximum size of 45 cm.

The saw-scaled viper is called by various names in the different states of India

State	*Name*
Uttar Pradesh	
Punjab	
Rajastan	Afai
Bihar	
Bengal	Bankoraj
Gujrat	
Maharashtra	Phoorsa
Kashmir	Gunas
Madras	Birianpankhu
Mysore	Kallu hawu

The viper is olive brown dorsally and has white pale brown wavy lines on its sides. The upper aspect of these lines blends in with an irregular brown spot in the vertebral region. The head in addition to being triangular shaped, has a pale brown arrow head mark (Fig. 8); this is the most characteristic feature of this snake. Pale brown marks are also found near the nasal and eye regions. The dorsal spots number from 25 to 31 and have a faint white border. The ventral scales have small black spots on their sides.

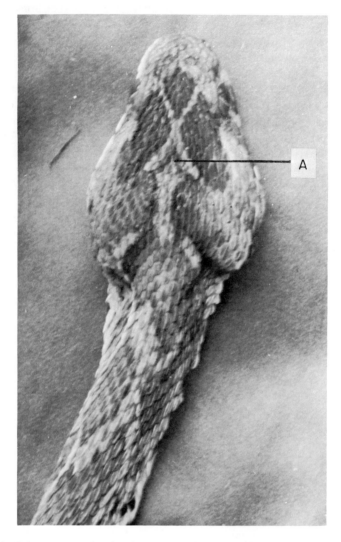

FIG. 8. *Echis carinatus,* showing the arrow mark on the head (at A). (Photo by N. E. Vad.)

Scalation:

Head: The head scales are small, imbricate, and keeled. Supraoculars 10-13; the fourth is the largest; the others are very narrow or broken up. The eye has an elliptical pupil and is ringed by 10–15 very small scales. One to two rows of scales intervene between the supralabials and the eye. Temporals are very small and keeled.

Intralabials: 10–12.

Body: The dorsal scales are in 25–31 rows, the outer rows being the largest. The scales beyond the neck are heavily keeled or serrated. In fact, these scales along with the keeled lateral scales on the body produce the noise when the snake moves sideways while rubbing on these serrations. Ventral scales 148–156, and subcaudals 25–30.

The females are heavier than the males. Five specimens of females ranging from 0.42 to 0.48 m in length weighed from 35 gm to 48 gm, while six specimens of males from 0.42 to 0.47 m weighed from 25 gm to 48 gm.

The penis is bifurcate and spiny; it extends from 12 to 18 scale lengths. The snake is viviparous and breeds during the monsoon months. In the laboratory, one female gave birth to over 30 young snakes at one time. The young shed their skin on the first, third, and seventh days and from the twenty-first day onward. The saw-scaled viper eats insects, slugs, scorpions, snakes, centipedes, and young lizards, the major portion of their diet being insects. The gut contents also reveal large numbers of parts of insects, particularly Dermaptera. Pollen grains from yellow flowers, red soil, and bits of twigs were also found in the gut. Snakes have been reared totally on earwigs. The viper has been observed to go inside tents for prey.

A. Habits

The viper locomotes sideways, rubbing the scales causing a hissing type of noise. It can attack while moving in this fashion and raises its head to an average height of about 25 cm. There are two other snakes, *Dipsas triagonata* and *Psammophis species* that resemble the viper but they do not locomote in this fashion nor are they poisonous. In the Ratnagiri District of Maharashtra as many as 6000 snakes have been collected in a year and transported to Haffkine Institute for their venom. The snakes are found in loose red soil and underneath stones. In former times a bounty was offered for the capture of vipers and 140,828 snakes were collected and destroyed in 1876. In 1890, 225,721 snakes were destroyed. One hundred and twenty-three people died as a result of viper bites that year.

B. Venom

The poison of this snake is vagotoxic. The average venom output per snake is 0.0046 gm while the lethal dose for an average-sized man is 8 mg. Consequently the usual bite is not fatal. The first reaction to a bite is intense swelling and burning pain. There is a continuous oozing from the puncture marks. Suppuration may start within 24 hours, and there is bleeding from the gums, kidney, or nose. Necrosis occurs at the site of bite. If sufficient anti-venin is not given, the secondary symptoms persist for days and the patient succumbs to them in time.

The venom when collected is thick and after centrifugation there is considerable debris. Venom collected from 467 *Echis* after centrifuging and lyophilizing gave 0.8122 gm of supernatant liquid and 0.0242 gm of debris. The latter after reconstitution was nontoxic. The MLD of this venom was 1/40,000.

Saw-scaled vipers have survived only 2 years in captivity.

REFERENCES

Chopra, R. N., and Chowhan, J. S. (1932). *Indian Med. Gaz.* **70**, 445.
Chopra, R. N., and Chowhan, J. S. (1938). *Indian Med. Gaz.* **73**, 720.
Deane, J. B. (1833). "Worship of Serpents." Rivington, London.
Deoras, P. J. (1957). *Everyday Sci.* **1**, 19.
Deoras, P. J. (1961). *J. Univ. Bombay* [N.S.] **30**, 50.
Deoras, P. J. (1963). "Venomous and Poisonous Animals and Noxious Plants of the Pacific Area." Pergamon Press, Oxford. p. 337.
Deoras, P. J. (1966). "Snakes of India." Natl. Book Trust, New Delhi.
Deoras, P. J., and Vad, N. E. (1955). *Haffkine Inst. Ann. Rept.* p. 60.
Deoras, P. J., and Vad, N. E. (1965). *J. Univ. Bombay* **34**, 41.
Deoras, P. J., and Vad, N. E. (1965–66). *J. Univ. Bombay* **34**, 57.
Deoras, P. J., and Vad, N. E. (1968a). *Proc. 11th PPSC, Tokyo,* **8**, 22.
Haffkine Institute. (1958). Annual Report, p. 5.
Indian Council of Medical Research. (1954). Technical Reports, p. 224; p. 215 (1955); p. 24 (1956).
"Mortality Due to Snake Bite." (1954–1958). Annual Report Director of Public Health, Government of Bombay.
Mukul, Dey. (1950). "My Pilgrimage to Ajanta and Bagh." Headley Bros., London.
Oldham, C. F. (1905). "The Sun and the Serpent." Constable, London.
Sarkar, B. B., Mitra, S. R., and Ghosh, B. N. (1942). *Indian J. Med. Res.* **30**, 453.
Smith, M. A. (1943). "The Fauna of British India, Ceylon, and Burma, *Reptilia* and *Amphibia*," Vol. III. Taylor & Francis, London.
Swaroop, S., and Grab, B. (1954). *Bull. World Health Organ.* **10**, 35.
Vidal, G. W. (1890). *J. Bombay Nat. Hist. Soc.* **5**, 64.
Vogel, J. P. (1926). "Indian Serpent Lore." Arthur Prosbsthain, London.
Wall, F. (1905–1919). *Bombay Nat. Hist. Soc.* **16**, 533; **17**, 1, 259, 857; **18**, 1, 227, 525, 711; **19**, 287, 555, 775; **20**, 65, 603, 933, and **21**, 1.

Chapter 23

The Venomous Snakes of Australia and Melanesia

HAROLD G. COGGER

AUSTRALIAN MUSEUM, SYDNEY, AUSTRALIA

I. INTRODUCTION

The snakes of Australia and the southwestern Pacific region contain a greater proportion of venomous species than those of any other part of the world, and include some of the most deadly snakes known. Fortunately, the really dangerous species form only a very small percentage of the total snake fauna, for most snakes in this region are not considered harmful to man. Snakebite is not a serious problem as it is in other parts of southeastern Asia. The annual death rate from snakebite is probably only 1 or 2 per million. At present, specific antivenins are available for all of the dangerous species likely to be commonly encountered.

The majority of venomous snakes in this area are poorly known; distributional data are sketchy for most species, while information on biology and life history is virtually nonexistent for all but a handful. The classification is confused, and frequently fails to indicate affinities; few modern systematic studies have been carried out.

The venoms of only a few species have been adequately studied; the venoms of the majority are virtually unknown.

The geographic limits implied by the term "Australia and Melanesia" are shown in Fig. 1. Unfortunately, Melanesia is defined essentially on ethnographic, rather than zoogeographic, grounds. Its fauna does not represent a discrete unit, but is rather a complex mixture of old endemic elements and more recent faunal exchanges. However, as a result of the distance of Melanesia from the continental masses of the New and Old Worlds, and its access to new terrestrial fauna only through the Indo-Malayan Archipelago, its fauna is sufficiently distinct to justify treatment as a single unit.

Within this area, venomous land snakes are found only in Australia (and its island state, Tasmania), New Guinea and its island archipelagos to the east —the Bismarck and Louisiade Archipelagos, and the Solomon Islands—and the Fiji Islands. New Zealand, which is considered ethnographically as part of Polynesia, has no snake fauna; the islands of New Caledonia and the New Hebrides have no venomous land snakes.

The true sea snakes (family Hydrophiidae) are not discussed in this chapter. However, the banded sea snakes of the genus *Laticauda* are frequently encountered on land on the islands of the southwestern Pacific, but may be readily identified by their paddle-shaped tails and banded bodies. The remaining sea snakes are strictly marine.

No mention is made of currently recognized geographic subspecies. Not only are these considered outside the scope of this contribution, but adequate studies of intraspecific geographic variation in the snakes of this region have never been carried out.

The treatment in this chapter is essentially regional, as this is both practical and convenient. In the regional lists, those species marked with an asterisk are dangerous to man.

The keys to the genera of venomous snakes (pp. 49, 54, and 55) require some qualifying comments. Free use has been made of "field" characters; i.e., those characteristics most obvious in fresh material, which require a minimum knowledge of special terminology, and which can be examined without need of dissection or use of a microscope. Keys using such characters as the number of teeth following the fang, relative shape and positions of the bones of the skull, etc., are readily available in a number of the references listed at the end of the chapter. Not only are such characters of secondary value to the field worker, but they are frequently less reliable than generally supposed. A number of such unreliable key characters have been perpetuated by succeeding generations of herpetologists; indeed, the most disconcerting

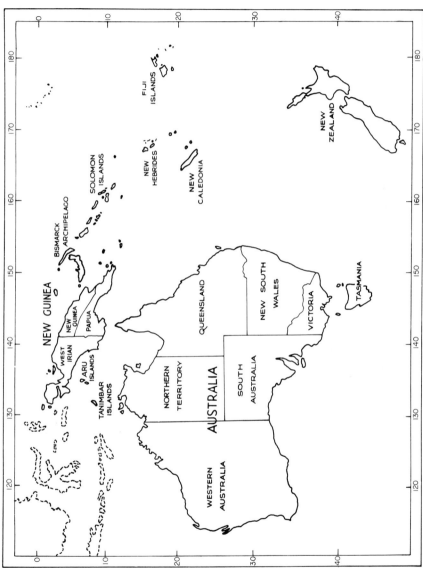

FIG. 1. Map showing the geographic limits of Australia and Melanesia (sometimes referred to as Australasia). The venomous snakes of the land areas with broken outlines are not dealt with in this chapter.

feature of most of the existing keys to the elapids is their emphasis on these unreliable characters. A number of more recent works on Australian reptiles have conveniently avoided this problem by omitting keys altogether. An attempt has been made in this chapter to overcome such deficiencies by use of artificial, regional keys.

All scale row counts given are those at midbody. Although it might seem to be self-evident, the distinction between "venomous" and "dangerous" snakes should be noted.

In Section VII the dangerous land snakes of the whole region are discussed in greater detail, but it should be emphasized that this section has been arranged to supplement the information available in a number of handbooks and articles dealing with the snakes of Australia and the southwestern Pacific (De Rooij, 1917; Kinghorn and Kellaway, 1943; Loveridge, 1945; Kinghorn, 1956; Slater, 1956; Worrell, 1961b, 1963; Werler and Keegan, 1963). Other important references to the herpetology of this region are those of Barbour (1912) and Brongersma (1934). The keys, however, are not available in any recent publication; they should be used, whenever possible, in association with the full descriptions of the species concerned, but it is hoped that they will be of special value to the field worker in this region.

II. THE CLASSIFICATION, ORIGINS, AND RELATIONSHIPS OF THE VENOMOUS LAND SNAKES OF AUSTRALIA AND MELANESIA

All venomous land snakes of this region are members of only two families, the Colubridae and the Elapidae.

The family Colubridae includes the bulk of the world's snakes, most of which are solid-toothed and harmless. A small number, however, possess venom glands and one or more pairs of enlarged, grooved fangs on the hind end of the maxillary bone. However, all Australian and Melanesian species have relatively small fangs, together with venom that is apparently of low toxicity, and all are considered harmless to man. No serious effects have ever been recorded from the bite of any of these species. With the exception of the brown tree snake *(Boiga irregularis)*, all rear-fanged or opisthoglyph colubrids in this area are aquatic or semiaquatic.

Although the rear-fanged colubrids are included in the regional lists, the absence of any dangerous species from the region does not warrant inclusion of these species in the keys.

The family Elapidae includes many of the world's deadliest snakes—such well known groups as the Afro-Asian cobras, the African mambas, the Indo-Malayan kraits, and the New World coral snakes. Nowhere, however, are the elapids so numerous and diverse as in the Australo-Papuan region, and it is this family to which *all* dangerous land snakes in Australia and Melanesia belong.

The elapids are characterized by the possession of a pair of enlarged, grooved fangs on the front of the maxillary bone. Unlike the mobile fangs of the vipers (Viperidae) and pit-vipers (Crotalidae), the fangs of the elapids are usually relatively short, and are more or less immovably fixed to the jaw. Though they may be functionally hollow, the fangs originate as a groove or canal in the front surface of the fang. In most species, however, the inward growth of calcium over the original canal forms a distinctly hollow fang, the only external sign of the original canal being a faint groove down the front surface. When the snake bites, venom passes from the venom gland, along a duct to an opening at the base of the fang, and then down the fang to an opening situated on the front surface of the fang near its tip. The fangs arise from paired sockets on each maxilla, and by the alternate use of the fang in each socket there is always a functional fang available when the older one is shed. Only rarely do both fangs function simultaneously, but as the teeth (including the fangs) are periodically replaced, it is not uncommon to find one or more "reserve" or "replacement" fangs arising near the base of the front ones on each side. The dentition and biting mechanism of Australian snakes has been fully discussed by Fairley (1929b).

In Fig. 2 are shown skulls of some elapids from the Australo-Papuan region.

The venoms of some of the elapids of this area are among the most potent known. The venoms of the dangerous species are principally neurotoxic, and death is generally caused by respiratory paralysis. Other important properties of the venoms of some species are discussed, if known, under the individual species. The venoms of the majority of these snakes have not been studied.

Most elapids of this region are of slender, elongate build, with head only slightly or not distinct from the neck. However, there is one notable and well-known exception—the death adder *(Acanthophis antarcticus)*, which resembles the viper in body shape. It has a relatively broad, arrow-shaped head, narrow neck, short stout body, and a short, sharply tapering tail (Fig. 3).

Within the Elapidae, classification has been based on numerous external and anatomical characters. All of the standard characteristics of scalation have been used, while such characters as the number and form of the maxillary teeth and the relative positions of various skeletal elements have frequently been employed.

Classification of the snakes of this region, especially at the generic level, is still in rather a chaotic state. A generic classification was first stabilized in the British Museum Catalogue of Boulenger (1896) and has remained with little change since that time. Indeed, the most recent checklists of De Haas (1950) and Klemmer (1963) show only minor deviation from Boulenger's generic classification. The greatest failing of this classification is that the genera were originally defined on only a few specimens, with little consideration of total variability and inter- and intraspecific variation. This resulted in genera so broadly defined that in many cases they embraced species that by "modern" concepts of classification belong to several genera. Recent work of Worrell (1961a, 1963) has attempted to rectify the shortcomings of the Boulengerian generic classification, but in doing so has so ignored the total morphology in favor of a handful of minor skull characters, as well as variation in these characters, that many new, virtually undefined genera have been erected.

Undoubtedly the true picture lies somewhere between these two extremes. However, because of the more balanced and erudite morphological studies of

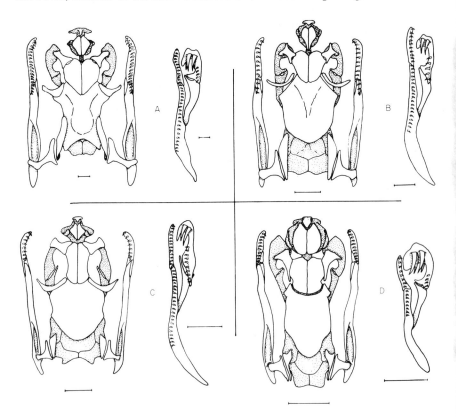

FIG. 2. Skulls of (A) tiger snake, *Notechis scutatus*; (B) mallee snake, *Denisonia suta*; (C) yellow-faced whip snake, *Demansia psammophis*; (D) bandy-bandy, *Vermicella annulata*.

FIG. 3. The death adder, *Acanthophis antarcticus*, showing the viperlike shape of the body and the distinctive tail and spine (arrow). (Photo: Howard Hughes/Australian Museum.)

the older British Museum school, .as well as their relative stability, the classification presented in this chapter largely reflects the older generic classification. Brongersma and Knaap-Van Meeuwen (1964) have described a new species of *Denisonia (D. boschmai)* from Merauke in West New Guinea. In the same paper, they rejected the multiplicity of generic names proposed by Worrell for species which have long been placed in the genus *Denisonia*.

Unfortunately it is not possible to distinguish a dangerous elapid from a harmless colubrid on external character alone. Positive identification can be made only by examining the open mouth for the presence of fangs. These are not always obvious in many of the smaller species, for the fangs may not only be very short, but each is normally enclosed in a fleshy sheath which generally hides all but the sharp tip of the fang. It is often necessary to use a needle or twig to turn back the sheath to reveal the fang.

Within this region, there is one character that can be of considerable value in determining whether or not a particular snake is dangerous. The loreal shield is a small scale present in some snakes between the eye and the tip of the snout. It is *absent* in *all* dangerous snakes of Australia and Melanesia. However, it is also absent in some harmless species. Hence its *absence* in any particular snake will reveal nothing of the snake's potential danger; its *presence,* however, wili immediately identify a *harmless* snake.

All dangerous land snakes in Australia and Melanesia are members of only one family; this facilitates identification, and has practical implications in the treatment of snakebite. Each of the antivenins for the dangerous snakes of this region gives a varying degree of cross protection against the bites of the others. In New Guinea, the presence of only five dangerous elapid species has allowed the development of a polyvalent serum that can be used for nearly all cases of snakebite in this region.

There seems little doubt that both the Colubridae and the Elapidae had their origins to the west, and entered the Australo-Papuan region through the Indo-Malayan Archipelago. There also seems little doubt that the elapids have had a much longer history in this area than have the colubrids.

The considerably earlier entry of the elapids is evidenced by their extensive radiation and specialization, their high degree of endemicity, and their geographic range and distribution through a broad spectrum of environments. Colubrids, on the other hand, are generally represented in this area by only a handful of species whose environmental limitations are broader than most other members of the family.

The present distribution of snakes in this region is largely a reflection of changes brought about during the Pleistocene period. Prior to this, each of the major land masses in this region had evolved distinctive endemic elapid faunas. The contemporary elapid fauna of Australia has developed from old endemic stock. The endemic elapid element is represented in New Guinea by *Apistocalamus* and its allied genera, and probably by *Micropechis*; in the Solomon Islands by *Denisonia par*; and in Fiji by *Ogmodon*. However, as a result of the accumulation of polar ice during the great ice ages of the Pleistocene period, the sea level was at times as much as 300 feet lower than at the present time. This resulted in either broad land connections (as between Australia and New Guinea), or a narrowing of the sea barrier between land masses (as between the Bismarck Archipelago and the Solomon Islands), enabling a number of forms to move into new areas. The last major land bridge is believed to have occurred about 17,000–20,000 years ago, during the Würm Glaciation (Gill, 1961). With the subsequent melting of the polar ice caps, land bridges were again severed, resulting in the disjunct distribution of many forms as seen today. The fact that most of these forms have only slightly differentiated from their parental stock, or not at all, together with their restriction even now to the parental habitat, is evidence of their recent separation.

Even during these Pleistocene times, exchanges took place that were almost entirely internal—an exchange of species between land masses within the Australo-Pacific region. There is little evidence of the entry of Indo-Malayan elements at that time, although a few Australian forms were able to penetrate into the archipelago to the west (e.g., the Australian death adder, *Acanthophis antarcticus*). This is not the place for a discussion of the hypotheses of Wallace and others regarding the faunal barrier between the Oriental and Australian regions, for this subject has been fully treated elsewhere (Darlington, 1957; Mayr, 1945). However, there is no doubt that a very effective barrier prevented the entry of most Indo-Malayan herpetofaunal elements into the Australo-Papuan region.

Typical of the exchanges that occurred during Pleistocene times is that between Australia and New Guinea, which are discussed in Sections III and IV.

III. THE VENOMOUS LAND SNAKES OF AUSTRALIA

A. Discussion

By far the largest land mass in this region, Australia, not unexpectedly, has the richest snake fauna. However, the most notable feature of the Australian herpetofauna is that nearly 70% (63 species) of its snakes are venomous members of the family Elapidae. This initially alarming figure falls into perspective when it is realized that only 13 of these species are harmful to man. Yet despite the fact that nowhere are the elapids so numerous and diverse as in the Australo-Papuan region, little is known of the biology and life history of the great majority of species.

Australia is essentially a flat country, largely devoid of major or extensive topographic features, and hence physiographic barriers to distribution or dispersal are relatively few. The dominant feature is the eastern highland region (including the Great Dividing Range), which extends from Cape York Peninsula along the eastern coast to western Victoria. These highlands vary considerably in height from Cape York, where they are little more than a complex series of low hills, to the southeastern corner of the continent where they attain their maximum altitude of nearly 7500 feet. These highlands are bordered on the eastern side by a narrow strip of well-watered coastland, and on the western side by the dry interior. The remainder of the continent is mostly flat, dry to arid country, with the exception of a northern coastal belt and the wet southwestern corner. Mountain ranges of moderate altitude, such as the Darling and Hammersley Ranges in Western Australia, the Flinders and Musgrave Ranges in South Australia, and the MacDonnell Ranges in the Northern Territory, are scattered throughout the continent. These frequently provide environmental conditions very different from the surrounding country, and by virtue of their distinctive environments they may act as faunal refuges. However, they are generally not sufficiently extensive to act as barriers to animal distribution.

Climatically, continental Australia ranges southward from the tropical north through a broad temperate region. With the exception of a relatively narrow, discontinuous coastal fringe, the continent is extremely dry; about one third receives less than 10 inches of rainfall per annum, while another third receives less than 20 inches. The central region consists largely of extensive tracts of desert and semidesert; salt lakes and claypans alternate with consolidated sand dunes and arid scrubs.

The vegetation, largely dependent on rainfall, follows a radiating pattern, from the desert grasslands and scrubs of the central regions through the savannah grasslands and woodlands typical of the 20 to 30 inch rainfall zone, to the tropical woodlands of the north and the sclerophyll forest of the high rainfall zones in the east, southeast, and southwest. Rain forest occurs in a series of isolated pockets along the eastern seaboard.

Despite the lack of distributional barriers in Australia, the snake fauna is rich and diverse; and despite the variety and extremes of environmental conditions, the majority of snakes, particularly the larger, dangerous species, are widely distributed through a range of these environments. Many species extend from the wet sclerophyll forests of the coastal belts to the arid interior, or may be as frequently encountered in the tropical north as in the southern latitudes. For this reason, patterns of distribution are difficult to correlate with major environmental factors.

The snake fauna of Australia is highly endemic; New Guinean or Indo-Malayan elements are few. Within the family Elapidae there are no species or genera that are clearly of recent northern origin. Although a number of species and genera of elapids are common to both northern Australia and New Guinea, there is little doubt that the New Guinean representatives are derived from Australian stock that entered New Guinea when the two countries were last in physical contact, at the height of Pleistocene glaciation (p. 42). Examples of such exchange of elapid snakes are the taipan *(Oxyuranus scutellatus)*, the death adder *(Acanthophis antarcticus)*, and the black snakes *(Pseudechis)*.

The colubrid snakes probably entered Australia contemporaneously with the exchange of elapid snakes. With the exception of a number of widely distributed species of aquatic or semiaquatic rear-fanged colubrids (which are found along the northern coast of Australia), only one terrestrial species has succeeded in entering Australia. This is the ubiquitous brown tree snake

(Boiga irregularis), which is found in the coastal areas of northern and eastern Australia.

A remarkable feature of the reptile fauna of Australia is the large proportion of species occurring in the arid interior. This can be clearly seen in Table I, in which the distributions of all Australian snakes are outlined briefly.

Tasmania, which climatically is rather similar to the southeastern corner of the mainland, has only three species of land snakes, all of which are venomous and all of which also occur on the mainland. These are the tiger snake *(Notechis scutatus)*, the copperhead *(Denisonia superba)*, and the white-lipped snake *(Denisonia coronoides)*. The latter grows to only 2 feet in length, and is not harmful to man. The copperhead is also known in Tasmania as the diamond snake. At the time when Australia and New Guinea were last in physical contact, so too were Australia and Tasmania.

Little is known of the biology of the Australian snakes. Few studies have been carried out regarding food preferences, growth, or breeding habits. Of the larger dangerous snakes, only two genera—*Demansia* and *Oxyuranus*—are oviparous. Members of the genera *Acanthophis, Notechis, Denisonia,* and *Tropidechis* are "live bearing" or ovoviviparous.

There is a definite and expected trend toward increase of the number of ovoviviparous species with increasing latitude.

Snakebite fatalities in Australia are few in relation to the population. The statistics in the first quarter of this century were reviewed by Fairley (1929a), while Trinca (1963) records only 45 fatalities over a 10-year period, from 1952 to 1961. This relatively low figure is the result of a number of factors; the most significant of which are that (a) all Australian snakes will avoid contact with man whenever possible; (b) the density of the human population is very low in most parts of Australia; (c) most Europeans wear clothing and footwear that offers protection from the relatively short fangs of most Australian snakes; (d) there is only a small native population; and (e) most of the Australian environment is sufficiently open to enable a large snake to be first seen from a safe distance, thus contact may be avoided. Finally, the ready availability of medical treatment, even in most remote areas, minimizes the chance of a fatal outcome from a bite.

In any one area in Australia, it is likely that only a small number of dangerous snakes will be found, so that it should not be difficult to become acquainted with the dangerous species of any one district. Indeed, this is advisable because one of the greatest deterrents to successful treatment of snakebite is ignorance of the species responsible for the bite.

In Tables I, II, and III the dangerous species are marked with an asterisk. Perusal of these tables will quickly indicate the dangerous species known or likely to occur in any one region.

TABLE I

THE VENOMOUS LAND SNAKES OF AUSTRALIA

Family Colubridae

Boiga irregularis (Merrem) Brown tree snake	Northern and eastern Australia
Cerberus rhynchops (Schneider) Bockadam	Coastal areas of northern Australia
Enhydris polylepis (Fischer) Macleay's water snake	Coastal areas of northeastern Australia
Enhydris punctata (Gray) Spotted water snake	Coastal areas of northern Australia
Fordonia leucobalia (Schlegel) White-bellied mangrove snake	Coastal areas of northern Australia
Myron richardsoni Gray Gray's water snake	Coastal areas of northern Australia

Family Elapidae

*Acanthophis antarcticus (Shaw) Death adder	Australia generally, except Tasmania; also in New Guinea
Aspidomorphus harriettae (Krefft) White-crowned snake	Central coastal parts of eastern Australia
Aspidomorphus kreffti (Günther) Dwarf crowned snake	Coastal parts of eastern Australia, north of Sydney
Aspidomorphus squamulosus (Duméril & Bibron) Golden crowned snake	Central coastal parts of eastern Australia
Brachyaspis curta (Schlegel) Bardick or desert snake	Southwestern Australia, through South Australia to western Victoria
Brachyurophis australis (Krefft) Australian coral snake	Inland Queensland and New South Wales
Brachyurophis campbelli (Kinghorn) Kinghorn's girdled snake	Queensland
Brachyurophis roperi (Kinghorn) Roper girdled snake	Northern Territory
Brachyurophis semifasciata Günther Half-girdled snake	Western Australia
Brachyurophis woodjonesii (Thomson) Thomson's girdled snake	Cape York Peninsula
Demansia acutirostris Mitchell Sharp-snouted brown snake	Lake Eyre, South Australia
*Demansia guttata Parker Spotted brown snake	Inland Queensland
Demansia modesta (Günther) Ringed brown snake	Inland New South Wales and Queensland to Western Australia
*Demansia nuchalis (Günther) Western brown snake	Western Australia, Northern Territory, South Australia and western parts of New South Wales and Queensland

* Asterisk indicates dangerous species.

TABLE I (*continued*)

Family Elapidae (*continued*)

*Demansia olivacea (Gray) Spotted-headed snake	Northern and northeastern Australia
Demansia ornaticeps (Macleay) Ornate whip snake	Far northern Australia
Demansia psammophis (Schlegel) Yellow-faced whip snake	Australia generally, except Tasmania
*Demansia textilis (Duméril & Bibron) Eastern brown snake	Eastern and southeastern Australia, excluding Tasmania
Demansia torquata Günther Collared whip snake	Coastal and inland regions of northern and northeastern Australia
Denisonia brevicauda Mitchell Short-tailed snake	Eyre Peninsula, South Australia
Denisonia brunnea Mitchell Mitchell's desert snake	Eyre Peninsula, South Australia
Denisonia carpentariae (Macleay) Carpentaria whip snake	Northeastern Australia
Denisonia coronata (Schlegel) Crowned snake	Southwestern to southeastern Australia
Denisonia coronoides (Günther) White-lipped snake	Coast and highlands of southeastern Australia, including Tasmania. Doubtfully recorded from southwestern Australia
Denisonia daemellii (Günther) Gray snake	Central Queensland to inland New South Wales
Denisonia devisii Waite & Longman De Vis' banded snake	Inland New South Wales and Queensland
Denisonia fasciata Rosen Rosen's snake	Southern part of Western Australia
Denisonia flagellum (McCoy) Little whip snake	Southeastern Australia, from eastern South Australia through Victoria to southeastern New South Wales
Denisonia gouldii (Gray) Black-headed snake	Australia generally, except the southeastern corner and Tasmania
Denisonia maculata (Steindachner) Ornamental snake	Coast and ranges of central Queensland
Denisonia mastersi (Krefft) Master's snake	Slopes and highlands of southeastern Australia
Denisonia nigrostriata (Krefft) Black-striped snake	Northeastern Australia
Denisonia pallidiceps (Günther) Australian small-eyed snake	Coastal and nearcoastal areas of northern and eastern Australia
Denisonia punctata Boulenger Little spotted snake	Inland or arid parts of Queensland, Northern Territory and northwestern Australia
Denisonia signata (Jan) Black-bellied marsh snake	Coast and tablelands of Queensland and New South Wales

* Asterisk indicates dangerous species.

TABLE I (*continued*)

Family Elapidae (*continued*)

Denisonia superba (Günther) Australian copperhead	Tasmania and southern Australia, from eastern South Australia to the tablelands of New South Wales
Denisonia suta (Peters) Myall or mallee snake	Inland Australia, reaching coastal areas only in drier regions
Elapognathus minor (Günther) Little brown snake	Southwestern Australia
Furina christeanus (Fry) Yellow-naped snake	Northern Australia
Furina diadema (Schlegel) Red-naped snake	Most parts of Australia
Glyphodon dunmalli Worrell Dunmall's snake	Southeastern Queensland
Glyphodon tristis Günther Brown-headed snake	Northern Australia and islands of Torres Strait
Hoplocephalus bitorquatus (Jan) Pale-headed snake	Central New South Wales to northern Queensland
Hoplocephalus bungaroides (Boie) Broad-headed snake	Coast and tablelands from central New South Wales to southeast Queensland
Hoplocephalus stephensi Krefft Yellow-banded snake	Coastal areas from central New South Wales to southeast Queensland
Notechis scutatus (Peters) Tiger snake	Southern Australia (except arid regions) and Tasmania. Southeastern Queensland
Oxyuranus scutellatus (Peters) Taipan	Northern and northeastern Australia, extending through inland Queensland to northeastern South Australia and western New South Wales
Pseudechis australis (Gray) Mulga snake	Inland Australia, extending to the coastal regions in the north and west
Pseudechis colletti Boulenger Collett's snake	Inland parts of central and southern Queensland
Pseudechis guttatus De Vis Spotted black snake	Central Queensland to central New South Wales excluding most coastal areas
Pseudechis porphyriacus (Shaw) Red-bellied black snake	Eastern Australia, from eastern South Australia to north Queensland
Rhinelaps approximans (Glauert) Glauert's banded snake	Western Australia
Rhinelaps fasciolatus Günther Narrow-banded snake	Southwestern Australia
Rhinelaps warro (De Vis) Black-naped burrowing snake	Northeastern Queensland
Rhinhoplocephalus bicolor Müller Müller's snake	Southwestern Australia

* Asterisk indicates dangerous species.

TABLE I (*continued*)

Family Elapidae (*continued*)

Rhynchoelaps bertholdi (Jan) Desert banded snake	Western, South and Central Australia
Tropidechis carinatus (Krefft) Rough-scaled snake	Coast (and parts of the highlands) of eastern Australia, from central New South Wales to north Queensland
Vermicella annulata (Gray) Bandy-bandy	Australia generally, except Tasmania
Vermicella bimaculata (Duméril & Bibron) Black-naped burrowing snake	Southwestern Australia
Vermicella calonota (Duméril & Bibron) Western black-striped snake	Southwestern Australia
Vermicella minima (Worrell) Miniature burrowing snake	Northwestern Australia
Vermicella multifasciata (Longman) Many-ringed bandy-bandy	Northern Australia

* Asterisk indicates dangerous species.

B. Key to Genera of the Elapids of Australia

1. Tail without spine 2
   Tail with spine (Fig. 3) . . . . *Acanthophis*
2. Internasals present 3
   Internasals absent . . . . *Rhinhoplocephalus*
3. Scales in 21 or more rows at midbody 4
   Scales in less than 21 rows at midbody 7
4. Subcaudals single 5
   Subcaudals divided 6
5. Scales smooth . . . . *Hoplocephalus*
   Scales strongly keeled . . . . *Tropidechis*
6. Anal single . . . . *Oxyuranus*
   Anal divided 12
7. All subcaudals single 8
   At least some subcaudals divided 11
8. Head only slightly distinct from neck; scales in 13 to 19 rows; if in 19 rows, ventrals more than 140 9
   Head very distinct from neck; scales in 19 rows; ventrals less than 140 . . . . *Brachyaspis*
9. Frontal shield longer than broad; where frontal is only slightly longer than broad, lower anterior temporal shield is much shorter than frontal 10
   Frontal shield scarcely or not longer than broad; lower anterior temporal shield about as long as or longer than the frontal . . *Notechis*

10. No teeth following the fang .... *Elapognathus*
    Three or more teeth following the fang .... *Denisonia*
11. All subcaudals divided....................................... 12
    Some subcaudals single, remainder divided .... *Pseudechis*
12. Diameter of eye about equal to or less than the distance from the
    mouth 13
    Diameter of eye much greater than the distance from the mouth *Demansia*
13. Rostral shield rounded..................................... 14
    Rostral shield sharp, projecting, shovel-shaped........ *Brachyurophis*
14. Diameter of eye about equal to or only slightly less than the distance
    from the mouth; body with or without cross bands.............. 15
    Diameter of eye much less than the distance from the mouth; body
    without cross bands .... *Glyphodon*
15. Body with or without cross bands; if unbanded, ventral surface
    immaculate white or cream 16
    Body without cross bands; uniformly dark brown in color dorsally;
    ventral surface colored or patterned .... *Aspidomorphus*
16. Nasal and preocular scales in contact 17
    Nasal and preocular scales widely separated 18
17. Ventrals more than 150 .... *Vermicella*
    Ventrals less than 150 .... *Rhynchoelaps*
18. Subcaudals more than 35 .... *Furina*
    Subcaudals less than 35 .... *Rhinelaps*

IV. THE VENOMOUS LAND SNAKES OF NEW GUINEA

A. Discussion

As in Australia, the rear-fanged colubrids are poorly represented. Indeed, with only two exceptions (see Tables I and II) the same species are shared by the two countries. None of the New Guinean opisthoglyph colubrids are harmful to man. The remaining venomous snakes are all members of the family Elapidae.

Topographically, New Guinea is dominated by a central, mountainous "spine", which rises in a number of places to more than 15,000 feet. This "spine" does not consist of a single mountain range, but rather of a complex of ranges forming a network that varies greatly in width, and is intersected by upland valleys that sometimes form broad highland plains.

To about 6500 feet the central mountains are clothed in a mixture of sub-tropical forests, except where the clearing and gardening activities of the native peoples have resulted in large tracts of grassland and secondary forest growth (bush fallow). Generally these latter areas include the valley floors and the adjacent mountainsides to about 7000 feet—occasionally to above 8000 feet. The subtropical forests are in turn replaced by montane forests of conifers and antarctic beech *(Nothofagus)* which extend to about 11,000 feet. The relatively small areas of alpine scrubs and grassland above this altitude are believed to be devoid of snakes.

Surrounding the central ranges is a lowland fringe of varying width. It is widest in southern New Guinea, where between the Digoel and Fly Rivers it extends more than 200 miles inland without attaining an altitude of more than 300 feet. Much of this enormous area of lowland is subject to periodic inundation, and consists of a mixture of mangrove forest (coastally), swamp forest, and swamp grassland. In the drained areas, rainfall patterns produce a mixture of rain forest, monsoon forest, and, in many areas along the southern coast, open savannah similar to that occurring in northern Australia.

Except for the great lowland valleys of the Sepik and Mamberamo Rivers in the north, the remaining lowlands are largely restricted to a narrow coastal fringe.

Most dangerous snakes in New Guinea occur only at altitudes below 2500 feet. Exceptions are the death adder *(Acanthophis antarcticus)* and the endemic small-eyed snake *(Micropechis ikaheka),* a forest-dwelling species that occurs up to nearly 5000 feet. The latter is capable of inflicting a fatal bite, and is much feared by the natives in many areas.

The remaining dangerous snakes of the genera *Demansia, Pseudechis,* and *Oxyuranus* are entirely restricted to the open savannahs and monsoon forests of the southern lowlands, and occur as far eastward as Milne Bay. These snakes, together with *Acanthophis,* are clearly of recent Australian origin, and all are congeneric—indeed most are conspecific—with Australian forms. They occur in an area that, as mentioned previously, has a climate and vegetation closely resembling those of Cape York Peninsula.

Endemic to the New Guinean region are the small cryptozoic elapids of the genera *Toxicocalamus, Ultrocalamus, Pseudapistocalamus,* and *Apistocalamus.* They have been considered among the most primitive members of the family Elapidae, but there is little doubt that they are also highly specialized. Most are known only from a handful of specimens, and it is not unlikely that future studies will relegate some currently recognized as separate species to synonyms of other species. Their distribution is particularly sketchy, but some species occur above 7000 feet.

In the Bismarck Archipelago, New Britain and New Ireland are mountainous, forest-covered islands with relatively narrow low coastal fringes. The mountains of New Britain attain an altitude in excess of 7000 feet, while those of New Ireland reach 4000 feet. Little is known of the snake faunas of these islands, or of the Admiralty Islands to the north. The only dangerous

TABLE II

The Venomous Land Snakes of New Guinea

Family Colubridae

Boiga irregularis (Merrem)	New Guinea and associated islands;
Brown tree snake	Bismarck Archipelago
Cantoria annulata Jong	Southern New Guinea
Cerberus rhynchops (Schneider)	Coast and rivers of New Guinea
Bockadam	
Enhydris enhydris (Schneider)	Coast and rivers of New Guinea
Enhydris polylepis (Fischer)	Coast and rivers of New Guinea
Macleay's water snake	
Enhydris punctata (Gray)	Coast and rivers of New Guinea
Spotted water snake	
Fordonia leucobalia (Schlegel)	Coast and rivers of New Guinea
White-bellied mangrove snake	
Heurnia ventromaculata Jong	Northwest New Guinea
Myron richardsoni Gray	Coast and rivers of New Guinea
Gray's water snake	

Family Elapidae

*Acanthophis antarcticus (Shaw)	New Guinea generally, Moluccas
Death adder	
Apistocalamus grandis Boulenger	New Guinea
Apistocalamus lamingtoni Kinghorn	Eastern New Guinea
Apistocalamus lonnbergi Boulenger	Northeastern New Guinea
Apistocalamus loriae Boulenger	New Guinea
Apistocalamus pratti Boulenger	Eastern New Guinea
Aspidomorphus mulleri (Schlegel)	New Guinea generally, Bismarck and Louisiade Archipelagos, d'Entrecasteaux Islands
Aspidomorphus schlegeli (Günther)	Western New Guinea
*Demansia olivacea (Gray)	Southern New Guinea
Spotted-headed snake	
Glyphodon tristis Günther	Southeastern New Guinea
Brown-headed snake	
*Micropechis ikaheka (Lesson)	New Guinea, Aru Islands
New Guinea small-eyed snake	
*Oxyuranus scutellatus (Peters)	Southern and southeastern New Guinea
Taipan	
*Pseudechis papuanus Peters & Doria	Southern and southeastern New Guinea
Papuan black snake	
Pseudapistocalamus nymani Lönnberg	Northeastern New Guinea
Toxicocalamus longissimus Boulenger	Eastern New Guinea, d'Entrecasteaux Islands
Toxicocalamus stanleyanus Boulenger	New Guinea
Ultrocalamus preussii Sternfeld	New Guinea

* Asterisk denotes dangerous species.

land snake recorded from any of these islands is the death adder (Kinghorn and Kellaway, 1943), which they reported only from New Britain. However, our knowledge is too inadequate to state categorically that dangerous snakes are absent from the remaining islands.

The snake faunas of Woodlark Island, the Entrecasteaux Islands, and the islands of the Louisiade Archipelago are also little known. Again, no dangerous land snakes have been recorded from any of these islands, but it is possible that one or more species may yet be found there.

There are many smaller offshore islands and island groups whose snakes are poorly known. For the most part, the death adder is the only dangerous species likely to occur on most of these islands, although *Micropechis ikaheka* is found on a number of the islands off the western tip of New Guinea, and also on the Aru Islands. Along the southern coast, too, many small offshore islands are inhabited by nearly all species living on the adjacent coast.

With the exception of the colubrids listed in Table II, there are no Indo-Malayan elements in the venomous snake fauna of the New Guinean region. Both the true vipers (Viperidae) and the pit vipers (Crotalidae), as well as the cobras and kraits within the family Elapidae, approach the western boundary of this region, but have failed to gain entry.

All of the dangerous snakes of New Guinea, except *Oxyuranus* and possibly *Micropechis,* are ovoviviparous. They display a wide range of behavior consistent with their diversity of structure and habitat. All but the death adder *(Acanthophis antarcticus)* are largely diurnal, although Slater (1956) reports that *Micropechis* is equally as active during the night as during the day. Most of the larger species of venomous snakes feed on a variety of small vertebrates, but the dietary habits of the majority of species are unknown.

Little is known regarding the occurrence of snakebite in New Guinea. Campbell and Young (1961) record 15 cases of elapid snakebite (only one fatal) treated at the Port Moresby District Hospital between November 1959 and August 1960. All cases came from an area within 100 miles of Port Moresby.

There is little doubt that the majority of bites occur within the southern coastal region of Papua, where the larger diurnal elapids are prevalent. In the case of all dangerous New Guinean snakes, the effects of envenomation are largely neurotoxic, and death usually results from respiratory paralysis. Specific antivenins are prepared by the Commonwealth Serum Laboratories in Melbourne for all of New Guinea's dangerous snakes except the small-eyed snake *(Micropechis ikaheka)*. In 1963 a polyvalent antielapid serum was used successfully for the first time.

An important range extension, from both the medical and biological viewpoints, is the confirmed presence of the eastern brown snake (*Demansia textilis*) on the northeastern coast of Papua (McDowell, 1967, and Australian Museum records from Dogura and Embogo).

B. Key to Genera of the Elapids of New Guinea

1. Tail without spine. 2
   Tail with spine (Fig. 3) . . . . *Acanthophis*
2. Preocular absent 3
   Preocular present 4
3. Internasals present . . . . *Toxicocalamus*
   Internasals absent . . . . *Ultrocalamus*
4. Nasal divided 5
   Nasal undivided . . . . *Pseudapistocalamus*
5. Anal divided 6
   Anal single. . . . . *Oxyuranus*
6. Scales in more than 15 rows 7
   Scales in 15 rows. 8
7. Scales in 17 rows. . . . . *Glyphodon*
   Scales in 19 or 21 rows . . . . *Pseudechis*
8. Diameter of eye about equal to or less than the distance from the
   mouth 9
   Diameter of eye much greater than the distance from the mouth *Demansia*
9. Frontal shield about as long as broad; no pattern of stripes or
   blotches on head or neck 10
   Frontal shield much longer than broad; head and neck usually with a
   pattern of distinctive stripes and blotches . . . . *Aspidomorphus*
10. Scales on body black- or brown-centered, broadly edged with yellow
    or cream . . . . *Micropechis*
    Scales on body mostly uniform black or brown . . . . *Apistocalamus*

V. THE VENOMOUS LAND SNAKES OF THE SOLOMON ISLANDS

A. Discussion

Topographically and climatically, the Solomon Islands are akin to those of the Bismarck Archipelago. Most of the larger islands that make up the Solomon group are very mountainous, and are covered by dense tropical and montane forests. The mountains often fall away steeply to the coast, leaving only a narrow fringing area of coastal lowland. The latter consists largely of a

TABLE III

THE VENOMOUS LAND SNAKES OF THE SOLOMON ISLANDS

Family Colubridae	
Boiga irregularis (Merrem)	Solomon Islands
Brown tree snake	
Family Elapidae	
Denisonia par (Boulenger)	Solomon Islands
*Micropechis elapoides (Boulenger)	Florida Island, Solomon Islands
Banded small-eyed snake	
Parapistocalamus hedigeri Roux	Bougainville, Solomon Islands

* Asterisk denotes dangerous species.

mixture of rain forest, grassland, secondary forest regrowth, and swamp forest, with coastal mangroves.

Although the Emperor Range on Bougainville exceeds 8500 feet (and has two active volcanoes), the mountains of most of the larger islands rise to between 4000 and 7000 feet. Many of the smaller islands that make up the archipelago are relatively low-lying, and are clothed in mixed lowland forests.

The snakes of the Solomon Islands are poorly known, and little has been recorded of their biology, habits, habitats, distribution, and venoms, or of the incidence of snakebite in the area. All of the elapids of the Solomon Islands are endemic; faunistically, these represent a long-isolated outpost of the New Guinean snakes. The similarity between *Denisonia par* of the Solomons and the Australian members of this genus is almost undoubtedly superficial and the result of convergence; future studies will probably show that this snake has only remote affinities with the Australian elapids, and that it is not congeneric with any Australian species.

No information is available on the occurrence of snakebite. It is unlikely that any but the small-eyed snake *(Micropechis elapoides)* could deliver a fatal bite, and this opinion is based on its size and the known potential of its close relative *Micropechis ikaheka* in New Guinea. *M. elapoides* is generally regarded as being confined to Florida Island, the type locality; however, specimens have come from Guadalcanal and Malaita Islands, and the species possibly occurs on other islands in the group.

B. Key to Genera of the Elapids of the Solomon Islands

1. Anal shield single 2
   Anal shield divided. . . . . *Denisonia*
2. Body uniform black or dark brown. . . . . *Parapistocalamus*
   Body with alternate bands of yellow and dark brown . . . . . . *Micropechis*

VI. THE VENOMOUS LAND SNAKES OF THE
FIJI ISLANDS

The Fiji Islands are a mixture of forested volcanic "high" islands and low, wooded coral or limestone islands.

The larger volcanic islands rise in places to more than 4,000 feet, and are divided into two quite distinct zones. The windward or rainy side of each island is clothed in relatively dense tropical forests; however, on the rain-shadowed lee side of each island, grasslands predominate, with only scattered trees and patches of rain forest. Large tracts of mangroves fringe the coasts.

Only one venomous land snake is found in the Fiji Islands—*Ogmodon vitianus,* a monotypic elapid genus endemic to these islands. Although venomous, it is not considered harmful to man. *Ogmodon vitianus* is a small snake, which attains a length of less than 2 feet. It is light to dark brown above, and cream-colored below with dark brown or black mottlings. The young have a light yellow bar across the back of the head. The latter is rather pointed, and is only slightly distinct from the neck. The scales are smooth, in 17 rows; the ventrals number 130–160; the anal is divided; the subcaudals are in 25–40 pairs.

Like the snakes of the Solomon Islands, little is known of the biology, habits, or venom of *Ogmodon.* Derrick (1957) records that it is known in Fiji as the bolo, that it is not aggressive, and that it is only rarely seen. Tippett (1958) states that "it is a mountain snake, and sometimes called 'Snake of the Mountain Ferns' *(gata ni balabala).*"

VII. THE DANGEROUS LAND SNAKES OF
AUSTRALIA AND MELANESIA

In the following section the dangerous species of snakes of the Australian and Melanesian regions are considered in greater detail. Within this geographical region, dangerous snakes are found only in Australia, New Guinea, and the Solomon Islands. The dangerous snakes occurring in each of these areas are listed in Table IV. Then follows more detailed descriptions and notes on the individual species, in which the diagnostic characters of each are shown in bold type.

To determine whether a particular snake is dangerous, an attempt should first be made to establish the genus to which it belongs, using the appropriate regional generic key. Once the genus has been determined, reference to Table IV will indicate whether the particular genus contains any dangerous species; if so, the species can be determined by comparing the diagnoses provided in the descriptions of individual species.

To confirm an identification made in this way, the full description, distribution, and biological notes should all be taken into account, and reference

TABLE IV

THE DANGEROUS SNAKES OF AUSTRALIA AND MELANESIA

Australia
  Acanthophis antarcticus (death adder)
  Demansia guttata (spotted brown snake)
  Demansia nuchalis (western brown snake)
  Demansia olivacea (spotted-headed snake)
  Demansia textilis (eastern brown snake)
  Denisonia superba (copperhead)
  Notechis scutatus (tiger snake)
  Oxyuranus scutellatus (taipan)
  Pseudechis australis (king brown or mulga snake)
  Pseudechis colletti (Collett's snake)
  Pseudechis guttatus (spotted or blue-bellied black snake)
  Pseudechis porphyriacus (red-bellied black snake)
  Tropidechis carinatus (rough-scaled snake)
New Guinea
  Acanthophis antarcticus (death adder)
  Demansia olivacea (spotted-headed snake)
  Micropechis ikaheka (small-eyed snake)
  Oxyuranus scutellatus (taipan)
  Pseudechis papuanus (Papuan black snake)
Solomon Islands
  Micropechis elapoides (banded small-eyed snake)

should be made wherever possible to other books listed in the references. However, it is believed that in the majority of cases the descriptions and keys in this chapter will enable identification of any dangerous species from this region with a relatively high degree of accuracy.

Some Australian snakes have not been included in Table IV; these are sometimes considered potentially dangerous species. Included in this category are the broad-headed snakes of the genus *Hoplocephalus* and some species of the genus *Denisonia (D. maculata, D. devisii,* and *D. fasciata)*. However, although severe reactions have resulted from the bites of some of these species, no fatalities or near-fatalities have ever been recorded. For this reason these species are not described in greater detail.

The designations given apply only within the major areas in which the particular species is known (Table IV). In the following descriptions the species are arranged in alphabetical order.

Acanthophis antarcticus   Death adder

Description: Body viperiform, with broad head, narrow neck, thick body, and short, thin tail ending in a spine (Fig. 3). Dorsal scales faintly, but distinctly keeled, in 21 or 23 rows. Ventrals 110–130. Anal single. Subcaudals

40–55, undivided anteriorly, divided posteriorly. Pupil vertical. Rust-colored, gray, brown or almost black above, with irregular darker cross bands. Belly cream or yellowish white, usually with irregular dark brown or black spots. Spine at tip of tail usually whitish or cream-colored. Grows to a little over 3 feet in length (Fig. 3).

Distribution: Australia generally, except the extreme southeastern corner and Tasmania; New Guinea mainland below about 3500 feet; islands west of New Guinea to Ceram; New Britain (Fig. 4).

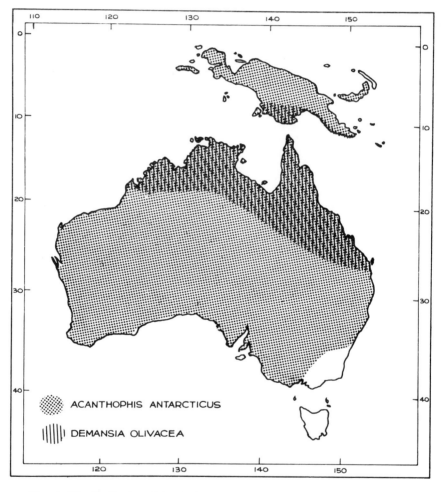

Fig. 4. The distribution of the death adder *(Acanthophis antarcticus)* and the spotted-headed snake *(Demansia olivacea)*.

Remarks: The death adder is found in a large range of habitats, from the desert regions of Central Australia to wet coastal forests. A number of subspecies have been described. It is usually found most abundantly in dry, open country, and is particularly common in the savannahs of northern Australia and lowland New Guinea. The death adder is not a particularly active snake; it is nocturnal, and spends much of its time lying quietly in sand or leaf litter. Its food consists principally of small vertebrates, and captive specimens have been observed to use the tail spine as a lure by twitching it at the approach of a lizard or small bird. When annoyed it usually adopts a flexed, "clock spring" posture (Fig. 3), and is able to strike with astonishing speed over a short distance. It is an ovoviviparous species, and produces up to 20 living young.

Venom: The venom of the death adder is extremely potent; before the production of a specific antivenin a fatality rate of 50% was recorded (Tidswell, 1906). The venom is powerfully neurotoxic; it has little coagulant effect and is only feebly hemolytic (Kellaway, 1938). The fangs are relatively large.

Demansia guttata   Spotted brown snake

Description: Body slender, head only slightly distinct from the neck. **Scales smooth, in 21 rows. Ventrals 200–210. Anal divided. Subcaudals 50–65, divided. Internasals present.** Pupil round. **Diameter of eye much greater than the distance from the mouth.** Color ranges from light to medium brown, with some dark-edged scales, and sometimes with a variable series of dark vertebral blotches. Belly white or cream, with scattered orange spots.

Distribution: Known only from inland Queensland (Fig. 5).

Remarks: Like other members of its genus it is oviparous, diurnal, and feeds on small vertebrates. Worrell (1963) reports that it is inoffensive.

Venom: Nothing is known of the venom of this species.

Demansia nuchalis   Western brown snake, gwardar, dugite

Description: Head and body slender, head only slightly distinct from the neck, scales smooth, in 17 or 19 rows. Ventrals 195–230. Anal divided. **Subcaudals 45–70, divided. Internasals present.** Pupil round. **Diameter of eye much greater than the distance from the mouth.** Color very variable, from rich reddish-brown to olive-green or gray, usually with a few scattered dark-brown scales on the nape. The scales on the body are frequently edged with dark brown, and some specimens have scattered darker scales. The head, particularly in the young, is sometimes dark brown or black. A number of subspecies have been described, and in one

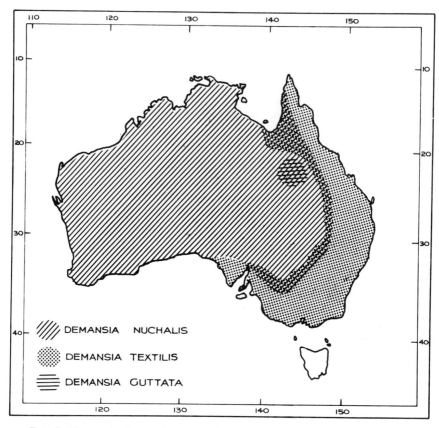

Fig. 5. The distribution of the spotted brown snake *(Demansia guttata)*, the western brown snake *(Demansia nuchalis)*, and the eastern brown snake *(Demansia textilis)*.

of these the dark-edged scales may form distinct though irregular cross bands. Grows to about 6 feet in length.

Distribution: Most parts of Australia except the extreme east, southeast, and Tasmania (Fig. 5). In some areas of western New South Wales it is sympatric with *D. textilis*.

Remarks: Similar in habits to the eastern brown snake *(Demansia textilis)*.

Venom: Little is known of the venom of this species, although it is believed similar in most respects to that of *D. textilis*.

Demansia olivacea   Spotted-headed snake

Description: Body slender, head only slightly distinct from neck. **Scales smooth, in 15 rows.** Ventrals 160–220. **Anal divided. Subcaudals 70–100, divided. Nasal divided. Preocular present.** Pupil round. **Diameter of eye much greater than the distance from the mouth.** Color very variable, from brown through steely gray to almost black. The individual scales are edged behind with black, giving a distinctly reticulated appearance in lighter specimens. Cream, olive-green or gray below, often reddish on the underside of the tail. **Head lighter with distinct dark spots or flecks** which sometimes coalesce to form irregular lines. Length has been recorded as up to 6 feet.

Distribution: Northern and northeastern Australia and the coastal lowlands of southern New Guinea (Fig. 4).

Remarks: Like the taipan, this species prefers the drier, open savannahs. It is normally a shy species, and is aggressive only when provoked. It is oviparous; up to 20 eggs have been recorded. Its food generally consists of small mammals and reptiles.

Venom: The venom of this species has not been studied quantitatively. It is doubtful whether it could inflict a fatal bite, but the relatively large size to which it grows warrants inclusion in this list.

Demansia textilis   Eastern brown snake

Description: Head and body slender, head only slightly distinct from the neck. **Scales smooth, in 17 rows.** Ventrals 185–240. Anal divided. **Subcaudals 40–80, usually all divided. Internasals present.** Pupil round. **Diameter of eye much greater than the distance from the mouth.** Color variable, from very light to very dark brown, although some specimens are gray or almost rust-colored. Belly usually yellowish, gray, or cream with brown or orange spots. Juvenile specimens are often strongly banded, or with a black head and black band across the nape. *Demansia textilis* only occasionally exceed 7 feet in length (Fig. 6).

Distribution: The eastern part of Australia from the Gulf of Carpentaria to Victoria and the eastern part of South Australia. Usually restricted to the coast, tablelands, and western slopes, but extending inland in a number of areas (Fig. 5).

Remarks: A quick-moving, nervous, diurnal snake, aggressive when provoked. Food consists largely of reptiles and small mammals. The brown snake is oviparous; it lays up to 30 or more eggs. It is found in a wide range of habitats, from the wet coastal areas to the drier interior.

Venom: Powerfully neurotoxic, and has a strong coagulant action; this coagulant principle is very diffusible, and death may result from thrombosis which causes hemorrhaging from the mucous membranes (Kellaway, 1938).

Denisonia superba   Copperhead

Description: Medium build, head distinct from neck. **Scales smooth, in 15 rows.** Ventrals 140–165. **Anal single. Subcaudals 35–55, single. Frontal shield much longer than broad, not much broader than the supraocular.**

FIG. 6. The eastern brown snake *(Demansia textilis)*. (Photo: R. D. Mackay.)

FIG. 7. The Australian copperhead *(Denisonia superba)*. (Photo: R. D. Mackay.)

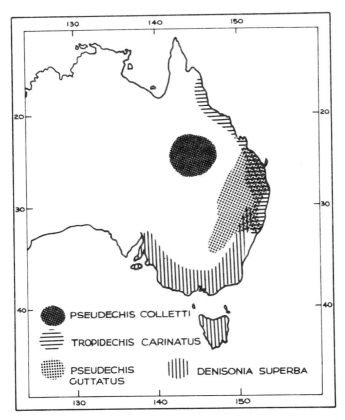

FIG. 8. The distribution of the copperhead *(Denisonia superba)*. Collett's snake *(Pseudechis colletti)*, the spotted black snake *(Pseudechis guttatus)*, and the rough-scaled snake *(Tropidechis carinatus)*.

Internasals present. Fang followed by more than 5 smaller teeth. Pupil round. Color very variable, from black to light brown or rich reddish brown above, many specimens with a dark vertebral stripe. **Lips usually alternately barred with dark brown and cream.** Ventrally bright yellow, cream or gray, with irregular darker variegations. Scales adjoining ventrals sometimes yellow or pink. Often a dark nuchal collar, particularly in the young, edged in front and behind with yellowish bands. Grows to about 5 feet in length (Fig. 7).

Distribution: Highland areas from northern New South Wales to Victoria and the Mount Lofty Ranges and Kangaroo Island in South Australia. Also Tasmania and some of the islands of Bass Strait (Fig. 8).

Remarks: The copperhead is essentially a cold climate species, and is often active and foraging at temperatures at which most other species are

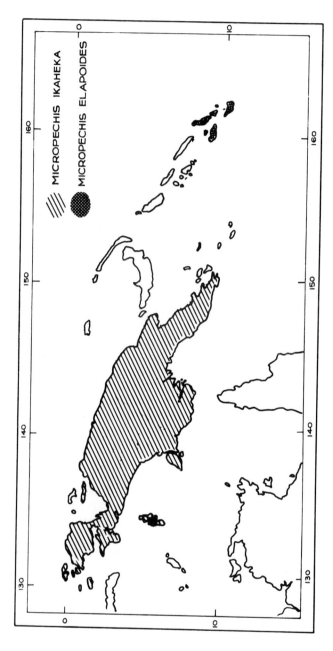

FIG. 9. The distribution of the banded small-eyed snake (*Micropechis elapoides*) and the small-eyed snake (*Micropechis ikaheka*).

torpid. It is one of only two species of snakes occurring above the winter snow line on Mount Kosciusko. It is largely diurnal, but during the warmer months of the year it is frequently nocturnal in its habits. It is not normally an aggressive species, and few bites have been recorded. Food consists of frogs, reptiles, and small mammals, and large numbers of snakes may congregate in swampy areas that provide a plentiful supply of food. The species is ovoviviparous; Kinghorn (1956) records up to 20 young in a brood.

Venom: Kellaway (1938) states that the venom of *Denisonia superba* is strongly neurotoxic, and death usually results from respiratory paralysis. The venom is also powerfully hemolytic and cytolytic, but lacks coagulant effect. Its vasodepressant action has been described by Kellaway and Le Mesurier (1936).

Micropechis elapoides   Banded small-eyed snake

Description: Medium build, head distinct from the neck. Scales smooth, in 17 rows. Ventrals 200–215. **Anal single.** Subcaudals 30–45, divided. Pupil round. Diameter of eye much less than the distance from the mouth. **Color cream or yellow both above and below, with 20–30 broad dark brown or black bands along the length of the body,** extending to but not meeting on the belly. Snout black or dark brown, with irregular dark markings on the remainder of the head. Grows to about 5 feet in length (Fig. 10).

Distribution: Known to occur on Florida, Malaita, and Guadalcanal Islands, but is probably also found on other islands in the Solomon group (Fig. 9).

FIG. 10. The banded small-eyed snake *(Micropechis elapoides)*.

Remarks: Nothing is known of the habits of this species. Probably similar to *Micropechis ikaheka* of New Guinea.

Venom: Nothing known.

Micropechis ikaheka   Small-eyed snake

Description: Medium build, head distinct from the neck. **Scales smooth, in 15 rows.** Ventrals 170–225. **Anal divided.** Subcaudals 35–60, divided. **Nasal divided. Preocular present.** Pupil round. **Diameter of the eye much less than the distance from the mouth. Frontal shield about as long as broad.** Color typically **yellowish brown above, the individual scales with dark brown centers and yellow edges.** Groups of scales with larger dark centers form irregular darker cross bands along the length of the body, although in some specimens the anterior part of the body is uniformly yellow. Head usually black or brown above, light brown below. Ventral surface creamy yellow, the ventral scales lightly edged with brown. Attains a length of about 5 feet (Fig. 11).

Distribution: Throughout New Guinea and the islands adjacent to the Vogelkop. Aru Islands (Fig. 9).

Remarks: Little is known of this species and its habits. Slater (1956) reports that it is normally timid, but very aggressive if provoked, that it inhabits the denser areas of rain forest, and that it is active by both day and night. It is known as a "bush" snake in the New Guinea highlands, and is much feared by the natives. It is believed to range to a higher altitude (about 5000 feet) than any other dangerous snake in New Guinea.

FIG. 11. The small-eyed snake *(Micropechis ikaheka)*. (Photo: K. R. Slater.)

Venom: Nothing is known of the venom of this species except that bites have been known to cause death in human beings.

Notechis scutatus   Tiger snake

Description: A medium to heavily built snake, the head distinct from the neck. **Scales smooth, in 15, 17 or 19 rows. Ventrals 140–190.** Anal single. **Subcaudals 35–70, single. Frontal shield scarcely or not longer than broad. Lower anterior temporal shield about as long as or longer than the frontal. Internasals present.** Pupil round. Color variable; above gray,

FIG. 12. Typical banded mainland tiger snake *(Notechis scutatus)*. (Photo: A. Holmes.)

olive-green, brown, reddish brown to black, with or without irregular darker or lighter cross bands. Creamy white to gray below. Young specimens usually more strikingly banded than adults. A number of subspecies have been described, including several mainland and insular melanotic forms. In eastern mainland Australia the tiger snake grows to about 5 feet in length; in southwestern Australia specimens may exceed 6 feet, but are usually much smaller. A melanotic form on Chappell Island in Bass Strait sometimes exceeds 7 feet in length (Figs. 12 and 13).

Distribution: From southeastern Queensland, through the eastern half of New South Wales, and through Victoria to Eyre Peninsula and Kangaroo

FIG. 13. Melanotic insular tiger snake *(Notechis scutatus)* from Bass Strait. (Photo: R. D. Mackay.)

Island in South Australia. Also Tasmania and many islands of Bass Strait. An isolate in southwestern Australia (Fig. 14).

Remarks: The tiger snake is found in a variety of habitats, but generally prefers poorly drained marshland, swampy areas or river flats, where it may occur in large colonies. At the northern limit of its range it inhabits rain-forest areas. It is typically diurnal, but like the copperhead may forage on warm nights. It is normally rather retiring, but may be particularly savage if provoked; the neck and body are flattened, and the striking stance consists of holding the body in a long, low curve (Fig. 13). It is ovoviviparous, and feeds largely on frogs and small mammals.

Venom: The venom of the tiger snake has received more attention than that of any other snake in the Australo-Papuan region. It is one of the most potent venoms known; it is strongly neurotoxic, with dominant effects on the central and peripheral nervous systems. It is feebly hemolytic, with some cytolytic action, and contains a coagulant principle, which is not diffusible like that of *Demansia textilis* (Kellaway, 1938). Morgan (1956) rates the venom of the tiger snake as only half as toxic as that of the taipan *(Oxyuranus scutellatus)*, although more recent work suggests that the difference in toxicity between the venoms of these two snakes is only very slight. Considerable differences in venom yield and toxicity have been found between different populations of *Notechis* (Kellaway and Thomson, 1932; Kellaway and Williams, 1935; Morgan, 1938; Worrell, 1961b). Prior to the availability of a specific antivenin, a mortality rate of 40% was recorded from bites of this species.

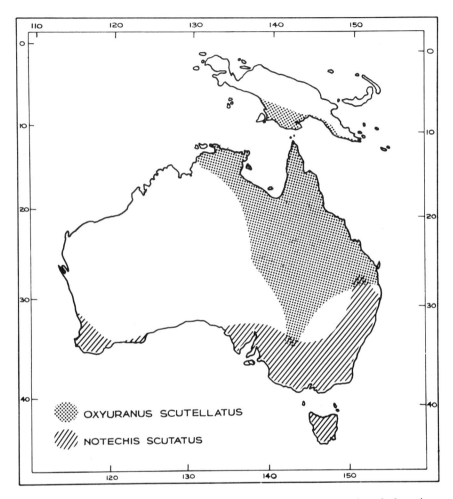

FIG. 14. The distribution of the tiger snake *(Notechis scutatus)* and the taipan *(Oxyuranus scutellatus)*.

Oxyuranus scutellatus   Taipan

Description: Slender to medium build; head distinct from neck. **Scales feebly keeled, in 21 or 23 rows.** Ventrals 230–250. **Anal single. Subcaudals 55–75, divided. Nasal divided. Preocular present. Internasals present.** Pupil round. Australian specimens are usually a relatively uniform medium or dark brown above, and cream or yellow below, sometimes with orange spotting. New Guinean specimens are usually steely gray to black, often with a reddish vertebral stripe. The taipan grows to a little more than 10 feet in length (Fig. 15).

Fig. 15. The taipan *(Oxyuranus scutellatus)*. An example of the New Guinean race. (Photo: K. R. Slater.)

Distribution: Northern and northeastern Australia, and parts of inland eastern Australia. The southern coastal lowlands of New Guinea, from Milne Bay in the east to West Irian (Fig. 14).

Remarks: Both in Australia and in New Guinea the taipan is most commonly encountered in drier, open savannah country, although it is also found in lowland forest areas and the arid interior of eastern Australia. It is normally very shy, but may be extremely aggressive when provoked. It is oviparous, and lays up to 13 eggs (Thomson, 1933; Slater, 1959). It feeds mainly on small mammals, although birds and reptiles are sometimes eaten.

Venom: The venom of the taipan is largely neurotoxic. Worrell (1963) obtained up to 400 mg of venom from a single milking, while Morgan (1956) has recorded its high toxicity (see comments on preceding species).

Pseudechis australis   Mulga or king brown snake

Description: A snake of medium to heavy build, head distinct from the neck. In very large specimens, the head is broad and heavily built. **Scales smooth, in 17 rows.** Ventrals 180–225. Anal divided, occasionally single. **Subcaudals 50–75, partly single, partly in pairs. Internasals present.** Pupil round. **Light brown to rich reddish brown above,** cream to

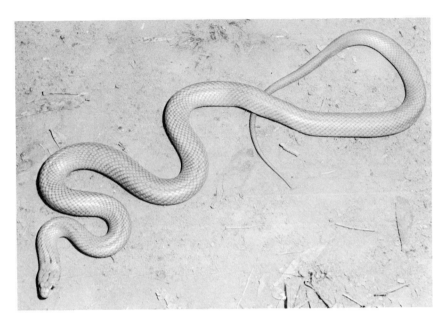

FIG. 16. The mulga or king brown snake *(Pseudechis australis)*. (Photo: R. D. Mackay.)

salmon-colored below, sometimes with orange blotches. May exceed 9 feet in length, but averages only about 6 feet (Fig. 16).

Distribution: Australia-wide except the eastern coast and tablelands. Does not occur in Tasmania. There is one doubtful record from Merauke in West Irian, but this record has not been included in Fig. 17.

Remarks: Usually diurnal, but sometimes nocturnal. Not normally an aggressive snake, but in common with most species, can be savage when aroused. It feeds largely on reptiles and small mammals. It is ovoviviparous.

Venom: Unlike its close relative *Pseudechis porphyriacus,* the venom of the mulga snake is anticoagulant. In other respects it is similar to that of the former species (Kellaway, 1938).

Pseudechis colletti   Collett's snake

Description: Medium build, head distinct from the neck. **Scales smooth, in 19 rows.** Ventrals 215–235. Anal divided. **Subcaudals 55–65, anterior ones single, remainder divided. Internasals present.** Pupil round. **Light to very dark brown above, with irregular pink or cream-colored scales that may sometimes form cross bands.** Belly white or cream, sometimes with darker blotches. Head may be dark gray or black.

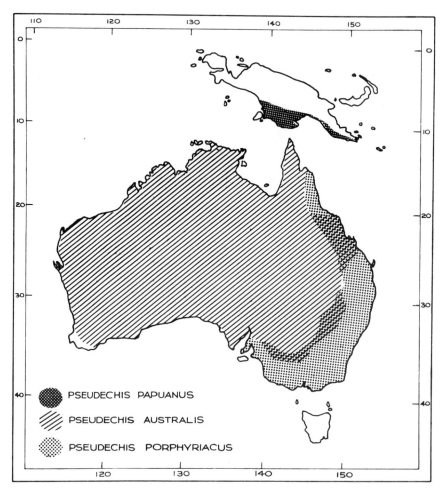

FIG. 17. The distribution of the mulga snake *(Pseudechis australis)*, the Papuan back snake *(Pseudechis papuanus)*, and the red-bellied black snake *(Pseudechis porphyriacus)*.

Distribution: Known only from central inland Queensland (Fig. 8).

Remarks: Little is known of the biology or habits of this species. It is ovoviviparous, and like other members of its genus feeds on small vertebrates.

Venom: Nothing is known of the venom of this species.

Pseudechis guttatus   Spotted or blue-bellied black snake

Description: Medium build, head only slightly distinct from the neck. **Scales smooth, in 19 rows.** Ventrals 180–205. Anal divided. **Subcaudals 45–65,**

anterior single, posterior divided. Internasals present. Pupil round. **From black to light brown above, sometimes with scattered cream-colored scales** that may be arranged roughly in cross bands. In extreme cases the scales may all be creamy, with hind edges tipped with black. Belly gray or blue-gray. Grows to about 6 feet in length (Fig. 18).

Distribution: Southeastern Queensland along the tablelands and western slopes to central New South Wales. Occasionally extends to coastal regions (Fig. 8).

Remarks: This species is found in a wide range of habitats, from the wetter, richly forested areas of the Great Dividing Range to the drier western slopes and plains. It is not normally aggressive; it is ovoviviparous, and feeds on small vertebrates.

Venom: Worrell (1963) states that the venom of this species is similar in action to that of *Pseudechis porphyriacus*.

Pseudechis papuanus   Papuan black snake

Description: Medium build, head distinct from the neck. **Scales smooth, in 19 (rarely) 21 rows.** Ventrals 215–230. **Anal divided.** Subcaudals 45–65, mostly single but a few posterior ones divided. **Nasal divided. Preocular present.** Pupil round. Color normally black above, gray or bluish gray below. Occasional specimens are brown above. Attains a length of about 7 feet (Fig. 19).

Fig. 18. The spotted or blue-bellied black snake *(Pseudechis guttatus)*. (Photo: Howard Hughes/Australian Museum.)

Distribution: Southern coast and lowlands of New Guinea (Fig. 17).

Remarks: Normally a shy snake, it becomes extremely savage if provoked. It is found in a number of lowland habitats, where it feeds on small mammals and reptiles. Like other members of its genus it is ovoviviparous, and largely diurnal.

Venom: Little is known of the properties of the venom of this species, but it is believed to be more toxic than that of any other member of the genus. It is largely neurotoxic.

FIG. 19. The Papuan black snake *(Pseudechis papuanus)*. (Photo: K. R. Slater.)

Pseudechis porphyriacus   Red-bellied black snake

Description: Medium build, head distinct from the neck. **Scales smooth, in 17 rows.** Ventrals 170–215. Anal generally divided, occasionally single. **Subcaudals 40–70, the first few single, the remainder paired. Internasals present.** Pupil round. **Color black above,** iridescent in life, belly usually bright pink, but may be almost white. **Lateral scales in contact with ventrals usually bright red or crimson.** Tail usually dark gray or black below. Although this species grows to more than 8 feet in length, it averages only 5 feet.

Distribution: Eastern Australia, from eastern South Australia to Cape York Peninsula. Largely restricted to the coast and ranges, but extends into the inland in various parts of New South Wales. Does not occur in Tasmania (Fig. 17).

Remarks: Normally a shy, inoffensive diurnal snake. It is most commonly encountered in swampy areas or near streams. Its food consists principally of frogs, but reptiles and small mammals are also eaten. The species is ovoviviparous, and produces up to 20 young.

Venom: The venom is powerfully hemolytic and also causes intravascular coagulation. It also has some neurotoxic action. Its coagulant principle is not diffusible like that of *Demansia textilis* (Kellaway, 1938). A bite in an adult is unlikely to be fatal.

Tropidechis carinatus   Rough-scaled snake

Description: Medium build, head distinct from the neck. **Scales strongly keeled, in 23 rows.** Ventrals 160–185. Anal single. **Subcaudals 45–55, single. Internasals present.** Pupil round. Color generally gray, olive-green, or brown above, usually with irregular darker cross bands, the latter more prominent in the young. Ventrally, yellowish olive. Grows to a little more than 3 feet in length (Fig. 20).

Distribution: Coast and parts of highlands of eastern Australia, from about Newcastle in New South Wales to beyond Cairns in Queensland (Fig. 8).

Remarks: Although a secretive, retiring snake, it may be extremely aggressive. It is active both by day and night. It is found in a range of "wet" habitats, from sclerophyll to rain forests, where it feeds on small vertebrates. It is believed to be ovoviviparous.

FIG. 20. The rough-scaled snake *(Tropidechis carinatus)*. (Photo: R. D. Mackay.)

Venom: Nothing is known of the venom of this species, except that its bite has been known to kill an adult man. The effects of the venom are largely neurotoxic.

REFERENCES

Barbour, T. (1912). *Mem. Museum Comp. Zool. Harvard* **44**, No. 1, 1–203.
Boulenger, G. A. (1896). *Brit. Museum Cat. Snakes* **3**.
Brongersma, L. D. (1934). *Zool. Mededeel.* **17**, 161–251.
Brongersma, L. D., and Knaap-Van Meeuwen, M. S. (1964). *Zool. Mededeel.* **39**, 550–554.
Campbell, C. H. (1966). *Med. J. Australia* **11**, 922–925.
Campbell, C. H. (1967). *Med. J. Australia* **1**, 735–740.
Campbell, C. H., and Young, L. N. (1961). *Med. J. Australia* **1**, 479–486.
Darlington, P. J. (1957). "Zoogeography: The Geographical Distribution of Animals," pp. 462–472, Wiley, New York.
De Haas, C. P. J. (1950). *Treubia* **20**, No. 3, 511–625.
De Rooij, N. (1917). "The Reptiles of the Indo-Australian Archipelago. II Ophidia." Brill, Leiden, Netherlands.
Derrick, R. A. (1957). "The Fiji Islands," Rev. ed., p. 158. Govt. Press, Suva.
Fairley, N. H. (1929a). *Med. J. Australia* **1**, 296–313.
Fairley, N. H. (1929b). *Med. J. Australia* **1**, 313–327.
Gill, E. D. (1961). *Proc. Roy. Soc. Victoria* (N.S.) **74** [2], 125–133.
Kellaway, C. H. (1938). *Med. J. Australia* **2**, 585–589.
Kellaway, C. H., and Le Mesurier, D. H. (1936). *Australian J. Exptl. Biol. Med. Sci.* **14**, No. 1, 57–76.
Kellaway, C. H., and Thomson, D. F. (1932). *Australian J. Exptl. Biol. Med. Sci.* **10**, No. 1, 35–48.
Kellaway, C. H., and Williams, F. E. (1935). *Australian J. Exptl. Biol. Med. Sci.* **13**, No. 1, 17–21.
Kinghorn, J. R. (1956). "The Snakes of Australia," 2nd ed. Angus & Robertson, Sydney, Australia.
Kinghorn, J. R., and Kellaway, C. H. (1943). "The Dangerous Snakes of the South-West Pacific Area." North Melbourne, Australia.
Klemmer, K. (1963). *In* "Die Giftschlanger der Erde," pp. 253–464. Elwert, Marburg/Lahn.
Loveridge, A. (1945). "Reptiles of the Pacific World." Macmillan, New York.
McDowell, S. B. (1967). *J. Zool. Lond.* **151**, 497–543.
Mayr, E. (1945). *In* "Science and Scientists in the Netherlands Indies" (P. Honig and F. Verdoorn, eds.), pp. 241–250. Board for the Netherlands Indies, Surinam and Curacao, New York.
Morgan, F. G. (1938). *Proc. Roy. Soc. Victoria* [N.S.] **50**, No. 2, 394–398.
Morgan, F. G. (1956). *In* "Venoms," Publ. No. 44, p. 359. Am. Assoc. Advance. Sci., Washington, D.C. (quoted in Werler and Keegan, 1963).
Slater, K. R. (1956). "A Guide to the Dangerous Snakes of Papua." Govt. Printer, Port Moresby.
Slater, K. R. (1959). *Australian Zoologist* **12**, No. 4, 306–307.
Storr, G. M. (1964). *Western Australian Naturalist* **9**, 89–90.
Storr, G. M. (1967). *J. Roy. Soc. West. Aust.* **50**, 80–92.
Thomson, D. F. (1933). *Proc. Zool. Soc. Lond.* [4], 855–860.

Tidswell, F. (1906). "Researches on Australian Venoms", Department of Public Health, New South Wales, Government Printer.

Tippett, A. R. (1958). *Trans. & Proc. Fiji Soc.* **5**, 115.

Trinca, G. F. (1963). *Med. J. Australia* **1**, 275–280.

Underwood, G. (1967). Publ. No. 653, pp. 1–179. Brit. Mus. Nat. Hist.

Werler, J. E., and Keegan, H. L. (1963). *In* "Venomous and Poisonous Animals and Noxious Plants of the Pacific Region" (H. L. Keegan and W. V. Macfarlane, eds.), pp. 219–325. Pergamon Press, Oxford.

Williams, E. E., and Parker, F. (1964). *Senckenberg. Biol.* **45**, 543–552.

Worrell, E. (1961a). *Western Australian Naturalist* **8**, No. 1, 18–27.

Worrell, E. (1961b). "Dangerous Snakes of Australia and New Guinea." Angus & Robertson, Sydney, Australia.

Worrell, E. (1963a). "The Reptiles of Australia." Angus & Robertson, Sydney, Australia.

Worrell, E. (1963b). *Aust. Rep. Park Rec.* No. **2**, 1–11.

ADDENDUM

Since this chapter was submitted for publication, the following important contributions to the taxonomy and biology of Australasian venomous snakes have appeared:

Campbell, C. H. (1966). *Med. J. Australia* **11**, 922–925.

Campbell, C. H. (1967). *Med. J. Australia* **1**, 735–740.

McDowell, S. B. (1967). *J. Zool. London* **151**, 497–543.

Storr, G. M. (1964). *Western Australian Naturalist* **9**, 89–90.

Storr, G. M. (1967). *J. Roy. Soc. West. Aust.* **50**, 80–92.

Underwood, G. (1967). *Brit. Mus. Nat. Hist.* Publ. No. 653, pp. 1–179.

Williams, E. E., and Parker, F. (1964). *Senckenbergiana Biol.* **45**, 543–552.

Worrell, E. (1963). *Aust. Rep. Park Rec.* No. **2**, 1–11.

An important range extension, from both the medical and biological viewpoints, is the confirmed presence of the eastern brown snake (*Demansia textilis*) on the northeastern coast of Papua [McDowell (1967) and Australian Museum records from Dogura and Embogo].

Chapter 24

The Pharmacology and Toxicology of the Venoms of the Snakes of Australia and Oceania

E. R. TRETHEWIE

UNIVERSITY OF MELBOURNE AND ROYAL MELBOURNE HOSPITAL,
MELBOURNE, AUSTRALIA

I. GENERAL INTRODUCTION

Oceania includes the islands of the Pacific area and notably Australia, Fiji, Hawaii, the shores of Japan and Malaya, New Guinea, New Zealand, and the Philippines. New Zealand is free of snakes. Asia contains almost every variety other than *Crotalus.* The Hawaiian Islands contain only one terrestrial snake, *Typhlops braminus,* which is harmless, and one sea snake, *Pelamis platurus* (yellow-bellied sea snake), which is relatively rare in the vicinity. The Philippines contain many of the Asian snakes, especially the cobra.

Classification of snakes of Oceania is not yet clear-cut. As far as the snakes of Australia are concerned, they may be classified as follows:

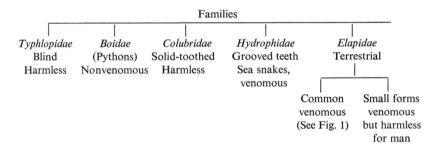

The family Typhlopidae is comprised of blind and harmless snakes. The pythons (Boidae) are also nonvenomous. Solid-toothed Colubridae are harmless. A family of Hydrophidae (grooved toothed) are sea snakes and are venomous, some powerfully so. Terrestrial forms (family Elapidae) have grooved teeth, though some (taipan and death adder) are almost completely canalized by deep grooving. This is usually a feature of vipers (enclosed groove teeth), but when they bite they hang on. The Elapidae, on the contrary, usually bite like a dog and let go, except for the taipan, which will chew. Thus the Australian snakes are like vipers in the clotting effects of their venoms, but correspond to snakes of the cobra and rattler type in their biting habits and pharmacological effects of venom. Characteristically, all the dangerous Australian snakes are elapids.

II. PHARMACOLOGICAL EFFECTS OF SNAKE VENOMS

Venom can produce pharmacological effects by direct effects of toxic constituents or release of pharmocologically active substances from the tissues, especially in relation to enzyme activity, though also from non-enzymic substances.

Numerous pharmacologically active substances are liberated from the lung and liver of various animals. For instance, histamine is liberated from the lung and liver of the dog by black snake venom (*Pseudechis porphyriacus*) (Figs. 1, 2). A muscle-stimulant substance slower and more prolonged in

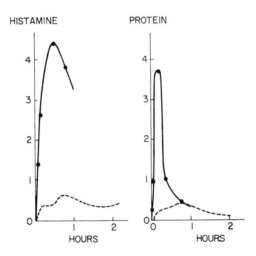

Fig. 1. Output of histamine and protein from the isolated perfused lung of the dog by the powerfully hemolytic black snake venom (continuous line) and the feebly hemolytic death adder venom (interrupted line). Output of histamine expressed as micrograms histamine per minute per microgram histamine per gram lung per milligram venom injected. Output of protein expressed as milligrams per minute for milligram venom injected. (From Trethewie, 1939.)

action is also released (Fig. 2) (now designated SRS). SRS is the term used for the muscle stimulant substance liberated in anaphylaxis (Kellaway and Trethewie, 1940a) and the generic term was then used for substances having a similar muscle-stimulant activity which were released in other conditions (Trethewie, 1941). It has been shown that adenyl compounds and a deaminating enzyme are released from liver and heart by cobra venom (*Naja naja*) (Fig. 3). It has also been shown that the release of histamine, SRS, and protein (Fig. 1) by the powerfully hemolytic black snake venom and the feebly hemolytic death adder venom (*Acanthophis antarcticus*) was

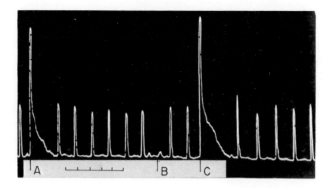

FIG. 2. Release of histamine and SRS (slow-reacting muscle-stimulating substance) from the isolated jejunum of the guinea pig suspended in Tyrode solution. The unlettered contractions are to 0.1 μg histamine. A and C: Responses to 0.4 ml of perfusate sample obtained from the hepatic veins after the intraportal injection of 10 mg black snake venom. Immediate sharp contraction due to histamine, slow relaxation due to SRS. B: Response to 0.4 ml perfusate obtained from another lobe of the liver injected with 10 mg death adder venom. Gut desensitized to both venoms. Time in minutes. (From Trethewie, 1939.)

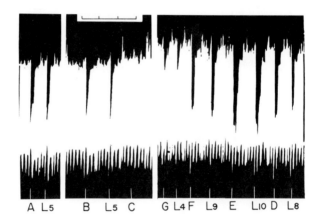

FIG. 3. Release of adenyl compound from the isolated Tyrode perfused liver of the rabbit. Responses are from the atropinized auricle of the intact desensitized guinea pig. Responses marked 5, 8, 10 are to 0.5, 0.8, 1.0 ml/liter in 100 Lacarnol. A, B: Responses to 10 mg liver before and 40 mg liver (uncorrected for edema) after injection of venom. C, F, and G: Responses to 1.0 and at D and E to 0.5 ml of perfusate samples obtained from the hepatic and veins, C obtained before and D, E, F, and G obtained 3, 7, 20, and 30 minute periods respectively after the intraportal injection of cobra venom. Heart block due to adenyl compound. Less adenyl compound released than accounted for by liver loss due to enzymic breakdown. Time in minutes. (From Kellaway and Trethewie, 1940b.)

proportional to the hemolytic power of the respective venom for the red cells of the dog from which the perfused organ was obtained (Fig. 4). This is considered to lend force to the suggestion that the release of histamine is due to the formation of lysolecithin by phosphatidases, found in all these venoms, which damages the cell and thereby effects its release. The release of histamine is considered in the case of black snake venom to be a contributory cause of death.

Clotting is another powerful effect of black snake venom, but its neurotoxic effect (in contradistinction to that of death adder) is feeble. Heparin, when used to treat mice injected with black snake venom, prolonged the life of animals significantly, but did not reduce the mortality. This indicates

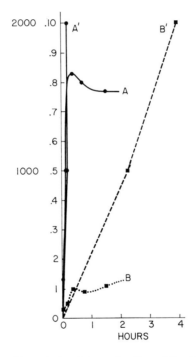

FIG. 4. Comparison of hemolysis of dog red cells and the release of histamine from two lobes of the liver of the same dog by the powerfully hemolytic black snake venom and the feebly hemolytic death adder venom. Continuous line A: histamine output in microgram per minute per microgram histamine content of lung per milligram venom (black snake) injected. Dotted line B: similar output for death adder venom. Continuous line A: time to complete hemolysis of reciprocal of black snake venom concentration (ordinate). Dotted line B: similar time for hemolysis for death adder venom. (From Trethewie, 1939.)

that factors other than clotting effect, e.g., histamine release, might be an important contributory cause of death. A similar prolongation only of survival time occurred with tiger snake venom (*Notechis scutatus*) (Fig. 5). Heparin reduces the mortality from Russell's viper and *Echis carinatus* (Ahuja *et al.,* 1946a,b). The combination of heparin and antihistamine significantly reduced the mortality of mice injected with black snake venom (Table II, p. 112, Chapter 25). A comparison of the effect of intravenously injected black snake venom on pulmonary and systemic blood pressure in the intact cat showed a fall in systemic pressure and a delayed rise in pulmonary pressure followed by a gradual return and later a fall below the original level (Fig. 6). This pulmonary vascular effect was neither abolished nor even reduced by antihistamine, and therefore it is unlikely that the effect is due to histamine.

The release of adenyl compounds from the perfused heart of the cat and rabbit, from the liver of the rabbit and dog, and from the kidney of the rabbit, is considered to account for some of the cardiovascular effects of venoms. Electrocardiographic studies in the intact cat, a species which shows dilation of the coronaries in response to venom (and incidentally adenyl compounds produce coronary dilatation), show first a bradycardia with lengthened P-R interval. This is followed by heart block, at first partial, e.g.,

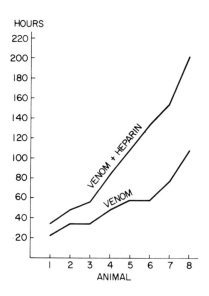

Fig. 5. Survival time (ordinate) of guinea pigs injected with the same dose of tiger snake venom (nine animals in each group, one of each survived). Upper line group, heparin (1 mg/kg) injected subcutaneously immediately after venom injection and daily thereafter. (From Trethewie and Day, 1948a.)

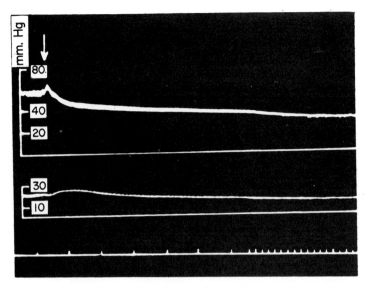

FIG. 6. Carotid (upper) and pulmonary artery (lower) pressure records from a heparinized cat injected with 1 mg/kg black snake venom intravenously (at the arrow). Time in minutes. The pulmonary pressure rise was not reduced following antihistamine injection. Time in minutes. (From Trethewie and Day, 1948b.)

2 to 1 block, and finally complete heart block (Fig. 7). At the same time, when adenyl compounds are released the appropriate deaminating enzyme is also released. M/40 cyanide will inhibit this enzyme activity *in vitro*. It was found, however, that the perfused kidney, even when perfused with M/20 cyanide, still showed a discrepancy in output in the perfusate compared with loss from the organ. This was thought to be due to unexpected cardiac effects and the subsequent discovery of renin suggested a probable explanation, namely, renin release, and this has now been proved. The demonstration that lysolecithin also can cause the liberation of adenyl compounds and of inactivating enzymes affords a hint that at least part of this action of releasing adenyl compound by venom may be attributable to the formation of lysolecithin.

Bradykinin, carboxypolypepidase, catalase, diastase cephalinase, cholinesterase, deoxyribonuclease, diaminoxidase, diastase, dipeptidase, endopeptidase, flavine adenine dinucleotide, hyaluronidase, invertase, L-amino acid oxidase, lecithinase, lipase, 5-nucleotidase, ophio-ATPase, phospholipase, phosphodiesterase, phosphomonesterase, polypeptidase, and protease are contained in venom and account for most of the pharmacological effects. Bradykinin may account for bradycardia in some instances.

The injection of black snake venom into the dog liver perfused with

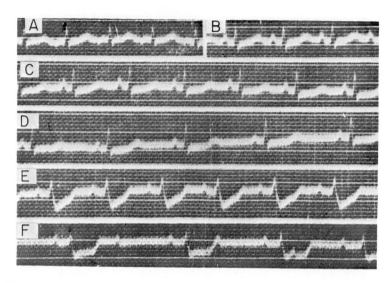

Fɪɢ. 7. Response of the ECG (lead 2) of the rabbit injected with 0.3 and then 0.4 mg cobra venom per kilogram intravenously 28 minutes later. A, ECG before, B, ECG after first injection; C,D,E,F, 3,4,12, and 19 minutes after the second injection. Note the bradycardia (C,D) lengthened P-R interval (E) with depressed S-T segment and 2:1 heart block (F). (From Kellaway and Trethewie, 1940b.)

diluted heparinized blood liberates an inactive conjugate of histamine, in addition to free histamine (Fig. 8). Histamine was freed from conjugate by hydrolysis.

Conjugation of histamine occurs following the injection of free histamine intraportally in the blood-perfused liver. In addition an "antihistamine-like" substance appears in the blood after the injection of venom into the blood-perfused liver. Potassium is released from the liver by venom, and heparin greatly inhibits the potassium released (Table III, p. 112, Chapter 25).

An attempt has been made to apply a rational form of treatment for snake venom poisoning following principles indicated from the above find-ings. These lines are considered supplementary to specific antivenin therapy, local measures, including cold applications, and general measures which should be applied vigorously and early. The mechanisms involved are (Fig. 9):

1. Enzymic release of histamine, SRS, adenyl compounds, deaminating enzyme, potassium, and probably many other substances. Phosphatidase present in the venom which is also responsible for hemolytic effect is implic-ated here. Proteinase may also be operative in releasing histamine and other substances. Oxidase, hyaluronidase, hydrogenase, and other enzymes, of course, also produce their specific effects.

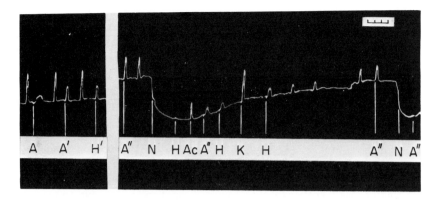

FIG. 8. Release of conjugated histamine by black snake venom from the isolated blood-perfused liver of the dog. Responses of the isolated jejunum of the guinea pig suspended in Tyrode solution. Left-hand panel: at A response to 0.4 ml alcohol eluate obtained from spun plasma after injection of venom. Elution made from charcoal to which conjugated histamine of plasma was adsorbed. At A′ response to 0.4 ml of similar eluate after treatment with HCl and neutralization. Unlettered contractions to 0.1 μg histamine. At H′ response to 0.04 μg histamine. Right-hand panel: at A″ response to 0.5 ml alcohol eluate after treatment with HCl, equal in response to 0.05 μg histamine (unlettered contractions) and to 0.06 μg acetylcholine and 10 mg potassium (not shown). At N 0.5 μg Neo-Antergan added to bath. At H inhibited response to 0.05 μg histamine; Ac, 0.06 μg acetylcholine; K, 10 mg potassium. Time in minutes. (From Trethewie and Day, 1949.)

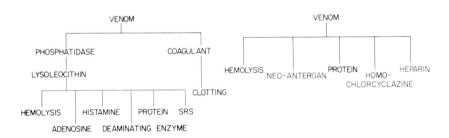

FIG. 9. Schema of toxic effect of venom. Left-hand figure: Enzyme activity plan. Right-hand figure: Antipharmacological agents can inhibit some effects but notably adenosine and deaminating enzyme are not yet countered.

2. Clotting and hemolytic effects of venom.

3. Neurotoxic effect and cardiac effects.

With regard to the first, antihistamine treatment of mice injected with black snake venom has been found ineffective (Table III, p. 112, Chapter 25). This might be due to coagulative effects which retard the chances of the appropriate therapeutic agent reaching the site of histamine release. Heparin alone is of value in viper venom poisoning, such as Russell's viper and *Echis carinatus*. Mention has already been made of the use of heparin, in both black snake venom poisoning and tiger snake venom poisoning. Survival time of guinea pigs was prolonged with heparin in both instances but no reduction in mortality occurred. Heparin, when given to mice after the injection of black snake venom, was followed by increase in mortality, due probably to inhibition of coagulation, allowing greater systemic release of histamine. It is probable that this mechanism of clot inhibition with further freer release of adenosine accounts for the disappointing results of anticoagulant treatment in cardiac infarction. The mortality is only slightly reduced by anticoagulants, while this might have been expected to be dramatically valuable in coronary thrombosis. Adenosine as well as deaminating enzyme is released in this condition (Trethewie and Thach, 1961) and an antiadenosine might be expected to be very useful therapeutically, when available, in reducing the 48-hour high mortality in severe cardiac infarction. When, however, heparin combined in large doses with antihistamine is given, there is a dramatic and a significant reduction in mortality for mice injected with black snake venom. This procedure should allow the antihistamine drug freer access to areas of injury, as the coagulative effect of the venom should be diminished. In addition, as pointed out above, heparin inhibits the release of potassium to a degree and also probably inhibits release of histamine. Neurotoxic and cardiac effects are unfortunately not influenced by this procedure.

More recently, therapy has been applied in rabbits against the neurotoxic effect of tiger snake venom along the lines suggested by methods used in management of anesthetic overdosage with curare. It was found, however, that artificial respiration, although prolonging life, had no effect on mortality, which was ascribed to cardiovascular effects of venom probably related in part to the liberation of adenyl compounds as already described; but it is also probably partly due to direct effects of cardiotoxin.

More will need to be known concerning pharmacological antagonists than at present to develop this line of argument, but it should be a fruitful line of investigation. Tazieff and Trethewie (1969) have reduced the effect of cardiotoxin from *N. nigricollis* with calcium.

Prolongation of life provides a longer time in which to administer antivenin therapy.

The first-aid procedures are washing or sucking off excess venom, local incision and administration of permanganate crystals or excision, injection of adrenaline or acid into the region of the bite, local ice-freezing and tourniquet application. Other snake-bite (rattler) is not improved by cold.

In the hospital, treatment is directed to shock. Stimulants and artificial respiration are applied if necessary and antivenin is injected intravously or intramuscularly after testing sensitivity. Small doses are given and gradually increased if intradermal tests indicate sensitivity, but unfortunately this test may not be related to systemic sensitivity. In Australia tiger antivenin is the serum of choice and in special instances where the snake type is known, taipan antivenin, brown snake antivenin and death adder antivenin are administered. A polyvalent antivenin and an antihistamine are now available (C.S.L.).

III. SPECIFIC TOXIC EFFECTS OF VENOM

The toxic action of venoms may be attributed to proteolytic enzymes, phosphatidase and other enzymes systems (see above), neurotoxin, cardiotoxin (this may be partly due to phosphatidase). The proteolytic activity accounts for the coagulating effect of venom where these effects are produced, and this is typical of Viperidae and also of the Australian Elapidae. The action is either (1) Conversion of prothrombin to thrombin and the coagulation of the blood, which proceeds along the classical lines of thrombin activating fibrinogen to form fibrin. Calcium and tissue extract influence this activity. (2) Coagulation of pure fibrinogen to fibrin by direct action.

It has been suggested that as with posterior pituitary gland hormone production, where one substance may have fringe molecules with differing activity, there is some evidence that the four toxic actions referred to above may overlap as, e.g., the rattlesnake venom of America has phosphatidase action and neurotoxin from a single protein constituent. This is probably not so in the case of Australian snake venoms, most of which like tiger snake, brown snake, copperhead, and death adder venoms, are powerfully neurotoxic, since this association has not been observed. In fact, Australian black snake venom, which is powerfully hemolytic in relation to phosphatidase activity, is only feebly neurotoxic.

Cardiotoxin, which was thought to be combined in either proteolytic enzymes, phosphatidase, or neurotoxin, is considered by some to be a separate fraction of snake venom, but this distinction has not been worked out in relation to Australian snake venom where some cardiotoxic actions have been demonstrated, and it has been shown that normal liver can reduce the impact of this. The pattern is further complicated by the fact, as mentioned above, that Australian and other snake venoms (notably the Indian cobra) release adenosine, which is a powerful cardiac depressant, in the heart.

IV. SPECIFIC AUSTRALIAN VENOMS

The distribution of the common venomous Australian snakes is shown
in Fig. 10.

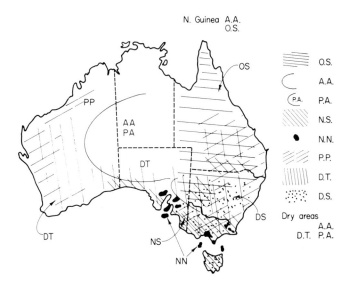

FIG. 10. Distribution of common venomous Australian snakes. O.S., *Oxyuranus
scutellatus;* A.A., *Acanthophis antarcticus;* P.A., *Pseudechis australis;* N.S., *Notechis
scutatus;* N.N., *Notechis nigra;* P.P., *Pseudechis porphyriacus;* D.T., *Demansia textilis;*
D.S., *Denisonia superba.*

A. *Notechis scutatus* (Australian Tiger Snake)

This snake is found in the southern parts of Australia, including Tasmania,
and a black variety occurs in the Bass Strait islands and Kangaroo Island.

The venom is very powerfully neurotoxic. The LD_{50} for the guinea pig
is 6.5/μg/kg. Death may be delayed up to 2 weeks with paralysis supervening
though it may occur in a few hours. Wasting occurs partly because of anorexia
and loss of ability to obtain food. The skin of the subject bitten may be still
highly toxic from contained venom for several days, e.g., skin removed from
a subject dying on the fourth day when extracted was highly lethal to mice.
The toxicity can be reduced by injection of the bitten area with dilute acid
or adrenaline and by ice cooling. For this reason some advocate excision of
the area bitten. The fangs are about 0.5 inches long and inject approximately
50 mg venom at a bite, which without treatment is fatal.

Antivenin developed for tiger snake venom will neutralize all the various venoms from other Australian snakes except that of the taipan (*Oxyuranus scutellatus*) and it is not complete against death adder. It is largely complete against brown snake but a specific antivenin has been prepared against this snake also.

Likewise, if an organ is desensitized to tiger snake venom, the organ is no longer sensitive to Indian venoms (e.g., cobra venom) though the reverse does not hold. Thus the antigenic structure of the Australian venoms includes that of the Indian venoms but the latter are not as complete antigenically as the Australian. This does indicate overlap in activity of Australian snake venom, which includes fringe groupings of elapid (cardio- and neurotoxic) and viperine (clotting) venoms. The snakes are probably distinct from those of India, indicating a very early separation of the Australian land mass in Oceania.

The survival time of tiger snake envenomation is increased by heparin but the mortality is quite unchanged. Likewise polyvinylpyrrolidone and tetrahydroaminoacridine, which reduce the toxicity of some other Australian snake venoms, are quite ineffective against tiger snake venom.

Tiger snake venom activates prothrombin but not fibrinogen. It also contains cholinesterase, hyaluronidase, L-amino oxidase, ophio-ATPase, phosphodiesterase, phosphomonoesterase, and protease.

Notechis scutatus venom, which contains phosphatidase A will protect mice against the lethal effect of staphylococcal toxin. A similar effect has been observed with cobra venom, which protects the animal to a degree against poliomyelitis.

Only venoms with phosphatidase A content produce this effect, but curiously the active principle in these venoms is relatively heat stable and its effectiveness is increased by toxoiding with formalin. Phosphatidase A and lysophosphatidase are found in normal tissues and are thought to play a part in the body's defenses.

Tiger snake venom also contains purines, notably guanosine, inosine, hypoxanthine, adenosine 3'-phosphate and adenosine. These compounds are believed to arise directly from the venom glands.

The neurotoxin can be separated into neurotoxin A and fraction B, both of which produce SRS gut responses and both increase capillary permeability, a function usually related to bradykinin. The neurotoxin can be produced free of enzyme activity and is a preformed toxic substance.

Symptoms of poisoning in man develop a half an hour or so after a bite, with nausea and vomiting. Sweating and coma may develop. Dulling of sensation with aphasia and dysphagia occur. The pupils dilate and respiratory paralysis follows, usually causing death; artificial respiration will not save the subject, partly due to effects of cardiotoxin or release of pharmacologically active substance, as has been observed with other Australian venoms.

Artificial respiration can prolong life and perhaps allow longer time for the use of antivenin and other supportive measures.

B. *Pseudechis porphyriacus* (Black Snake)

This snake is found especially in swampy regions and all around the eastern belt of Australia. A desert form, without a red stripe, occurs in the central dry areas of Australia (P.A., Fig. 10). This venom is powerfully hemolytic but only feebly neurotoxic. It is also powerfully coagulant. It causes the release of histamine, also acetylated histamine, and SRS (slow-reacting substance) from the tissues.

Heparin increases the toxicity of the venom, probably because it allows the release of pharmacologically active substances which are otherwise trapped in the tissues made *fixé* by coagulation. It is believed to both coagulate fibrinogen and activate prothrombin. It contains both L-amino oxidase and lecithinase. The LD_{100} for mice is approximately 3.5 gm/kg by subcutaneous injection.

Hemolysis, which reaches a maximum activity in black snake venom, but occurs with all the common venomous Australian snakes is of two types —direct and indirect. Direct hemolysis occurs with washed red cells without added lecithin, and indirect hemolysis depends particularly on the addition of lecithin which is converted to the hemolytic agent lysolechithin by the phosphatidase A. Phosphatidase A will protect mice against the lethal effect of staphylococcal toxin. This is due to the effect of unsaturated acids. Phosphatidase A and probably other enzymes take part in a natural defense mechanism against bacteria.

Following biting, local pain and swelling are produced with vomiting, and hematuria, hematemesis, and prostration follow. Children and babies may die from the bite but adults recover. Tiger snake antivenin is specific.

C. *Acanthophis antarcticus* (Death Adder)

This snake is found throughout Australia and parts of New Guinea and adjacent islands. The venom of this snake is highly neurotoxic but only feebly hemolytic. The LD_{100} (approx.) for mice is 0.5 to 0.7 mg/kg subcutaneously. It is a clotting venom, being of the family Elapidae and is thought both to coagulate fibrinogen and activate prothrombin.

It contains cholinesterase, hyaluronidase, L-amino acid oxidase, lecithinase, and ophio-ATPase. Protein, histamine, and SRS are liberated from perfused organs but not nearly so powerfully as by black snake venom. While hyaluronidase may account for the spreading factor, the venom is related to that of a genus which causes spreading in the absence of hyaluronidase and this is thought to be related possibly to copper and/or zinc compounds.

There is a latent period of about an hour following biting before symptoms appear, followed by faintness, drowsiness, and sweating. Dullness of sensation follows, with staggering gait followed by dysphagia and dilatation of the pupils. There is peripheral circulatory failure, and hemorrhagic phenomena occur. Death finally results from respiratory failure. A specific death adder antiserum is now available, which is preferable to tiger snake antivenin.

D. *Denisonia superba* (Copperhead)

This snake is found especially in the drier areas in the near south of Australia and in the eastern portion of the continent. The venom coagulates fibrinogen and is considered by some to activate prothrombin. It is certainly a clotting venom. Cholinesterase, hyaluronidase, L-amino acid oxidase, lecithinase, and opio-ATPase are among the many enzymes present. The LD_{100} for mice is given as 1–2 mg/kg. This, of course, is not an accurate parameter but the LD_{50} is not recorded.

Copperhead venom liberates histamine from the lungs of animals and from the liver of the dog. It also liberates SCS (a slow-contracting substance). The venom contains neurotoxin and following envenomation rapid loss of muscle tone and consciousness occurs, and peripheral circulatory failure supervenes. Tiger snake antivenin is a specific antitoxin.

E. *Demansia textilis* (Brown Snake)

This snake is found in the drier areas of the southern parts of Australia. The venom is neurotoxic and produces clotting. It is not known whether the clotting mechanism acts via the coagulation of fibrinogen or activation of prothrombin. Similar venom is produced by *Demansia affinis* and *Demansia australis*.

Enzymes present include esterase, L-amino acid oxidase and ophi-ATPase. The LD_{100} (approx.) for mice is 0.25 mg/kg.

Symptoms may be delayed for 12 hours following biting, and are accompanied by headache, dizziness, abdominal pain, and vomiting. Weakness supervenes, there is peripheral circulatory failure accompanied by respiratory and cardiac failure. Thromboses occur, hemoglobinuria appears, and the temperature is subnormal. The bite from a 6-foot (2-m) snake is commonly fatal.

Tiger snake antivenin is not very useful in protecting against brown snake bite and a specific antivenin to brown snake venom is now available.

F. *Oxyuranus scutellatus* (Taipan)

This snake is found in the north and northeastern parts of Australia, and is gradually moving southward. It is also prevalent in New Guinea. The

LD_{50} for guinea pigs is approximately 0.020 mg/kg and for mice 0.12 mg/kg (subcutaneous injection).

The venom is said to contain "thrombase" but it is not known whether this activates prothrombin or coagulates fibrinogen directly. It is only feebly hemolytic, but powerfully neurotoxic. There is some protection afforded by tiger snake antivenin, but a specific taipan antivenin is now available.

The bite of this snake is one of the most deadly in Australia and some species measure 3 m in length. Neurotoxic paralytic effects with flaccid paralysis of the limbs are the cause of death, with respiratory paralysis from the bulbar involvement. The fangs are 1 cm long in a 1.5 m snake, and 500 mg venom are injected with one bite.

V. SEA SNAKES OF OCEANIA

There are many varieties of sea snake (Hydrophidae) found in Australian, Malayan, and Pacific waters generally. They are exceedingly venomous.

A. *Enhydrina schistosa* (Common Sea Snake)

This snake is found on the shores of the Pacific Ocean, also from the Persian Gulf to Japan, the Australian coast, and Southern Pacific Islands.

The lethal dose for a man is considered to be about one third of an average bite dose and is approximately 0.05 mg/kg. The bite feels like a pin prick while bathing or fishing in the water and there is no local pain subsequently. In approximately 1 hour paresis develops, usually in the legs and involves the trunk, arms, and neck muscles in about 2 hours. Aching, stiffness, and pain precede the muscular weakness, which is usually of the flaccid variety, but may be spastic and then hyper-reflexia is present. Bulbar paralysis usually follows with severe trismus. Ptosis also develops early. Thirst, a feeling of coldness, and sweating supervene and the pupils dilate. Vomiting, and nasal regurgitation commonly occur.

The muscles are tender and respiratory paralysis with terminal hypertension precede death in cyanosis. Failing vision is said to indicate a fatal outcome. Albuminaria and hemoglobinuria occur and the blood concentrates with a mild leukocytosis. Cobra antivenin is not protective. Artificial respiration and probably neostigmine are valuable supportive measures.

B. *Pelamis platurus* (Yellow-Bellied Sea Snake)

Pelamis platurus is an equally toxic sea snake, and is found even more extensively than *Enhydrina schistosa,* extending throughout the Pacific to the shores of tropical America and southeast Africa. Its venom produces pharmacological effects similar to *Enhydrina schistosa*, with respiratory paralysis.

C. Other Sea Snakes

Less common forms of sea snakes are *Hydrophis* which is less toxic than cobra venom in small animals; *Lapemis curtus* and *Laticauda colubrina*, which are of equal and double toxicity respectively, and *Lapemis hardwickii,* which is also extremely toxic for small animals.

VI. SNAKES OF NEW GUINEA

A. *Oxyuranus scutellatus canni* (Papuan Taipan)

This snake is found in the coastal regions of New Guinea. It is more venomous than the Australian taipan and the toxic effects are only exceeded by the king cobra and then only if the cobra is fully grown at 18 feet.

B. *Acanthophis antarcticus laevis* (Papuan Death Adder)

This death adder is found in the forest and is active at the end of the wet season. The pharmacological effects of the venom correspond with the Australian death adder.

C. *Demansia olivacea* (Black Whip Snake)

This snake is common in New Guinea but little is known of the pharmacology or biochemistry of its venom.

D. *Pseudechis papuanus* (Papuan Black Snake)

This snake is found in the coastal regions of New Guinea and the venom corresponds to the King Brown Snake of the Australian mainland. This snake causes more human deaths than any other Australian or New Guinea snake. A specific antivenin is available.

E. *Pseudechis australis* (King Brown Snake)

This snake, which is found in most of Australia except the southwest, is possibly to be found in coastal New Guinea areas.

F. *Pseudechis australis* (*Naja australis*)

This snake is slightly more dangerous than *Pseudechis porphyriacus*. The large size, measuring 2–3 m, can cause human deaths.

G. *Hydrophidae**

These snakes are found in New Guinea waters as in Australian waters.

* See addendum on page 101.

VII. SNAKES OF THE PHILIPPINES

A. Terrestrial Snakes

The most important of the venomous terrestrial snakes are of the cobra type corresponding to the Indian cobra. These are *Naja naja philippinensis, Naja naja samarensis, Naja naja moilepis,* and *Ophiophagus hannah* (king). While it is believed the venom of these snakes corresponds pharmacologically to that of the Indian snakes, discussed elsewhere, antivenin to the Indian cobra does not protect against the Philippine cobra, nor does antivenin to the Philippine cobra protect against the Indian cobra. There are four to six deaths from snake bite annually in the Philippines.

B. *Hydrophidae*

The sea snakes in Philippine waters have already been discussed with the Australian sea snakes.

VIII. SNAKES OF HAWAII AND OTHER PACIFIC ISLANDS

A. Terrestrial Snakes

Typhlops braminus (family Typhlopidae). This snake is a blind snake, is harmless, and comprises the only snake found in this land.

B. *Hydrophidae*

Pelamis platurus (Yellow-bellied sea snake). This snake, previously described, is venomous and is relatively rare in the vicinity of the Hawaiian Islands.

C. Snakes of Malaya

Many of the snakes of Malaya such as the king cobra, naja naja, and Russell's viper are the same as those found in India. However, the Malayan viper *Agkistrodon rhodostoma* (Boie) is deserving of special mention. Less than half the subjects bitten develop serious poisoning. This is because the snake rarely injects a large amount of venom. If local swelling is very slight the bite is not serious. The main pharmacological effect is a clotting defect and this prime characteristic is present in 40% of cases. In 15% of victims the hemorrhagic syndrome develops, hematemesis occurs, and generalized ecchymoses appear. The coagulating defect persists and in the absence of specific treatment is still present after 3 weeks. The prolonged coagulation defect is described as a defibrinating syndrome and this is similar to what is found in other defibrinating syndromes. The clot quality is similar. Blood transfusion and transfusion with human fibrinogen is disappointing. Specific antivenin is rapidly curative.

The following venomous snakes occur in the Pacific Islands as listed. The pharmacological effects of the venom, where known, have already been described in previous sections.

COLUBRIDAE (family)
  Boiginae (subfamily)
    Boiginae irregularis is a slightly venomous terrestrial rear-fanged snake. Its distribution is the Bismarck Archipelago, D'Entrecasteau group, New Guinea and the Solomon Group.

ELAPIDAE (family)
  Denisonia (genus)
    Denisonia par. This snake occurs in the Solomon group.
    Denisonia woodfordii occurs in the Solomon group also.
  Micropechis (genus)
    Micropechis elapoides. This snake is very rare and occurs in the Solomon group. It also occurs in Papua.
  Pseudelaps
    Pseudelaps mülleri occurs in northern Papua and New Guinea and Moratau Island of the D'Entrecasteau group.

HYDROPHIDAE (family)
  Laticauda (genus)
    Laticauda colubrina occurs in Fiji, the New Hebrides, The Solomons, and the Tonga Islands.
  Pelamis (genus)
    Pelamis platurus occurs in the eastern Pacific along the American coast from the coast of Ecuador to the Gulf of Panama and many localities on the Oriental side of the Pacific, as listed previously.

APPENDIX: DANGEROUS VENOMOUS ANIMALS OF AUSTRALIA, OTHER THAN VENOMOUS SNAKES

A. *Synancja trachynis* (Stonefish)

This fish inhabits tropical reefs and estuaries throughout Oceania. It is found around the north, east, and west coasts of Australia, as far south as Morton Bay and Houtands Brolhos. The fish is covered with 18 spines and fusiform venom glands are attached to the upper parts of the spines. Envenomation occurs by treading on or handling the fish. Approximately 50–90 mg. of dried venom is obtained from each fish. The venom is largely protein in nature, is unstable to freezing, heating to 50°C, oxidizing agents, and acid and basic dyes. The LD_{50} for mice is 0.5 mg/kg by intraperitoneal injection and 0.35 mg/kg by intravenous injection. The venom has powerful myotoxic action, causing muscular paralysis which involves skeletal, cardiac, and involuntary muscle. Contraction of the peripheral vasculature causes initial hypertension followed by hypotension from involuntary muscle paralysis; myocardial damage also lowers blood pressure; capillary permeability is increased. Atrioventricular block is produced; notwithstanding this, the heart continues to beat for a considerable time after respiration has ceased. Death results from respiratory paralysis; conduction block is considered due to slow depolarization of the muscle.

The clinical effects of the venom are pain, skin necrosis, irregular respiration, pulmonary edema, muscular incoordination, followed by weakness and paralysis, bradycardia and hypertension. Stonefish antivenin has been prepared.

In treatment relief of pain is an urgent problem and local chilling of the area with excision of the area if possible is the treatment of choice, including antivenin.

B. *Hapalochlaena maculosa* (Ringed Octopus)

This octopus, which measures approximately 20 cm across when fully grown and weighs 12 gm on the average, contains two venom glands weighing approximately 0.15 gm each and is found commonly in Australian coastal waters. The venom is injected into the skin by a beaklike apparatus, on handling. The venom causes neuromuscular block of rat diaphragm and later depression of the response of the muscle to direct stimulation; the effect is reversible. The isolated jejunum of the guinea pig responds with immediate contraction followed by a typical SRS delayed relaxation. The respiration in the intact animal ceases in the case of the cat, within 15 minutes of the intravenous injection of one-eighth of the gland content; and although the isolated heart is unaffected by the venom, in the intact animal the heart fails from anoxia soon after the failure of respiration (Trethewie, 1965).

In man, a prick on the skin, following handling, is followed in 1–3 hours by nausea, fear, and failure of respiration. In the absence of artificial respiration death may ensue in 3–4 hours. Prostigmin should be used in treatment for as yet no antivenin is available. Controlled respiration is required.

C. Poisonous Spiders

There are two spiders in the Australian area which produce toxic venom, namely *Latrodectus hasselti* and *Atrax robustus*.

I. *Latrodectus hasselti*

On hot days the toxicity of the venom is greater than on cool days, but at near zero temperatures the toxicity increases again. The LD_{50} for mice, by intravenous injection, is about five spiders per kilogram. Following injection, symptoms include lachrymation, eye closure, watery discharge from the nose and mouth, and paralysis in 24–48 hours. There is spasticity and rapid loss of weight occurs. Pulmonary edema develops, and 20 gm. mice die approximately within 6 hours of being bitten. Babies may die following the spider bite; a specific antivenin is available.

2. Atrax robustus

There are eight species of the genus *Atrax* present in the Australian zone, all of which are dangerous to man. Death has occurred following the bite of *A. robusutus* and *A. formidabilis*. This spider requires adequate room for biting since the biting mechanism is like that of an axe. The red-backed spider *(L. hasselti)* does not require this space since its fangs open and shut like a pair of scissors. Death in small animals may occur within a few seconds, and the venom may be injected intracerebally in small animals. Monkeys bitten become comatose in $1\frac{1}{2}$ hours and may die in 24–48 hours. Generally animals either die within a few hours or recover completely, no delayed effects being observed. Following biting the animal may cry, becomes restless, with watery discharge from nose and mouth, lachrymation and eye closure, respiratory distress, muscular incoordination and tremor followed by convulsions, paralysis, and coma. Death occurs from respiratory failure. Hypothermia is produced. Sneezing is produced in guinea pigs, possibly due to histamine release. At autopsy the lungs are congested and emphysematous and hemorrhagic pulmonary edema is produced and the auricles are dilated with dark blood. A specific antivenin is available for treatment in man.

REFERENCES

Ahuja, M. L., Brooks, A. G., Veeraraghavan, N., and Menon, I. G. K. (1946a). *Indian J. Med. Res.* **34**, No. 2.

Ahuja, M. L., Veeraraghavan, N., and Menon, I. G. K. (1946b). *Nature* **158**, 878.

Austin, L., Cairncross, K. D., and McCallum, I. A. N. (1961). *Arch. Intern. Pharmacodyn.* **81**, 339.

Buckley, E. E., and Porges, N. (1956) *In* "Venoms," Publ. No. 44, p. 243. Am. Assoc. Advance. Sci., Washington, D.C.

Burt, C. E., and Burt, M. D. (1932). *Bull. Am. Museum Nat. Hist.* **63**, Art. 5.

Butcher, D. A. (1959). Fisheries Circ. No. 2. Fisheries and Game Dept., Victoria, Australia.

Calmette, A. (1907). "Venoms, Venomous Animals and Antivenomous Serum-Therapeutics." Bale & Danielsson, London.

Campbell, A. (1969). A Clinical Study of Snakebite in Papua. M.D. Thesis, University of Sydney.

Doery, H. M. (1956). *Nature* **177**, 381.

Doery, H. M. (1957). *Nature* **180**, 799–800.

Doery, H. M. (1958). *Biochem, J.* **70**, 535.

Doery, H. M., and North, E. A. (1960). *Brit. J. Exptl. Pathd.* **41**, 243.

Doery, H. M., and Pearson, J. E. (1961a). *Biochem. J.* **78**, 820.

Doery, H. M., and Pearson, J. E. (1961b). *Brit. J.* **28**, 620.
Feldberg, W., and Kellaway, C. H. (1937). *Australian J. Exptl. Biol. Med. Sci.* **15**, 81.
Feldberg, W., and Kellaway, C. H. (1938). *J. Physiol.* (*London*) **94**, 187.
Feldberg, W., Holden, H. F., and Kellaway, C. H. (1938). *J. Physiol.* (*London*) **94**, 232.
Gitter, S., Amiel, S., Gilat, G., Sonnino, T., and Welwart, Y. (1963). *Nature* **197**, 383.
Glavert, L. (1950). "A Handbook of the Snakes of Western Australia," 2nd ed. (rev.). Western Australian Naturalists' Club, Perth, Australia.
Hunt, R. A. (1947). *Victorian Naturalist, (Melbourne)* **64**.
Kaire, G. H. (1968). *Med. J. Australia* (in press).
Kellaway, C. H. (1937). *Bull. Johns Hopkins Hosp.* **60**, 1.
Kellaway, C. H. (1939). *Ann. Rev. Biochem.* **8**, 541.
Kellaway, C. H., and Feldberg, W. (1938). *J. Physiol.* (*London*) **94**, 187.
Kellaway, C. H., and Trethewie, E. R. (1940a). *Quart. J. Exptl. Physiol.* **30**, 121.
Kellaway, C. H., and Trethewie, E. R. (1940b). *Australian J. Exptl. Biol. Med. Sci.* **18**, 63.
Kinghorn, J. R. (1956). "The Snakes of Australia." Angus & Robertson, Sydney, Australia.
Kinghorn, J. R., and Kellaway, C. H. (1943). "The Dangerous Snakes of the South West Pacific Area." Army Handbook.
Krefft, G. (1869). "Snakes of Australia," 1st ed. Richards, Sydney, Australia.
Loveridge, A. (1945). "Reptiles of the Pacific World." Macmillan, New York.
McPhee, D. R. (1959). "Snakes and Lizards of Australia." Jacaranda Press, Brisbane, Australia.
Mitchell, F. J. (1900). "Harmless or Harmful?" South Australian Museum Handbook, Adelaide, Australia.
Noguchi, H. (1909). "Snake Venoms." Carnegie Inst. Washington, D.C.
North, E. A., and Doery, H. M. (1953). *Nature* **161**, 1542.
Ogilvie, Sir H., Thomson, W. A. R., and Garland, J. (1959). *Practitioner* **183**, 354.
Oliver, J. A., and Shaw, C. E. (1953). *Zoologica* **38**, Part 2, No. 5.
Proceedings. (1957–1958). *Proc. Roy. Zool. Soc. N. S. Wales,* 1955–1956.
Reid, H. A. (1955). *Trans. Roy. Soc. Trop. Med. Hyg.* **49**, 4.
Reid, H. A. (1956a). *Brit. Med. J.* **II**, 73.
Reid, H. A. (1956). *In* "Venoms," Publ. No. 44, p. 367. Am. Assoc. Advance. Sci., Washington, D.C.
Reid, H. A. (1957). *Trans. Roy. Soc. Trop. Med. Hyg.* **50**, 517.
Reid, H. A. (1959). *Practitioner* **183**, 530.
Reid, H. A. (1961a). *Brit. Med. J.* **I**, 1281.
Reid, H. A. (1961b). Lancet **II**, 399.
Reid, H. A. (1962). *Brit. Med. J.* **II**, 576.
Reid, H. A., and Lim, K. J. (1957). *Brit. Med. J.* **II**, 1266.
Reid, H. A., and Marsden, A. T. H. (1961). *Brit. Med. J.* **I**, 1290.
Russell, F. E. (1959). *U.S. Public Health Serv., Publ.* **74**. 855.
Russell, F. E., and Emery, J. A. (1961). *Am. J. Med. Sci.* **241**, 135.
Russell, F. E., and Truman, E. L. (1959). *In* "Myasthenia Gravis" (H. R. Viets, ed.), p. 101. Thomas, Springfield, Illinois.
Schenberg, S. (1959). *Science* **129**, 1361.
Spector, W. S. (1956). "Handbook of Biological Data." Saunders, Philadelphia, Pennsylvania.
Taylor, E. E. (1919a). *Philippine J. Sci.* **14**, 105.
Taylor, E. E. (1919b). *Philippine J. Sci.* **21**, 161 and 257.
Tazieff, P., and Trethewie, E. R. (1969) (to be published).
Trethewie, E. R. (1939). *Australian J. Exptl. Biol. Med. Sci.* **17**, 145.

Trethewie, E. R. (1941). *Australian J. Exptl. Biol. Med. Sci.* **19**, 175.
Trethewie, E. R. (1947). *Australian J. Exptl. Biol. Med. Sci.* **25**, 291.
Trethewie, E. R. (1956a). *Med. J. Australia* **II**, 8.
Trethewie E. R. (1956b). *In* "Venoms," Publ. No. 44, p. 243. Am. Assoc. Advance. Sci., Washington, D.C.
Trethewie, E. R. (1963). Report of Research and Investigation, University of Melbourne, Melbourne University Press, Australia.
Trethewie, E. R. (1965). *Toxicon* **3**, 55.
Trethewie, E. R., and Day, A. J. (1948). *Australian J. Exptl. Biol. Med. Sci.* **26**, 37.
Trethewie, E. R., and Day, A. J. (1948b). *Australian J. Exptl. Biol. Med. Sci.* **26**, 153.
Trethewie, E. R., and Day, A. J. (1949). *Australian J. Exptl. Biol. Med. Sci.* **27**, 385.
Trethewie, E. R., and Thach, W. T. (1961) *Med. J. Australia* **II**, 550.
Trinca, G. F. (1963). *Med. J. Australia* **I**, 275.
Waite, E. R. (1898). "Australian Snakes," 1st ed. London.
Waite, E. R. (1929). "The Reptiles and Amphibians of South Australia." South Australian Branch of the "British Science Guild."
Whitley, C. P. (1943). C.S.I.R., *Bull.* **159**.
Wiener, S. (1956a). *Med. J. Australia* **I**, 739.
Wiener, S. (1956b). *Med. J. Australia* **II**, 331.
Wiener, S. (1957). *Med. J. Australia* **II**, 377.
Wiener, S. (1958). *Med. J. Australia* **II**, 219.
Wiener, S. (1959a). *Med. J. Australia* **I**, 420.
Wiener, S. (1959b). *Med. J. Australia* **II,** 629.
Wiener, S. (1959c). *Med. J. Australia* **II**, 715.
Wiener, S. (1961a). *Med. J. Australia* **I**, 449.
Wiener, S. (1961b). *Med. J. Australia* **II**, 41.
Wiener, S. (1961c). *Med. J. Australia* **II**, 44.
Worrell, E. (1952). "Dangerous Snakes of Australia." Angus & Robertson, Sydney, Australia.
Worrell, E. (1958). "Song of the Snake." Angus & Robertson, Sydney, Australia.
Worrell, E. (1961). "Dangerous Snakes of Australia and New Guinea." Angus & Robertson, Sydney, Australia.

* Addendum to page 95:

In Papua most snake bites (about 50%) occur in the savanna woodland, especially in Port Moresby; 10% of the victims develop symptoms of snake bite. On the average there are eight deaths a year. P. black snake is the commonest culprit. In 10% of the cases the mark is not obvious and in 40% of the cases first aid procedures have obliterated evidence. Local pain and swelling are uncommon. Pain in regional lymph nodes and flaccid paralysis is common. Death adder antivenin rapidly reverses the paralysis from this snake. Asphyxia is a common cause of death because of obstruction. Proteinuria is a common sign. Blood coagulation occurs, but there is a slight defibrinating bleeding effect. Hypotension is common. Treatment includes antivenin, tracheotomy, and artificial respiration. The fatality rate is about 7%. Very little is known about W. N. Guinea (W. Irian) snakes.

Chapter 25

The Pathology, Symptomatology, and Treatment of Snake Bite in Australia

E. R. TRETHEWIE

UNIVERSITY OF MELBOURNE AND ROYAL MELBOURNE HOSPITAL,
MELBOURNE, AUSTRALIA

I. INTRODUCTION

The pathology of snake bite by Australian snakes is related to the effects of hemolysis, blood clotting, neurotoxic and cardiotoxic effects and the release of pharmacologically active substances such as histamine, slow-reacting substance, (SRS) adenosine-inactivating enzyme and numerous other substances and enzymes.

The symptomatology is a reflection of these underlying pathological processes. Treatment follows three lines:

1. First aid
2. Hospitalization
   a. Local measures
   b. General supportive therapy

3. Specific therapy
    a. Antivenin
    b. Antipharmagological agents against active released substances

The commonly dangerous terrestrial snakes are *Notechis scutatus* (tiger snake), *Acanthophis antarcticus* (death adder), *Denisonia superba* (copperhead), *Demansia textilis* (brown snake), *Oxyuranus scutellatus* (taipan) and *Pseudechis porphyriacus* (black snake). The common sea snakes in Australian waters are *Enhydrina schistosa* (common sea snake) and *Pelamis platurus* (yellow-bellied sea snake); both are extremely toxic. Bites by these latter snakes are relatively rare.

First-Aid Procedures

Immediately after the subject is bitten a tourniquet should be applied firmly to occlude the arterial supply over the first single bone proximal unit, e.g., base of phalanx for finger, upper arm for hand (and finger) or forearm and *mutatis mutandis* for the leg. The surface venom is then sucked off, or washed off if water is available. The tourniquet is released for 30 seconds after 20 minutes and this is repeated at 15-minute intervals, the tourniquet not being left on for more than 2 hours in all. Some advise venous obstruction only. It is to be remembered that more limbs were lost during the World War I because of the application of a tourniquet for too long than from other cause. If a finger is bitten by a certainly known very deadly snake (e.g., a taipan or tiger snake) 6 feet (2 m) or more long, it is probably wisest if facilities are available to cut the finger off proximal to the bite immediately. This has been done on a number of occasions with an axe with success. If a snake-bite outfit is carried (which is occasionally the case, e.g., with hikers) surface venom is washed off or sucked off, a cruciate incision is made over the fang marks, the area is sucked, and Condy's crystals (potassium permanganate) sprinkled into the area. If ice is available it should be liberally applied to the area in the attempt to freeze it. This has been performed on occasions with marked success. It both inhibits enzyme activity and reduces blood flow to curtail toxin absorption. A tourniquet can still be applied with effect for the first time 1 hour, or even longer, after biting. If the bite is not on a limb, but on the body or the adnexae, e.g., vulva (e.g., being bitten when squatting to micturate) first aid is limited to washing off venom, incision, sucking and application of an oxidizing agent (e.g., $KMnO_4$) if available and especially the use of ice if available, which should be replenished repeatedly. This may be available from a nearby house. The subject is then transported or walks to the nearest doctor's surgery or is conveyed to hospital. It is probably wise to keep him awake with stimulants such as coffee or by hurried walking or even slapping the subject. Many now object to first-aid incision.

On arrival at the surgery, if no antivenin is available, the area immediately outside the bite is injected with noradrenaline and the area excised deeply under local anesthesia. If antivenin is available, this is given first (for details see later) and the excision is then performed.

II. PATHOLOGY

A. Bite Area

The snake having bitten "like a dog," the two puncture marks produced by the fangs are 1–2 cm apart. Sometimes only one mark is seen, sometimes three or four. Reserve fangs produce the other punctures, if present. Sometimes scratches may occur from the jaw pressure. Commonly, with Australian snakes, there is remarkably little local reaction immediately; in fact it may be difficult to detect the puncture marks. Sometimes there is bleeding which may be profuse. In 1–2 hours, sometimes earlier, local erythema develops with variable edema and bruising. Centripetal vessels may show thrombosis and hemorrhage. Sometimes remarkably little reaction is noted and this makes it difficult to assess a story, especially with children or a nervous adult. It is best to err on the side of safety and assume snakebite has occurred.

B. General Pathology

A hemorrhagic zone commonly develops submucously in the second part of the duodenum. There is general spoiling of organs in severe cases which could probably be shown up well by intravenous dye injection (e.g., Evan's blue 1824) especially in the liver and probably in the nervous system. This is the case intraperitoneally in animals. The lungs may show hemorrhage and thrombosis. The kidneys may show hemoglobin breakdown products in the tubules and the urine show hemoglobinuria. There may be quite marked intravascular hemolysis.

The findings in the case of a 40-year-old male dying 5 days after being bitten on the forearm by a tiger snake were as follows: The blood was unclottable; red and white cells were generally intact but red cell lysis was evident. Fluid from the wound was lethal for guinea pigs. The forearm was swollen and discolored. The axillary lymph glands were not enlarged. Blood escaping from the neck veins was very thick. The lungs were emphysematous and showed congestion posteriorly with thick viscid blood. There was little staining of the endocardium. The lining of the aorta showed slight coffee-colored staining. The bile was viscid. The spleen was of normal texture and dark with viscid blood. The pancreas, small intestine, and other organs were normal. There was moderate excess of cerebrospinal fluid. The meshes of the pia-arachnoid were full of clear fluid that dripped away. There was no excess of fluid in the ventricles, but the brain appeared wet.*

* This is a précis of an autopsy report kindly supplied by Professor J. B. Cleland, Emeritus Professor of Pathology, Adelaide University.

III. SYMPTOMS

These usually appear in 15–60 minutes, though occasionally, when vein puncture has occurred, death may take place in 2–3 minutes. Sometimes symptoms are delayed for 2 or more hours. Rarely, as for example occasionally with brown snake, they may be delayed for 12 hours.

Initially there is faintness and sweating with headache, nausea, and often vomiting, and the vomiting may continue with vomit of blood-stained material. Pains are frequently felt in the chest over the precordium and these may be severe. Diarrhea is common. Nervousness may play a part in this occurrence. Pain occurs in the chest and across the back with dyspnea.

Nervous symptoms include dizziness, apathy, and drowsiness with muscular weakness, The pupils dilate. Syncope may then follow with the development of coma.

With tiger snake bite, dulling of sensation, aphasia, and dysphagia occur. The pupils dilate and respiratory paralysis follows. Dullness of sensation with dysphagia also occurs with death adder, and a staggering gait supervenes. Nervous symptoms are marked in the case of taipan bite, quite significant with brown snake and less evident with copperhead.

Peripheral circulatory failure and respiratory failure are common, and while the former is generally regarded as solely peripheral, sometimes it may be central in addition and central cardiac failure may be a marked feature.

In the case of black snake bite, cardiac and neurotoxic effects especially are minimal, but hematuria, hematemsis, and prostration are common. Also, local pain and swelling are noted at the site of the bite. Children may die from black snake bite, but adults nearly always survive.

IV. SIGNS

General Appearance

Shortly after being bitten the subject, while not physically ill, may appear very frightened and apprehensive. Sweating is prominent and this is probably both neurogenic and toxic. In the later stages it is largely toxic. Sometimes, as with sea snake bite, the subject has not really known he has been bitten and is quite unconcerned—the slight prick felt while in the water is ignored. Again, when a snake fancier is bitten he is often quite unconcerned, but such a subject may die. He has been bitten before and recovered, thinks he is safe, but if the interval since the last bite is too long or this is a more venomous snake, he can then die.

Vomiting, abdominal pain, nausea, and sweating often usher in more severe symptoms.

Cardiovascular collapse is indicated by a rapid thready pulse and the failure is both peripheral and central. Capillary circulation (on blanching) is poor in return. Shock is severe and sometimes there may be bradycardia.

Paralysis of the lower motor neuron type is noted with flaccidity and absent deep reflexes. This may become wide-spread after 2 hours, but often vigorous treatment instituted within 1 hour may abort its appearance. Spastic paralysis may be observed with sea snake bite and then the deep reflexes are increased. Respiratory paralysis with cyanosis and lack of diaphragmatic movement is a well-marked feature in severe cases and this may precede extensive general paralysis. Dilated pupils are also prominent in severe cases and can precede any extensive paralysis.

Hemorrhagic phenomena, with ecchymoses, blood in the urine, and the vomiting of blood are observed in cases of severe snake bite in Australia and the effects of hemolysis may be extensive with black snake bite while neurotoxic effects are minimal. Thrombosis is well marked with Australian snakes, is considerable with black snake and also marked with brown snake. In the case of death adder, the neurotoxic effects, including a staggering gait, are marked, while hemolytic effects are minimal. With tiger snake bite paralysis is severe and with taipan flaccid paralysis is extensive. Brown snake and copperhead bite is not so severe. (See Table I.)

V. DIAGNOSIS

Clinically the diagnostic patterns are four.
1. A subject is bitten by a snake, this is clearly stated, significant puncture marks are seen, some description is given of the snake and the patient presents for examination before any marked symptoms appear. There may or may not be considerable apprehension.
2. An apprehensive subject (adult), or a child whose mother is worried, presents a confused story and there is significant doubt that a snake was actually encountered. However, if definite fang marks are observed this again does not present difficulty, though if they are doubtful extreme difficulty in diagnosing is experienced.
3. A subject is bitten by a snake, there is a clear story, and the patient is presented *in extremis* or with varying degrees of collapse, some hours after being bitten. Staggering gait, difficulty with respiration, vomiting, hematemesis, cardiovascular collapse, and even coma may be present.
4. The patient is found in coma and no story is available.

There are gradations between each of these clinical types and the fourth is rare in Australia.

The diagnostic problem is usually twofold: (1) Was the patient bitten by a snake? (2) What was the nature of the snake?

TABLE I

Clinical Effects of Australian Snake Bite

Snake	Alimentary symptoms	Neurotoxic (paralytic)	Cardiotoxic	Peripheral circulatory	Vascular: clotting and hemolytic
Tiger snake	Pain and vomiting	Dulling of sensation, aphasia, dysphagia, respiratory paralysis	Significant toxic myocarditis	Circulatory collapse, sweating	Present
Black snake	Nausea and vomiting	Minimal	Minor	Some prostration	Hematuria and hematemesis
Death adder	Vomiting	Drowsiness, dulling of sensation, staggering gait, dysphagia, pupil dilatation, respiratory failure	Faintness	Sweating, peripheral circulatory failure	Present
Copperhead	Vomiting	Rapid loss of muscle tone, coma	Present	Peripheral circulatory failure	Present
Brown snake	Pain and vomiting	Headache, dizziness, weakness, respiratory failure. Symptoms may be delayed several hours	Cardiac failure	Peripheral circulatory failure	Thromboses, hemoglobinuria
Taipan	Vomiting	Flaccid paralysis bulbar paralysis respiratory paralysis	Cardiac failure	Peripheral circulatory failure	Thromboses, hemolysis slight
Sea snake (common, yellow bellied)	Vomiting, nasal regurgitation	Paresis, aching, stiffness. Flaccid paralysis, sometimes spastic. Respiratory paralysis	Cyanosis with hypertension	Sweating	Albuminuria, hemoglobinuria

In 10–15% of cases of Australian snake bite, the snake is not seen. In the case of a snake, such as the taipan, which chews, this is less likely. Commonly with sea snake bite, a prick is felt and the snake is not seen. Most snakes seen in fresh water in Australian rivers are terrestrial forms swimming from bank to bank. If a subject says he was bitten by a snake he did not see, and puncture marks 1–2 cm apart are present, the subject should be treated for snake bite. Sometimes the patient is unaware of the bite, marks not being recognized,

and concludes the snake did not bite, but an hour or so later obvious marks and redness are observed. Symptoms commonly follow about this time. Treatment should also be instituted here even in the absence of general symptoms.

Fifty percent of bites are on the leg below the knee and the vast majority are peripheral on the limbs so that a tourniquet can be applied.

An apprehensive patient may think biting has occurred when in effect it has not, or the marks were not produced by a snake. If no marks whatever are seen, the subject should be examined again in 1 hour and if they are still absent, the subject probably was not bitten, but should be observed for 12 hours to watch for the onset of symptoms such as pain, vomiting, and weakness. If there are marks of a bite or any general symptoms appear, treatment should be instituted at once.

Nature of the Snake

In Australia almost all snake bites are produced by venomous snakes. Treatment should, therefore, be instituted at once, and first-aid treatment follows the lines already outlined. The choice of antivenin is between tiger snake, taipan, brown snake, and death adder. If it is impossible to tell from the description of the snake, as is often the case, especially with children, tiger snake antivenin should be used. The choice of multiple serum follows the regional distribution outlined below. A taipan chews and if snake bite has occurred in the taipan area and the snake was large and chewed (holding on), taipan antivenin should be used because tiger snake antivenin is not very suitable in such cases. Tiger snake antivenin can be given as well in cases of doubt. Before antivenin is administered, 100 mg of an antihistamine preparation such as Anthisan or the corresponding dose of a variant should be given intravenously over 1–2 minutes and then the antivenin is given in the appropriate dosage intravenously. At the same time 3–7 minims (0.2–0.4 ml.) of adrenaline (1 in 1000) should be given subcutaneously. If homochlorcyclazine is available, it should be given at the same time as the antihistamine. It possesses the further advantages of being also anti SRS and antitryptaminic. Ordinarily one can expect about 8–10% of subjects to develop symptoms of sensitivity to intravenous antivenin varying from mild urticaria, to angioneurotic edema, to bronchospasm (mild to severe), to severe anaphylactic collapse. Death from anaphylaxis is very unlikely if the above procedure is followed. Even in the absence of such protective treatment very severe symptoms are not common.

The tiger snake (when fully grown being 3 feet or 1m long) has dark bands over a lighter body color of light gray or orange or green. In the Bass Strait Islands and Tasmania (Fig. 10, Chapter 24) unstriped black and brown forms occur. All have 15–19 dorsal scales. The taipan is often 6–9 feet (2–3 m) long, is brown with a spotted belly, with 23 slightly ridged dorsal scales. The death adder is up to 3 feet (1 m) long, with a short banded body and a tail, with 21–23 dorsal scales. The copperhead is 3–6 feet (1–2 m) long, yellow-brown to black in color with orange or red lateral scales. The black snake, found throughout eastern Australia, is usually 3–6 feet (1–2m) long, occasionally 7 feet. The skin is glossy black and lateral and ventral scales are red. There are 17 dorsal scales. The king brown or mulga snake *(Pseudechis australis)*, found on the Australian mainland generally, is coppery brown in color, is not spotted like *Demansia*, and has 17 dorsal scales.

Naturally owing to difficulties in determining the snake type from the description of the victim or others present, who are usually not well versed in snake lore, a polyvalent serum is the serum of choice, and this is now available. We expect soon to have a simple test, however, to determine the snake responsible within reasonable time clinically. This test employs cut-out plates yielding antigen-antibody reactions (Trethewie and Rawlinson, 1966, 1967). This has been improved (Trethewie and Thomas, 1969) (see Fig. 2). Russell (1967) found the serum of patients negative for antibody after biting.

The Commonwealth Serum Laboratories (Australia) recommend initially 3,000 units of their tiger snake antivenin for mainland tiger snake bite, 6,000 units for the Tasmanian tiger snake bite and 12,000 units for the Bass Strait Islands bites. For taipan 12,000 units taipan antivenin should be used and for brown snake 600 units brown snake antivenin. For death adder, 6,000 units death adder antivenin are given and for copperhead and black snake 3,000 units of tiger snake antivenin are advisable. For the king brown snake either 12,000 units tiger snake antivenin or 15,000 units Papuan black snake antivenin are suitable.

Where the snake type is not known, the schedule as outlined below listing the antivenins suitable for the various geographic areas indicated provides a suitable basis for antivenin therapy.

VI. ANTIVENIN SCHEDULE WHERE THE TYPE OF SNAKE IS NOT KNOWN

For all cases of snake bite on the mainland of Australia, except north of Rockhampton in Queensland, give 3,000 units tiger snake antivenin. In Tasmania and the Bass Strait Islands give 6,000 units and 12,000 units tiger snake antivenin, respectively. In addition, for all snake bites throughout the mainland of Australia give 500 units brown snake antivenin. In northern New South Wales, the Northern Territory, and Queensland, give also 12,000 units taipan antivenin. Throughout the mainland of Australia, except Victoria, give also 6,000 units death adder antivenin. In Victoria 3,000 units of

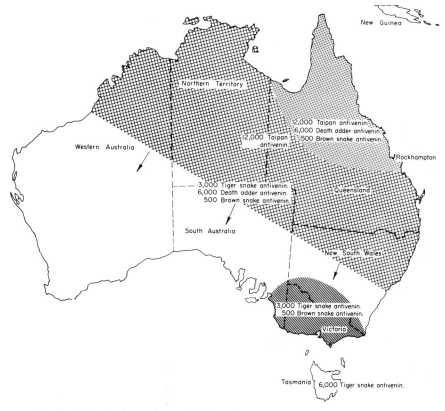

FIG. 1. Antivenin dosage by area in Australia where identification of the snake has not been made.

Tiger Snake antivenin and 500 units of Brown Snake antivenin are given. This distribution is shown in Fig. 1. Alternatively polyvalent antivenin, which is now available, may be given.

VII. THE TREATMENT OF SHOCK

In severe cases of snake bite, treatment for shock is necessary. While the heart is impaired in these cases, supportive treatment with a small blood transfusion of 300 ml, repeated if necessary, can be given with benefit. An exchange transfusion can be permitted also. It is to be remembered that in experimental cardiac infarction in dogs with shock blood transfusion greatly reduces mortality.

Physicians are not yet courageous enough to apply this therapy to man in cardiac infarction, but in view of the results of blood transfusion in all other cases of shock such treatment is probably advisable. Transfusion will improve the coronary blood flow and this is a critical factor in shock. Cardiac failure is a cause of death here and in heart-lung preparations; if the blood pressure is not maintained by mechanical means, the coronary flow falls, and the heart fails. In a further attempt to improve the coronary blood flow, Aramine, in a dose of 10 mg intravenously, and repeated as required, may be life saving.

Respiratory paralysis requires mechanical aid of a respirator, and body fluid and electrolyte balance must be maintained.

Hemolysis, where extensive, can raise blood potassium to dangerous levels and intravenous glucose (50 ml of 25 %) can reduce the impact of this, as can exchange transfusion.

Tracheotomy may be required in severe cases with bulbar and respiratory paralysis and close nursing attention as with tetanus intoxication may be necessary. However, in snake bite there are no gross spasms.

TABLE II

EFFECT OF SUBCUTANEOUS HEPARIN AND ANTIHISTAMINE ON MORTALITY OF MICE INJECTED PREVIOUSLY WITH BLACK SNAKE VENOM SUBCUTANEOUSLY[a]

Experiment	Venom, mg/kg	Heparin, mg/kg	Neo-Antergan, mg/kg	Died	Lived	Mortality, %
a	1.1	—	—	14	26	35
b	1.1	—	50	13	7	65
c	1.35	—	—	32	32	50
d	1.35	13.5	—	17	7	71
e	1.35	13.5	50	1	19	5
f	1.8	—	—	17	3	85
g	1.8	3.6	—	9	1	90
h	1.8	12	—	9	1	90

[a] From E. R. Trethewie and A. J. Day, *Australian J. Exptl. Biol. Med. Sci.* 26, 153 (1948).

TABLE III

INHIBITION OF OUTPUT OF POTASSIUM AND HISTAMINE AFTER INJECTION OF VENOM

Dog No.	Tyrode perfusion		Heparinized blood perfusion (spun plasma) (μg)
	Histamine output (μg)	Potassium increase (mg %)	
T_4	81.8	28.1	19.0
T_5	239.2	28.5	47.1
T_6[a]	18.8	8.4	21.4

[a] Heparinized perfusion.

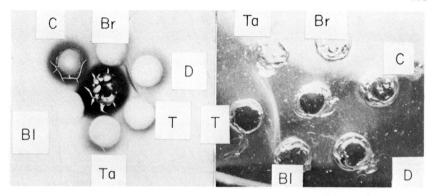

FIG. 2. Gel diffusion. Left-hand test: center well material squeezed from bite (guinea pig bitten by *Acanthophis antarcticus*), peripheral: D, death adder; Br, brown; C, copperhead; Bl, black; Ta, taipan; T, tiger antivenin. Positive: death adder.

Right-hand test: center well material obtained with saline-filled syringe from bite (cut) (guinea pig bitten by *Notechis scutatus*), peripheral: same notation. Positive: tiger snake.

REFERENCES

Russell, F. E. (1967). *Toxicon* **5**, 147.
Trethewie, E. R., and Rawlinson, P. A. (1966). *Mem. Inst. Butantan Sao Paulo* **33**(1) 235.
Trethewie, E. R., and Rawlinson, P. A. (1967). *Med. J. Aust.* **2**, 111.
Trethewie, E. R., and Thomas, A. L. (1969). *Toxicon* **7**, no. 3.

Chapter 26

Classification, Distribution, and Biology of the Venomous Snakes of Northern Mexico, the United States, and Canada: Crotalus and Sistrurus

LAURENCE M. KLAUBER*

SAN DIEGO SOCIETY OF NATURAL HISTORY AND
SAN DIEGO ZOOLOGICAL SOCIETY, SAN DIEGO, CALIFORNIA

I. INTRODUCTION

The rattlesnakes comprise a group of venomous snakes characterized by the presence at the end of the tail of a series of loosely articulated rings. These rings are composed of a hornlike material, which, when the tail is

* Deceased.

115

vibrated, elicits a sharp hissing sound. All rattlesnakes possess this appendage; no snake without one can accurately be called a rattlesnake, no matter how much it may resemble a rattlesnake in other physical characteristics.

Rattlesnakes occur only in the New World, where they are found from southern Canada southward to central Argentina. They have proliferated into a large number of species and subspecies, which differ extensively in size, color, pattern, scalation, and details of morphology, as well as in venom toxicity and other characteristics not superficially apparent. Although the majority of species prefer arid areas, some are found in marshes as well as forests. They are present from sea level to altitudes close to the timber line. In many of the areas inhabited they are the most dangerously venomous snakes occurring there, and they are often the most numerous as well.

Rattlesnakes belong to the family Crotalidae, the pit vipers, so-called because the members of this family differ from their nearest relatives, the family Viperidae, in having, on each side of the head between the eye and nostril, a small pit or aperture, which is the seat of a temperature discriminating organ.

Rattlesnakes are divided into two genera, *Crotalus* and *Sistrurus,* which differ in the size and arrangement of the scales on the crown of the head, those in *Sistrurus* being regularly arranged plates, while in *Crotalus* they are more subdivided and irregular in disposition. In these characteristics, *Crotalus* more nearly resembles *Bothrops* and *Lachesis,* two other members of the family Crotalidae found in the New World, as well as *Trimeresurus* in the Old World; whereas *Sistrurus* resembles *Agkistrodon,* a crotalid occurring in both hemispheres.

Altogether at this time 27 species (62 subspecies) are recognized in the genus *Crotalus,* and 3 species (7 subspecies) in *Sistrurus.*

II. CHECKLIST OF THE RATTLESNAKES

The following checklist presents a schedule of the species listed alphabetically, with the subspecies also listed alphabetically under each species. To avoid duplication the name of the type or nominate subspecies is given as representative of the species; that is, the name of the describer and date after the nominate subspecies refers to the original describer of the species, and not to the worker who first split it into subspecies.

The checklist also includes the English common name most usually applied to each snake and a brief summary of its range. Although this chapter covers only the rattlesnakes as far south as Mexico, for the sake of completeness the few rattlers found in Central and South America have been included in the checklist.

A CHECKLIST OF THE RATTLESNAKES

(Valid subspecies are listed alphabetically under each species.)

Genus *Crotalus* Linné, 1758

Crotalus adamanteus Beauvois, 1799
Eastern diamondback rattlesnake. Southeastern and Gulf states, North Carolina to Louisiana.

Crotalus atrox Baird and Girard, 1852
Western diamondback rattlesnake. Arkansas and Oklahoma, south to central Mexico and west to southeastern California.

Crotalus basiliscus basiliscus Cope, 1864
Mexican west coast rattlesnake. Sonora to Michoacán.

   C. b. oaxacus Gloyd, 1948
   Oaxacan rattlesnake. Oaxaca.

Crotalus catalinensis Cliff, 1954
Santa Catalina Island rattlesnake or rattleless rattlesnake. Santa Catalina Island, Gulf of California, Mexico.

Crotalus cerastes cerastes Hallowell, 1854
Mojave Desert sidewinder. Mojave Desert.

   C. c. cercobombus Savage and Cliff, 1953
   Sonoran Desert sidewinder. Sonoran Desert in Arizona and Sonora.

   C. c. laterorepens Klauber, 1944
   Colorado Desert sidewinder. Colorado Desert in southeastern California, southwestern Arizona, and northwestern Mexico.

Crotalus durissus durissus Linné, 1758
Central American rattlesnake. Southeastern Mexico and Central America.

   C. d. culminatus Klauber, 1952
   Northwestern Neotropical rattlesnake. Southwestern Mexico.

   C. d. terrificus Laurenti, 1768
   South American rattlesnake or Neotropical rattlesnake. Colombia to Argentina.

   C. d. totonacus Gloyd and Kauffeld, 1940
   Totonacan rattlesnake. Northeastern Mexico.

   C. d. tzabcan Klauber, 1952
   Yucatán Neotropical rattlesnake. Yucatán Peninsula.

   C. d. vegrandis Klauber, 1941
   Uracoan rattlesnake. Eastern Venezuela.

Crotalus enyo enyo Cope, 1861
Lower California rattlesnake. Southern two-thirds of Baja California, Mexico.

   C. e. cerralvensis Cliff, 1954
   Cerralvo Island rattlesnake. Cerralvo Island, Gulf of California, Mexico.

C. e. furvus Lowe and Norris, 1954
Rosario rattlesnake. Northwestern Baja California, Mexico.

Crotalus exsul Garman, 1883
Cedros Island diamondback rattlesnake. Cedros Island off Pacific Coast of Baja California, Mexico.

Crotalus horridus horridus Linné, 1758
Timber rattlesnake. Northeastern and north central United States.

    C. h. atricaudatus Latreille, 1790
    Canebrake rattlesnake. South Altantic and Gulf states, and lower Mississippi Valley.

Crotalus intermedius intermedius Troschel, 1865
Totalcan small-headed rattlesnake. East central Mexico.

    C. i. gloydi Taylor, 1941
    Oaxacan small-headed rattlesnake. Mountains of southwestern Mexico.

    C. i. omiltemanus Günther, 1895
    Omilteman small-headed rattlesnake. Central Guerrero, Mexico.

Crotalus lepidus lepidus Kennicott, 1861
Mottled rock rattlesnake. Southern Texas and New Mexico, and northeastern Mexico.

    C. l. klauberi Gloyd, 1936
    Banded rock rattlesnake. Southeastern Arizona, southwestern New Mexico, and northwestern Mexico.

    C. l. morulus Klauber, 1952
    Tamaulipan rock rattlesnake. Tamaulipas, Mexico.

Crotalus mitchelli mitchelli Cope, 1861
San Lucan speckled rattlesnake. Southern Baja California, Mexico, and adjacent islands.

    C. m. angelensis Klauber, 1963
    Ángel de la Guarda Island speckled rattlesnake. Ángel de la Guarda Island, Gulf of California, Mexico.

    C. m. muertensis Klauber, 1949
    El Muerto Island speckled rattlesnake. El Muerto Island, Gulf of California, Mexico.

    C. m. pyrrhus Cope, 1866
    Southwestern speckled rattlesnake. Southwestern United States and northwestern Mexico.

    C. m. stephensi Klauber, 1930
    Panamint rattlesnake. East central California and southwestern Nevada.

Crotalus molossus molossus Baird and Girard, 1853
Northern black-tailed rattlesnake. Southwestern United States and northern Mexico.

    C. m. estebanensis Klauber, 1949

San Esteban Island rattlesnake. San Esteban Island, Gulf of California, Mexico.

C. m. nigrescens Gloyd, 1936
Mexican black-tailed rattlesnake. Tableland of Mexico, from southern Sonora to Puebla.

Crotalus polystictus Cope, 1865
Mexican lance-headed rattlesnake. Tableland of Mexico, from southern Zacatecas to central Veracruz.

Crotalus pricei pricei Van Denburgh, 1895
Western twin-spotted rattlesnake. Mountains of southeastern Arizona and northwestern Mexico.

C. p. miquihuanus Gloyd, 1940
Eastern twin-spotted rattlesnake. Mountains of northeastern Mexico.

Crotalus pusillus Klauber, 1952
Tancitaran dusky rattlesnake. Mountains of Michoacán and Jalisco, Mexico.

Crotalus ruber ruber Cope, 1892
Red diamondback rattlesnake. Southern California and northern Baja California, Mexico, including several adjacent islands.

C. r. lucasensis Van Denburgh, 1920
San Lucan diamondback rattlesnake. Southern Baja California, Mexico.

Crotalus scutulatus scutulatus Kennicott, 1861
Mojave rattlesnake. From the Mojave Desert of California southeast to south central Mexico.

C. s. salvini Günther, 1895
Huamantlan rattlesnake. South central Mexico.

Crotalus stejnegeri Dunn, 1919
Long-tailed rattlesnake. Mountains of Sinaloa and Durango, Mexico.

Crotalus tigris Kennicott, 1859
Tiger rattlesnake. South-central Arizona and northwestern Mexico.

Crotalus tortugensis Van Denburgh and Slevin, 1921
Tortuga Island diamondback rattlesnake. Tortuga Island, Gulf of California, Mexico.

Crotalus transversus Taylor, 1944
Cross-banded mountain rattlesnake. Vicinity of Mexico City.

Crotalus triseriatus triseriatus Wagler, 1830
Central Plateau dusky rattlesnake. Mountains of southwestern Mexico.

C. t. aquilus Klauber, 1952
Queretaran dusky rattlesnake. Mountains of central Mexico.

Crotalus unicolor van Lidth de Jeude, 1887
Aruba Island rattlesnake. Aruba Island off the coast of Venezuela.

Crotalus viridis viridis Rafinesque, 1818
Prairie rattlesnake. The Great Plains from long. 96° W. to the Rocky Mountains, and from southern Canada to extreme northern Mexico.

    C. v. abyssus Klauber, 1930
Grand Canyon rattlesnake. Grand Canyon, Arizona.

    C. v. caliginis Klauber, 1949
Coronado Island rattlesnake. South Coronado Island, off northwest Baja California, Mexico.

    C. v. cerberus Coues, 1875
Arizona black rattlesnake. Mountains of northern and central Arizona.

    C. v. concolor Woodbury, 1929
Midget faded rattlesnake. Southwestern Wyoming, western Colorado, and eastern Utah.

    C. v. helleri Meek, 1905
Southern Pacific rattlesnake. Southern California and northern Baja California, Mexico.

    C. v. lutosus Klauber, 1930
Great Basin rattlesnake. The Great Basin from the Rocky Mountains to the Sierra Nevada.

    C. v. nuntius Klauber, 1935
Hopi rattlesnake. Northeastern and north central Arizona.

    C. v. oreganus Holbrook, 1840
Northern Pacific rattlesnake. Pacific Slope, east of the Cascades and west of the Sierra Nevada, from British Columbia south to central California.

Crotalus willardi willardi Meek, 1905
Arizona ridge-nosed rattlesnake. Mountains of southern Arizona and northern Sonora.

    C. w. amabilis Anderson, 1962
Del Nido ridge-nosed rattlesnake. North central Chihuahua, Mexico.

    C. w. meridionalis Klauber, 1949
Southern ridge-nosed rattlesnake. West central Mexico.

    C. w. silus Klauber, 1949
West Chihuahua ridge-nosed rattlesnake. Southwestern New Mexico to southwestern Chihuahua.

Genus *Sistrurus* Garman, 1883

Sistrurus catenatus catenatus Rafinesque, 1818
Eastern massasauga. Western New York to Nebraska and Kansas.

    S. c. edwardsi Baird and Girard, 1853
Desert massasauga. West Texas to southeastern Arizona, and northern Mexico.

S. c. tergeminus Say, 1823
Western massasauga. Southwestern plains.
Sistrurus miliarius miliarius Linné, 1766
Carolina pigmy rattlesnake. North Carolina to Georgia and Alabama.
S. m. barbouri Gloyd, 1935
Southeastern pigmy rattlesnake. South Carolina to Florida and south-eastern Mississippi.
S. m. streckeri Gloyd, 1935
Western pigmy rattlesnake. Mississippi to southwestern Tennessee and eastern Texas.
Sistrurus ravus Cope, 1865
Mexican pigmy rattlesnake. Central Mexican plateau.

III. KEYS TO THE RATTLESNAKES

In preparing keys for the identification of the rattlesnakes, I have facilitated their use and made determinations more accurate by segregating the coverage into two areas, that is, the United States and Canada in one key, and Mexico in the other. Further, the keys will only identify the snakes down to the species level; users wishing to identify subspecies are referred to the more complete keys contained in another work by the present author.

KEY TO THE RATTLESNAKES OF THE UNITED STATES AND CANADA

(Species only; for subspecies see Klauber, 1956, pp. 99 and 101.)

1a. Top of the head with large, symmetrically arranged plates anteriorly, usually 9 in number; a pair of large parietals in contact with each other.
. . . . Genus *Sistrurus* 2
1b. Top of the head with small scales anteriorly; parietals, if enlarged, not in contact. . . . . Genus *Crotalus* 3
2a. Upper preocular usually in contact with the postnasal; anterior sub-ocular usually contacts the fourth and fifth supralabials; 11 or more dorsal scale rows at the center of the tail; no red or orange in the inter-blotch spaces on the middorsal line. . . . . *Sistrurus catenatus*
2b. Upper preocular not in contact with the postnasal; anterior subocular usually contacts the third and fourth supralabials; 10 or fewer dorsal

scale rows at the center of the tail; usually with red or orange between blotches on the middorsal line. . . . . *Sistrurus miliarius*

3a. Outer edges of the supraoculars extended into raised and flexible horn-like processes distinctly pointed at the tip. . . . . *Crotalus cerastes*

3b. Outer edges of the supraoculars not extended into pointed hornlike processes. 4

4a. Tip of the snout and the anterior canthus rostralis raised into a sharp ridge. . . . . *Crotalus willardi*

4b. Tip of the snout and the anterior canthus rostralis not raised into a sharp ridge. 5

5a. Postnasal in broad contact with the first supralabial; anterior subocular in contact with the third and fourth supralabials; midbody scale rows plus supralabials on both sides of the head total 41 or fewer; a dorsal pattern usually comprised of two rows of small brown spots on opposite sides of the middorsal line, but some or many of which spots may be joined together middorsally. *Crotalus pricei*

5b. Postnasal usually separated from the first supralabial, or, if in contact, only to a small extent; anterior subocular not in contact with the third and fourth supralabials; midbody scale rows plus supralabials on both sides of the head total 42 or more; dorsal pattern not comprised of two rows of small brown spots on opposite sides of the middorsal line. . . 6

6a. Prenasals usually separated from the rostral by small scales or granules*; or, at least, the front edges of the prenasals chipped and sutured. . . . . *Crotalus mitchelli*

6b. Prenasals contacting the rostral, and their front edges not chipped or sutured. 7

7a. Upper preocular usually split vertically, the anterior section being higher than the posterior and curved over the canthus rostralis in front of the supraocular. . . . . *Crotalus lepidus*

7b. Upper preocular not split vertically; or if split, the anterior section not conspicuously higher than the posterior and not curved over the canthus rostralis in front of the supraocular. 8

8a. A vertical light line on the posterior edges of the prenasals and first supralabials. . . . . *Crotalus adamanteus*

8b. No vertical light line on the posterior edges of the prensals and first supralabials. 9

---

* About 17% of the specimens of *C. viridis* from central Arizona have this rostral-prenasal separation, which is here used to identify *C. mitchelli*. If the specimen is from Arizona and has a black, dark-brown, or dark-gray ground color rather than cream, tan, pink, light-gray, or light-brown, then it is *C. viridis* and not *C. mitchelli,* even if the prenasals are separated, or almost separated, from the rostral by the interposition of a row of small scales.

9a. More than two internasals—that is, scales between the nasals and in contact with the rostral, regardless of their size or position*. . . . .
. . . . *Crotalus viridis*

9b Two internasals. 10

10a. Supraoculars pitted, sutured, or with outer edges broken. . . . .
. . . . *Crotalus mitchelli***

10b. Supraoculars unbroken. 11

11a. Tail rings absent or only obscurely distinguishable from the ground color owing to lack of contrast; tail often very dark-brown, dark-gray, or entirely black. 12

11b. Light and dark alternating tail rings clearly evident, usually black or brown on a light-gray or tan ground color. 13

12a. Ventrals rarely fewer than 180; supralabials rarely 15 or fewer; usually 8 or fewer large flat scales in the internasal-prefrontal area; usually a straight dividing line or suture between the scales in the frontal and prefrontal areas; scales in the anterior part of the frontal area larger than those behind; body pattern comprising diamonds, which posteriorly are open laterally to form rings. . . . . *Crotalus molossus*

12b. Ventrals rarely more than 179; supralabials rarely more than 15; usually more than 8 scales in the internasal-prefrontal region; no definite division or straight suture between the scales of the frontal and prefrontal areas; scales in the anterior part of the frontal area not conspicuously larger than those behind; body pattern comprising irregular crossbands or chevrons, in some specimens obscured by a black suffision. . . . . *Crotalus horridus*

13a. A definite division between the scales of the prefrontal and frontal areas; minimum scales between the supraoculars rarely more than 2; each supraocular bordered inwardly and posteriorly by a large flat crescentic scale. . . . . *Crotalus scutulatus*

13b. No definite division between the scales of the prefrontal and frontal areas; minimum scales between the supraoculars rarely less than 4; no large, flat crescentric scales behind the supraoculars. 14

---

* Some specimens of *Crotalus viridis* have only two internasals and thus would fail to key out properly by the use of this character. To avoid taking the wrong course, one may reinforce the decision by proceeding to 11 regardless of the number of internasals, if the snake is from central or northern California west of the Sierra Nevada, from Oregon, Washington, Canada west of the 100th meridian, Idaho, northern Nevada, Utah (except the extreme southerwestern corner), Colorado, Wyoming, Montana, the Dakotas, western Nebraska, or western Kansas.

** The *C. mitchelli* subspecies *C. m. stephensi,* which occurs in the Death Valley region of California and Nevada, lacks the rostral-prenasal separation characteristic of most specimens of *C. mitchelli,* but has irregularities in the supraoculars instead.

14a. Tail usually of alternating brown and tan rings not in sharp contrast with the posterior body color. . . . . *Crotalus tigris*

14b. Tail of alternating black and light ash-gray rings, both colors in sharp contrast with the posterior body color. 15

15a. Body color predominantly cream, buff, gray, or gray-brown,* with dark dots conspicuous in the body blotches; scale rows most often 25; minimum scales between supraoculars usually 5 or less; first infralabials rarely divided transversely**. . . . . *Crotalus atrox*

15b. Body color predominately pink, red, brick-red, or red-brown, without conspicuous dark dots in the body blotches; scale rows most often 29; minimum scales between supraoculars usually 6 or more; first infralabials usually divided transversely . . . . *Crotalus ruber*

KEY TO THE RATTLESNAKES OF MEXICO

(Species only; for subspecies see Klauber, 1956, pp. 109 and 115.)

1a. Top of the head with large plates anteriorly (usually 9 in number) including a single frontal and a pair of large, symmetrical (but sometimes sutured) parietals in contact with each other. . . . . . . Genus *Sistrurus* 2

1b. Top of the head with scales of varying sizes; more than one scale in the frontal area; parietals, if enlarged, not in contact or symmetrical. . . . . Genus *Crotalus* 3

2a. Upper preocular not in contact with the postnasal; rostral curved over the snout; canthus rostralis rounded; dorsal body blotches longer than wide or color black. . . . . *Sistrurus ravus*

2b. Upper preocular in contact with the postnasal; rostral not curved over the snout; canthus rostralis sharply angled; dorsal body blotches square or wider than long. . . . . *Sistrurus catenatus*

3a. Outer edges of the supraoculars extended into raised and flexible hornlike processes distinctly pointed at the tip. . . . . *Crotalus cerastes*

3b. Outer edges of the supraoculars not extended into pointed hornlike processes. 4

4a. Subcaudals in males more than 40, and in females more than 35. . . . . *Crotalus stejnegeri*

---

* May be pink or red in New Mexico, but *C. ruber* does not occur in that state, or anywhere else in the United States except in southern California.

** Many California specimens of *C. atrox* have divided first infralabials, which is unfortunate for the herpetologist, since this is the only state in which a key is necessary to separate *C. atrox* and *C. ruber,* for the latter is found in no other state.

4b. Subcaudals in males fewer than 40, and in females fewer than 35. 5

5a. Tip of the snout and the anterior canthus rostralis raised into a sharp ridge. . . . . *Crotalus willardi*

5b. Tip of the snout and the anterior canthus rostralis not raised into a sharp ridge. 6

6a. Rattle matrix shrunken; no loose rattle segment. . . *Crotalus catalinensis*

6b. Rattle matrix normal; almost always at least one or more loose rattle segments subsequent to the juvenile (button) stage. 7

7a. Prenasals separated from the rostral by small scales or granules. . . . . *Crotalus mitchelli*

7b. Prenasals contacting the rostral. 8

8a. Body pattern comprises a series of 35 or more crossbands composed of conspicuous dark-gray or brown dots on a buff, pink, or light-gray background; dorsoventral width of proximal rattle contained in the head length less than $2\frac{1}{2}$ times. *Crotalus tigris*

8b. Body pattern comprises diamonds, hexagons, rectangles, ovals, or ellipses; or, if bands, not made up of conspicuous dots; dorsoventral width of proximal rattle contained in the head length more than $2\frac{1}{2}$ times. 9

9a. Anterior subocular contacts one or more (usually 2) supralabials. . . . 10

9b. Anterior subocular fails to reach any supralabial. 16

10a. Each supraocular transversely crossed by a thin, black-bordered light line that usually bends backward outwardly; clearly outlined round or oval dark blotch immediately below and touching the eye; dorsal pattern usually comprised of longitudinal ellipses; usually a pair of slim intercanthals, each about twice as long as wide. . . . . *Crotalus polystictus*

10b. No thin, black-bordered transverse lines on the supraoculars; no clearly outlined round or oval blotch immediately below the eye; intercanthals, if paired, not long and slim. 11

11a. Midbody scale rows plus the supralabials on both sides of the head total 41 or fewer. 12

11b. Midbody scale rows plus the supralabials on both sides of the head total 42 or more. 14

12a. A dorsal pattern of transverse bars, undivided middorsally. . . . . *Crotalus transversus*

12b. A dorsal pattern of blotches or spots. 13

13a. Most of the dorsal blotches located on the middorsal line. . . . . *Crotalus intermedius*

13b. Dorsal pattern comprised largely of two rows of small blotches on opposite sides of the middorsal line. . . . . *Crotalus pricei*

14a. Prefrontals (canthals) paired and in contact, and with even but convex posterior edges. . . . . *Crotalus pusillus*

14b. More than 2 scales in the prefrontal area. 15

15a. Upper preocular usually split vertically, the anterior section being higher than the posterior and curved over the canthus rostralis in front of the supraocular; dorsal body blotches occupy less longitudinal space than the interspaces; primary dorsal body blotches or crossbands seldom exceed 24 except in Tamaulipas. . . . . *Crotalus lepidus**

15b. Upper preocular not split vertically; or, if split, the anterior section not conspicuously higher than the posterior and not curved over the canthus rostralis in front of the supraocular; dorsal body blotches occupy more longitudinal space than the interspaces; primary dorsal body blotches usually exceed 24. . . . . *Crotalus triseriatus*

16a. More than two internasals—that is, scales between the nasals and in contact with the rostral, regardless of size or position. *Crotalus viridis*

16b. Two internasals. 17

17a. Upper preocular usually split vertically, the anterior section being higher than the posterior and curved over the canthus rostralis in front of the supraocular; dorsal body blotches occupy less longitudinal space than the interspaces; pattern usually of blotches or crossbars seldom exceeding 24 in number. . . . . *Crotalus lepidus*

17b. Upper preocular not split vertically; or, if split, the anterior section not conspicuously higher than the posterior and not curved over the canthus rostralis in front of the supraocular; dorsal body blotches occupy more longitudinal space than the interspaces; pattern of diamonds, hexagons, rectangles, or ellipses usually exceeding 24 in number. 18

18a. Ventral scales 161 or fewer. . . . . *Crotalus triseriatus*

18b. Ventral scales 162 or more. 19

19a. Tail rings strongly contrasting, alternating white or light-gray with brown or black. 20

19b. Tail black, brown, or gray, with rings, if present, only moderately evident, because of lack of contrast between the light and dark areas. . 24

20a. Light and dark tail rings of approximately equal widths; the postocular light stripe, if present, reaches the supralabials 1 to 3 scales in advance of the angle of the mouth; minimum scales between the supraoculars rarely less than 4; no flat crescentric scale bordering each supraocular posteriorly. 21

20b. Dark tail rings narrower than the light; the postocular light stripe, if present, passes backward above the angle of the mouth; minimum scales between the supraoculars usually 2, rarely more than 3; usually a flat crescentic scale bordering each supraocular on the posterior-

---

* It has been necessary to double-key *C. lepidus,* as the anterior subocular sometimes contacts one or more labials and sometimes does not. The same is true of *C. triseriatus* although this contact fails in the latter form much less often than in *C. lepidus.*

inward side. . . . . *Crotalus scutulatus*

21a. First infralabials usually divided; basic body color pink, red, or red-brown; dark punctations not conspicuous in dorsal blotches. 22

21b. First infralabials usually undivided; basic body color cream, tan, buff, or gray; dorsal blotches conspicuously punctated with dark dots. . . . 23

22a. A pair of intergenials usually present; generally no contact between the prenasal and the first supralabial; dark tail rings often interrupted laterally. . . . . *Crotalus exsul*

22b. Intergenials usually absent; prenasal usually contacts the first supralabial; dark tail rings rarely interrupted laterally. . . . . *Crotalus ruber*

23a. Upper preocular usually in contact with the postnasal, or such contact prevented by an upper loreal. . . . . *Crotalus atrox*

23b. Upper preocular usually not in contact with the postnasal and no upper loreal present. . . . . *Crotalus tortugensis*

24a. Scales in the frontal and prefrontal areas mostly small, rough, and knobby. . . . . *Crotalus enyo*

24b. Scales in the frontal and prefrontal areas mostly large and flat. 25

25a. No paired dark vertebral stripes on the neck; or if present, not extending posteriorly as much as 1 head length before they meet the first dorsal blotches. 26

25b. On the neck a pair of quite regular dark stripes 1 to 3 scale rows wide, separated by a single light middorsal stripe 2 or 3 scale rows wide, these stripes extending from 1 to 4 head lengths behind the head before they meet the first dorsal blotches. . . . . *Crotalus durissus*

26a. A black or dark-brown bar, bordered before and behind with cream or buff, crossing the head between the anterior points of the supraoculars.
. . . . *Crotalus durissus* (subspecies *totonacus* only)

26b. No transverse bar in the prefrontal areas as above described. 27

27a. Tail usually black or very dark-brown with light crossbars seldom in evidence posteriorly; rattle matrix usually black; subcaudals usually fewer than 28 in the males and 23 in the females. . . . . *Crotalus molossus*

27b. Tail usually gray, with light-gray crossbars in evidence posteriorly; rattle matrix usually gray; subcaudals usually 28 or more in the males and 23 or more in the females. . . . . *Crotalus basiliscus*

Note: The species *catalinensis, enyo, exsul, ruber,* and *tortugensis* occur only on the Peninsula of Baja California, the adjacent islands, or both; they do not occur on the Mexican mainland.

IV. MORPHOLOGY
A. Size

The sizes attained by rattlesnakes are of more than passing interest, since size influences several factors that are likely to affect the gravity of a bite, including venom quantity, depth of injection, penetration through protective clothing, height of the bite if on the legs, distance reached in a strike, and similar variables. Only the unit toxicity of the venom, and the size and natural resistance of the victim are as important in determining the outcome of an accident as are the factors that vary primarily with the size of the snake.

The several species of rattlesnakes differ widely in the adult size attained. There are some that occasionally reach 7 feet (or very rarely 8 feet) in length, whereas there are others that seldom exceed 20 inches. In general it is found that most subspecies of rattlesnakes are fullgrown at lengths of between $2\frac{1}{2}$ and 4 feet.

The differences in bulk or weight between the smaller and larger species are even more impressive than their differences in length; a large rattlesnake 3 times as long as a small one will weigh about 37 times as much. It can readily be appreciated how much more formidable the large snake will be in its ability to reach and to drive its fangs deeply into a human victim.

The largest of rattlesnakes is the eastern diamondback *(Crotalus adamanteus)* of the southeastern states of the United States. This snake occasionally reaches a length of 7 feet, and there are a few records indicating that very rarely an individual approximating 8 feet, give or taken an inch or so, has been encountered. These measurements, of course, refer to snakes "in the round"; the measurement of skins, which are subject to considerable stretching in drying for preservation, gives greatly exaggerated figures.

The largest eastern diamondbacks probably occur in Florida, but here the rapid increase in the human population has resulted in a great reduction of wilderness areas in which the snakes can live to reach their maximum growth, so that today 6-foot snakes of this species are by no means common.

The next to the largest rattlesnake is the western diamondback *(Crotalus atrox)* of the southwestern United States and Mexico, a snake occasionally reaching 7 feet, although most adults are considerably shorter than this. The western diamondback, plentiful in extensive agricultural and grazing areas, is probably a greater hazard to man and his domestic animals, than any other snake in North America.

Other species that are known to reach 6 feet, at least occasionally, are the Mexican west coast rattlesnake *(Crotalus basiliscus)*, the timber and canebrake rattlesnakes *(Crotalus horridus* subspecies), and the Neotropical rattlesnake *(Crotalus durissus)*, particularly the subspecies *C. durissus durissus* in Central America.

The smallest rattlesnakes belonging to the genus *Crotalus* are *C. intermedius, C. pricei, C. transversus,* and *C. willardi* which seldom or never exceed a length of 22 inches. These are all mountain forms; it is characteristic of the rattlesnakes that those species inhabiting high altitudes tend to be smaller than the lowland forms.

The species of the genus *Sistrurus* are all relatively small. Even the largest, the massasauga *(S. catenatus)* seldom attains a length of 30 inches.

People encountering rattlesnakes in the wild almost always exaggerate the lengths of the snakes they have seen. Even a person of experience, who has no particular fear of snakes, cannot estimate the length of a coiled rattler with any accuracy, nor, indeed, even one outstretched in a crawling position. Sight records of length are quite valueless.

There is a considerable body of folklore, as well as unauthenticated written records, describing rattlesnakes much larger than any I have mentioned, but, to whatever extent the bases for these stories can be traced, they are found to be quite unreliable. For example, there is a story current in a number of places in the southeastern states concerning a rattler of huge size, 15 to 20 feet or more in length, with rattles as large as teacups, that was run over by a train. Sometimes the skin of this creature has been saved, but never the rattles, which, unfortunately, are always stated to have been destroyed in the accident that killed the snake. When the skin is investigated by some curious herpetologist, it is always found to be that of a python, a boa, or some other exotic creature, as can be easily determined by noting the smooth, instead of the keeled scales of a rattlesnake, and the number of rows of dorsal scales which always exceed 33, the maximum in any rattlesnake species.

In general, male rattlesnakes grow slightly larger than females; it is probable that the average size of the adult males in a large population exceeds the females by about 12%. The one exception is the sidewinder *(Crotalus cerastes)* in which the largest females slightly exceed the comparable males.

In size at birth, compared with the size attained as adults, the smaller species have a lesser proportionate growth than the larger. The smaller species grow to be about $3\frac{1}{2}$ times their lengths at birth, whereas this ratio is increased to about 5 times, or a little higher, in the larger species. Island and other forms that are stunted, grow proportionately less from birth to maturity than do their full-sized mainland relatives.

Rattlesnakes are stout, bulky snakes as are many snakes of the families Crotalidae and Viperidae. Although some of the African vipers are relatively stouter than the rattlesnakes, it is probable that *Crotalus adamanteus* grows to be the heaviest of all venomous snakes, although by no means the longest. In weight *C. adamanteus* may occasionally reach 30 pounds or slightly more.

Rattlesnakes become somewhat stouter or more bulky as they age, the equation representing the weight in terms of length having an exponent of about 3.3, whereas 3 would indicate a constant degree of stoutness.

Like most crotalids and viperids, rattlesnakes have rather large heads. This is a characteristic often inaccurately believed to be found in all venomous snakes, but the idea is quite erroneous, since the family Elapidae, containing some of the most dangerous snakes in the world, have heads which, in size and shape, resemble those of the harmless Colubridae.

As is the case with so may animals, rattlesnakes have relatively larger heads at birth than at maturity. At birth the head length ranges from about $6\frac{1}{2}$ to $8\frac{1}{2}\%$ of the body length overall; in the adult state from 3.3 to 4.7%. *C. adamanteus, C. cerastes,* and *C. triseriatus* are species with proportionately large heads; whereas *C. durissus, C. scutulatus,* and *C. tigris* have small heads. *C. enyo* and *C. polystictus* are characterized by narrow heads; *C. cerastes* and *C. mitchelli* by broad.

Because of the presence of the rattle and the resulting foreshortening of the tail, rattlesnakes are relatively short-tailed snakes. In the males the ratio of the tail length to overall length varies from about 7 to $12\frac{1}{2}\%$, in the females from 5 to 10%. *Crotalus stejnegeri,* the long-tailed rattlesnake, has relatively the longest tail, as its name implies; rattlesnakes with proportionately short tails are *C. ruber,* some subspecies of *C. mitchelli,* and *C. viridis.*

B. Longevity

The ages attained by rattlesnakes in the wild are not known with accuracy since the marking, release, and recapture of these dangerous snakes has not been extensively practiced. Judging by size and some rough indications given by the rattles, it is presumed that they occasionally attain an age of 12 to 14 years. At the San Diego Zoo several species of captive rattlesnakes have been kept alive for from 18 to 20 years. Large species live longer than the small.

Although it is true that snakes, unlike mammals and birds, do not reach a period of complete cessation of growth, the growth of a fully adult rattlesnake is much slower than during its preadult years. Thus unusually large rattlesnakes of a species are not necessarily unusually old, the difference being more likely an individual variation.

C. Squamation

The several species of rattlesnakes differ considerably from each other in squamation, that is in the number and arrangement of certain series of scales, and in the position of various scales and their contacts with each other. So far as is known, these details of squamation remain unchanged from birth to death despite the frequent shedding of the skin, and this contancy in each individual, and a moderate degree of constancy within each subspecies, render squamation an exceedingly important criterion in all taxonomic work. The differentiation of species that have quite similar characteristics of pattern and color cannot be effected without using differences in squamation.

From the taxonomic standpoint the most important scale series on the body are the number of dorsal scale rows, the number of ventral plates, and the number of subcaudal scales.

In the rattlesnakes the dorsal scale rows at midbody vary from 19 to 33. The variation within a subspecies seldom exceeds 4 rows, although 8 is reached in a few subspecies. Sexual dimorphism in scale rows is virtually absent in the rattlesnakes. The number of scale rows is strongly correlated with size, that is, average subspecific size, not individual size; small subspecies are likely to have 19 to 23 rows, large subspecies 25 to 31. Normally the dorsal rows are odd in number because of the presence of a single middorsal row. The dorsal scales on a rattlesnake are strongly keeled with the exception of the 1 to 3 lowest lateral rows on each side, which are usually smooth. Two apical scale pits are often present on the posterior ends of the dorsal scales, particularly on the middle rows toward the tail.

The ventral scales, or plates or scutes as they may be called, are quite useful in classification, despite the fact that there is considerable intrasubspecific variation, and also sexual dimorphism, since the females of each subspecies usually average about 5 more ventral scales than the males. The number of ventrals also is sharply correlated with average body size, with a low of about 130 scales in the sidewinder *(C. cerastes)* to a high of over 200 reached in *C. basiliscus* and *C. ruber.* The pigmy rattler *(S. miliarius)* has fewer ventral scutes than any other rattler, sometimes with as few as 122 ventrals. The ventral-size correlation is by no means a perfect one since the averages are often affected by specific and intersubspecific relationships. Within each subspecies the coefficient of variation in the number of ventrals in each sex ranges from about 1 to $3\frac{1}{2}\%$, with an average of 2%.

The subcaudal scales are found to be less important in classification than the other two body series, since many scales in this series are so irregular as to lead to uncertainties in the counts. There is considerable sexual dimophism in the number of subcaudals; the usual difference between the sexes comprises about 5 more scales in the males than the females. Tail proportionality is also important; short-tailed snakes have fewer scales in this series.

The greatest aids to the taxonomist are to be found in the head scales, where there are not only fundamental differences in subdivisions, such as those that separate the 9-plate prefrontal and frontal pattern that distinguishes *Sistrurus* from the greater proliferation of the scales in these areas in *Crotalus,* but also a wide variety in the sizes, subdivisions, and contacts between certain scales on the head in the many species of *Crotalus.* The use of these difference in classification can be readily noted by consulting the keys (pp. 121–125).

In general it will be found that scale contacts in head scales are of more service in taxonomic problems than scale numbers; the most readily countable scales, the supralabials and infralabials, usually vary in number by 2 to 4 scales within a subspecies; but scale contacts and mutual arrangements tend to have a greater intrasubspecific constancy. A few subspecies are characterized by scale shapes; of these the most important are the peaked supraoculars of the sidewinder *(C. cerastes)* which have earned this snake its alternate name of horned rattlesnake; and the ridged snout produced by tipped up internasals and canthals on the ridge-nosed rattler *(C. willardi)*.

As examples of the types of differences found in other species of rattlesnakes, there are the 4, instead of the normal 2 internasals in *C. viridis,* the lack of contact between the prenasal and rostral in most specimens of *C. mitchelli,* and the split upper preocular in *C. lepidus.* It is evident from this discussion that anyone who wishes to make accurate identification of rattlesnakes must become familiar with the nomenclature of squamation. A reference that will be found useful is contained in the present author's enlarged work on the rattlesnakes, particularly the glossary and diagrams on pp. 92–99 of Volume 1.

D. Color and Pattern

The colors and patterns of the many kinds of rattlesnakes differ considerably and it would be of value to both professional and amateur naturalists if it were possible to segregate the various forms by these prominently evident characters. Unfortunately this is not possible, for although some species are so conspicuously different from all others as to be readily distinguishable, this is not true of many important species, and there are further uncertainties that result from individual and intrasubspecific variations.

Mainly rattlesnakes are blotched snakes having patterns comprised of dorsal diamonds or hexagons with one or more series of smaller lateral blotches on the sides. Blotches are characteristic of slow-moving snakes compared with the longitudinal lines or stripes that most frequently adorn slim, fast-moving species. The typical rattlesnake blotch is a multicolored area involving several different hues in its parts. There is, first, the central area of each blotch, then the part surrounding the center, then the part next to the light outer border, then the light border itself, and, finally, the background color between blotches, usually somewhat darker middorsally than laterally. This multiplication of colors makes it difficult to describe a rattlesnake pattern succinctly, hence tabular descriptions of rattlesnakes based on color and pattern are not particularly useful, being further complicated by individual and intrasubspecific variation.

Fortunately there are some rattlesnake species characterized by divergences from the mode which render them readily distinguishable. For example, there are several species with bands or crossbars rather than the more common middorsal blotches, including *C. horridus, C. lepidus, C. tigris,* and *C. transversus.* However, it should be noted that, even in the blotched snakes, the posterior blotches often engage the lateral series to become, in effect, crossbars, as, for example, in *C. molossus.*

Among other conspicuous deviants from the usual pattern are *C. pricei,* in which the dorsal blotch series has been split into twin series of spots on each side of the middorsal line; *C. polystictus,* whose dorsal pattern comprises two or more series of longitudinal ellipses; *C. willardi* in which the side borders of the blotches have become indefinite or have disappeared entirely; and *C. durissus* (except the subspecies *C. d. totonacus*) wherein the anterior blotches have been modified into longitudinal lines extending from one to four head lengths along the neck.

The number of blotches (or crossbars) between head and tail varies from about 15 in *C. lepidus* to 60 or more in some of the small mountain species. The coefficient of variation in the body blotches in a subspecies ranges from 4 to 12%, with an average of 6%.

The tails of most rattlesnakes are ringed, usually with colors that correspond with those of the posterior body blotches, but sometimes with rings whose alternating colors differ from those of the body blotches. Such sharply contrasting tails are particularly evident in *C. atrox* and *C. ruber,* in which the tails look as if they had been attached to the wrong snake. In other species the tails are virtually without pattern, this being particularly evident in *C. molossus* and *C. horridus,* in which the tails are generally black.

Although rattlesnakes of almost every color are to be found, including red, orange, yellow, green, and blue, the colors are seldom brilliant, that is, there is some admixture with gray; there is seldom the pure color that characterizes, for example, the green of some tree snakes.

Yellow is present in *C. v. concolor,* orange in *C. v. abyssus* and *C. m. pyrrhus* from certain areas, and red or pink in the latter form from Arizona. Green is evident in *C. molossus, C. scutulatus,* and *C. lepidus,* and blue in *C. lepidus* and *C. pricei.* But most rattlesnakes run to more neutral grays and browns, with color combinations that are beautiful in their contrasts. A few such as *C. horridus* and *S. catenatus* have melanistic phases, and albinistic aberrants rarely occur in many species.

Despite the usefulness of color and pattern in distinguishing species, these criteria have in the past led to incorrect determinations when assigned too much importance. For instance, for a long time there was confusion between *C. atrox* and *C. scutulatus* resulting from some of their superficial likenesses in color and pattern, an important error in view of the differences lately found between the toxicity and physiological effects of the venoms of these two species. If one should depend too much on color and pattern, there will be confusion between the reddish specimens of *C. atrox* found in some parts of New Mexico with *C. ruber,* a quite different rattlesnake that occurs only in southern and Baja California. *C. triseriatus* is difficult to segregate from *C. intermedius,* if one uses only pattern and color to identify them, but scale arrangements indicate that they are quite different species.

In most rattlesnakes a close examination of the dorsal pattern will disclose the fact that the darker elements of the pattern are composed of a multiplicity of dark dots or punctations. This effect is particularly evident in *C. atrox* and its relatives, in *C. mitchelli,* and in *C. tigris.*

The head of most rattlesnakes is adorned with a variety of light and dark marks, including light supraocular crossbars, and light preocular and postocular stripes, with a dark ocular stripe between. The presence and direction of these marks are often of some service in reaching taxonomic decisions. For example, in *C. adamanteus* there is a vertical light line on the posterior edges of the prenasals and first supralabials which is distinctive of the species; in *C. atrox,* the postocular light line intersects the supralabials, whereas in *C. scutulatus* it passes backward above the angle of the mouth; and in *C. polystictus* the supraocular crossbars curve backward outwardly.

Whatever element of the pattern is accepted as describing the color of any species of rattlesnake, whether the major blotch color, or the ground color between blotches, the intrasubspecific variations are found to be extensive. Most of these show distinct tendencies to favor protective coloration, whereby the snake tends to merge inconspicuously into its background. This is particularly evident in *C. cerastes,* in which, within each subspecies, there will be found color variations that tend to match the sand on which the snake lives; and in *C. viridis* in which the several subspecies have become modified to match the colors of the rocks or soil on which they have become most prevalent. This tendency is so universal among the rattlers that certain generalities may be presented. Mountain species are usually dark, lowland forms light. Disruptive coloration, with its contrasts, is to be found in subspecies that live in forested areas where shadows are most prevalent. However, procrypsis is not always the determining factor in fixing the snake's color, as witness the brick red of *C. ruber,* which contrasts so decidedly with the green of the cactus or the gray of the granite boulders among which the snake is so often found.

Rattlesnakes have some power of color modification following temperature changes, although this never approaches the extensive changes that are so apparent in many species of lizards. However, it has been observed in *C. cerastes* and *C. viridis,* and, without doubt it occurs in others. The change results in a lightening of the color as the temperature increases. The effect is produced by a contraction of the melanophores in the skin with a rising temperature.

Many species of rattlesnakes show an ontogenetic change. The young are more brightly colored and with more distinct patterns than the adults; indeed the adults of some forms, of which *C. v. abyssus* is an example, almost lose their blotches entirely with age.

Sexual dimorphism in color is not evident in most species of rattlesnakes. However, in *C. horridus* the yellowish individuals are nearly always females, whereas the black specimens are generally males.

E. Senses

Of the rattlesnake senses, that of smell is probably the most acute and useful. The sense of sight is moderately good, but has relatively narrow spatial limitations. Hearing is absent but is replaced in some degree by an acute sensibility to vibrations of the ground upon which the snake rests. In addition the rattlesnake, like all pit vipers, has a sense organ of which man has no counterpart; this is the facial pit, which permits the snake to sense the presence and direction of an object having a slightly higher temperature than that of its surroundings, a useful adjunct for a creature whose principal prey comprises small mammals.

The rattlesnake vision is not particularly acute. Tests indicate that a snake can distinguish objects within a range of about 15 feet, particularly if they are in motion. The snake has no supersensitive spot on the retina such as the fovea of a mammal. To a large extent the vision of each eye is independent of the other and there is binocular vision only within a narrow angle. Hence the snake seldom makes an attempt to orient the head toward the object seen, unless it does so as a part of its threatening posture. Distance accommodation is effected by movement of the lens toward or away from the retina.

Rattlesnakes, like all other snakes, are without eyelids. The eye is covered or protected by a transparent domelike spectacle or brille, which is a part of the skin that the snake periodically sheds. The snake seems to have confidence in the protective value of the spectacle and does not flinch when an object is brought near the eye.

A short time before the skin is to be shed a milky fluid accumulates between the old spectacle and the inner one that will replace it. As a result the snake becomes partly blind. This opacity disappears a few days before the skin is shed.

The most notable feature of the rattlesnake's eye is the vertical pupil, a characteristic of certain families or genera of snakes, particularly those that are largely nocturnal, for the vertical pupil is more efficient in closing out surplus bright light than a round or oval pupil. The iris is almost completely closed in bright light, or when the snake is sleeping. Although rattlesnakes are by no means exclusively nocturnal, the vertical pupil does improve their vision at night.

Folklore has attributed qualities to the rattlesnake's eye that it really does not have, such as the ability to enlarge and turn red when the snake is enraged, and also the power to fascinate its prey.

Rattlesnakes, like all snakes, have no external ear openings, and the rudiments of the auditory apparatus seem to be ineffective in transmitting the impacts of airborne waves to the brain. At least, despite the fact that certain nerves have the ability to transmit the effects of such waves to the brain, the snake shows no apparent reaction thereto. Thus the long-held belief that snakes are deaf to airborne sounds remains valid. However, rattlesnakes, like many other snakes, are extremely sensitive to tremors in the ground on which they rest, and to these tremors they do exhibit an evident response. Rattlesnakes have shown by their reactions that they can sense the tread of a horse at a distance of 100 feet or more.

The rattlesnake's sense of smell is acute; indeed it is accentuated by the presence of two separate avenues whereby the impact of odors may be carried to the brain. There is first the tongue, which need not touch a foreign object to assess its character but does so, much more frequently, by picking up gaseous, liquid, or minute solid particles from the air and conveying them to two pits, known as Jacobson's organs, in the roof of the mouth whence nerves carry the effect to the brain for analysis.

The knowledge of how the tongue is used has been a comparatively recent discovery. Prior to this it was assigned many incorrect purposes, such as for licking prey or as a stinger, and many people still confuse the tongue with the snake's fangs. Often a snake, seeking to avoid detection through procrypsis, by the use of its tongue gives the only evidence of being on the alert, for it continually everts and retracts the tongue. When it is apparent to the snake that it has been discovered, the tongue may also become a part of the threatening pose. In the rattlesnake this is done by everting the tongue to the limit and then alternately erecting it vertically upward and then depressing it downward, with the bifid tips spread as widely apart as possible. This habit definitely adds to the snake's dangerous appearance.

In addition to the tongue and Jacobson's organs, the rattlesnake has nostrils, olfactory bulbs, and a sense of smell of the more customary type.

Besides the usual senses with which man is familiar because of his own possession of similar devices, the rattlesnake, like all pit vipers, has a novel, additional sense organ. This is the facial pit, a deep depression on each side of the head slightly below a line from the eye to the nostril. This pit is a temperature-difference detector, whereby the snake is able to determine the presence, through its interception of infrared waves, of any object differing in temperature from its surroundings to the extent of only a few degrees. This is valuable to an animal whose prey is largely comprised of small warm-blooded creatures like rodents. Indeed, because there is a pit on each side of the head, through the resulting stereoscopic effect the snake becomes aware of the direction and distance of the prey, which facilitates the accuracy of the snake's strike. The pit is essentially a short-range detector, limited to objects about 15 inches away.

F. Lungs

In an animal as attenuated as a snake, it might be assumed that some of its organs would be considerably modified in shape so as to be accommodated in the slim body. Of all the major organs probably the lungs have been the most extensively changed.

Actually there is but one functional lung; in the Crotalidae and many other families, only one active lung remains, the other having been suppressed to a useless remnant. In the rattlesnakes and most other genera, the right lung is the one remaining, and it has become so enlarged that it extends over about three fourths of the snake's body.

The rattlesnake's lung comprises three parts, a tracheal section, a bronchial section, and a posterior, nonvascularized, air-storage sac having no function for blood aeration. The tracheal section has respiratory alveoli along its upper, inner surface, and, as it is prevented from collapsing by transverse cartilaginous incomplete rings, permits the snake to breathe while engulfing food. The tracheal alveoli increase the functional lung area and thus compensate in some degree for the absence of the other lung. The bronchial lung furnishes the major part of the aerating surface, and is the part of the lung most nearly resembling the lungs of other animals. Posterior to this is the third section of the lung, a large, thin-walled, bladderlike extension, which, having no alveoli, has no aerating function, but is merely useful for air storage. It gives the snake buoyancy when in the water, furnishes an extra air supply when the snake is swallowing food, reduces the rapidity of breathing, accentuates the ability to hiss, and also, in some degree, may have a temperature-equalizing function.

G. The Rattle

Initially I stated that the rattle is an appendage peculiar to the rattlesnake; that no snake without a rattle can be considered a rattlesnake and that none with one is not a rattlesnake. The first of these statements is subject to slight modification; rarely, three kinds of adult rattlesnake are encountered without rattles, or with partial rattles that are soundless.

There is, first, a snake that has lost its rattles in an accident, usually removed by some man who, thinking he had killed the snake, cut off the rattles as a souvenir. But subsequently the snake survived. Such a snake can be recognized by its blunt tail. A second kind of rattleless rattlesnake that is rare indeed is one on which, through some malformation of the tail at birth, no rattle at all is formed. And, finally, the rattlesnakes *(C. catalinensis)* inhabiting a single small island (Santa Catalina) in the Gulf of California, Mexico, apparently have an inheritable defect in the matrix upon which the rattle is formed, such that only the rattle surrounding the matrix is retained, each prior rattle segments being detached and lost when the snake sheds its skin.

With these exceptions, every rattlesnake has a rattle, although during its earliest youth it is soundless, since it has but one rattle segment. There must be at least two segments to make a noise, for the sound is produced by two or more segments striking against each other when the tail is vibrated.

Although a great many purposes have been attributed to the rattle, including a poisonous quality of its own, a mating call, a call for help from other nearby rattlesnakes, a lure for prey, or, contrarily, a warning to prey, all of which were suggested long ago, these have now been abandoned in favor of the theory most consistent with the actual use of the rattle, namely a warning to creatures that might injure the snake. For the strident sound of the rattle— a sharp hiss like that of escaping steam—has a definitely frightening impact upon the hearer, whether man or beast. Any theory of use dependent on the rattles being heard by any other rattlesnake is invalid since it is probable that no rattlesnake can hear the sound of a rattle, its own or that of a companion. And such an intricate device would never have been perfected for the obviously self-injurious purpose of warning prey; or the altruistic warning of other creatures for their good rather than to benefit the snake. Many other kinds of snakes, both venomous and harmless, vibrate their tails when threatened or annoyed, and the primitive rattle must have been developed from this ancient and widespread habit, which is observed among some species of snakes in every continent.

The rattle is a highly intricate device, intricate in the relation of its parts, whereby successive segments or rings are loosely interlocked so that they clash against each other when the tail is vibrated, yet with such an effective grip that each segment retains its hold on the one representing the succeeding skin shedding, unless the grip be interrupted by breakage. Furthermore, the conformation of each rattle is such that the entire rattle string tends to tip upward posteriorly, thus reducing wear from dragging on the ground. Since the rattle intricacy is the same in the two genera *Crotalus* and *Sistrurus,* it may be presumed that the appendage was developed before the evolvement of the characters that differentiate the two genera.

A rattlesnake is born with a prebutton, a rattle so flexible that it is shed when the youngster first sheds its skin at an age of 8 to 10 days. Subsequently the young rattler attains a button, the first segment that is permanently retained, meaning by permanently, at least until the button is lost by breakage. In adult rattlesnakes it is quite easy for one with experience to tell whether or not the button remains, and thus whether the string is complete. Rattlesnakes acquire a new rattle each time the skin is shed, which is about 4 times per year in its first 2 years, and about twice in its subsequent years. Hence rattle strings numbering more than 12 segments would often be found in fully adult snakes at an age of 8 years or more, were it not for wear and breakage. But such strings are rather infrequent since the loss of rattles through wear and accident is quite normal; so excessively long strings are rarely found on adult snakes; indeed they would be inefficient vibrators. As a noisemaker a string of seven to ten segments is more effective.

The rattles not only conform to the size of snake, that is, to its age and the resulting thickness of its tail, as each segment is formed, but also to the adult size of the species. Thus, large rattlers, such as *C. adamanteus* and *C. atrox* have correspondingly large and strident rattles audible at distances of 100 feet or more, while the small species, such as *C. intermedius, C. pusillus,* and *S. miliarius,* have rattles so small as to be almost inaudible when one is only a few feet distant from the snake. Furthermore, there are the dimensional differences, such as the size of the button and the growth increments between successive segments during the adolescent growth of the snake, that are characteristic of the several species and subspecies of rattlesnakes, which are to some extent independent of adult body size. These differences, when known, are useful in taxonomy. For example, *C. durissus* and its subspecies start life with a small button, after which the increments in the sizes of successive segments are relatively large until the rattles are appropriate to large snakes. On the other hand, *C. m. mitchelli* starts life with the largest of all buttons, but with the next two or three segments, loses this supremacy to larger rattlesnakes such as *C. adamanteus* and *C. atrox.*

The speed with which a snake vibrates its rattle is partly dependent on temperature. Within temperatures in which the snake is likely to be fully active, the speed varies from about 35 cycles per second at a temperature of about 60°F to 85 cycles at a temperature of about 95°F.

The rattle is composed of keratin, the albuminoid substance that is the basis of such tough animal parts as horn, hair, nails, and feathers. It is virtually impervious to moisture or oxygen, and withstands wear without further nourishment.

Despite the early myths to the effect that male rattlesnakes carry their rattles with the flat side vertically and females at right angles thereto, actually all rattlesnakes carry them with the major axis vertical. The direction of vibration is transverse, that is, the snake vibrates its tail to the right and left, not up and down.

Although rattlesnakes usually sound their rattles when annoyed or disturbed, since this is a characteristic feature of their defensive or threatening posture, it would be dangerous to assume that they will always do so. In traveling about in rattlesnake-infested country, one must always be on the alert to observe the snake before it is approached within striking range, and not depend upon the warning rattle. Rattlesnakes are quite capable of striking or biting without sounding the rattle at all.

One of the earliest and most indelible myths concerning the rattle is that the number of segments accurately indicates the age of the snake. This is not true since the snake acquires a new rattle with each shedding of its skin, which may be 3 to 5 times per annum during its first two or three years when growth is most rapid, and twice per year thereafter; besides which, in the adult stage, an indeterminate number of rings has usually been lost through wear and breakage.

Another myth is that the snake keeps its rattles above water when swimming. This is not true; the snake propels itself in water by waving the tail from side to side with the rattles fully immersed. When wet, the rattles will make no sound.

V. HABITS AND ECOLOGY

Like all other snakes, rattlesnakes are greatly influenced by external temperatures. Unlike mammals and birds, which have internal means for heating and cooling and thus for keeping their body temperatures at an optimum favorable to their body processes and needs, snakes are dependent on external temperature effects, such as the sun's radiation, the air temperature, and the temperature of the ground on which they rest, to secure a body temperature within limits favorable to their requirements for activity. At temperatures too low the snake can only move slowly; and if there is a further drop in temperature, the creature is completely immobilized and is at the mercy of the weather or its enemies. Similarly, in the other direction, too high a temperature thickens the blood and the snake quickly dies, and this at a temperature not particularly uncomfortable to man.

Thus the snake's activities, whether daily or seasonal, are circumscribed by its temperature requirements. Its primary necessities of food, reproduction, and protection must be sought in periods of favorable weather and temperature; and it employs various expedients which enable it to take advantage of variations in microclimate, whereby it can extend the periods during which its body temperature will be maintained at satisfactory levels. Thus, when the sun is too hot it seeks the shade; when the ground becomes dangerously warm it takes refuge in rock crevices or down mammal holes. In the summer season it becomes largely nocturnal.

When below optimum temperatures are to be met it basks in the sun. This is likely to be its habit during the first warm sunny days of spring or the last in the fall. In the temperate zone, where winter temperatures fall below those that the snake can endure with safety and which in any case will render the creature immobile for an extended period, it seeks a safe winter seclusion deep in some hole or rocky cleft below the frost line. Here it enters a den in which it will live in safety but in torpor for the entire winter season, as long as 5 or 6 months, depending on the latitude and altitude. Usually such a den for hibernation is used year after year by the snakes of the contiguous area; once found to be safe, successive generations of rattlers follow their elders into the same refuge that experience has proved secure and adequate.

Hibernating dens of this character are especially utilized by the various subspecies of *C. viridis* in the western United States, and *C. horridus* in the east, it being by no means unusual for from 100 to 500 snakes to gather at a single den. Various harmless snakes may likewise use the same haven. There is no enmity between them at this season.

At these dens the snakes do not all go underground to stay on a single day, nor do they all emerge simultaneously. On the contrary, they first seek refuge for the night hours, and emerge each day to take advantage of the last warm sunlight of autumn. Exactly the same conditions exist in reverse in the spring. These habits are taken advantage of seasonally by men who know the location of the dens; raids are made either to kill or capture large numbers of rattlesnakes each year. A single day in spring or fall will be more fruitful than many days of hunting rattesnakes in the summer. Suitable trapping methods have also been perfected so that the dens need not be visited daily.

The duration of hibernation varies with the climate, which in turn depends on latitude and altitude. In areas of mild climate, California and Florida, for example, the rattlers are more likely to seek individual seclusion, rather than gathering together in large numbers, and when the weather is propitious for a day or so in winter they are likely to emerge and bask in the sun. Further toward the tropics their annual activity requires no seasonal interruption.

Throughout the temperate zone spring is the season of maximum activity, for, after issuing from hibernation, the snakes are hungry; also this is the mating season. So at this season there is a concentration of activity even more intensified then seems to be required by weather alone. In the southwestern United States, May is the month in which the snakes are most active. Even in the tropics, where the weather does not require the cessation of activity, the snakes are more active in some months than in others.

A. Locomotion

Rattlesnakes employ four methods of locomotion: horizontal undulatory, rectilinear, concertina, and sidewinding. None of these is peculiar to the rattlesnakes, although sidewinding is less widespread in other genera. Horizontal undulatory progression is the usual method of crawling employed by most snakes. It is the method in which the snake's body is disposed in several waves on either side of a central axis which points in the direction toward which the snake is crawling. The motion is produced by the loops of the body pushing backward against objects on the ground or irregularities in the surface. The ventral plates are not used ratchetwise in this method of crawling.

Rectilinear progression is a method extensively used by large, thick-bodied snakes, and is therefore appropriate for the larger rattlers, particularly when a rapid advance is not necessary, as, for example, when they are stalking prey or investigating some suspicious object. The term "caterpillar crawling" is sometimes applied to this method, and also "rib-walking," although the latter is quite inappropriate, since the snake does not use its ribs to facilitate the motion. In this method the snake's skin and surface muscles move relative to the interior body column, so that there is a reciprocating motion between the two. First a section of the skin slides forward relative to the central body mass. The snake then contracts, or bunches, several adjacent ventral scutes and presses them against the ground, where they are anchored by the rear edges of the plates. Then the body itself advances relative to the skin. Several of the skin sections move independently but simultaneously, so that the skin compression waves appear to flow backward along the body, as the body itself advances. In this method of crawling the posterior edges of the ventral scutes do have some of the ratchet effect with which they are commonly credited but the snake does not walk on its ribs. In rectilinear crawling, the snake leaves a straight track.

In concertina progression the central part of the body is alternately gathered into two or more side waves and then restraightened. As the waves are formed the head and neck are stationary, but as the waves are straightened, the head and neck are thrust forward, while the tail is stationary. Thus, by alternately forming and straightening the central curves, the snake slowly advances. The concertina method, if used at all, is likely to be employed when a slow, hesitating advance is desired, as, for example, when a snake is investigating some strange object. This method has the advantage that the head advances in a straight line, and further that there are short periods during which the head is motionless and therefore the senses may be more deliberately exercised.

The fourth method of crawling is called sidewinding and is a method appropriate to a snake seeking to advance on a surface such as loose sand against which a transverse thrust cannot be effectively applied. The locomotion is called sidewinding since the snake advances at an angle with a line drawn from head to tail. It is practiced by the sidewinder or horned rattlesnake, C. cerastes, and likewise it has been developed independently by the small viperine snakes inhabiting the deserts of the Old World, such as the Sahara, the Kalahari of South Africa, and the deserts of Asia between the eastern Mediterranean and western India. In the sidewinding method the snake applies vertical rather than transverse forces to the ground. This it does by raising the anterior part of the body slightly above the sand and advancing it to a new anchorage at the side. Then, with the head anchored, the body is laid down behind it in a straight line, after which the head is advanced to a new anchorage, and again the body is advanced. The resulting track is a series of straight lines, each equal to the length of the snake, advancing in echelon toward the objective. It is surprising with what rapidity a snake, adept at sidewinding, can progress. It appears to advance with a smooth, flowing motion diagonally across the sand. It is probable that *Crotalus cerastes* can move faster (up to 3 miles per hour) than any other rattlesnake applying any type of progression.

B. Warning and Defensive Methods

Rattlesnakes are very seldom aggressive unless they are deliberately annoyed or injured. If their privacy is encroached upon, they tend to lie quiet and to escape discovery by procrypsis or their blending coloration. But if this fails, they usually seek safety by crawling away toward some refuge; and if this, in turn, fails to eliminate the attention of some threatening trespasser, then they assume a defense posture which serves both to disuade the enemy from attack and at the same time prepares them to use their really potent weapon, namely, the strike.

The rattler having been prevented from escaping, and caused to stand his ground, will, by his actions and appearance, leave no doubt in the mind of any creature having the most elemental desire for safety that here is an animal that should be avoided. For the snake raises up in a loose S-shaped spiral, about half the body resting on the ground in a ring that serves as a base of operations, while the anterior part of the body rears up into a loose S-shaped spiral, which, when straightened out, becomes the forward lunge that constitutes the strike. The head of the snake faces the enemy; the tongue is alternately protruded and retracted, and, when protruded, is raised vertically upward, and then dropped downward, with the tips widely separated. The snake exhales in a violent hiss, and superimposed on this the rattle is continuously sounded with its maximum stridency. The snake does not have to approach its enemy in order to accentuate its obvious threat. If the warning is successful and the intruder retreats, after a short time the snake will crawl toward some refuge; if this fails it can maintain its threatening pose indefinitely, always ready to strike if the enemy approaches within range.

It should be mentioned that the defensive pose presupposes a sufficient time for the snake to assume it. A rattlesnake accidently stepped on can bite a man's leg without the necessity of coiling or striking, and is likely to do so without sounding its rattle.

C. Climbing and Swimming

Rattlesnakes are stout, heavy-bodied snakes and are not natural tree climbers, as are so many slim snakes that seek their prey aloft. Nevertheless they can climb and those that live in forests are occasionally found up in trees. They climb, as do all snakes, not by wrapping themselves around a trunk or branch, but by forcing their bodies into cracks or irregularities in the bark, or by looping themselves over knots or branches.

Similarly, although rattlers are not addicted to water as are such water seekers as snakes of the genera *Natrix* or *Thamnophis,* or the water moccasin *Agkistrodon piscovorus,* nevertheless rattlesnakes have no hesitancy in taking to water if necessary to reach some goal. They are good swimmers and are so buoyant that the head can be kept well above the water surface to see nearby objects.

Not only do rattlesnakes cross lakes or streams but they are not infrequently found at sea, distant as much as 20 miles from land, particularly in the relatively warm waters of the Gulf of Mexico and off the South Atlantic states. Here many of the offshore islands are populated by *C. atrox* or *C. adamanteus.*

In the colder waters of the Pacific they are not so frequently encountered, nevertheless they have populated many of the islands. This is particularly true of the islands of the Gulf of California, Mexico, where the water is warmer. Almost every island of this Gulf is populated by one or more species of rattlesnake, but here the landings may be less frequent than they are on the Atlantic islands. Gene flow from the mainland species that produced the first colonists has been sufficiently infrequent and intermittent to permit the differentiation into endemic island species or subspecies, such as, for example, *C. tortugensis* and *C. catalinensis*; as well as the subspecies *C. enyo cerralvensis, C. mitchelli angelensis, C. mitchelli muertensis,* and *C. molossus estebanensis.* Not all the colonists have differentiated to an extent warranting taxonomic recognition by any means; many islands are occupied by mainland forms. On the Pacific Coast only two forms have been so separated from their mainland congeners to be so recognized; *C. exsul* on Cedros Island, a derivative of the mainland *C. ruber ruber,* and *C. viridis caliginis* of South Coronado Island, a descendant of *C. viridis helleri* of the adjacent mainland.

D. Food and Hunting

The principal food of rattlesnakes consists of small mammals such as mice and rats, gophers and ground squirrels, chipmunks, and tree squirrels. Obviously the size of the prey depends on the size of the snake; the largest individuals of such species as *C. adamanteus* or *C. atrox* can engulf a full-grown cottontail rabbit. Birds are occasionally eaten, particularly ground-nesting birds by *C. viridis* on the prairies. A few rattlers of species that live in well-watered areas subsist, at least in part, on frogs or toads. Probably the most important food supply for the small species of rattlesnakes, particularly in the Southwest, comprises lizards. These, by reason of their attenuated shape, are especially suitable food for the small species of rattlers and the juveniles of the larger kinds that might have difficulty in swallowing a full-grown mouse.

Field observations indicate that rattlesnakes often secure their prey by lying in wait along the trails where it may travel, and, aided by sight or the facial pits, striking it as it passes by. Also, mammals are followed into their holes, particularly in the summer when the snakes take refuge below ground to escape the heat.

When a rattlesnake strikes a small mammal, it seldom retains hold of the creature, which may run for a short distance before succumbing to the venom, soon deadly to so small an animal. The snake follows by scent, and then swallows the prey at leisure, often a lengthy process because of the size of the prey, compared with the size of the snake's mouth. In swallowing, the snake may use its fangs to draw the prey into its mouth, but more often employs only its small, solid pterygoid and mandibular teeth. Food is always swallowed whole; a snake has no means for masticating or dismembering prey.

The venom is useful in two ways: first it avoids the necessity of holding on to a struggling creature that might injure the snake; and, second, the venom accelerates digestion.

Rattlesnakes in zoos are fed dead prey, which they take quite readily. They are known to do the same in the wild when they find them, including animals already partly decompsed.

Although rattlesnakes in the early days were credited with the ability to secure prey by fascination, particularly by means of a baneful glance from their wicked eyes, it is now known that they neither have this power nor need it.

Studies of captive rattlesnakes show that they will thrive on a meal of suitable size every 14 to 18 days. In the wild, where they must subsist on what they can catch, the irregularity in the size of the prey produces a greater variation in the frequency of feeding. Also the season affects their feeding. They feed most frequently in the spring, when their fat supply has been depleted by hibernation; and in the fall when they are building their fat reserves in preparation for the winter.

A snake eats less, proportionately, than warm-blooded creatures, not only because of its decreased heat loss, but also feeds less frequently because of the greater average weight of the prey compared to its own weight. For example, the prey for a single meal may range from 5 to over 100% of the snake's own weight; studies have shown that the average is about 40%, a far higher proportion than is found in the feeding schedule of any carnivorous mammal.

Although rattlesnakes have been kept in captivity for a year or more without food, they will not survive this long without water. While it has been found that most rattlesnakes will drink water, from time to time, when it is available, some of the desert species drink infrequently, and in the wild may depend on the moisture content of their prey. Milk of course, is not a natural food for any species of rattlesnake, and they will not drink it in captivity, unless water is denied them.

E. Reproduction

Rattlesnakes usually mate in the spring and the young are born in the autumn, the actual dates being dependent on the local climatic conditions. They are ovoviviparous, that is, the young are retained in the body of the mother and issue in thin-walled, ready-to-hatch eggs from which they break out immediately, slitting the parchmentlike sac with an egg tooth, a temporary specialized tooth in the front of the upper jaw. Young rattlers are able to fend for themselves when born. They do not stay with the mother and she exhibits no protective interest in them. If the brood and mother are found together, it is only because they have temporarily sought the same refuge. That the mother may swallow the young for protection when danger threatens is a persistent and worldwide myth applied to many kinds of ovoviviparous snakes.

The male organs of rattlesnakes are extruded from the tail. They are paired; and each of the pairs in turn is deeply bifurcated. Proximally they are covered with spines and distally with fringes or flounces. There are specific differences in shape and attenuation, in the number of spines and flounces, and in the presence of mesial spines. The spines per lobe vary from 20 in *C. triseriatus* to 200 in *C. adamanteus*; and the flounces from 15 to 60. There is also a generic difference, in that there is a gradual transition from spines to flounces in *Sistrurus* but a more sudden shift in *Crotalus*.

The sexes are born in equal numbers, but there is a slightly higher female mortality; also because of the reduced activity of females carrying young, males are more often encountered in the field. Only during a short spring mating season are pairs found together. The idea that they roam about together throughout the year for mutual protection is mythical.

One of the interesting sexual activities indulged in by male rattlesnakes is the combat dance. In this ritual two male rattlers rear up and each endeavors to press his opponent to the ground with the forepart of his body. The dance may last 15 minutes or more, depending on the success of one combatant over the other. Neither snake is injured in this contest, but usually the one defeated withdraws from further contact. The sparring match may be revived on several successive days. Although the dance was long confused with mating, it is now known that the patterns are quite different. It is presumed that the dance is a sexual exhibition, yet no female need be present to induce males to indulge in it.

In the northern latitudes rattlesnakes usually mate in the spring, shortly after issuing from hibernation. There is evidence that the males trail the females by scent. The females generally give birth to their first young at the age of 3 years; however in the more severe climates of the north, with a resulting short active season, they cannot bring the young to term by the ensuing autumn, and, as a result, there is a biennial, instead of an annual reproductive cycle. A single species, if it has a wide enough latitudinal range, may have a one-year cycle in the south and a two-year in the north. It is believed that the rattlesnakes along the southern border of the United States and from there southward into Mexico, bear young annually.

The length of gestation in rattlesnakes having an annual cycle is probably about 4 months. Most of the young Nearctic rattlesnakes are born in the last half of August or the first part of September. Where the snakes have a long period of hibernation, it is important that they be successful in finding food before entering their dens, for those that fail are not likely to survive.

One of the complications that make it difficult to ascertain rattlesnake gestation schedules is the fact that female snakes are provided with a means of sperm storage. This produces an irregularity between mating dates and dates of birth. Off-season matings are by no means infrequent; and it is possible that two broods may occasionally result from a single mating, or a single brood from several matings.

The number of young rattlesnakes in a brood is dependent on several variables, such as the species of snake and the size of the mother. Young mothers and those that are smaller than the species average produce fewer young than their larger and older sisters; and some species are definitely more prolific than others. Small species such as *C. lepidus* and *C. pricei* usually have about 5 to 8 young; medium-sized species such as *C. viridis* about 7 to 10; and large kinds like *C. adamanteus* about 10 to 15. The semi-tropical rattlesnakes *C. basiliscus* and *C. durissus* are apparently more prolific than the more northern kinds, and broods of 40 or more are not unknown; in fact, a brood of 60 *C. basiliscus* has been recorded.

F. Enemies

Rattlesnakes have many natural enemies, including mammals, birds, and other snakes, but it is doubted whether all of them together are as important in controlling the rattlesnake population as are wind and weather, and the food supply. Besides, man himself is the greatest enemy of the rattlesnake, not necessarily by design, but because of the number killed by automobiles on the highways.

As for most of the natural enemies of rattlesnakes, the adults are not much subject to predation, since they are dangerous and the creatures that would prey on them learn to avoid them. But the juvenile rattlers have less ability to defend themselves. They can be attacked and eaten by smaller creatures than could cope with an adult, so that many of them are destroyed, although, even so, their inability to capture food no doubt accounts for a greater destruction than enemies.

Some rattlesnake enemies prey upon the snakes for food. Among the wild animals with this purpose are coyotes, foxes, wild cats, and badgers; and among the domestic animals hogs, dogs, and cats. Hogs are said to be particularly effective in clearing their ranges of rattlesnakes; a bite to them is less hazardous because of their protective layer of fat. The ungulates, both wild and domestic, such as deer, antelope, sheep, goats, horses, and cattle, all seem to have a propensity, for reasons that are not entirely clear, for killing rattlesnakes by stamping on them. Deer especially have been seen many times to kill rattlesnakes by this means.

Dogs often become adept rattlesnake killers. They have a stereotyped method, which they employ with success in other parts of the world to kill snakes, whether venomous or harmless. The dog circles the alert, coiled snake until he wears outs its patience, causing it to stretch out and try to crawl away. He then rushes in, seizes the snake in his teeth, executes a quick bite and, releasing his hold, tosses the snake in the air. A few successful rushes of this kind soon render the snake helpless; and, unless the dog becomes careless in attempting to show off before his master, he will usually escape without being struck.

Eagles and hawks, especially red-tailed hawks, prey quite regularly on rattlesnakes and have often been seen carrying them aloft. Roadrunners *(Geococcyx californianus)* are reputed to be great destroyers of rattlers, but probably most of their victims are juveniles.

Kingsnakes *(Lampropeltis)*, racers *(Mastocophis)*, black snakes *(Coluber constrictor)*, and indigo snakes *(Drymarchon)* all attack rattlesnakes of sizes that they can eat. Kingsnakes, especially, are supposed to be natural enemies of rattlesnakes, seeking to kill them on every occasion when they meet. This is hardly accurate; they have no altruistic desire to eliminate a creature dangerous to man. However the kingsnake does include snakes, both harmless and venomous, in its diet, and when hungry will not hesitate to attack any rattler, usually a juvenile, that it can successfully engulf. It has a highly stereotyped method of attacking a rattler, and, although almost invariably bitten in the process, suffers no damage from the venom of its prey.

Rattlesnakes, both in the wild and in captivity, suffer from various bacterial and protozoan parasites. One arthropod, the mite *Ophionyssus* is particularly troublesome in snake exhibits. Tongue worms or linguatulids are found to afflict many rattlers in the wild. Round worms are also serious parasites.

But of all the enemies of rattlesnakes, the automobile, quite inadvertently, has proved to be the most devastating. Unfortunately it is equally destructive of useful, harmless snakes. In these days of smooth-surfaced, fast-traffic, multiple-lane highways, it is unusual for any snake to cross in safety. For, at best, his progress is featured by inefficient whipping of his body from side to side, with little forward motion. The counting of live or run-over snakes on the highways today, as compared with their numbers 20 years ago, has shown how effectively the density of the snake population along our highways has been reduced.

G. Control

As people move out into the suburbs, they naturally encounter more snakes, including rattlesnakes, then before. This is particularly true when the new garden contains, or did contain, brush and rocks. As a result, our scientific institutions receive many inquiries asking how this new menace is to be met, particularly if there are young children in the family.

If the area to be protected is relatively small, such as a children's yard, we usually advise snakeproof fences of quarter-inch mesh, high enough to discourage heavy-bodied snakes like rattlers from crawling over, and sunk deep enough below the surface to prevent burrowing rodents from digging holes through which the snakes might gain entrance. Shrubbery must be kept back from the fence on the outer side. Gates must be kept closed. Other recommendations are the elimination of hiding places around the grounds, such as piles of rocks, lumber, or rubbish; the destruction of the rattlesnake's food supply, such as mice, rats, gophers, and other small rodents; and the presence of a lively dog, which may sound the alarm if he comes upon a rattlesnake. But particularly we suggest that the residents learn to distinguish harmless from venomous snakes, and that they do everything to protect and encourage the presence of the harmless snakes. For the harmless snakes are the natural competitors of the rattlesnakes for food; the more there are the fewer rattlers can live around the premises. Furthermore, some of the harmless snakes, such as kingsnakes and racers, may kill and eat small rattlesnakes.

Various chemical repellents have been marketed but apparently have not met with success.

Where stock is to be protected on the open range, the best means of reducing the rattlesnake population is to find the dens in which they hibernate and to institute systematic campaigns for killing them on the days when they are seeking seclusion in the fall or emerging in the spring. Traps may be employed so that continuous vigilance at the den is unnecessary. Where the climate is such that the rattlers do not hibernate together in large numbers, spring hunts may prove effective, although this will require the help of many people to produce much effect on the snake population. By suitable publicity, the offering of prizes for the most or largest snakes, these spring campaigns can be made to attract hunters from distant places.

The payment of bounties for the control of rattlesnakes has been practiced in many places but has not been particularly effective. It sometimes results in bringing in snakes from distant areas or the saving of females until the young are available for the claim of bounties.

The rattlesnake's feeding schedule is so irregular that poisoning or trapping (except at dens) is likely to be ineffective, and is attended by danger to people as well.

With respect to the advisability of rattlesnake destruction in areas distant from human habitation, there is some question. Rattlesnakes, from a purely economic standpoint, are of definite value to the farmer or stock raiser, because of the destruction of great numbers of injurious rodents. Against this favorable characteristic must be balanced the chance of people or stock being bitten. It has been well stated that rattlesnakes have no place in a settled country, no matter how beneficial their food habits may be. The obvious solution to this uncertainty is to encourage the presence of harmless snakes, especially the larger kinds, such as gopher or bull snakes, rat snakes, king-snakes, racers, and similar rodent eaters, for these have all the favorable attributes of the rattlesnakes, but none of the hazards to man and his domestic animals that characterize the venomous snakes.

Aside from their rodent-destroying propensities, rattlesnakes have little economic value. True, there is some trade in live snakes with zoos and carnival shows; dried venom is necessary for the preparation of antivenin and other scientific purposes; there is a desultory sale for ornamental objects made from rattlesnake skins; and a small demand for canned rattlesnake meat as a food fad. But all of these support only a few commercial institutions along the southern border of the United States, and they cannot be viewed as being of major economic importance.

VI. AVOIDANCE OF RATTLESNAKE BITE

The treatment of rattlesnake bite is the subject of discussion in another chapter in this book by another author. However, I shall give here some advice on how to avoid being bitten.

In the United States, rattlesnake bite is no longer, statistically speaking, an important hazard. No one knows how many people are bitten annually in this country, since nonfatal bites need not be reported to the health authorities. We may guess that between 6000 and 7000 persons are bitten per annum, not a large figure in a population of about 200 million. Our knowledge of the number of bites will soon be improved through the efforts of some active research workers, now engaged in gathering the necessary statistics by inquiries made of hospitals and medical societies.

The annual fatalities from rattlesnake bite in the United States now average about 10 per year (98 in 10 years). So we may conclude that the mortality rate is somewhat below 1%, even if we estimate the number of bites at 1,000 per year.

The number of people bitten by rattlesnakes would be greatly reduced were they to adhere to certain simple precautions. These may be recommended to any persons whose employment or avocation takes them into areas where rattlesnakes occur.

Wear protective clothing. It need not be especially heavy; ordinary trousers and stout boots afford considerable protection to the legs, where most bites occur. Keep the leg-protective clothing on when walking around your camp at night; remember that rattlers become nocturnal in the heat of the summer.

When hiking amid grass, brush, cactus, or rocks, keep to cleared paths as much as possible. Be alert at all times; watch where you put your hands and feet and avoid spots where a snake may lurk.

Remember that rattlers are found at higher altitudes than generally supposed—up to 11,000 feet in the western United States and up to 14,000 feet in Mexico. Although they are much less plentiful in these high places than lower down, still it will be safer to be on the lookout for them.

When hunting and watching for game to run or fly, keep an eye on the trail to avoid stepping on or near a rattler.

Step on a log or rock, not over it; this will permit you to see whether a rattler lies concealed on the far side. Also examine the surroundings before sitting down to rest on a log or rock.

Seek a cleared spot when crawling under or through a barbed-wire fence.

Do not lift a stone, plank, log, or any other object with the hand, for there might be a rattler concealed beneath. Use a stick or hook of some kind to turn the object over.

Do not reach into a hole for wounded game or any other purpose.

Firewood should be gathered in the daytime when the surroundings are clearly evident; one should not walk around camp in darkness without a flashlight wherewith to light the way. Bedrolls should be prepared during daylight, avoiding proximity to brush, rocks, or rubbish.

In climbing up a rocky cliff, do not seek a handhold above your head or elsewhere beyond your sight.

Learn something about the snakes in the area where you live, hunt, or fish, so that you may distinguish the harmless from the venomous kinds. Remember that local residents sometimes deny that rattlesnakes are to be found in the vicinity, in order not to discourage or frighten visitors.

When a rattlesnake sounds off suddenly close to your feet, ascertain where the snake is so that you won't step toward him rather than away. One should not forget that there may be two snakes present, only one of which is rattling. No rattlesnake will chase a person; a man can easily walk faster than it can crawl, so don't panic.

Injured or dead rattlesnakes should not be handled; the head of a decapitated snake should not even be touched. If the wish to secure the rattles of a snake you have killed is irresistible, keep one foot on the snake's head while you apply your knife to its tail.

Keep the remains of a dead rattler away from children and dogs. It is safest to bury the remains, handling the body with a stick or some long-handled tool. Never play practical jokes or frighten people with a dead snake.

VII. LEGEND AND FOLKLORE

A vast body of folkore and myths has grown up concerning rattlesnakes. Much of it is interesting, some is amusing, yet some of the beliefs thus spread and perpetuated can be dangerous to those who accept them as dependable fact.

The New World rattlesnake, through its possession of the unique noise-making device on the end of its tail, is so noteworthy, so distinctive from any Old World snake, that it has always been a cause of wonder and surprise. For this reason it has become the subject of many myths, legends, and tall stories. There have been three principal sources of these beliefs; (1) actions attributed to Old World snakes before the discovery of America, and then brought across the Atlantic and applied to rattlesnakes by the explorers and colonists; (2) myths invented by the Indians before Columbus; and (3) tales devised by explorers and settlers after their arrival in the new continent. Of course at this late date it is quite impossible to determine where many of the myths originated, except that any entailing the rattle itself could not fall in the first of these categories, for no European could have imagined the rattle before 1492.

Rattlesnake myths are of all kinds. Many have to do with the treatment of rattlesnake bite, an obvious objective of great importance to people living in primitive areas. Of these treatments, many employed plants or botanical products; as such they were merely a part of the backwoods pharmacopoeia, a field of medicine still having a wide adherence today. Time has not proved any of these cures for snake bite to be effective.

The Indian treatments often were based on the activities of medicine men or shamans who were credited with being specialists skilled in the treatment of snake bite. Their cures were nearly always surrounded by mysticism and were performed with elaborate rituals, which were believed as important as any physical materials that they employed in their treatments.

Many traditions that attributed queer habits or accomplishments to rattlesnakes were deliberately invented by humorously inclined guides to amaze and shock innocent travelers and city folk. Such tales were too readily accepted and subsequently were included in the printed narratives the travelers wrote on their return.

I have the space to list only a few of the more persistent exaggerations and misinterpretations that have transported this rather simple, elemental creature, the rattlesnake, into the realm of fantasy. Here they are:

Rattlesnakes are particularly vindictive and aggressive toward man, exercising a calculated hate in seeking to injure him. They will chase people to bite them, crawling faster than a man can run. Mates travel together for mutual protection; and, if one is killed, the other will never cease seeking out the man responsible for the foul deed to avenge its mate, sometimes traveling great distances to find the culprit.

On the other hand, rattlesnakes will not bite children. Pregnant women are immune to the bite; indeed, a powdered rattle, taken internally, will facilitate childbirth.

A rattlesnake's breath is deadly. Its fiery glance fascinates its prey, so that the doomed creature cannot escape, and, indeed may be drawn, helpless, toward the fatal jaws. It is dangerous for a man to look a rattlesnake in the eye; he may be hypnotized and fall upon the snake. The rattles when shaken give off a poisonous dust. A rattlesnake will not rattle at night, but the rattles are always sounded before the snake strikes.

Rattlesnakes have a strong odor, whereby they may be detected over a considerable distance; it resembles the odor of cucumbers. Rattlers are completely blind just before they shed their skins; they shed at particular seasons of the year, at which times they are especially dangerous. Rattlesnake venom can be used to cure its own bite. (This is a misunderstanding of how antivenin is prepared.) Rattlesnakes never bite their prey, for if they did so, they would poison themselves. A rattler removes its venom glands before drinking water. It must drink water after biting anything, otherwise the venom left in its mouth would poison it. If a hornet finds a dead rattlesnake, it drinks some of the venom, thereby greatly increasing the potency of its own sting. Rattlesnake venom, taken internally, is a cure for epilepsy and various other diseases; and eating the flesh will cure tuberculosis.

Rattlesnakes dislike blue and are more likely to bite anyone dressed in that color. A camp may be protected by laying a hair rope around it, for the hairs prevent the snake from crossing, otherwise the camper may find one in his bedroll.

A rattlesnake skin around the throat will cure a cold, or, worn as a hatband, it will cure a headache. It is also effective against toothache, especially in children. The skin, placed in a violin or guitar will improve the instrument's tone.

If you wish to shoot a rattler you need not aim the gun for the snake will line up its head with the shot; indeed, it can strike an oncoming bullet.

Rattlesnakes remove their fangs when courting to avoid injuring their mates. When some enemy threatens young rattlers they seek refuge down the mother's throat and crawl out again when the danger is passed.

A rattlesnake hung on a fence will bring rain; if a rattler is seen moving in the heat of the day it will rain; rattlesnakes know what the weather will be in advance, and will seek refuge if it will be particularly bad.

As long as a man is drunk he is immune to the effects of rattlesnake bite; liquor is a definite antidote for rattlesnake venom.

The copperhead is a female rattlesnake. There are snakes known as pilot snakes that guide rattlers toward their prey. Rattlers lick their prey and cover it with saliva before swallowing it. Rattlesnakes like milk and will milk cows to get it.

Rattlesnakes can puncture auto tires by biting them and auto mechanics have been killed by the venom left on the point of a puncturing fang. Rattlesnake venom taken internally in small quantities is a cure for numerous diseases. Rattlesnake oil is a cure for rheumatism; it can also be used by acrobats to increase their elasticity.

When threatened with inevitable destruction, rattlers commit suicide by biting themselves. Rattlesnakes mate with bull snakes to produce particularly dangerous offspring—dangerous because they have no warning rattle. Rattlesnakes jump clear off the ground when they strike and can reach an enemy twice as far away as the snake is long. A rattlesnake fatally bit a man through his boot; later his son wore the boot and died from the dried venom on the point of the fang sticking through the leather. When a rattler bites some wooden object the results may be quite surprising; for example, a hoe handle that was bitten swelled to such an extent that 20 five-room houses were built with the lumber that resulted. Roadrunners kill rattlesnakes by building around them a corral made of cactus lobes. When the snake tries to climb over the cactus wall it is stuck by the spines and is so angered that it bites itself and dies.

Certain Indian tribes used various repellents, which, when carried in the field, prevented a rattlesnake from coming near enough to strike. Some Indian tribes refused to kill rattlers, fearing the vengeance of their kin. But they invited white men to kill any in their camps, not caring particularly what might happen to the killer.

One of the oldest and most persistent cures for rattlesnake bite is the application of the flesh of a freshly killed, or dying, animal (usually a chicken) to the wound. A related method is the use of some part of the offending snake, such as its head, heart, or liver, either taken internally or applied to the wound.

People in different areas have great faith in the curative powers of certain substances, either taken internally or applied as a poultice to the bite. Among these may be mentioned milk, eggs, onions or garlic, vinegar, oils or fats, ammonia, salt, tobacco, and iodine. Among the substances only used externally are kerosene, turpentine, gunpowder (black), mud, the saliva of a fasting man, or snake stones.

Probably the earliest, as well as the most persistent, rattlesnake myth is that one can tell the snake's age from the number of its rattles, since it acquires one additional rattle each year. Actually, the snake gets a new rattle each time it sheds its skin, which is oftener than once a year; besides which adult rattlers have usually lost an indeterminate number of rings through breakage and wear.

Rattlesnakes are used in certain religious ceremonies. Of the Indian rituals the most famous is the Hopi Snake Dance in which a group of dancers carry live rattlesnakes dangling from their mouths. In the southeastern states there is a sect of Christian religious enthusiasts who make a practice of handling live rattlesnakes to demonstrate their faith by conforming to the biblical admonition "They shall take up serpents . . . and it shall not hurt them."

BIBLIOGRAPHY

(Suggested Accessory Reading Material)

Anonymous. (1961). "Antivenin (Crotalidae) Polyvalent (North and South American Antisnakebite Serum)," pp. 1–44. Wyeth Lab. Div., Philadelphia, Pennsylvania.

Aymar, B. (1956). "Treasury of Snake Lore." New York.

Bellairs, A. d'A. (1957). "Reptiles," pp. 1–195. London.

Bogert, C. M. (1959). *Sci. Am., 200,* No. 4, 105–120.

Buckley, E. E., and Porges, N. (1956). In "Venoms," Publ. No. 44, pp. xii–467. Am. Assoc. Advance. Sci., Washington, D.C.

Bullock, T. H., and Diecke, F. P. J. (1956). *J. Physiol. (London)* 134, No. 1, 47–87.

Conant, R. (1956). *Copeia* W, No. 3, 172–185.

Conant, R. (1958). "A Field Guide to Reptiles and Amphibians of the United States and Canada East of the 100th Meridian." Boston, Massachusetts.

Cowles, R. B., and Phelan, R. L. (1958). *Copeia* No. 2, 77–83.

Dullemeijer, P. (1961). *Koninkl. Ned. Akad.* **C64**, 383–396.

Gloyd, H. K. (1940). *Chicago Acad. Sci., Spec. Publ.* No. 4.

Goin, C. J., and Goin, O. B. (1962). "Introduction to Herpetology." San Francisco, California.

Klauber, L. M. (1952). *Bull. Zool. Soc. San Diego* **26**, 1–143.

Klauber, L. M. (1956). "Rattlesnakes: Their Habits, Life Histories, and Influence on Mankind," 2 vols. Univ. of California Press, Berkeley, California.

Klemmer, K. (1963). "Die Giftschlangen der Erde," pp. 253–464.

La Barre, W. (1962). "They Shall Take Up Serpents: Psychology of the Southern Snake-Handling Cult," pp. 1–208. Univ. of Minnesota Press, Minneapolis, Minnesota.

Oliver, J. A. (1955). "The Natural History of North American Amphibians and Reptiles." Princeton, New Jersey.

Oliver, J. A. (1958). "Snakes in Fact and Fiction." New York.

Parrish, H. M. (1959). *Arch. Internal Med.* **104**, 198–207.

Parrish, H. M. (1963). *Am. J. Med. Sci.* **245**, No. 2, 35–47.

Pope, C. H. (1955). "The Reptile World." New York.

Russell, F. E. (1961). *J. Am. Med. Assoc.* **177**, No. 13, 903–907.

Schmidt, K. P., and Davis, D. D. (1941). "Field Book of Snakes of the United States and Canada." New York.

Schmidt, K. P., and Inger, R. F. (1957). "Living Reptiles of the World," pp. 1–287. Garden City, New York.

Shaw, C. E. (1962). *Copeia* No. 2, 438.

Smith, H. M., and Taylor, E. H. (1945). *U.S. Natl. Museum, Bull.* **187**, 1–239.

Stebbins, R. C. (1954). "Amphibians and Reptiles of Western North America." New York.

Werler, J. E. (1963). "Poisonous Snakes of Texas and First Aid Treatment of Their Bites," Bull. 31, pp. 1–62. Texas Game and Fish Commission.

Wright, A. H., and Wright, A. A. (1957–1962). "Handbook of Snakes," 3 vols. Ithaca, New York.

Chapter 27

The Coral Snakes, Genera Micrurus and Micruroides, of the United States and Northern Mexico

CHARLES E. SHAW

SAN DIEGO ZOOLOGICAL GARDENS, SAN DIEGO, CALIFORNIA

I. INTRODUCTION

The coral snakes of the New World are members of the family Elapidae which includes some 41 genera widely distributed in Africa, Asia, and Australia. The elapids are distinguished from other dangerously venomous snakes, with the exception of the sea snakes (family Hydrophiidae), in having

157

the short, permanently erect fangs situated at the front of the upper jaw that characterize the proteroglyph snakes. The coral snake genera *Micrurus* and *Micruroides* are confined to the western hemisphere.

Micruroides is a monotypic genus occurring in northwestern Mexico and in southwestern United States in Arizona, New Mexico, and, possibly, extreme western Texas. *Micruroides* is distinguished generically from *Micrurus* by the presence, on the maxillary bone, of one or two small teeth behind the venom-conducting fang. *Micrurus* is a much more widely ranging genus; its nearly 50 species occur along the Atlantic coast from North Carolina to Florida, westward along the Gulf coast to Texas, and as far north as Arkansas. The genus then ranges southward through Mexico and Central America to Uruguay and northern Argentina in South America.

The coral snakes found in the United States are, like the majority of their relatives, glossy in appearance and brightly colored with alternate rings of red, yellow or white, and black. To the uninitiated, the accurate identification of the coral snakes is complicated by their resemblance to several harmless species. Within the limits of distribution of the eastern coral snake (*Micrurus fulvius*), the scarlet snake (*Cemophora*) and scarlet kingsnake (*Lampropeltis doliata*) may be confused with the venomous coral snake. In contrast to both these harmless snakes, the eastern coral snake has a black snout as well as adjacent red and yellow rings that encircle the body. Identification based on color may be a further source of confusion in that both albinism (Hensley, 1959) and melanism (Gloyd, 1938) have been recorded for the Texas coral snake (*Micrurus fulvius tenere*). In the southwestern United States, harmless snakes such as the shovel-nosed snake (*Chionactis*), Arizona mountain kingsnake (*Lampropeltis pyromelana*), and long-nosed snake (*Rhinocheilus*) may be confused with the western coral snake (*Micruroides euryxanthus*). Again, the western coral snake also has a black snout, and the red rings encircling the body have white- or cream-colored borders.

The coral snakes are usually quite secretive, even fossorial in habit, and both nocturnal and diurnal. Their inoffensive nature, coupled with their attractive coloration, has, on many occasions, led to their being handled as pretty playthings, particularly by children, and often with impunity. However, at least with the larger eastern coral snake (*Micrurus fulvius*), envenomation as the result of the bite can be fatal.

II. A CHECKLIST OF THE CORAL SNAKES OF THE UNITED STATES AND NORTHERN MEXICO

The coral snakes of the United States, which also range into northern Mexico, comprise two species, each with three subspecies. These are:

Arizona coral snake (*Micruroides euryxanthus euryxanthus*) (Kennicott)

Sonora coral snake (*Micruroides euryxanthus australis*) Zweifel and Norris

Sinaloa coral snake (*Micruroides euryxanthus neglectus*) Roze

Eastern coral snake (*Micrurus fulvius fulvius*) (Linnaeus)

South Florida coral snake (*Micrurus fulvius barbouri*) Schmidt*

Texas coral snake (*Micrurus fulvius tenere*) (Baird and Girard)

Key

1. The first ring behind the light-colored ring on the back of the head is black; a single tooth, the venom-conducting fang, is at the front of the maxillary bone (2). The first ring behind the light-colored ring on the back of the head is red; one or two teeth occur at the rear of the maxillary bone in addition to the venom-conducting fang (4).

2. No black in the dorsal scales of the red rings, *Micrurus fulvius barbouri*. Black spotting or streaking in the dorsal scales of the red rings (3).

3. Irregular streaking and spotting of black in the red rings, *Micrurus fulvius tenere*. Black in the red rings tends to form two large spots, *Micrurus fulvius fulvius*.

4. Red scales middorsally number 93 or less, *Micruroides euryxanthus euryxanthus*. Red scales middorsally number 100 or more (5).

5. Ventral scales number 217 or more; white- or cream-colored rings two and one-half to four scales wide, *Micruroides euryxanthus australis*. Ventral scales number less than 217; white- or cream-colored rings one-half to two scales in width, *Micruroides euryxanthus neglectus*.

III. WESTERN CORAL SNAKE (*Micruroides euryxanthus*)

Within the past few years, partially because of road development and the employment of the automobile as an adjunct to reptile collecting, especially in desert areas, a relatively large series of specimens of *Micruroides* have become available for study. As a consequence *Micruroides euryxanthus* has been found to comprise, in addition to the nominate race *M. e. euryxanthus* two additional races or subspecies extralimital to the United States,

---

* Duellman and Schwartz, 1958, p. 314, place *Micrurus f. barbouri* in the synonymy of *Micrurus f. fulvius*. Subsequently, other authors have continued to recognize *M. f. barbouri* as a race of *M. fulvius*.

M. e. australis and *M. e. neglectus*. A summary of the nomenclature of the western coral snake follows:

Arizona Coral Snake (*Micruroides euryxanthus euryxanthus*) (Kennicott)

1860 *Elaps euryxanthus* Kennicott, *Proc. Acad. Natl. Sci., Philadelphia* **12**, 337. Type locality: "Sonora." Restricted to Guaymas, Sonora, Mexico. Smith and Taylor, 1950, *Univ. Kansas Sci. Bull.* **33**, 344.

1917 *Micrurus euryxanthus* Stejneger and Barbour. Check List of North Amer. Amph. and Reptiles, p. 106.

1928 *Micruroides euryxanthus* Schmidt. *Bull. Antiv. Inst. Am.* **2**, 63.

1955 *Micruroides euryxanthus euryxanthus* Zweifel and Norris, *Am. Midland Naturalist* **54**, 246.

Distribution: Possibly extreme western Texas,* New Mexico, and Arizona southward to northern Sonora and northwestern Chihuahua, Mexico; also Tiburon Island, Gulf of California, Mexico. Records of many years standing for Swan Falls on the Snake River, southwestern Idaho, and St. George Canyon, southwestern Utah, have yet to be confirmed by additional specimens.

Sonora Coral Snake (*Micruroides euryxanthus australis*) Zweifel and Norris

1955 *Micruroides euryxanthus australis* Zweifel and Norris, *Am. Midland Naturalist*, **54**, 246. Type locality: Guirocoba, Sonora, Mexico.

Distribution: Western Chihuahua and southeastern Sonora, Mexico. An intergrade with *M. e. euryxanthus* from approximately central Sonora (30 miles north of Hermosillo) is reported by Zweifel and Norris (l.c.).

Sinaloa Coral Snake (*Micruroides euryxanthus neglectus*) Roze

1967 *Micruroides euryxanthus neglectus* Roze, Amer. Mus. Novitates, No. 2287. Type locality: 16.3 miles north-northwest of Mazatlán, Sinaloa, Mexico.

Distribution: Known only from the type specimen and a paratype collected 20 miles north of Mazatlán, Sinaloa, Mexico.

A. Morphology

Micruroides euryxanthus is a much smaller species than the coral snake of the eastern United States (*Micrurus fulvius*). Most adult western coral

---

* Brown (1950) inclines to the belief that *Micruroides euryxanthus* occurs in extreme western Texas though its presence is admittedly debatable. On the other hand, Werler (1964) lists only *Micrurus fulvius tenere* as occurring in Texas.

snakes are 15 to 18 inches in length, but exceptional individuals up to 22 inches in length have been reported. In *Micruroides* the body is sub-cylindrical with a short tail. The snout is blunt and the head is small, without a distinct neck. The eye is small with a round pupil. The rostral plate is normal and nearly as high as it is broad. The normal 9 plates cover the top of the head; these consist of two internasal scales followed by a pair of prefrontals, a small frontal, a single supraocular on each side, and a pair of parietals. There is no loreal scale. There is 1 preocular scale and 1 or 2 posto-cular scales followed by 1, then 2, temporal scales. There are 7 supralabial scales and 6 or 7 infralabial scales. There are 2 pairs of chin shields. The scales on the body are smooth and in 17–15–15 rows. The anal scale is divided.

Zweifel and Norris (1955) have provided the fullest information regarding the range of variation in ventral and subcaudal scales as well as sexual dimorphism in these characters of *Micruroides euryxanthus*. The ventral scales of *Micruroides* vary from 206 to 244 with males of *M. e. euryxanthus* averaging 223.37 and females 233.06. Males of *M. e. australis* have a mean ventral scale count of 221.3 and females 226.0. In subcaudal scale count, males of *M. e. euryxanthus* average 27.75 and females 24.06. In *M. e. australis* the subcaudal scales of the males average 25.6 and females 24.5. These authors also noted that a single specimen of *M. e. euryxanthus*, a male, from Tiburon Island, Gulf of California, Mexico, possessed 250 ventral scales, thus exceeding the maximum number reported for any mainland specimen.

The two known specimens of *Micruroides e. neglectus*, both males, have 206 and 207 ventral scales and 25 and 26 subcaudal scales. The reduction in the number of ventral scales together with yellow rings $\frac{1}{2}$ to 2 scales long are the features used by Roze (1967) to distinguish the southernmost race, *neglectus*, from *M. e. australis* which has a greater number of ventral scales and also yellow rings that are $2\frac{1}{2}$ to 4 scales in length.

The characteristic that Zweifel and Norris (1955) found most constant in segregating *M. e. australis* from *M. e. euryxanthus* was the greater width of the red rings encircling the body of the former race. This difference is expressed by counting the total number of red scales along the middorsal line of the snake's body exclusive of the tail. Expressed numerically, *M. e. australis* has from 100 to 107 red scales with a mean of 104.4, while the northern race *M. e. euryxanthus* has from 42 to 93 red scales with a mean of 72.2. According to Roze (1967), *M. e. neglectus* more nearly resembles *M. e. australis* with regard to the width of the red rings, the two known specimens both having a total of 106 red scales middorsally on the body.

B. Ecology and Life History

Despite human knowledge of *Micruroides euryxanthus* for better than 100 years, most of the details of its ecology, life history, and habits remain unknown. Presumably the two southern races of the species found in Mexico have life histories and habits similar to the nominate subspecies, although they dwell in an area of somewhat different vegetation.

In Arizona, *Micruroides e. euryxanthus* is an inhabitant of the semiarid areas along the foot of the southern plateau, having been recorded from Signal, Mohave County, in the northwest to Grant County, New Mexico, in the southeast. Its distribution does not appear to include the more extreme desert areas of western Arizona or northwestern Sonora, Mexico. *Micruroides e. euryxanthus* is said (Stebbins, 1966) to occur from sea level to 5800 feet in a variety of situations such as farmland, grassland, woodland, thorn scrub, and brushland. It is evidently partial to rocky desert areas with sandy soil. It is normally a secretive snake, concealing itself either by burrowing beneath the ground surface, beneath rocks or litter of various kinds, or in crevices of one sort or another. It is generally out at night although individual snakes have been found wandering about during the day particularly on overcast days after a rain.

Nothing seems to have been recorded concerning the life history of the races of *Micruroides euryxanthus*. Even reports of the presence and number of oviductal eggs in preserved specimens are lacking. It may be assumed, however, that mating takes place in the spring with egg deposition in the late spring or early summer. Hatching of the eggs probably occurs during the early fall months, but to date no hatchling seems to have been recorded.

A peculiar reaction of *Micruroides euryxanthus*, when it is molested or annoyed, and one which it shares with the hook-nosed snakes of the genus *Ficimia*, is the habit of everting, to a greater or lesser extent, its cloacal lining with an accompanying audible and sharp popping sound.

Vorhies (1929) noted that *Micruroides*, like the other coral snakes, feed upon snakes, and that captive specimens have been observed to eat *Tantilla* and *Leptotyphlops*. There is little doubt that any snake or lizard small enough to be subdued and engulfed by *Micruroides* would be an acceptable food item.

That these snakes are difficult to keep in captivity is attested by the fact that there seem to be no published records of their captive longevity. At the San Diego Zoological Garden, most specimens live a few days to several months, usually without feeding. One exceptional individual of *M. e. euryxanthus* did survive 3 years and 8 months from the time it was collected in the foothills of the Catalina Mountains, Pima County, Arizona. During its captive life it fed regularly on small local lizards of five species and small snakes of four species.

C. Bites and Envenomation

In spite of such statements, as the one made by Fowlie (1965) that: "This extremely poisonous snake fortunately accounts for few fatalities due to its placidity and small jaw size which make it unable to bite humans except on the ear lobe or on the interdigital web," there are, so far as I am aware, no recorded human fatalities from the bite of *Micruroides euryxanthus*. Lowe and Limbacher (1961) state that, to their knowledge, there has never been a case of a person bitten by an Arizona coral snake.

Evidently the first recorded bite of a human by *Micruroides e. euryxanthus* is recorded by Oliver (1958). While handling an adult Arizona coral snake (length not stated) in an attempt to pose it for photographs, Dr. Oliver was bitten on the outside of the little finger with a chewing motion of the snake's jaws. The bite went unnoticed by the victim who, while restraining the snake with one hand, had his attention distracted from it. A bystander called Oliver's attention to the fact that he was, indeed, being bitten. Oliver pulled the snake from his finger and observed that the very short fangs had not penetrated the skin. This bite was completely asymptomatic.

More recently, Russell (1967) published the first reports of envenomation of humans by *Micruroides*, describing four cases that came under his own expert observation during the period 1955–65. All these cases had in common the fact that the victims, all adults, were handling the snakes at the time they were bitten. In one case the victim was proving to a friend that the coral snake's fangs could not penetrate the skin. When the snake did bite, the victim rapidly tore it from his arm. No significant symptoms developed from this bite.

In the other three cases reported by Russell (1967) the victims were bitten by adult *Micruroides e. euryxanthus* of 40, over 45, and 55 cm, respectively. The latter measurement represents the maximum known length for a *Micruroides*.

In one of the three coral snake bites resulting in envenomation, only one fang penetrated the skin. In the other two instances both bites showed two puncture wounds. In one of these the skin punctures were about 4 mm apart, and the other, involving the 55-cm snake, showed puncture wounds about 3.3 mm apart.

All three victims reported the following common symptoms: immediate, but not severe, pain at the site of the bite, with the pain persisting from 15 minutes to several hours. Nausea, weakness, and drowsiness occurred several hours following envenomation. Paresthesia was also common to all three victims, being limited to the bitten digit in one case and spreading to the hand and wrist in the other two instances. All victims were asymptomatic within 7 to 24 hours after being bitten, with the exception of one who appeared not to become asymptomatic until four days following the bite. The latter case not only involved the largest coral snake, but, also, the victim was an Arizona physician whose avocation was herpetology.

As might be expected, the physician's personal observation and notes concerning his bite and its symptoms are the most complete of any given by Dr. Russell. The victim is reported to have removed the snake from his finger almost as soon as it had bitten. About 3 hours after the bite there was paresthesia over the bitten finger and hand. There followed a progressive deterioration of handwriting ability over a period of about 6 hours and at the end of this time the victim's handwriting appeared as that of a 5-year-old child. He experienced headache, nausea, and, especially, difficulty in focusing his eyes. There was some slight ptosis. His inability to accommodate visually was manifested by his walking into doors on several occasions. Though he was not certain, he thought he experienced some photophobia and increased lacrimation. He also complained of weakness and drowsiness and, in trying to recall the experience, he stated that he could remember very little about it. In common with the other cases reported by Russell (1967), there was no change in cardiac rate and no respiratory difficulty.

In the absence of an antivenin specific for *Micruroides*, and in view of his clinical experience, Dr. Russell (1967) has suggested the following measures subsequent to envenomation by *Micruroides*: "The wounded area should be scrubbed thoroughly. The depth to which these snakes bite is so limited that it may be possible to remove part of the venom by the mechanics of the washing procedure. Incision and suction seem unwarranted. Surgical excision of the area might be indicated in a child bitten by an extremely large specimen, but this would seem a radical and unnecessary procedure for most cases. Oxygen or positive pressure should be used if a respiratory deficit develops. Intravenous fluids, antiemetic agents and CNS stimulants might be indicated in some cases. The appropriate antitetanus agent should be given, but routine use of antibiotics is not recommended."

D. Venom

No analysis of the venom of *Micruroides euryxanthus* appears to have been made.

E. Antivenin

No specific antivenin for the bite of *Micruroides euryxanthus* is available. The circular accompanying the antivenin prepared by Wyeth Laboratories, Inc. for *Micrurus fulvius* specifically states that *M. fulvius* antivenin *will not* neutralize the venom of *Micruroides euryxanthus*.

IV. EASTERN CORAL SNAKE (*Micrurus fulvius*)

A summary of the nomenclature of the coral snakes of the eastern United States is as follows:

Eastern Coral Snake (*Micrurus fulvius fulvius*) (Linnaeus)

1766   *Coluber fulvius* Linnaeus, *Syst. Nat.* [*Ed.* 12] **1**, 381. Type locality: Restricted to Charleston, South Carolina. Schmidt, 1953, Check List of North Amer. Amph. and Reptiles, p. 223.

1896   *Elaps fulvius* Boulenger (part), *Cat. Snakes Brit. Museum* **3**, 422.

1928   *Micrurus fulvius fulvius* Schmidt, *Bull. Antiv. Inst. Am.* **2**, 64.

Distribution: North Carolina southward along the Atlantic and Gulf coastal plain to Mississippi.*

South Florida Coral Snake (*Micrurus fulvius barbouri*) Schmidt

1928   *Micrurus fulvius barbouri* Schmidt, *Bull Antiv. Inst. Am.* **2**, 64.

Distribution: The extreme south of Florida.

Texas Coral Snake (*Micrurus fulvius tenere*) (Baird and Girard)

1853   *Elaps tenere* Baird and Girard, *Cat. North Am. Reptiles* **1**, 22, 156. Type locality: Restricted to New Braunfels, Comal County, Texas. Smith and Taylor, 1950, *Univ. Kan. Sci. Bull.* **33**, 361.

1936   *Micrurus fulvius tenere* Schmidt, *Field Mus. Nat. Hist.,* Zool. Series **20**, 40.

Distribution: Louisiana, Arkansas and Texas; Coahuila, Tamaulipas and San Luis Potosí, Mexico.

A. Morphology

In *Micrurus* the body is subcylindrical with a short tail (*Micrurus* means short tail). The snout is short and blunt, and the head is small, without a distinct neck. The eye is small with a round pupil. The rostral plate is about as high as it is wide. There are 9 regular plates at the top of the head including 2 internasals, 2 prefrontals, a frontal, 2 supraoculars, and paired parietals. There is no loreal scale. There are 1 preocular scale and 2 postocular scales followed by 1 or 2 temporal scales. There are 7 supralabial and 7 infralabial scales. There are 2 pairs of chinshields, with the first pair separated from the mental scale by the first infralabials. The body scales are smooth and in 15 rows at midbody. The anal scale is divided.

In the eastern coral snake the first ring behind the yellow ring on the back of the head is black. From the neck to the origin of the tail, the body is patterned with alternate black and red rings separated by a narrow yellow ring. There are no red rings on the tail. Depending on the subspecies involved,

* Extralimital records of *Micrurus fulvius* for Illinois are discussed by Smith (1961), for Indiana and Ohio by Link (1951), and for Ohio by Conant (1951). Smith believes that the Illinois record is based on a specimen transported from further south. Conant also believes Ohio records to be based on accidentally transported specimens particularly since 2 subspecies are involved.

the red rings may be immaculate (*Micrurus f. barbouri*), contain 2 more or less discrete paired black spots (*Micrurus f. fulvius*), or have irregular streaking or spotting of black (*Micrurus f. tenere*).

Specimens of *Micrurus f. fulvius* up to 47.5 inches in length have been previously reported (Oliver 1955), but recently McCollough and Gennaro (1963) have placed a 51-inch specimen on record. Most adult *M. f. fulvius*, however, vary from about 2 to 2.5 feet in length. Neill (1957) believes that there may well be some geographical variation in size and notes that *M. f. barbouri* from the extreme south of Florida is generally under 30 inches in length, the maximum recorded specimen being 37.5 inches. Werler (1964) records the maximum length of *M. f. tenere* at nearly 42 inches.

B. Ecology and Life History

Carr (1940) states that *Micrurus f. fulvius* is found in areas of upland and mesophytic hammock as well as high pine. This author also mentions seeing several individuals in and near the water, one specimen being found on a mass of floating vegetation in a shallow pond. Neill (1957) believes the optimum habitat in central Florida is the plant association known as turkey oak and longleaf pine (*Quercus laevis* and *Pinus australis*) and equates this association with the "high pine" of Carr (1940). Two other situations to which *Micrurus* is partial, according to Neill (1957), are scrub and live oak hammock. He believes these snakes to be most common in dry places where the ground is brush covered, open, and not heavily vegetated. Carr (1940) characterizes the habitat of *Micrurus f. barbouri* of southern Florida as tropical hammock or glade land, and notes that it is occasionally ploughed up on muckland farms. The habitat of *Micrurus f. tenere* is said by Conant (1958) to be lowland as well as the cedar brakes, rocky canyons, and hillsides of the Edwards Plateau of Texas.

Carr (1940) described *M. f. fulvius* as partly fossorial and frequently found in leaf mold and decaying vegetation. Werler (1964) has said that *M. f. tenere* is found in or under decaying logs, or other trash, especially in damp areas. He further noted that in the San Antonio area the Texas coral snake is often found under flagstones in residential areas.

Wright and Wright (1957) have described *Micrurus f. fulvius* as being mainly crepuscular in its activities. Neill (1957) is emphatic in stressing that *M. f. fulvius* is principally diurnal in its activities in Georgia and Florida and is most active from sunrise to about 9:00 A.M. He cites the collecting circumstances of 121 coral snakes only one of which was encountered at night during a 10-year period. He notes further that in Florida, at least, the failure to find coral snakes at night is not a reflection of the diurnality of collectors since most snake hunting in the state is done at night.

Most authorities writing of coral snakes have generally stressed their secretiveness, and, from this, one seems led to assume that they are not abundant. That in some areas, at least, coral snakes are not uncommon is well illustrated in a report by Beck (1939). He reviewed an attempt by authorities of Pinellas County, Florida to reduce their venomous snake population by offering bounties for the coral snake and the diamondback rattlesnake (*Crotalus adamanteus*). Rewards were offered over a period of 39 months and, at first, amounted to 2 dollars, later the sum was reduced to 1 dollar. During this 39-month period, a total of 9529 snakes were redeemed by county officials and of these 1958 (20.5%) were coral snakes. Records covering a 5-month period from March 1 to July 31, 1938, also furnish some additional information on the seasonal incidence of coral snake in this county, the yield of specimens for this period being: March, 56; April, 118; May, 60; June, 53; July, 30.

Stickel (1952) observed that "little is known of the reproductive habits of the coral snakes, except that the eastern species lays 2 to 12 eggs, usually fewer than 6." Neill (1957) mentioned an *M. f. fulvius* that laid 13 eggs, while Telford (1955) reported an *M. f. fulvius*, 819 mm in total length, that laid 7 creamy-white eggs. Five of these adhered in one group and 2 in another. These eggs ranged in length from 36 to 46 mm; 5 eggs were 14 mm, and 2 were 13 mm in width. These eggs did not hatch. Ditmars (1920) reported a large specimen of *Elaps* (*Micrurus*) *fulvius* that laid 7 eggs on June 29. These were incubated in damp wood pulp. One egg was opened, apparently immediately after being laid, and was found to contain a transparent embryo about 2 inches in length. A second egg was opened on August 6, and it contained an unpatterned embryo 5.25 inches long. The 5 remaining eggs hatched on September 26 and 28 after a 90- and 92-day incubation. These young coral snakes were described as about 7 inches in length (one measured 7.75 inches in length). Ditmars noted that the young had a paler coloration than the adult, the red rings being a pale brick color. Neill (1957) described the eggs of *M. f. fulvius* as whitish, soft-shelled, usually less than six in number, and deposited in rotting logs or debris. He further noted that the newly hatched young are 7 to 8 inches in length, require a 2- to 3-month incubation before hatching, and are paler in color than the adults. Werler (1964) recorded 2 to 9 eggs as the clutch complement of *M. f. tenere*.

The eastern coral snake, when annoyed or molested, will thrust its tail up with the tip curled as if to direct its tormentor's attention to this end of the snake rather than its head.

All three races of the eastern coral snake appear to be short-lived in captivity; at least no noteworthy records of their length of life in captivity seem to have been published.

C. Bites and Envenomation

Insofar as the incidence of bite from the eastern coral snake (*Micrurus fulvius*) is concerned, these snakes evidently present relatively little hazard unless actually handled or restrained by unwitting victims. Carr (1940), however, described two instances in which *Micrurus f. fulvius* apparently struck without provocation. In one of these Carr himself was struck on the leg as he was walking through a hammock, and, in the other case, the sole of a man's shoe was bitten as he stepped from a boat at the edge of a lake.

Varying degrees of toxicity have been attributed to the venom of *Micrurus fulvius*. These range from the remark of Cope (1900) that the bite of the smaller species of coral snake (presumably he included *Micrurus fulvius* in this group) was innocuous to man and larger animals, to the comment of McCollough and Gennaro (1963) that the venom of *M. f. fulvius* is the most toxic per milligram of dried weight of any snake in the United States. Such disagreement perhaps may be based on the shortness of the coral snake's fangs and the consequent inability of this venom-conducting tooth to invariably penetrate even human skin to a depth at which envenomation might take place. Minton (1957) gave the fang length of a 2- to 2.5-foot *Micrurus fulvius* as 2 mm. Bogert (1943) gave the fang length of 3 specimens of *Micrurus fulvius*, ranging in total length from 700 to 817 mm (ca. 28 to 33 inches), as 2.1 to 2.5 mm. It should be noted here that the lower lumen or orifice through which venom is ejected by the snake is not situated at the fang tip, but somewhat above the tip on the front face of the fang; this presents an added handicap to short-fanged coral snakes. A "scratching" bite, or a bite in which the snake was immediately pulled from the victim before being allowed to embed its fangs fully with the typical chewing motion characteristic of elapid snakes, would probably result in little or no envenomation. In the case of such comparatively small snakes as coral snakes, with their correspondingly short fangs, the variation in individual skin thickness, or callous, of the victim would, in many instances, seem to be a factor involved in the severity of the bite. Adequate protective apparel such as gloves for gardeners as well as shoes for those walking in areas where coral snakes are known to occur, combined with a suitable caution in disturbing woodpiles, logs, rocks, and stones would undoubtedly reduce the incidence of coral-snake bite. Parrish and Kahn (1967), and others, have observed that most coral-snake bites are inflicted on fingers, hands, and arms as well as toes and feet.

That the bite of the eastern coral snake (*Micrurus fulvius*) can be fatal, though not invariably, is certainly well documented now. Stickel (1952) stated that estimates of mortality from coral-snake bite vary from 20% to as high as 75%, with 20% the more nearly correct figure. Of 20 cases of coral-snake bite, not all of which were attributable with certainty to this species, reported by Neill (1957), four of the victims succumbed, a mortality of 20%. McCollough and Gennaro (1963) noted that of 277 venomous snake bites reported to the Florida State Board of Health in 1962, only seven were from coral snakes and none of these was fatal. Parrish and Kahn (1967) report that only one (9%) of 11 cases discussed by them died. These authors further estimate that 20 ± 5 persons are bitten by coral snakes annually in the United States and that the probable mortality rate is about 10%.

In spite of past or present opinions or statistics, coral-snake bites, if definitely identifiable as such, should be viewed seriously since severe and even fatal envenomation may occur without any significant local tissue reaction. It should be noted that McCollough and Gennaro (1963) found that 10 individuals of the eastern coral snake ranging in length from 1 foot 8 inches to 3 feet, when stripped or milked, produced a venom yield from 1 to 13 mg, there being little correlation between the size of the snake and the venom yield.

The eastern coral snake bites by first seizing the victim and then commencing a chewing motion in an attempt to embed its short fangs or by biting repeatedly. Summarizing their findings concerning the symptoms of envenomation by the eastern coral snake, McCollough and Gennaro (1963) state: "Immediate findings were one, two, or more tiny puncture wounds, dependent on the number of bites, with pain usually insignificant or only in proportion to the size of these wounds. From one to seven and one-half hours after the bite, the following appeared: Apprehension, giddiness, dyspnea, nausea, salivation, vomiting, and weakness were among the initial complaints; convulsions occurred to two of six fatal cases; paralysis made its appearance as late as four to seven and one-half hours post-bite and was usually of a bulbar type involving the cranial motor nerves, death ensuing within an hour or two from respiratory paralysis. Once symptoms appeared, progression was alarmingly rapid. In those patients placed in a respirator, or in those surviving the initial bulbar paralysis, peripheral paralysis occurred, and in three instances, this became complete prior to death. The sensorium remained clear in all patients as long as oxygen exchange was maintained by respirator or other means."

Parrish and Kahn (1967) state that of the 11 coral-snake bites discussed by them none of the victims bitten by the Texas coral snake (*Micrurus f. tenere*) showed signs of venom poisoning. Stimson and Engelhardt (1960) generally confirmed these observations in 9 cases, in the remark that "not one of these was as serious as the least of the pit viper cases." These authors did not state whether envenomation occurred in any of the cases noted by them or not. They did, however, say that "Extensive correspondence with hospital and physicians within the territory where *Micrurus* [presumably, they are referring to the territory of *M. f. tenere*] exists has resulted in but one report of a death caused by a coral snake." Neill (1957) recorded a bite from a Texas coral snake in which the victim, a boy, age 13, developed signs of respiratory difficulty and occasional spasmodic twitching of the entire body about 4.5 hours after the bite.

D. Grades of Bites

Parrish and Kahn (1967) categorize coral snake bites in three grades as follows.

I. Grade 0: No Venenation

This type of bite results in tooth or fang scratches or actual fang punctures and resultant minimal local swelling. There are no systemic symptoms of envenomation within 36 hours subsequent to the bite.

2. Grade I: Moderate Venenation

Caused by scratch marks from the snake's teeth or fangs or actual fang punctures or both. There is slight to moderate local swelling and one or more systemic symptoms such as euphoria, nausea, vomiting, excessive salivation, paresthesia in the bitten extremity, ptosis, weakness, abnormal reflexes, motor paralysis, depression, dyspnea; respiratory paralysis is not complete within the first 36 hours after the bite.

3. Grade II: Severe Venenation

As in Grade I, but complete respiratory paralysis occurs within the first 36 hours following the bite.

McCollough and Gennaro (1963) point out, in particular, that "In Florida and other areas of the South, unexplained nausea, vomiting, convulsions and paralytic symptoms particularly involving the muscles supplied by the cranial nerves and accessory nerves of respiration, in an infant or child who is young enough that an accurate history is unobtainable and who has been ambulatory in wooded areas, pose the additional consideration of coral snake bite in the differential diagnosis."

E. Treatment

Parrish and Kahn (1967) recommend that any victim who has a break in the skin as the result of a coral-snake bite should be hospitalized for at least 48 hours. The bitten area should be cleansed with germicidal soap and water to remove any venom on the skin, and the victim's vital signs should be checked frequently.

In the presence of fang punctures, Parrish and Kahn (1967), as well as McCollough and Gennaro (1963), recommend the use of a tourniquet, incision of fang punctures, and suction, even though there is no evidence that these measures are of value. Parrish and Kahn (1967) also recommend, in the presence of one or more fang punctures, and even in the absence of symptoms, the immediate intravenous administration of coral-snake antivenin. Symptoms may progress rapidly, and victims have died as the result of respiratory paralysis within 4 hours of being bitten. It is also advisable to have immediately available the services of an anesthesiologist, a mechanical respirator, an oxygen supply, and facilities and equipment for a tracheotomy.

F. Antivenin

Antivenin of equine origin for the bite of the eastern coral snake (*Micrurus fulvius*) has been recently produced by Wyeth Laboratories Inc. Mrs. Eleanor E. Buckley of Wyeth Laboratories informs me (*in litt.*) that this antivenin will be made available to physicians free of charge, as a Wyeth service, for victims of the bite of *Micrurus fulvius*. A supply of this antivenin will be maintained at the U.S. Communicable Disease Center in Atlanta, Georgia. Public Health departments as well as certain poison control centers in the states in which the eastern coral snake is found will also be furnished with *Micrurus fulvius* antivenin.

REFERENCES

Beck, W. M. (1939). *Florida Naturalist* **12**, 94.
Bogert, C. M. (1943). *Bull. Am. Museum Nat. Hist.* **81**, 285.
Brown, B. C. (1950). "Baylor Univ. Studies," p. 208. Baylor Univ. Press.
Carr, A. F. (1940). *Univ. Florida Publ. Biol. Sci.* Ser. **3**, 1.
Conant, R. (1951). "The Reptiles of Ohio." 2nd ed. Notre Dame Univ. Press.
Conant, R. (1958). Field Guide to Reptiles and Amph. of the Eastern United States. Houghton, Boston, Massachusetts.
Cope, E. D. (1900). The Crocodilians, Lizards, and Snakes of North America. *Rept. U.S. Natl. Museum for 1898*, 1120.
Ditmars, R. L. (1920). "The Reptile Book." Doubleday, New York.
Duellman, W. E., and Schwartz, A. (1958). *Bull. Florida State Museum* **3**, 181.
Fowlie, J. A. (1965). "The Snakes of Arizona." Azul Quinta Press, Fallbrook, California.
Gloyd, H. K. (1938). *Herpetologica* **1**, 121.
Hensley, M. (1959). *Publ. Museum Mich. State Univ. Biol. Series* **1**, 135.
Link, G. (1951). *Nat. Hist. Miscellanea*, **92**.
Lowe, C. H., and Limbacher, H. P. (1961). *Arizona Medicine* **18**, 128.
McCollough, N. C., and Gennaro, J. F. (1963). *J. Florida Med. Assn.* **49**, 968.
Minton, S. A., Jr. (1957). *Sci. Am.* **196**, 114.
Neill, W. T. (1957). *Herpetologica* **13**, 111.
Oliver, J. A. (1955). "The Natural History of North American Amphibians and Reptiles." Van Nostrand, Princeton, New Jersey.

Oliver, J. A. (1958). "Snakes in Fact and Fiction." Macmillan, New York.

Parrish, H. M., and Kahn, M. S. (1967). *Am. J. Med. Sci.* (new series) **253**, 561.

Roze, J. A. (1967). *Am. Museum Novitates* 2287.

Russell, F. E. (1967). *Toxicon.* **5**, 39.

Schmidt, K. P. (1953). "Check List of North American Amphibians and Reptiles, 6th Ed.

Schmidt, K. P., and Davis, D. D. (1941). "Field Book of Snakes of the U.S. and Canada."
Putnam, New York.

Smith, H. M., and Taylor, E. H. (1950). *Univ. Kansas Sci. Bull.* **28**, 298.

Smith, P. W. (1961). *Illinois Nat. Hist. Surv. Bull.* **28**(1). 18.

Stebbins, R. C. (1966). "Field Guide to Western Amphibians and Reptiles." Houghton,
Boston, Massachusetts.

Stickel, W. H. (1952). U.S. Fish and Wildlife Serv. Wildlife Leaflet **339**.

Stimson, A. C., and Engelhardt, H. T. (1960). *J. Occupational Med.* **2**, 163.

Telford, S. R. (1955). *Copeia* **3**, 258.

Vorhies, C. T. (1929). *Bull. Antiv. Inst. Am.* **2**, 98.

Werler, J. E. (1964). *Texas Park and Wildlife Dept. Bull.* **31**.

Wright, A. H., and Wright, A. A. (1957). "Handbook of Snakes of the United States and
Canada." Cornell Univ. Press, Ithaca, New York.

Zweifel, R. G., and Norris, K. S. (1955). *Am. Midland Naturalist* **54**, 246.

THE CHEMISTRY, TOXICITY, BIOCHEMISTRY, AND PHARMACOLOGY OF NORTH AMERICAN SNAKE VENOMS

Chapter 28

The Chemistry, Toxicity, Biochemistry, and
Pharmacology of North American Snake Venoms

ANIMA DEVI

DEPARTMENT OF PHYSIOLOGY, UNIVERSITY OF CALCUTTA, CALCUTTA, INDIA

I. OCCURRENCE AND DISTRIBUTION OF SNAKES
IN NORTH AMERICA

Nearly all the snakes of North America belong to three distinct genera
(Minton, 1956, 1957; Porges, 1958; Parrish, 1958; Klauber, 1956): (1)

175

Crotalus, (2) *Agkistrodon* and (3) *Sistrurus.* There are more than forty different species on this continent. One or more of them are generally found in every part of North America (not true for Canada); in some areas, like Central America, southern United States and the Pacific coast, they are found in abundance. Relatively few of the North American species and subspecies of *Crotalus* are a menace to mankind (Klauber, 1956). The most dangerous are *C. adamanteus* and *C. atrox. Agkistrodon* is more widely distributed, is well represented in Eastern Asia, and is probably relatively recent in North America (Minton, 1956, 1957). Members of this genus are considered less dangerous to mankind (Cochran, 1943; Pope, 1944) and are common to the United States. *A. bilineatus,* the Mexican moccasin, is found along the coast of Mexico.

Snakes of the genus *Sistrurus* (ground rattlers) are small and much less aggressive than *Crotalus* and *Agkistrodon. Sistrurus milarius* and *S. catenatus* are the commonest species of this genus (Klauber, 1956; Conant, 1958). The Pacific rattlesnakes that are generally found from British Columbia southward to lower California and eastward into Idaho have been divided into three races: *C.v. oreganus* in the north, *C.v. helleri* in the south, and *C.v. cerberus* in Arizona and extreme western New Mexico (Klauber, 1956).

A review by Parrish (1966) indicates that previously published estimates of the incidence of snake bite accident in the United States are much too low.

II. TOXICITY OF VENOMS

The venoms of North American snakes in general are less toxic than those of tropical snakes, particularly the Elapidae. The toxicity of a venom depends on many factors, seasonal conditions, size and source of the specimen (whether taken from wild or captive animals), amount of venom collected, frequency of milking, and the like (Klauber, 1956).

A. Dependence of Toxicity on the Nature of the Animal Employed

The toxicity of a venom varies with the nature of the animal employed for the purpose, e.g., rabbits are twice as susceptible to cobra venom as are dogs, and 25 times as sensitive as the mongoose (Elliot, 1933, 1934). Similarly, it takes six times as much venom of *C. atrox* (North American snake) per unit of body weight to kill a rat as is required for a guinea pig or rabbit (Billing, 1930). The pigeon, which has been widely used in the past to determine the toxicity of a venom, is unusually sensitive to snake venoms (Minton, 1956, 1957; Ghosh *et al.,* 1941). At the other extreme lies the white mouse, which is now more frequently used to test the toxicity of venom. The white mouse is remarkably resistant to snake venoms (Minton, 1956, 1957; Macht, 1937;

Githens and Wolff, 1939). It is one twelfth as susceptible as the pigeon to the venoms of rattlesnakes (Minton, 1956, 1957).

The toxicity (lethal dose, LD_{50}) of a venom can be reduced by various means. When a dilute solution of *A. piscivorus* venom is exposed to 18,375 rad of X-irradiation, it loses more than 80% of its toxicity if tested in the mouse and 95% of its phospholipase A activity (Flowers, 1963). A 10–100 mg dose of heparin can delay the lethal effects of *Vipera russelli* and tiger snake venoms in the rabbit, if administered intravenously either along with or just after injection, but is unable to prevent death from the delayed toxic effects of the venoms. It can neither enhance nor reduce the toxicity of *A. piscivorus* venom. The venoms of the true or pit vipers lose their toxicity if heated to 70°C, and the venoms of elapids lose their toxicity at 95°C if heated for 1 hour.

B. Amount of Venom Administered in a Single Bite

How fatal the snake bite will be depends as much on the toxicity of the venom as on the amount of venom administered during the bite, e.g., cobras use 62% of their venom in a single bite (Acton and Knowles, 1914a) and saw-scaled vipers eject 71% (Acton and Knowles, 1914b), while rattlesnakes hold 25–75% of the original contents of the glands in reserve (do Amaral, 1930); according to King (1941), it is 67%. This has been attributed largely to the structure of the gland and the muscles that discharge the venom. The amount of venom ejected varies also with the size and species (Klauber, 1956; Minton, 1957).

After giving due consideration to all pertinent questions, it is possible to arrive at certain conclusions. The South American rattlesnake *(C. d. terrificus)* is the most dangerous of all the rattlesnakes found in the Western Hemisphere. The venom of *S. c. catenatus* exhibits the highest toxicity among all the venoms of North American snakes (Githens and Wolff, 1939). Next to this are the toxicities of venoms of *Crotalus viridis viridis* and *C. adamanteus* (Githens and Wolff, 1939). The venoms of *C. atrox* and *A. p. piscivorus* do not differ significantly in their toxicities. The venom of *C. atrox,* according to Githen and Wolff (1939), is much less toxic than the venom of *C. adamanteus* but appears to be more potent than the former from the results of toxicity tests carried out with different species of rattlesnakes by Klauber (1956).

The eastern diamondback *(C. adamanteus)* is the largest and most dangerous snake in North America and is considered as one of the world's deadliest snakes. This is due as much to the toxicity of the venom as to high venom delivery. *Crotalus atrox* is more responsible for serious and fatal accidents than any other North American snake. Its venom is much less toxic than that of *C. adamanteus,* but the quantity discharged in a single bite is very high. *Crotalus viridis viridis* and *C. h. horridus,* particularly their larger sub-species, and *C. molossus,* are also considered dangerous snakes, but deaths inflicted by their bites are seldom encountered in North America.

III. BIOCHEMISTRY OF NORTH AMERICAN SNAKE VENOMS

Venom is a mixture of several proteins. Relatively few protein components have been separated until recently. The various components can be separated by employing such procedures involving starch gel electrophoresis, paper electrophoresis, and anion and cation exchange chromatography. Nearly all venoms, depending upon the species of snakes, show the presence of at least five to ten well-characterized components.

The toxicity of a venom may be due to the action of one or more such components present therein; e.g., in the case of venoms of elapids, toxicity appears to be due to a single component: neurotoxin (Devi and Sarkar, 1959; Master and Rao, 1961). Cobra venom contains seven other distinctly separable components (Master and Rao, 1961). If the venom is heated at different temperatures to destroy one or more of the various components (activities), it retains its lethal toxicity (MLD) until all but one (the neurotoxin) is destroyed (Devi and Sarkar, 1968). Although the toxicity of the venom is due to neurotoxin, the toxic manifestations are greatly augmented by the presence of other less toxic constituents of the venom. The venom of *Vipera russelli* shows the presence of eight different components when subjected to starch gel electrophoresis. The toxicity appears to be due to the combined actions of four such different components (Master and Rao, 1961). The toxicity of the venom of *Bothrops jararaca* is also due to combined action of several components. When a dilute solution of *Crotalus* venom in veronal buffer, pH 8.6 and ionic strength 0.5, is subjected to electrophoresis of 300 V for 30 minutes at 21°C in a microelectrophoretic cell, supported on cellulose acetate, it reveals the presence of several components and displays homogeneity of protein components in snakes from a single population and similar patterns in snakes in polytypic species (Laviton *et al.,* 1964), *Crotalus atrox* venom shows the presence of ten different components when subjected to starch gel electro-phoresis in borate buffer, pH 8.9 (Jimenez-Porras, 1961), while two-dimen-sional starch gel electrophoresis carried out in sodium acetate buffer, pH 4.1, ionic strength 0.02, reveals the presence of 20 zones in *C. adamanteus* venom (Neelin, 1963).

Of the various protein components that have been separated and tested, all but one or two exhibit enzyme activity. Cobra venom, as has been already pointed out, owes its toxicity to only one of its eight components, which does not appear to have any enzyme activity. All other components, however, exhibit enzyme properties. The activities of these enzymes differ from venom to venom collected from the same snake at different times of the year. The activity of any particular enzyme also differs from species to species.

Generally speaking, the venoms of nearly all North American snakes exhibit activities of several enzymes, e.g., the various fractions isolated from the venom of *C. atrox* by starch gel electrophoresis, when tested for different enzyme activities, revealed the presences of 5′-nucleotidase, nicotinamide-adenine dinucleotidase, adenine 5′-triphosphatase, deoxyribonuclease, ribonuclease, phosphodiesterase, phospholipase A, L-amino acid oxidase, proteases, and plasma clotting inhibitor (Jimenez-Porras, 1961). This venom does not show the activities of acid or alkaline phosphomonoesterase, acetylcholinesterase, or direct hemolysin. Not all venoms contain plasma clotting inhibitor; rather many of them contain enzymes capable of coagulating blood, releasing bradykinin from plasma globulin and also enzymes capable of destroying bradykinin.

The occurrence of neurotoxins and cardiotoxins in certain venoms has also been reported. The presence of the latter was first demonstrated in cobra venom by Sarkar (1948), who isolated it from the same venom (Sarkar, 1947). In the following pages, the occurrence, properties, and functions of some important enzymes found in the venoms of North American snakes have been described. The enzymes in snake venoms have been recently reviewed by Devi and Sarkar (1968) in Chapter 7, Vol. I, p. 167, of this treatise.

A. Proteolytic, Esterase, and Clotting Activities

The ability of snake venoms to hydrolyze gelatin was extensively investigated by Eagle and Harris as early as 1937, a property which is also shared by the venoms of a large number of North American snakes. The venoms can also digest casein and hemoglobin. Some of the crotalid venoms, particularly those belonging to the *Agkistrodon* (copperhead) genus, exhibit very high proteolytic activity. Among the venoms tested, the one of *A. piscivorus* has the highest activity, even though it is only 5% of that of crystalline trypsin (Deutsch and Diniz, 1955). The venom of *C. d. terrificus* (white) exhibits very feeble proteolytic activity, viz., 0.017% of that of the venom of *A. piscivorus*. The venoms of *C. atrox* and *C. horridus* show, on the other hand, appreciable activity, while the venom of *C. adamanteus* exhibits only 0.16% of that of crystalline trypsin, although it is severalfold higher than that of the venom of *C. d. terrificus* (Deutsch and Diniz, 1955).

The venoms of all North American snakes can hydrolyze such synthetic amino acid esters as benzoylarginine ethyl ester. The esterase activity of venoms is not as low as their proteolytic activity if compared to the corresponding activities of crystalline trypsin, e.g., the esterase activity of the venoms of *A. p. piscivorus, C. adamanteus, C. t. terrificus* (white) are 14, 4, and 2% and their proteolytic activity, are 5, 0.16, and 0.0008%, respectively, of that of trypsin. It should be noted that the venom of *A. piscivorus,* which exhibits highest proteolytic activity among the venoms of all species tested, does not show the highest esterase activity. On the other hand, the venom of *A. contortrix,* which exhibits less than half of the proteolytic activity (only 43%) of the venom of *A. p. piscivorus,* can hydrolyze BAEE more efficiently than the latter venom (Deutsch and Diniz, 1955).

The proteolytic and esterase activities of venoms are not inhibited by soybean trypsin inhibitor and ovomucoid although they both strongly inhibit such properties of trypsin. Ethylenediaminetetracetate (EDTA), on the other hand, inhibits the digestion of hemoglobin by venom proteolytic enzymes, but is unable to suppress the action of trypsin. EDTA is also unable to inhibit the action of *C. t. terrificus* (yellow and white) venoms on hemoglobulin (Deutsch and Diniz, 1955). None of the venoms can hydrolyze acetyl-L-tryosine ethyl ester, an excellent substrate for chymotrypsin. On the basis of these observations, it has been suggested that the proteases present in the venoms are similar to trypsin but not identical to it as both trypsin and venom enzyme behave differently toward the same inhibitors (Deutsch and Diniz, 1955; Devi *et al.,* 1960; Sarkar and Devi, 1964). It has been recently noted that when *C. atrox* venom is subjected to chromatographic separation in diethylaminoethyl cellulose (DEAE) three fractions having proteolytic activity are revealed. This activity is due to three separate proteolytic enzymes that are differentiated by their pH optimum, specificity toward substrates and the effects of ions on their activity (Pfleiderer and Sumyk, 1961). The pH optimum for α- and γ-enzymes in tris HCl buffer is between 8.8 and 9.0 when casein is used as the substrate for their activity measurements, while the pH optimum for β-enzyme is 8.0 to 8.2. α- and β-enzymes are activated by Ca^{++} and Mg^{++}. These ions do not have any effect on γ-enzyme. The activity of all the three enzymes is strongly inhibited by EDTA, cyanide, and sulfide ions as well as by Hg^{++}. The inhibition by EDTA cannot be reversed by $NaNCO_3$ or phosphate (Goucher and Flowers, 1963).

B. Clotting Activity of Venoms

The venoms of many snakes of North America, particularly *Crotalus,* can coagulate blood (Eagle, 1937). They do so by virtue of their ability to convert fibrinogen to a fibrin clot, thus simulating the action of thrombin or or papain. However, it is not known whether the venom acts in the same way as thrombin does. The mode of action is different from that of the venom of *Vipera russelli* (true viper), which also coagulates blood but does so by activating Stuart's factor (factor X), which in turn converts proaccelerin (factor V) to its activated form; the latter then converts prothrombin to thrombin (Davie and Ratnoff, 1965). Trypsin and papain are also known to enhance the clotting of blood (Eagle, 1937; Sarkar, 1960). Soybean trypsin inhibitor and ovomucoid cannot block the clotting activity of venoms, but they can inhibit the corresponding activity of trypsin and papain (Deutsch and Diniz, 1955; Devi *et al.*, 1960; Sarkar and Devi, 1964).

A comparison of the clotting and proteolytic activities of venoms does not indicate a close parallelism between the two activities as the venoms of *A. piscivorus* of *Agkistrodon* genus and of *C. atrox* of *Crotalus* genus with high and moderate proteolytic activity cannot coagulate blood, whereas the venoms of *C. t. terrificus* (yellow and white) with very feeble proteolytic activity can do so. The venom of *C. adamanteus*, which has low proteolytic activity (0.16% of that of crystalline trypsin), exhibits comparatively high clotting activity (Deutsch and Diniz, 1955).

The venoms of *C. atrox, A. contortrix*, and *A. p. piscivorus* can digest fibrinogen much faster than the venom of *Bothrops jararaca* (South American snake), but cannot convert fibrinogen to fibrin clot as *B. jararaca* can (Deutsch and Diniz, 1955). The failure of these venoms to coagulate blood is presumed to be due to their ability to hydrolyze simultaneously other peptide bonds in the fibrinogen molecule, thus rendering the latter uncoagulable, since addition of thrombin to this partially degraded fibrinogen fails to produce the clot (Sarkar and Devi, 1964). The venom of *A. p. piscivorus*, like trypsin or plasmin, can also digest fibrin. Its fibrinolytic activity is considerably less than that of the plasmin but much higher than that of trypsin when plasma clot is used as substrate. Like trypsin, it can hydrolyze lysine ethyl ester but its activity is much less (Davie and Ratnoff, 1965). None of these activities of venoms can be inhibited by soybean trypsin inhibitor, ovomucoid, or ϵ-amino caproic acid. These inhibitors strongly block the fibrinolytic and fibrino-genolytic activities of plasmin and trypsin (Devi, 1968). A detailed discussion by Sarkar and Devi on the relationships between proteolytic, esterase, clotting, fibrinogeolytic, and fibrinolytic activities of venoms can be found in the proceedings of the Ninth International Congress on Hematology (Sarkar and Devi, 1964).

IV. ENZYMES IN VENOMS CAPABLE OF RELEASING BRADYKININ FROM PLASMA GLOBULIN AND DESTROYING IT

The release of a hypotensive and smooth muscle-stimulating peptide, bradykinin, from plasma by some snake venoms was first reported by Rocha e Silva *et al.* (1949). Venoms also contain enzymes capable of destroying bradykinin (Deutsch and Diniz, 1955). Not all venoms, however, exhibit both of these activities, for example, the venom of *A. contortrix* is free from the bradykinin-destroying enzyme while the venoms of *C. atrox* and *C. d. terrificus* show the activities of both enzymes. The activities of the two enzymes are quite low in *C. d. terrificus* venom, and highest in *C. atrox*. In general, venoms with high proteolytic and esterase activities are most active in releasing and destroying bradykinin, although this is not true of every venom, e.g., the venom of *C. adamanteus* with very low proteolytic activity can release bradykinin very effectively. It is interesting to note that the venoms that cannot release bradykinin are also unable to destroy it (Deutsch and Diniz, 1955).

Recently Webster and Piene (1963) noted the formation of two pharmacologically active polypeptides when the venom of *A. contortrix* is incubated with purified kallidinogen, an α_2-globulin present in plasma. They have been identified as kallidin-9 and kallidin-10 from their R_f values. The former retains most of the biological activity. Crude preparations of bradykinin and kallidin appear to be chemically and pharmacologically the same substance (Werle *et al.*, 1950; Gaddum and Horton 1959). The same conclusion was also reached by Van Arman and Miller (1961), who further suggested that there are two bradykinins and kallidins.

The venom possibly contains two enzymes: one is responsible for the release of kallidin-10 and the other for kallidin-9. Venom is unable to convert kallidin-10 to kallidin-9. The former is converted to the latter by the plasma aminopeptidase.

Neither soybean trypsin inhibitor nor ovomucoid can prevent the formation or destruction of bradykinin by the venom, while they can strongly inhibit the release and destruction of bradykinin by trypsin. Bradykinin formation by venom is suppressed by the addition of arginine esters but not by benzoylarginine amide (Habermann, 1961).

The hydrolytic activity of *C. adamanteus* venom, its ability to release bradykinin, and its ability to coagulate fibrinogen are all inhibited by diisopropyl fluorophosphate (DFP) as are the hydrolysis of amino acid esters and release of kallidin from kalli by kallikrein (Habermann, 1961). These results, according to the author, indicated the presence of several DFP-sensitive enzymes in venoms. These enzymes can be separated by electrophoresis and distinguished by their pharmacological and biochemical properties (Habermann, 1961).

Since the abilities of venoms to hydrolyze peptide bonds in proteins, ester bonds in synthetic amino acid esters (like BAEE), to coagulate blood, to release bradykinin and to destroy it cannot be inhibited by soybean trypsin inhibitor and ovomucoid, it may be inferred that these different properties of venoms are due to a group of very similar enzymes which are probably collectively esterolytic-active proteins and not to a single enzyme. They resemble trypsin in many ways but are not identical to it, as venoms and trypsin behave differently toward the same inhibitors.

A. Phospholipases in Snake Venoms

Nearly all venoms hitherto tested for their ability to hydrolyze phospholipids, particularly phosphatidal ethanolamine and phosphatidal choline, show invariably the presence of a very active enzyme capable of hydrolyzing L-α-lecithin or cephaline at the β or 2 position (Saito and Hanahan, 1962). α'-Acyl-L-α glycerophosphorylcholine (lysolecithin) and free fatty acids are formed (Tattrie, 1959). High phospholipase A activity is observed in the venoms of *S. catenatus* and *A. piscivorus* while moderate activity is noted in the venoms of *C. horridus, C. atrox, C. ruber,* and *A. bilineatus.* The venom of *C. adamanteus* cannot hydrolyze the intact red cell (erythrocytes) membranes of man, rabbit, pig, and cow but is active against intact red cells of guinea pigs. Sonication of red cells renders them susceptible to the action of venom phospholipase A. Hemolysis is also observed when rabbit erythrocytes are treated with deoxycholate (Heemskera and Van Deenen, 1964).

Vogel *et al.* (1964) found an increase in the degradation of phosphatidal ethanolamine by phospholipase A of *C. adamanteus* venom if phosphatidal choline, in the form of an aqueous emulsion, is added to the assay system. Only a small amount of lysophosphatidal choline is formed. If phosphatidal choline and phosphatidal ethanolamine are dried together and subsequently used in the assay system, more lysophosphatidal choline is produced. Dauvillier *et al.* (1964) employed γ-elaidyl-β-stearyl-9-10-t$_2$-L-α-glycerophosphorylcholine, isolated from eggs, as substrate in their studies on the degradation of this compound by *C. adamanteus* venom phospholipase A in calcium borate–ether system. Complete hydrolysis of the substrate was achieved in 1 hour. Fatty acids and traces of β-elaidyl-L-α-glycerophosphorylcholine were the reaction products; appearance of lysophosphatidal choline could not be detected. γ-Lysolecithin according to the authors, cannot act as the substrate for the venom phospholipase A.

Complete inhibition of the synthesis of cholesterol esters by *C. adamanteus* venom was recently observed by Shah *et al.* (1964) when lecithin, labeled with fatty acids such as oleic acid, linoleic acid, or palmitic acid in β-position, and labeled free cholesterol, were incubated in the presence of extracts of acetone powder of rat plasma serving as the source of enzymes concerned. The esterification of cholesterol appears to be catalyzed by a fatty acid transferase, and in this process transesterification of fatty acid at the β-position of lecithin takes place in the reaction. The inhibitory action of *C. adamanteus* venom is associated with its phospholipase A activity, as due to its action, lecithin is converted to lysolecithin which without the fatty acid at the β-position is unable to participate in the synthesis of cholesterol ester. These results support the conclusion arrived at by previous workers that the action of phospholipase A of venoms is to remove the fatty acids from β-position of phospholipids.

Slotta and Frankel-Conrat (1938) obtained crotoxin, a crystalline protein, from the venom of C. *durissus terrificus* (Brazilian snake) in an electrophoretically homogeneous state. The crystalline crotoxin was also obtained by these workers from the venoms of other species, including also the venoms of many North American snakes (Slotta, 1956). It has a molecular weight 30,000. It exhibits hemolytic (phospholipase A) and neurotoxic activities which are due to two different constituents. Crotoxin possibly contains three other enzymes. It inhibits the succinate cytochrome c-reductase system. It does not, however, show any 5'-nucleotidase and coagulating activities (Nygaard and Sumner, 1952).

Recently Saito and Hanahan (1962) isolated phospholipase A from the venom of C. *adamanteus* in a state of very high purity, using a procedure involving pH change, heat treatment, and subsequent chromatography on DEAE cellulose. They obtained two protein fractions, I and II, with phospholipase A activity. Both of them have similar sedimentation constants but different isoelectric points and electrophoretic mobilities. Their specific activities are also different, but the nature of fatty acids liberated by their actions on native ovolecithin is the same. Their molecular weight ranges from 30,000 to 35,000. Both these fractions (proteins I and II) are free from other enzymes, viz., protease, nucleotidase, phosphodiesterase and phosphomonoesterase, present in the crude venom. The presence of two hemolysins (phospholipase A) in C. *adamanteus* venom was also reported by Michl and Kiss (1959). Two zones with hemolytic activity were also noted by Neelin (1963), when C. *adamanteus* venom was subjected to two-dimensional starch gel electrophoresis in sodium acetate buffer, pH 4.1. Phospholipase A purified from crotalid and colubrid venoms loses its activity on dialysis (Bjork, 1961; Neumann and Habermann, 1956). Particles with a molecular weight of 30,000 should not normally be dialyzable unless degraded; but in C. *atrox* and C. *adamanteus* venoms, both the active forms of the enzyme appeared in the dialyzate in their original proportions (Bjork, 1961; Saito and Hanahan, 1962). Since proteases occur in crude venom as well as electrophoretic and chromatographic fractions of phospholipase A, it is possible that the active dialyzable components are produced either by dissociation or degradation of the parent enzyme (Neelin, 1963).

The phospholipase A activity of venoms is heat resistant. Its optimum pH is 7.0–7.2 and for maximum activity, lecithin in ether solution should be used. Addition of NaCl, $CaCl_2$, and EDTA are also necessary for maximum activity (Saito and Hanahan, 1962).

Phospholipase B activity in snake venoms has never been detected before; the failure may be attributed to the unfavorable pH chosen for the reaction. The pH chosen for the reaction was 3.5–4.4. or 6.0–6.5, as phospholipase B activity in mold and animal tissues is maximum at 3.5–4.4 to 6.0–6.5, respectively. Recently Doery and Pearson (1964) determined the phospho-lipase B activity of a number of venoms at pH 8.5–10.0 and found considerable activity of this enzyme in venoms. The phospholipase B activities of *C. adamanteus* and *A. piscivorus* are 55 and 10% of that of *Pseudechis porphyriacus* (Australian snake) venom.

B. L-Amino Acid Oxidase in Snake Venoms

L-Amino acid oxidase, discovered by Zeller and Maritz (1944) and found to be a flavoprotein, occurs in almost all venoms of snakes belonging to different species of *Agkistrodon, Crotalus,* and *Sistrurus* genera. The activity differs from species to species. The L-amino acid oxidase in snake venoms is more active than its counterpart occurring in animal tissues, e.g., the turnover for the purified enzyme from the venom of *A. piscivorus* is 3,100 (Singer and Kearnary, 1950), while for the same enzyme purified from animal tissue is only 6 (Blanchard *et al.,* 1965). Marburg (1964) determined the L-amino acid oxidase activity of venoms of different species using leucine as substrate and found the MAO activity in the following order: *C. atrox* > *Naja naja* > *Vipera ammodytes,* a finding which does not appear to be compatible with the observation of Morozova (1966) who noted practically the same activity of this enzyme in the venoms of crotalids, viperids and elapids.

It is evident that venoms of many North American snakes contain very active L-amino acid oxidase. Wellner and Meister (1960) succeeded in crystallizing this enzyme from the venom of *C. adamanteus,* following a procedure involving selective heat denaturation, adsorption on and elution from calcium phosphate gel, ammonium sulfate fractionation, and dialysis against water. The crystalline enzyme is homogeneous in the ultracentrifuge but gives rise to two to three components when subjected to electrophoresis. All the components are equally active. The molecular weight of venom L-amino acid oxidase is approximately 130,000 and contains two moles of flavin adenine dinucleotide per mole of enzyme. The optimum pH is 7.5. It is stable in reduced form (Wellner and Meister, 1960).

The reaction catalyzed by venom L-amino acid oxidase can be represented by the classical equation:

$$RCH\,(NH_2)\,COOH + O_2 \rightarrow RCOCOOH + H_2O_2 + NH_3$$

Since there is no catalase in the venom, the hydrogen peroxide produced decarboxylates the L-keto acids oxidatively. A large number of natural amino acids and their analogs can be converted to their corresponding keto acids by the venom enzyme (Meister, 1956). An account of the extensive study on substrate specificity of this enzyme, made by Zeller (1951, 1963) and by many others (Bender and Krebs, 1950; Frieden *et al.,* 1951; Suzuki and Wanagâ, 1960; Radda, 1964; Zeller *et al.,* 1965) can be found in the reviews of

Zeller (1951) and Devi and Sarkar (1968). Zeller (Zeller *et al.*, 1965) found that glycine is not a good substrate for the L-amino acid oxidase of snake venoms unless a large quantity of the venom is used. When a benzene ring is introduced into glycine or alanine, the reaction velocity (V) is slightly increased while the Michaelis-Menten constant (K_m) is greatly depressed. They determined V and K_m for phenylalanines that were substituted in every ring position with groups of various size and reactivity, and noticed a decrease in V_m (maximum velocity) when a halogen residue (except fluorine) is introduced in orthoposition of phenylalanine, the 3-position of the imidazole nucleus of histidine, and the 4-position of the indole nucleus of tryptophan. Within halogen series the effects are more marked with increasing size of the residue, and the rate of reaction induced by various substituents follows the pattern $m > p > O$.

A large number of heterocyclic amino acids containing simple or fused rings have been found by Zeller *et al.* (1965) to be excellent substrates for venom L-amino acid oxidase. The introduction of a second aliphatic amino group (as in ornithine) into an otherwise suitable amino acid renders it unsuitable for the enzyme, but acetylation of this amino acid residue (as in arginine) restores the lost property. Radda (1964) found that the rate-limiting step in this reaction is the hydrolysis of α-imino acid rather than the dissociation of the product from the enzyme or the loss of α-H of the substrate. He used L-aminophenylacetic acid and its derivatives substituted in the aromatic nucleus as substrates and the venoms of *C. adamanteus* and Malayan pit viper, *A. rhodostoma,* as enzyme source in his studies.

Since the decrease in V_m caused by the introduction of orthohalogens increases with the increase in the size of the residue, it is evident that the size of the ring introduced has an effect on V_m, which on the other hand, does not appear to be influenced by the chemical nature of the substituents. The latter, having opposite effects on electron distribution, induced either by induction or resonance, exert the same effect.

C. Nucleases in Snake Venoms

I. Deoxyribonuclease and Ribonuclease in Venoms

The ability of snake venoms to hydrolyze yeast and thymus nucleic acids was first noted by Delezenne and Morel (1919) and subsequently by Taborda *et al.* (1952a,b). Gulland and Walsh (1945) also found RNase activity in Russell's viper venom which was used by them to study the composition of yeast RNA. The DNase activity of *C. d. terrificus* venom according to Taborda *et al.* (1952a) is highest among the venoms of snakes of *Bothrops* and *Crotalus* genera tested by them. They did not include in their studies the venom of *C. adamanteus*. The DNase activity of this venom is higher than that of the venoms of *B. atrox, C. atrox,* and *A. piscivorus* (Richard *et al.,* 1965). Recently Budovskii (1962) compared the DNase activity of venoms of different species of snakes. His results also indicate higher activity in the venoms of *Crotalus* genus. Laskowski *et al.* (1957) suggested that *C. adamanteus* venom possibly contains more than one DNase on the basis of two pH optima they noted with this venom. The pH optimum at 5.0 was also previously noticed by Haessler and Cunningham (1957). Partially purified venom phosphodiesterase is not active at pH 5.0 (Laskowski *et al.,* 1957).

These endonucleases hydrolyze the sensitive diester phosphate bonds in the interior of DNA or RNA chains. As the reaction proceeds, the oligonucleotides that are formed in the initial stages from the enzyme action undergo further degradation until all the phosphate bonds susceptible to the venom DNase's (or RNase's) action are hydrolyzed.

The venom *(B. jararaca)* DNase is more stable to heat than the venom RNase, since at 70°C the latter loses all its activity while only 20% of the DNase activity is destroyed at this temperature; complete loss of DNase activity occurs at 80°C (Taborda *et al.,* 1952a,b). Formaldehyde, fluoride, cyanide, cupric ions and cysteine inhibit the venom DNase and RNase activities. Mg^{++}, Mn^{++}, and Hg^{++} also inhibit the DNase activity but are less effective than the former group of inhibitors (Taborda *et al.,* 1952a,b). Mg^{++} and Mn^{++}, on the other hand, enhance the RNase activity of venoms.

D. Phosphodiesterase in Snake Venoms

Snake venoms contain in addition to endonucleases at least one more enzyme capable of splitting phosphodiester bonds in nucleic acids. The presence of such an enzyme in snake venoms, capable of hydrolyzing diphenylphosphate was first reported by Uzawa (1932). Gulland and Jackson (1938a) also investigated this activity of venoms of many species of snakes using diphenylphosphate as substrate. The latter compound is resistant to phosphomonoesterases but susceptible to phosphodiesterases of other origin (Hilmore, 1959). The venom phosphodiesterase hydrolyzes stepwise the (polynucleotide chains of nucleic acids from the end of the chain carrying a 3' hydroxyl end group (Hilmore, 1959; Razzell and Khorana, 1959a). The venom phosphodiesterase can completely hydrolyze highly polymerized DNA as well as DNA synthesized enzymically *in vitro* (Laskowski *et al.,* 1957; Boman and Kaletta, 1956; Adler *et al.,* 1958). In the presence of sufficient amount of purified venom phosphodiesterase, calf thymus DNA is completely hydrolyzed to 5' nucleotide units (Razzell and Khorana, 1959a; Schmidt and Laskowski, 1961). Langen (1961) also noted complete degradation of highly polymerized DNA to mononucleotides by venoms, previously treated to destroy 5' nucleotidase activity. The venom phosphodiesterase can hydrolyze both DNA and RNA but deoxyribonucleotides are attacked faster than the ribonucleotides (Razzell and Khorana, 1959a,b). Cyclic oligonucleotides, lacking an end group, are only very slowly hydrolyzed by the venom phosphodiesterase (Razzell and Khorana, 1959a).

The first attempt to purify the venom phosphodiesterase from crude *C. adamanteus* venom was made by Koerner and Sinsheimer (1957), but the twofold purified material obtained by these workers was heavily contaminated with 5'-nucleotidase and nonspecific phosphatases. A 70-fold purification of the enzyme from the same venom was achieved by Razzell and Khorana (1959a). The procedure involved acetone precipitation and chromatography on DEAE cellulose. The purified venom phophodiesterase was not, however, free from 5'-nucleotidase and nonspecific phosphatases. It is interesting to note that Sulkowski *et al.* (1963) recently succeeded in obtaining venom phosphodiesterase free from 5'-nucleotidase activity from *B. atrox* venom.

The pH optimum for the hydrolytic reaction is 8.9 to 9.8. Its action is inhibited by reducing agents and also by such chelating agents as EDTA. The action of EDTA can be reversed by higher concentrations of Mg^{++}, Mn^{++}, Zn^{++} and Ca^{++}. The purified venom phosphodiesterase can hydrolyze polynucleotide chain even in the absence of any added Mg^{++}; only a slight stimulation is observed by its addition, which suggests that the purified enzyme contains sufficient amount of Mg^{++}. Phosphate acts as an inhibitor only at high concentration (Razzell and Khorana, 1959a).

The purified venom phosphodiesterase can hydrolyze a variety of phosphoesters such as di-*p*-nitrophenyl-phosphate, *p*-nitrophenyl-uridine-5'-phosphate, *p*-nitrophenyl-thymidine-phosphate, thymidylyl-(5'–3')-thymine, thymidylyl-(5'–3')-thymidylic (-5')-acid and the corresponding cyclic dinucleotides. The rate of hydrolysis varies from substrate to substrate, e.g., *p*-nitrophenyl-thymidine-5'-phosphate is hydrolyzed 1,000 times faster than di-*p*-nitrophenyl-phosphate. Razzell and Khorana (1959a,b) studied the ability of a large number of different types of oligonucleotides to act as competitive inhibitors, using *p*-nitrophenyl-thymidine-5'-phosphate as substrate and purified venom phosphodiesterase as the enzyme. They all inhibit the venom phosphodiesterase, the value of K_i decreases as the chain length of the oligonucleotides is increased. Oligonucleotides carrying 3'-phosphate at the end of the chain are the most powerful inhibitors.

E. 5'-Nucleotidase in Snake Venoms

Gulland and Jackson (1938a) studied the ability of venoms obtained from different species of snakes that inhabit geographically separated parts of the world, to hydrolyze mono- and diphenylphosphates. All the venoms tested by them were capable of hydrolyzing diphenylphosphate (diesterase activity), but only a few of them would hydrolyze monophenylphosphate. Some of these venoms were also capable of hydrolyzing nucleoside-5'-monophosphate (Gulland and Jackson, 1938b).

Among the venoms of *Crotalus* snakes tested for 5'-nucleotidase activity, that of *C. adamanteus* exhibits highest activity (Richard *et al.*, 1965; Devi *et al.*, 1966). The phosphodiesterase and nonspecific phosphatase activities of this venom are also considerably less than those of *B. atrox* (Laskowski *et al.*, 1957). For this reason the venoms of *C. adamanteus* and *B. atrox* have been widely used in preference to other venoms to purify phosphodiesterase (Razzell and Khorana, 1959a; Sulkowski *et al.*, 1963) and 5'-nucleotidase (Bjork, 1963, 1964). The activities of all these four enzymes are less marked in the venom of *C. atrox*, although its 5'-nucleotidase activity compared to other three activities is considerably higher and is only slightly less than that of the venom of *B. atrox*, which renders this venom an excellent source for the purification of 5'-nucleotidase. The lowest activity is noted in *Bothrops jararaca* venom. The K_m values for the venoms of *C. adamanteus* and *C. atrox* are 3.4×10^{-3} and 2.4×10^{-3} M respectively (Devi *et al.*, 1966).

The 5'-nucleotidase activity in venoms is higher than that has been reported for animal tissues, bull sperm, microorganisms, etc. (Schmidt and Laskowski, 1961). The venom 5'-nucleotidase removes orthophosphate from adenosine-5'-monophosphate (AMP), liberating adenosine as one of the reaction products. It cannot hydrolyze *d*AMP more than 40% (Sulkowski *et al.*, 1963; Devi *et al.*, 1966). Other deoxy compounds such as *d*GMP, *d*CMP, and *d*TMP are also hydrolyzed by venom 5'-nucleotidase. *d*TMP is more sensitive than the other three deoxycompounds. Flavin mononucleotide which does not contain a ribose but a ribitol group is not attacked (Bjork, 1964). Appreciable hydrolysis of other nucleoside-5'-monophosphates like GMP, CMP, UMP, and IMP has been observed (Sulkowski *et al.*, 1963). It cannot hydrolyze 3'- or 2'-3'-nucleotides, but its activity and thermal stability are greatly enhanced by the addition of these nucleotides. Amino acids such as leucine can also augment the 5'-nucleotidase activity of the venom but have no effect on its thermal stability (Bjork, 1964).

Highest activity of this enzyme is observed with dialyzed venoms when measured in the presence of Mg^{++} and EDTA. Zn^{++} and Ni^{++} are strong inhibitors (Devi *et al.*, 1966; Kaye, 1955). Co^{++} at lower concentrations stimulates the enzyme activity while at higher concentrations it inhibits (Devi *et al.*, 1966; Zeller, 1951). Addition of sulfhydryl compounds such as cysteine and glutathione (reduced forms) enhances the activity. Orthophosphate and pyrophosphate strongly supress the hydrolysis of AMP by venom 5'-nucleotidase. NaF is less effective than orthophosphate or pyrophosphate (Master and Rao, 1961).

The pH optimum for venom 5′-nucleotidase differs from species to species. This difference in pH optimum disappears when purified 5′-nucleotidases of these venoms are used. The pH optimum for the purified enzyme is 8.5 to 8.6 (Bjork, 1964).

F. Enzymes in Snake Venoms Capable of Hydrolyzing Adenosine-5′-triphosphate

Although cobra venom and other venoms of the Elapidae family have been more widely used than any other venom for the study of the enzyme that hydrolyzes ATP, venoms of North American snakes belonging to *Crotalus, Agkistrodon* and *Sistrurus* (Zeller, 1948, 1950) are also capable of hydrolyzing adenosine-5′-triphosphate. Venoms of *C. d. terrificus* and *C. adamanteus* are more active than the venom of *Naja flava* (Radomski and Deichmann, 1958). ATPase activity of *C. atrox* venom is less than its 5′-nucleotidase and NADase activities, but much higher than its nuclease activity (Jimenez-Porras, 1961).

Maximum hydrolysis occurs at pH 8.3–9.5. The venom enzyme, like other ATP-hydrolyzing enzymes of animal tissue origin, is sensitive to heat. It loses more than 90% of its activity on standing at 38°C for several hours (Zeller, 1951). It requires bivalent cations for activity. Mg^{++} appears to be the best activator among them. Ca^{++} is less active than Mg^{++} but more effective than other divalent metals while Co^{++}, Ni^{++}, and Mn^{++} are comparatively much less active; Zn^{++}, Hg^{++} and Cd^{++}, are strong inhibitors (Zeller, 1951).

The products of hydrolysis are adenylic acid and pyrophosphate (Johnson *et al.,* 1953). The former undergoes further hydrolysis to adenosine and orthophosphate by the action of 5′-nucleotidase also present in the venom (Cochran, 1943). Cobra venom cannot hydrolyze ADP (Johnson *et al.,* 1953). Venoms of other species possibly do not have this enzyme either. Thiamine pyrophosphate, β-glycerophosphate, and hexose phosphates are not attacked by the venom ATPase (Zeller, 1950). It is also not known whether venom contains more than one enzyme capable of hydrolyzing ATP, as is found in mitochondria. Rat liver mitochondria contains three such enzymes; each has a different pH optimum (Green, 1959).

G. Enzymes in Snake Venoms Capable of Hydrolyzing Adenosine Diphosphate Nucleotide

The presence of an enzyme in the venoms of North American snakes capable of splitting pyrophosphate linkage of NAD has been also reported by several investigators (Jimenez-Porras, 1961; Radomski and Deichmann, 1958). This enzyme was first detected in black tiger snake *(Notechis scutatus)* venom (Chain, 1939). Its presence in cobra venom was subsequently reported (Ghosh and Bhâttacharjee, 1951). This enzyme is, however, different from the NAD-nucleosidases which liberate nicotinic acid amide from NAD

(Korenberg and Fuller, 1950). In freshly collected venom of the western diamondback rattlesnake, *C. atrox,* the NADase activity is less than its 5'-nucleotidase activity but is considerably higher than the activities of those enzymes involved in the hydrolysis of nucleic acids and adenosine-5'-triphosphate. The NADase activities of *C. d. terrificus* and *C. adamanteus* venoms are higher than that of *Naja flava* venom (Radomski and Deichmann, 1958).

V. PHARMACOLOGY OF NORTH AMERICAN SNAKE VENOMS

Although the various components present in the venoms of different species of snakes, their nature, properties, and mode of actions are known, the cause of death of animals (or humans) bitten by snakes is not well understood. Of the various components, only a very few are highly toxic. It is generally believed that death is caused by the combined actions of all the components present therein; one is possibly more responsible than others in inflicting death.

The various components present in the venoms are enzymes of diverse nature; the only component known to be toxic but not an enzyme is neurotoxin (Devi and Sarkar, 1968; Devi, 1968). Neurotoxin is the main toxic principle of the venoms of elapids but not of the venoms of crotalids or viperids. The ten different components separated from freshly collected *C. atrox* venom by starch gel electrophoresis are all enzymes (Jimenez-Porras, 1961), and interestingly enough they are, with one exception, hydrolytic enzymes. The pharmacological actions of snake venoms should be a reflection of the actions of the enzymes present therein, or substances released by their actions.

Many snake venoms contain powerful enzymes (such as proteases and phospholipase A) that produce necrosis and are responsible for releasing bradykinin (kallidin), and histamine, respectively. The release of the latter compound may be due to the alteration in the cell permeability brought about by lysolecithin or lysocephalin, formed by the action of venom phospholipase A on phospholipids of cell membranes. The venom therefore appears to exert its toxic effects indirectly as well. In the following pages the various physiological and pharmacological effects of venoms of North American snakes have been described, avoiding as far as possible the details.

Our knowledge about the site of action of snake venoms in the animals is also very much limited. Only recently the distribution of radioiodine-labeled snake venoms of *C. d. terrificus* and *A. piscivorus,* following their injections in mice, was studied using whole body autoradiography techniques. The results indicate their rapid metabolism and excretion by various routes. The venom of *C. terrificus* is mostly localized in the renal cortex and the venom of cottonmouth moccasin in the liver of the animal (Sumyk and Kashin, 1964).

'Greatest localization of cottonmouth venom in the lung and least in the brain of mouse were recorded by Gennaro and Ramsey (1959) following the injection of radioactive venom into the animal.

The various components of venoms are proteins. Their actions on different organs of the body depend largely on their access to the cytoplasmic constituents of cells. Very little is known about their permeability through the cell membrane. Only recently Johnson and Eldridge studied the action of *C. atrox* venom on the permeability of D-glucose-6-^{14}C in human lung epithelial cells by incubating ^{14}C for 60 to 240 minutes with a known number of cells (58,000 cells/ml), in the presence of different amounts of venom (0.3, 0.6, or 0.9 mg/ml). These authors noted a change in the rate of permeability of glucose-6-^{14}C which varied with venom concentration and time of incubation. The increased permeability observed by the authors may be attributed to the action of venom phospholipase A on phospholipids of cell membrane. The actions of venom proteases might also be responsible for the increased permeability observed by the authors.

A. Cardiovascular Effects of Venoms

Changes in blood pressure, respiratory failure, and cardiac arrest in animals following the intravenous injection of venoms of many North American snakes have been recorded by several authors. It is interesting to note that vascular effects of venoms cannot be abolished nor can they be reduced by antihistamine (Hadidan, 1956; Russell *et al.,* 1960; Devi and Sarkar, 1966). An account of these effects of venoms is summarized below:

Intravenous injection of *A. piscivorus* venom (1 ± 0.3 mg/kg body weight) into an anesthetized dog causes a rapid fall in blood pressure, bringing it to 30 to 40 mm Hg (Essex, 1932); in some cases respiratory failure and death of the animal are also observed (Brown, 1941). If the animal tolerates the initial effect of the venom, even then the blood pressure continues to decline but does so slowly until death occurs; in some cases it tends to rise slowly. If, on the other hand, the animal receives an initially smaller dose (0.02 to 0.05 mg/kg of body weight), a moderate fall in blood pressure (30 to 50 mm Hg) is noted; if the animal receives three or four larger doses (0.2 to 0.5 mg/kg of body weight) thereafter at regular intervals, no significant changes can be noticed, and incidence of respiratory failure is seldom encountered in such cases (Hadidan and Murphy, 1955). Animals which die 1 hour or so after receiving the initial large dose (1 ± 0.3 mg/kg of body weight) of the venom, when examined often show gross hemorrhage (Taube and Essex, 1937); but where death occurs earlier than 1 hour, hardly any visible sign of hemorrhage can be noticed (Hadidan, 1956; Essex, 1932; Brown, 1941).

An immediate fall in systemic arterial pressure and a concomitant increase in venous and cisternal pressure in anesthetized (phenobarbital hypnotized) adult cats were recorded by Russell *et al.* (1960) following the injection of venoms (2.5 mg/kg of body weight) of *C. atrox, C. ruber, C. cerastes, C. viridis,* and *C. mitchelli.* Application of artificial respiration at this point fails to restore the blood pressure to normal level; contrary to this, the blood pressure continues to fall, the ensuing respiration becomes more and more shallow and rapid until the animal dies. A profound and sustained fall in blood pressure in anesthetized dogs and rabbits has been also observed by Radomski and Deichmann (1957) after the administration of the venom of *C. adamanteus*; the blood pressure continues to decline until it reaches zero mm Hg; at this point respiration ceases and the heart stops beating.

Electrocardiograms show disturbances in rhythm and ischemic patterns following the injection of venoms of *C. atrox,* etc. S-T segment is depressed and T wave is deviated from normal. In many experiments no such significant changes could be detected (Russell *et al.,* 1960). Following the injection of a sublethal dose of the venoms of *C. atrox, C. ruber, C. viridis, C. mitchelli,* and *C. cerastes,* electroencephalographic recordings show flattened tracings (Russell *et al.,* 1960). Radomski and Deichmann (1957) did not, however, observe any adverse effect on the electrocardiograms in dogs following the administration of *C. adamanteus* venom.

Russell *et al.* (1963) observed differences in the signs and behavior elicited in mammals following the intravenous injections of lethal and sublethal doses of crude venoms of *C. adamanteus, C. atrox, C. viridis* and *C. horridus horridus* and partially purified venom phosphodiesterase prepared from these venoms by employing the method of Koerner and Sinsheimer (1957). The venom phosphodiesterase has not been considered to be responsible for the more deleterious effects of the crude venoms that have been recorded, following their injection. The enzyme appears to be responsible for production or stimulation of the primary pharmacological changes that in turn bring about the profound fall in systemic arterial pressure. Venom phosphodiesterase has, however, no deleterious effect on neuromuscular transmission in the mammals.

The protective action of hydrocortisone was noticed by Deichmann *et al.* (1958) in dogs which had received it either immediately or 2 or 4 hours after the intramuscular injection of a lethal dose of *C. adamanteus* venom. Cluxton (1951) observed marked clinical improvements in a patient bitten by a copperhead after treatment with ACTH. Allam *et al.* (1956) could not demonstrate any such protecting action of cortisone on the death rate of dogs, following the injection of lethal doses of *C. atrox* and *C. adamanteus.* Similar negative results were also reported by Schottler (1954) with ACTH, cortisone, or hydrocortisone when they were used against the venoms of *C. d. terrificus.* in mice. The results of Russell and Emery (1961) also corroborate the findings of Allam *et al.* (1956). These investigators found that methylprednisolone and hydrocortisone when used in the range of 5 mg to 500 mg/kg of body weight, were unable to protect the animal from the effect of lethal dose of the venom of *A. contortrix*. They noted, however, that if 100 mg/kg of body weight was used, the venom dose had to be raised by 25% to cause the death of the animal, indicating some protective action of these drugs. Administration of hydrocortisone does not prevent the death of the animal but can counteract the arterial hypotension resulting from the action of *C. adamanteus* venom in dogs, but delays the death of the animal (Morales *et al.*, 1963). The intravascular thrombosis that occurs after the injection of *C. adamanteus* venom into animals can be prevented by ε-amino caproic acid (Vick *et al.*, 1963). The authors suggested that ε-amino caproic acid acts by blocking the activation of an enzyme system responsible for producing the vascular changes (Vick *et al.*, 1963). ε-Amino caproic acid is known for its powerful inhibitory action on the activation of profibrinolysin (Alkjaersig *et al.*, 1959; Ablondi *et al.*, 1959) and on clot lysis (Alkjaersig *et al.*, 1959; Ablondi *et al.*, 1959; Sarkar and Devi, 1960). Since it has no action on the clotting of blood, and prevents clot lysis, it is difficult to understand how it exerts its beneficial effects.

B. Effects of Venoms on Nerves and Muscles

When an isolated frog sciatic nerve is kept immersed in a venom solution *(A. piscivorus)* prepared in Ringer-Lock solution, its action potential is not appreciably changed in 1 to 2 hours at 20°C. If the venom solution is replaced by Ringer's solution at this point and tested after 2 hours, a considerable loss of its action potential is noted and after 12 hours, the nerve fails to respond any more to electrical stimulation (Truant and Hadidan, 1956). Nerves exposed to only Ringer's solution maintain their excitability for several days under similar conditions. These authors also found an increased penetration of procaine if the nerve is first exposed to venom solution and to the drug thereafter, as is evident from the rate of loss of action potential of venom-treated and untreated nerves in procaine solution. The venom-treated nerve loses its action potential much faster than the untreated one.

Contraction of virgin guinea pig or rabbit uterus, rabbit ileum, and bronchioles of isolated guinea pig lungs is observed in the presence of *A. piscivorus* venom solution (Essex, 1932; Hadidan and Murphy, 1955). The effect on bronchioles of isolated guinea pig lungs can be abolished by antihihistamine drugs (Hadidan and Murphy, 1955). The venom does not exhibit any curarelike effect but can produce neuromuscular block (Hadidan, 1956). It is interesting to note that cobra venom is very active in paralyzing neuromuscular junction of frog nerve-muscle preparation as reported by Sarkar and Maitra (1950). The muscle also does not respond to any direct or indirect stimulation (through the nerve), when kept immersed in venom solution, but its ability to respond to stimulation through nerves is lost earlier than its ability to respond to direct stimulation (Sarkar and Maitra, 1950). The direct effect of cobra venom on frog skeletal muscle was also reported many years ago by Devi *et al.* (1954). The ability of muscle to respond to direct or indirect stimulation is not restored even after the exposure of the nerve-muscle preparation to Ringer's solution alone, suggesting that the effect of the venom is irreversible. Hadidan (1956) noted curarelike action of *A. piscivorus* venom in only one of his many experiments, in which the response to nerve stimulation disappeared first. Venoms from species of *Crotalus* genus, viz., of *C. atrox, C. ruber, C. viridis viridis, C. cerastes* can, however, depress the activity at the neuromuscular junction in the phrenic nerve-diaphragm preparation. The least active venom is that of *C. atrox* which requires 22 minutes to produce complete neuromuscular block (Russell *et al.*, 1960). Russell and Long (1960) found only a weak neuromuscular blocking activity in *C. atrox* venom. Pretreatment of squid joint axons with moccasin venom or *C. adamanteus* venom renders them susceptible to the action of α-tubocurarine ($1.4 \times 10^{-4}M$) and acetylcholine ($10^{-3}M$) which are, in the absence of venom, inactive in blocking conduction (Rosenberg and Podleski, 1962). Dimethylcurare also becomes active to squid joint axons after their treatment with venoms. Venoms, through the action of their phospholipase A on the phospholipid component of the membrane, produce lysophosphatides, which are directly or indirectly responsible for the reduction of barriers surrounding the axonal membrane. *A. piscivorus* venom also depolarizes the squid axon (Rosenberg and Podleski, 1962).

C. Effect of Venoms on Isolated Frog Heart

The action of various snake venoms on isolated frog heart has been studied by many workers (Magenta, 1922; Elliot, 1901; Cushny and Yogi 1918; Gunn and Epstein, 1933; Devi and Sarkar, 1956). While many investigations were carried out with cobra venom, Magenta (1922) used the venom of *A. piscivorus,* among other venoms, in his study. Essex and Markowitz (1930) and Essex (1932) also used the water moccasin venom to study its effects on rabbit hearts. Brown (1940) found that the effect of venom (1:1,000 to 1:125,000) on an isolated perfused frog heart depends on the concentration of the venom used, e.g., when a venom solution 1 in 1,000 is used, the heart stops beating in 18 minutes. If the heart is perfused with a venom solution 1 in 10,000, an increase (47%) in amplitude of contraction, lasting for 7 minutes, is first observed; the heart stops eventually after 24 minutes. If the venom solution is 1 in 1,000,000 dilution, an 11 to 12% increase in amplitude is noted in the initial stages of perfusion, and then the heart continues to beat with normal amplitude until the perfusion is discontinued at 180 minutes. The results suggest that the venom exerts a stimulating effect when used at lower concentration and depressing action at higher concentration. In all cases (except in the case of very dilute solution, viz., 1 in 1,000,000), the heart eventually stops beating. Muscular irregularities with local regions of contracture in the ventricle are frequently observed. The venom, according to the author, acts on vagal endings or the ganglion cells of the heart and thereby produces the slowing of the heart.

D. Necrotizing Effect of Venoms, Their Lytic and Hemolytic Activities and Their Effects on Clotting Time and Prothrombin Time

The proteases of snake venoms act in many ways upon the victim. In human victims who survive snake bites, deep necrosis down to the bones of a foot or hand has been often reported. Many of the venoms of the North American snakes exert a strong necrotizing effect. This effect of venoms was carefully studied by Minton (1956) by measuring the necrotic area developed after the intradermal injection of 0.5 mg of the venom into shaved belly skin of a guinea pig. The results are not as accurate as they should be because of the limited number of animals employed for each sample of the venom, irregular shape of the area of necrosis, and the difficulty of comparing deep, circumscribed necrosis with superficial wide necrosis. Giving due consideration to the limitation of the method employed, an inverse relationship was noted by the author to exist between the lethal toxicity (LD_{50}) and ability of the venom to produce local necrosis. There are some exceptions to this generalization. The results suggest that among all the venoms tested, the ones of *C. adamanteus* and *C. ruber* are most powerful in producing necrosis. The venom of *S. catenatus,* which has high lethal toxicity, shows low local necrotizing effect. The venom of *S. miliarius,* which exhibits low lethal toxicity, on the other hand, has considerably higher local necrotizing effect.

Flowers and Goucher (1963) studied the protecting action of (disodium salt) EDTA on tissue damage caused by the injection of the venom of *A. piscivorus* in a chimpanzee's fingers and feet. The disodium salt of EDTA was subcutaneously administered directly into the area immediately after the injection of the venom (5 and 10 mg, respectively). The EDTA-treated areas showed much less swelling, necrosis, hemorrhages, etc., than the untreated areas. Goucher and Flowers (1963) noticed marked inhibition of proteolytic activity of *A. piscivorus* venom by EDTA. Marked reduction of necrotic activity of EDTA-treated venom was also noted by these authors in rabbit skin. It is evident that the beneficial effect of EDTA is due to its ability to inhibit proteolytic activity of the venom.

The venoms of many North American snakes are capable of producing lysis of *Paramecium*. These venoms contain a lysing factor. The venoms of three species of *Agkistrodon* are moderately active. The venom of *A. piscivorus* is, however, particularly active in producing rapid lysis of *Paramecium*. The ability to hemolyze rabbit erythrocyte has been detected in all venoms of

North American snakes when tested *in vitro*, but complete lysis *in vivo* is seldom observed, even with venoms of high hemolysin content, viz., the venoms of *S. catenatus* and *A. piscivorus* (Parrish, 1966). The latter venom can produce hemolysis of red cells *in vitro*, but no hemolysis of red cells can be detected even after 4 hours following the administration of the venom of *A. piscivorus* into rabbits and dogs (Hadidan, 1956). *C. adamanteus* venom can produce leukocytosis, alter the clotting time, prothrombin time, and produce profound hemorrhages in the poisoned animals (Radomski and Deichmann, 1957).

VI. SUMMARY AND CONCLUSION

Although the number of deaths that occur every year from snake bites in North America is insignificant compared to total deaths that are encountered world-wide, a great many biochemical and pharmacological investigations of venoms of North American snakes have been carried out. There are many species of snakes (pit vipers), but the venoms of *C. atrox* and *C. adamanteus* have been more frequently used than any other venom of North American snakes for such studies. The venoms of other North American pit vipers are not very much different from those of *C. atrox* and *C. adamanteus* except in the degrees of their toxicity and activities of the enzymes present therein. Only in recent years has the separation of the various components present in the venoms of North American snakes been seriously undertaken, and with the advent of starch gel electrophoresis, chromatography, or DEAE-cellulose, etc., it has now been possible to resolve the various components of the venoms of *C. atrox* and *C. adamanteus*. On the basis of the results obtained with these venoms, it is evident that they possibly contain ten or more components. Except for one or two, all of them are enzymes having different properties. They are mostly degradative enzymes; hydrolyse peptide bonds in proteins, ester bonds in synthetic amino acid esters, split phosphodiester bonds in nucleic acids, pyrophosphate linkage in NAD, remove the terminal phosphate (orthophosphate) from nucleoside-5'-phosphate, pyrophosphate from adenosine triphosphate, and fatty acids from β or 2' positions of phospatidal choline or ethanolamine. The only enzyme present in the venom which has no hydrolytic activity appears to be L-amino acid oxidase, the enzyme which converts L-amino acids to their corresponding keto acids. Their clotting activity and ability to release and destroy bradykinin also appear to be due to their proteolytic activity. The enzymes which hydrolyze proteins, synthetic amino acid esters (BAEE), produce clots and release bradykinin, possibly belong to a family of similar but not identical enzymes. They resemble trypsin in many ways but also differ from it in other important respects. The venom of *C. adamanteus* contains three diphosphopyridine-sensitive proteolytic enzymes, which can be distinguished from pH optimum, effects of ions on this activity, etc.

Snake venoms contain, besides proteases, very active phospholipase A and ATPase. They are also rich sources of endo and exonucleases and 5'-nucleotidase. The latter enzymes have been often used in the study of the structure of nucleic acids, to distinguish nucleoside-5'-phosphate from nucleoside-3'-phosphate and 2', 3'-nucleotides. Venom L-amino acid oxidase has been used to prepare a large number of α-keto acids from their corresponding amino acids, and in the study of its specificity toward substrates. Snake venom (phospholipase A) has been used in the study of the structure of phospholipids and the nature of fatty acids at their β-positions. The venoms of *piscivorus* and *C. adamanteus* also exhibit phospholipase B activity, and the latter venom has also been shown to contain two protein fractions with equal phospholipase A activity.

The venoms have a profound effect on circulation, causing a fall in blood pressure, cessation of respiration, and cardiac arrest following injection into anesthetized animals. They also have stimulating (increase in amplitude of contraction) and depressing action on isolated perfused frog hearts, depending upon the concentration used; but in all cases, except when extremely dilute solution is used, the heart finally stops beating, and goes to systolic contracture. *C. adamanteus* venom can produce necrosis, leukocytosis, alter clotting time, and cause profound hemorrhage in the poisoned animals. The cardiovascular effects can neither be abolished nor reduced by antihistamines.

The present article emphazises the importance of the isolation of various components in relatively pure forms, and the study of their nature, biochemical properties, and pharmacological actions separately as well as collectively. Such comprehensive studies are believed to be more useful and significant in the understanding of the cause of deaths from snake bites.

REFERENCES

Ablondi, F. B., Hagon, J. J., Philipis, M., and De Renzo, E. C. (1959). *Arch. Biochem. Biophys.* **82**, 153.

Acton, H. W., and Knowles, R. (1914a). *Indian J. Med. Res.* **1**, 388–413.

Acton, H. W., and Knowles, R. (1914b). *Indian J. Med. Res.* **1**, 414–424.

Adler, J., Lehmann, I. R., Bessman, M. J., Simons, E. S., and Korenberg, A. (1958). *Proc. Natl. Acad. Sci. U.S.* **44**, 641.

Alkjaersig, N., Fletcher, A. P., and Sherry, S. (1959). *J. Biol. Chem.* **234**, 832.

Allam, M. W., Weiner, D., and Lukens, F. D. W. (1956). *In* "Venoms," Publ. No. 44, p. 393. Am. Assoc. Advan. Sci., Washington, D.C.

Bender, A. E., and Krebs, H. A. (1950). *Biochem. J.* **46**, 210.

Billing, W. M. (1930). *J. Pharm. Exptl. Therap.* **38**, 173–196.

Bjork, W. (1961). *Biochim. Biophys. Acta* **49**, 195.

Bjork, W. (1963). *J. Biol. Chem.* **238**, 2487.

Bjork, W. (1964). *Biochim. Biophys. Acta* **89**, 483.

Blanchard, M., Green, D. E., Nocito, V., and Ratner, S. (1965). *J. Biol. Chem.* **161**, 583.

Boman, H. G., and Kaletta, U. (1956). *Nature* **178**, 1394.

Brown, R. V. (1940). *Am. J. Physiol.* **130**, 613.

Brown, R. V. (1941). *Am. J. Physiol.* **134**, 202.

Budovskii, E. F. (1962). *Vopr. Med. Khim.* **8**, 73.

Chain, E. (1939). *Biochem. J.* **33**, 407.

Cluxton, H. E., Jr. (1951). *Proc. 2nd Clin. ACTH Conf., 1951,* Vol. 2, p. 445. McGraw-Hill (Blakiston), New York.

Cochran, D. M. (1943). "Poisonous Reptiles of the World," War Background Studies, No. 10. Smithsonian Inst., Washington, D.C.

Conant, R. (1958). "A Field Guide to Reptiles and Amphibians." Houghton, Boston, Massachusetts.

Condie, R. M., Staab, E. V., and Good, R. A. (1964). *Proc. Soc. Exptl. Biol. Med.* **116**, 696.

Cushny, A. R., and Yogi, S. (1918). *Phil. Trans. Roy. Soc. London* **B208**, 1.

Dauvillier, P., de Hass, G. H., Van Deenen, L. L. M., and Raulin, J. (1964). *Compt. Rend.* **259**, 4867.

Davie, E. W., and Ratnoff, O. D. (1965). *In* "The Proteins" (H. Neurath, ed.), Vol. 3, p. 359. Academic Press, New York.

Delezenne, C., and Morel, H. (1919). *Compt. Rend.* **168**, 241.

Deutsch, H. F., and Diniz, C. R. (1955). *J. Biol. Chem.* **216**, 17.

Devi, A. (1968). *In* "Venomous Animals and Their Venoms" (W. Bücherl, V. Deulofeu, and E. E. Buckley, eds.), Vol. I, p. 119. Academic Press, New York.

Devi, A., and Sarkar, N. K. (1956). *In* "Venoms," Publ. No. 44, p. 189. Am. Assoc. Advance. Sci., Washington, D.C.

Devi, A., and Sarkar, N. K. (1966). *Mem. Inst. Butantan Simp. Intern.* **33**(2), 573.

Devi, A., and Sarkar, N. K. (1968). *In* "Venomous Animals and Their Venoms" (W. Bücherl, V. Deulofeu, and E. E. Buckley, eds.), Vol. I, p. 167. Academic Press, New York.

Devi, A., Maitra, S. R., and Sarkar, N. K. (1954). *Ann. Biochem. Exptl. Med. (Calcutta)* **13**, 23.

Devi, A., Banerjee, R., and Sarkar, N. K. (1960). *Proc. 7th Intern. Congr. Hematol., Rome, 1958,* Vol. II, p. 1190. I. L. Pensiero Scientifica, Rome.

Devi, A., Asghar, S. S., and Sarkar, N. K. (1966). *Mem. Inst. Butantan Simp. Intern.* **33**(3), 943.

Diechmann, W. B., Radomski, J. L., Farrell, J. J., MacDonald, W. E., and Keplinger, M. L. (1958). *Am. J. Med. Sci.* **236**, 204.

do Amaral, A. (1930). *Mem. Inst. Butantan (São Paulo)* **5**, 195–205.

Doery, H. M., and Pearson, J. E. (1964). *Biochem. J.* **92**, 599.

Eagle, H. (1937). *J. Exptl. Med.* **65**, 613.

Eagle, H., and Harris, T. N. (1937). *J. Gen. Physiol.* **20**, 543.

Elliot, R. H. (1901). *Phil. Trans. Roy. Soc. London* **B197**, 361.

Elliot, R. H. (1933). *Blackwood's Mag.* **233**, 359–365.

Elliot, R. H. (1934). "The Myth of the Mystic East," p. 210. Edinburgh and London.

Essex, H. E. (1932). *Am. J. Physiol.* **99**, 681.

Essex, H. E., and Markowitz, J. (1930). *Am. J. Physiol.* **90**, 317 and 705.

Flowers, H. H. (1963). *Toxicon* **1**, 131.

Flowers, H. H., and Goucher, C. R. (1963). *U.S. Army Med. Res. Lab., Fort Knox, Ky., Rept.* **596**, 8.

Frieden, E., Hsu, L. T., and Dittmer, K. (1951). *J. Biol. Chem.* **192**, 425.

Gaddum, J. H., and Horton, E. W. (1959). *Brit. J. Pharmacol.* **14**, 117.

Gennaro, J. F., Jr., and Ramsey, H. W. (1959). *Nature* **184**, 1244.

Ghosh, B. N., and Bhattacharjee, K. L. (1951). *Sci. Culture* (Calcutta) **2**, 272.

Ghosh, B. N., De, S. S. and Chowdury, D. K. (1941). *Indian J. Med. Res.* **29**, 367.

Githens, T. S., and Wolff, N. D. (1939). *J. Immunol.* **37**, 33.

Goucher, C. R., and Flowers, H. H. (1963). *U.S. Army Med. Res. Lab., Fort Knox, Ky., Rept.* **599**, 17.

Gralen, N., and Sevedberg, T. (1938). *Biochem J.* **32**, 1375.

Green, D. E. (1959). *Recent Advan. Enzymol.* **21**, 73.

Gulland, J. M., and Jackson, E. M. (1938a). *Biochem. J.* **32**, 590.

Gulland, J. M., and Jackson, E. M. (1938b). *Biochem. J.* **32**, 597.

Gulland, J. M., and Walsh, E. D. (1945). *J. Chem. Soc.* 173.

Gunn, J. W. C., and Epstein, D. (1933). *Quart. J. Pharm. Pharmacol.* **6**, 182.

Habermann, E. (1961). *Arch. Exptl. Pathol. Pharmakol.* **240**, 352.

Hadidan, Z. (1956). *In* "Venoms," p. 205. Publ. No. 44, Am. Assoc. Advanc. Sci., Washington, D.C.

Hadidan, Z., and Murphy, M. M. (1955). *J. Gen. Physiol.* **39**, 185.

Haessler, H. A., and Cunningham, L. (1957). *Exptl. Cell Res.* **13**, 304.

Heemskera, C. H. T., and Van Deenen, L. L. M. (1964). *Koninkl. Ned. Akad. Wetenschap. Proc.* **B67**, 181.

Hilmore, R. J. (1959). *Ann. N. Y. Acad. Sci.* **81**, 660.

Jimenez-Porras, J. M. (1961). *J. Exptl. Zool.* **148**, 257.

Johnson, B. D., and Bertke, E. M. (1964). *Toxicon* **2**, 197.

Johnson M., Kaye, M. A. G., Hems, R., and Krebs, H. A. (1953). *Biochem. J.* **54**, 625.

Kaye, M. A. G. (1955). *Biochim. Biophys. Acta* **18**, 456.

King, W. A., Jr. (1941). *Reptile Reptr.* p. 44.

Klauber, L. M. (1956). "Rattlesnakes, Their Habits, Life Histories and Influence on Mankind." Univ. of California Press, Berkeley, California.

Koerner, J. F., and Sinsheimer, R. L. (1957). *J. Biol. Chem.* **228**, 1049.

Korenberg, A., and Fuller, H. S. (1950). *J. Biol. Chem.* **128**, 763.

Langen, P. (1961). *Biochem. Z.* **334**, 65.

Laskowski, M., Hagerty, G., and Laurila, U. R. (1957). *Nature* **180**, 1181.

Laviton, A. E., Meyers, G. S., and Grunbaum, B. W. (1964). *Taxonomic Biochem. Serol.* p. 667.

Macht, D. I. (1937). *Proc. Soc. Exptl. Biol. Med.* **36**, 499.

Magenta, M. A. (1922). *Compt. Rend. Soc. Biol.* **87**, 834.

Marburg, A. G. (1964). *Behringwerk. Mitt.* **43**, 293.

Master, R. W. P., and Rao, S. S. (1961). *J. Biol. Chem.* **236**, 1986.

Meister, A. (1956). *In* "Venoms," Publ. No. 44, p. 295. Am. Assoc. Advance. Sci., Washington, D.C.

Michl, H., and Kiss, G. (1959). *Monatsh. Chem.* **90**, 604.

Minton, S. A., Jr. (1956). *In* "Venoms," Publ. No. 44, p. 145. Am Assoc. Advance. Sci., Washington, D.C.

Minton, S. A., Jr. (1957). *Sci. Am.* **196**, 114.

Morales, F., Bhanganada, K., and Perry, J. E. (1963). *Proc. 10th Symp. Venomous Poisonous Animals, Noxious Plants, Pacific Reg.,* p. 385. Univ. Hawaii.

Morozova, V. F. (1966). *Dokl. Akad. Nauk Uz. SSR.* **21**, 52.

Neelin, J. M. (1963). *Can. J. Biochem. Biophys.* **41**, 1073.

Newmann, W., and Habermann, E. (1956). *In* "Venoms," Publ. No. 44, p. 174. Am. Assoc. Advance. Sci., Washington, D.C.

Nygaard, A. P., and Sumner, J. B. (1952). *J. Biol. Chem.* **200**, 723.
Parrish, H. M. (1958). *Vet. Med.* **53**, 197.
Parrish, H. M. (1966). *Public Health Rept.* (*U.S.*) **81**, 269.
Pfleiderer, G., and Sumyk, G. B. (1961). *Biochim. Biophys. Acta* **51**, 482.
Pope, C. H. (1944). "The Poisonous Snakes of the New World." N.Y. Zool. Soc., New York.
Porges, N. (1958). *Science* **117**, 48.
Radda, G. K. (1964). *Nature* **203**, 936.
Radomsky, J. L., and Deichmann, W. B. (1957). *Federation Proc.* **16**, 328.
Radomski, J. L., and Deichmann, W. B. (1958). *Biochem. J.* **70**, 293.
Razzell, W. E., and Khorana, H. G. (1959a). *J. Biol. Chem.* **234**, 2105.
Razzell, W. E., and Khorana, H. G. (1959b). *J. Biol. Chem.* **234**, 2114.
Richard, G. M., du Vair, G., and Laskowski, M., Sr. (1965). *Biochemistry* **4**, 501.
Rocha e Silva, Beraldo, M. W. T., and Rosenfeld, G. (1949). *Am. J. Physiol.* **156**, 261.
Rosenberg, P., and Podleski, T. R. (1962). *J. Pharmacol. Exptl. Therap.* **137**, 249.
Russell, F. E., and Emery, J. A. (1961). *Am. J. Med. Sci.* **241**, 507.
Russell, F. E., and Long, T. E. (1960). *Proc. 2nd Intern. Symp. Myasthenia Gravis, 1961.* p. 101. Thomas, Springfield, Illinois.
Russell, F. E., Emery, J. A., and Long, T. E. (1960). *Proc. Soc. Exptl. Biol. Med.* **103**, 737.
Russell, F. E., Buess, F. W., and Woo, M. Y. (1963). *Toxicon* **1**, 99.
Saito, K., and Hanahan, J. (1962). *Biochemistry* **1**, 521.
Sarkar, N. K. (1947). *Indian J. Chem. Soc.* **24**, 227.
Sarkar, N. K. (1948). *Ann. Biochem. Exptl. Med.* **8**, 11.
Sarkar, N. K. (1960). *In* "Enzymes in Health and Diseases" (D. M. Greenberg and H. Harper, eds.), p. 359. Thomas, Springfield, Illinois.
Sarkar, N. K., and Devi, A. (1960). *Proc. 7th Intern. Congr. Hematol., Rome, 1958* Vol. II, p. 1079. I. L. Pensiero Scientifica, Rome.
Sarkar, N. K., and Devi, A. (1964). *Proc. 9th Intern. Congr. Hematol., Mexico City, 1962* Vol. II, p. 315.
Sarkar, N. K., and Maitra, S. R. (1950). *Am. J. Physiol.* **163**, 208.
Schmidt, G., and Laskowski, M., Sr. (1961). *In* "The Enzymes" (P. D. Boyer, H. Lardy, and K. Myrbäck, eds.), 2nd ed., Vol. 5, p. 3. Academic Press, New York.
Schottler, W. H. A. (1954). *Am. J. Trop. Med. Hyg.* **3**, 1083.
Shah, S. N., Lossow, W. J., and Chaikoff, I. L. (1964). *Biochim. Biophys. Acta* **84**, 176.
Singer, T. P., and Kearnary, E. B. (1950). *Arch. Biochem.* **27**, 348.
Slotta, K. (1956). *In* "Venoms," Publ. No. 44, p. 253. Am. Assoc. Advance. Sci., Washington, D.C.
Slotta, K., and Frankel-Conrat, H. (1938). *Chem. Ber.* **71**, 1076.
Sulkowski, E., Bjork, W., and Laskowski, M. (1963). *J. Biol. Chem.* **238**, 2477.
Sumyk, G. B., and Kashin, P. (1964). *Inst. Ind. Technician, Chicago.* p. 94.
Suzuki, T., and Wanaga, S. I. (1960). *J. Pharm. Soc. Japan* **80**, 1002.
Taborda, A. R., Taborda, L. C., Williams, J. N., Jr., and Elvehjem, C. A. (1952a). *J. Biol. Chem.* **194**, 227.
Taborda, A. R., Taborda, L. C., Williams, J. N., Jr., and Elvehjem, C. A. (1952b). *J. Biol. Chem.* **195**, 207.
Tattrie, N. H. (1959). *J. Lipid Res.* **19**, 535.
Taube, H., and Essex, H. E. (1937). *A.M.A. Arch. Pathol.* **24**, 43.
Truant, A. P., and Hadidan, Z. (1900). See Hadidan (1956).
Uzawa, T. (1932). *J. Biochem.* (*Tokyo*) **15**, 19.
Van Arman, C. G., and Miller, L. M. (1961). *J. Pharmacol. Exptl. Therap.* **131**, 366.

Vick, J. A., Blanchard, R. J., and Perry, J. F., Jr. (1963). *Proc. Soc. Exptl. Biol. Med.* **113**, 841.

Vogel, C., Koppel, J. L., and Olwin, J. H. (1964). *J. Lipid Res.* **5**, 385.

Webster, M. E. and Piene, J. V. (1963). *Ann. N. Y. Acad. Sci.* **104**, 91.

Wellner, D., and Meister, A. (1960). *J. Biol. Chem.* **235**, 2013.

Werle, E., Kehl, R., and Koebke, K. (1950). *Biochem. Z.* **320**, 372.

Zeller, E. A. (1948). *Experientia* **4**, 184.

Zeller, E. A. (1950). *Helv. Chim. Acta* **33**, 821.

Zeller, E. A. (1951). *In* "The Enzymes" (J. B. Sumner and K. Myrbäck, eds.), Vol. 1, Part 2, p. 986. Academic Press, New York.

Zeller, E. A. (1963). *Biochem. Z.* **339**, 13.

Zeller, E. A., and Maritz, A. (1944). *Helv. Chim. Acta* **27**, 1888.

Zeller, E. A., Ramchandra, G., Fleisher, G. A., Ishimaru, T., and Zeller, V. (1965). *Biochem. J.* **95**, 262.

Chapter 29

Comparative Biochemistry of Sistrurus Miliarius Barbouri and Sistrurus Catenatus Tergeminus Venoms

CARLOS A. BONILLA,* WAYNE SEIFERT, AND NORMAN HORNER

DEPARTMENT OF BIOLOGY, MIDWESTERN UNIVERSITY, WICHITA FALLS, TEXAS

I. INTRODUCTION

Two representative species of the genus *Sistrurus* (*miliarius* and *catenatus*) are found in the United States. Morphologically, these rattlesnakes are divided into various subspecies on the basis of general coloration, blotch patterns, scale rows, and other criteria. Little is known, however, about the protein composition of the venoms of these closely related species. In this chapter, we shall present some biochemical and pharmacological properties of crude and partially purified *Sistrurus* venoms.

*Present address: Department of Physiology and Biophysics, Colorado State University, Fort Collins, Colorado.

II. MATERIALS AND METHODS

A. Venom

Lyophilized venom of *Sistrurus miliarus barbouri* (dusky pigmy rattle-snake) was purchased from the Sigma Chemical Company. *Sistrurus catenatus tergeminus* (western massasauga rattlesnake) venom was collected in our serpentarium, stored at 2°C, and processed no later than 15 minutes after collection. The venom was then cleared by centrifugation at 48,000 g for 30 minutes, and the supernatant fluid was lyophilized immediately.

B. Analytical Electrophoresis

Electrophoresis in acidic polyacrylamide gels was carried out as described by Jordan and Raymond (1967) under the following experimental conditions: voltage, 250 V (constant); current 100 ma; prerun, 90 minutes; electrolyte, 0.37 M glycine-citric acid buffer, pH 2.9; load, 60 μ liter (3.0–4.5 mg protein); standard, lysozyme (0.90 mg) 3X crystallized (Pierce Chemical Co.). Separations were carried out at room temperature but coolant (4°C) was circulated through the cooling plates at all times. In all instances, a 13.8% cyanogum-41 gel was used.

The gels were stained with Amido Schwarz for 16 hours and destained electro-phoretically (40 minutes) with a mixture of methanol-acetic acid-water (5:1:5).

C. Gel Filtration

Sephadex G-25 coarse (bead form 100–300 μ) and Sephadex G-75 fine (bead form 40–120 μ) were purchased (Pharmacia Fine Chemicals Co., Inc.). All analytical procedures were carried out at 2°C unless otherwise stated.

Dry Sephadex G-75 was suspended in a 1% NaCl solution and allowed to swell for 98 hours, after which the fine particles were removed by repeated decantations. The columns (2.5 × 50 cm) were packed according to the method of Flodin (1962) and equilibrated for 48 hours with 0.1 M Tris-HCl-0.1 M NaCl buffer (pH 8.6). For maximum chromatographic resolution, columns with a minimum void space in the outlet were utilized (gel filtration columns; Pharmacia Fine Chemicals Co., Inc.). Fractions found to contain material absorbing at 280 mμ were pooled and concentrated either by the addition of coarse dry Sephadex G-25 (Flodin *et al.*, 1960) or by lyophili-zation after a 24-hour dialysis against distilled water. Protein concentration was determined by the procedure of Lowry *et al.* (1951), using bovine serum albumin as the standard.

D. Biological Assay

Adult (25–40 days old) male and female Swiss albino mice (*Mus musculus*) from a homogenous stock were used in all experiments. The diet consisted of Purina Laboratory Chow and tap water *ad libitum*. Intraperitoneal administration with a constant volume of 0.01 ml/gm of body weight was used. Crude venoms or their partially purified fractions were dissolved in physiological saline prior to bioassay.

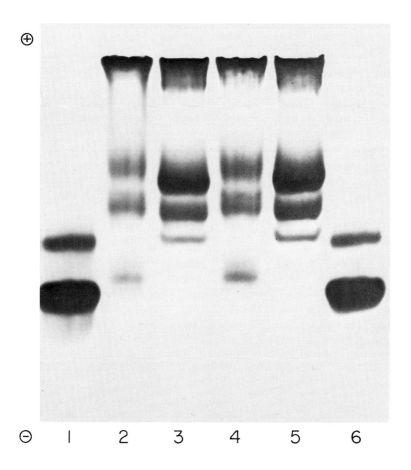

FIG. 1. Electrophoretogram of crude venom from *Sistrurus* snakes. (1) lysozyme (0.90 mg); (2) *S. catenatus tergeminus* (3.0 mg); (3) *S. miliarius barbouri* (3.0 mg); (4) *S. catenatus tergeminus* (4.5 mg); (5) *S. miliarius barbouri* (4.5 mg); (6) lysozyme (0.90 mg).

III. RESULTS AND DISCUSSION

A. Venom

Lyophilized *Sistrurus* venoms were very soluble (over 95%) in both physiological saline and 0.1 M Tris-HCl buffer (pH 8.6), but less soluble in distilled water (approximately 80%). The crude venoms contained a small amount of insoluble material of unknown composition which could be removed easily by centrifugation. In Tris-HCl buffer, *S. miliarius barbouri* venom was light yellow, probably due to the presence of small amounts of riboflavin, but it became colorless after dialysis against distilled water. *S. catenatus tergeminus* venom was colorless.

B. Electrophoretic Patterns

Electrophoretic patterns of crude *S. catenatus tergeminus* (slots 2 and 4) and *S. miliarius barbouri* (slots 3 and 5) venoms are shown in Fig. 1. Each of these secretions contained at least 3 different components migrating toward the cathode at pH 2.9. In both cases, a quantity of material remains at the origin indicating that under the stated conditions for electrophoresis some components are not being resolved. Although similar, the electrophoretic patterns are not identical and can be used, when in doubt, to differentiate these 2 species of rattlesnake. Unfortunately, the small amount of partially purified fractions obtained has so far precluded their electrophoretic analysis.

C. Gel Filtration Patterns

Comparative gel filtration patterns of *S. miliarius barbouri* and *S. catenatus tergeminus* lyophilized venoms are shown in Fig. 2. The peak tubes were pooled, dialyzed for 24 hours against distilled water, and concentrated by lyophilization. Protein recoveries (percent from total) for the partially purified fractions are seen in Table I. The lyophilized powder was dissolved in a given amount of physiological saline solution and tested for biological activity.

TABLE I

PERCENT PROTEIN RECOVERY AFTER PARTIAL PURIFICATION
OF *Sistrurus* VENOMS

Fraction	*S. miliarius barbouri*	*S. catenatus tergeminus*
I	24.7	20.5
II	46.5	28.2
III	15.8	13.3

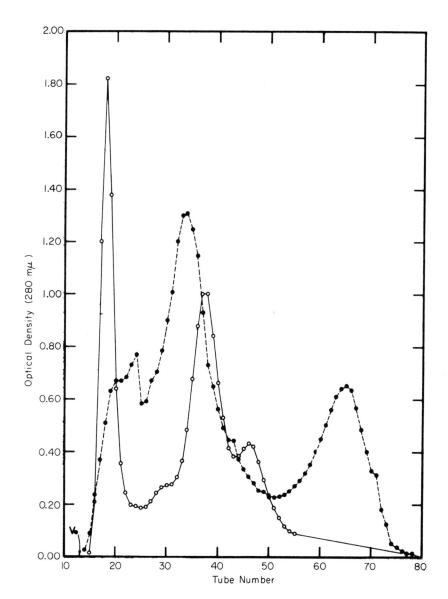

Fig. 2. Gel filtration of crude *Sistrurus* venoms on Sephadex G-75. The column (2.5 × 50 cm) was equilibrated and eluted with 0.1 *M* Tris-HCl-0.1 *M* NaCl buffer, pH 8.6: ○——○, *S. catenatus tergeminus* (62.0 mg protein); ●——●, *S. miliarius barbouri* (63.5 mg protein).

D. Biological Activity

The LD_{50} for *S. catenatus tergeminus* and *S. miliarius barbouri* by the intraperitoneal route were 0.76 and 0.28 mg/20 gm mouse. These results are in contrast with Minton's data (1956) and cannot be explained at this time. However, the age of the snakes (Minton, 1966) and the use of commercial (pooled) venom can drastically affect the lethality of crude venoms (Bonilla and Horner, 1968).

As shown in Table II, most of the *in vivo* lethal toxicity of *S. catenatus*

TABLE II

LETHALITY OF CRUDE AND PARTIALLY PURIFIED
Sistrurus catenatus tergeminus VENOM

Fraction	Dose (μg/gm)	Lethality (%)
Whole venom	9.50	0
	19.00	20
	38.00	50
I	12.30	25
	36.90	78
II	8.50	28
	17.00	91
	34.00	100
III	8.00	0
	24.00	0
	48.00	0

tergeminus venom resided in fraction II; fraction I was lethal only at extremely high concentrations, and fraction III had no lethal activity even at maximal concentrations (1 mg/gm body weight). The same results were obtained with partially purified fractions from *S. miliarius barbouri*. The LD_{100} for *S. miliarius barbouri* fraction II was 0.590 mg/20 gm while that of *S. catenatus tergeminus* fraction II was 0.680 mg/20 gm mouse.

No neurotoxic activity could be ascribed to either the whole venom or partially purified fractions obtained by gel filtration on Sephadex G-75. Crude *Sistrurus* venoms or fraction II caused severe peritoneal hemorrhages with large amounts of free fluid in the abdominal cavity and vascular collapse was always observed. However, intravascular hemolysis or clotting were completely absent, and the hematocrit values were normal at all times. Histological examination of liver and lung tissue was mainly negative, with the exception of small foci of degeneration and necrosis present in the liver of animals receiving fraction II. Thus, as in the case of viper envenomation, the pharmacological action of *Sistrurus* venoms can best be explained on the basis of hemorrhagic (probably protease) factors.

The data presented herein warrant further investigation of *Sistrurus* venoms. An attempt is being made at this time to isolate pure proteins with hemorrhagic activity. It is hoped that this will lead to a better understanding of hemorrhagic principles in snake venoms.

REFERENCES

Bonilla, C. A., and Horner, N. (1968). Unpublished observations.

Flodin, P., Gelotte, B., and Porath, J. (1960). *Nature* **188**, 493.

Flodin, P. (1962). "Dextran Gels and Their Applications in Gel Filtration." Pharmacia Fine Chemicals, Inc., Rochester, Minnesota.

Jordan, E. M., and Raymond, S. (1967). E-C Bulletin Supplement # 1034, E-C Apparatus Corp., Philadelphia, Pa.

Lowry, O. H., Rosebrough, N. J., Farr, A. L., and Randall, R. J. (1951). *J. Biol. Chem.* **193**, 265.

Minton, S. A., Jr. (1956). *In* "Venoms," Publ. No. 44, p. 295. Am. Assoc. Advance. Sci., Washington, D.C.

Minton, S. A., Jr. (1966). *In* "Animal Toxins" (F. E. Russell and P. R. Saunders, eds.), p. 211. Pergamon Press, Inc., New York.

Chapter 30

Neotropical Pit Vipers, Sea Snakes, and Coral Snakes

ALPHONSE RICHARD HOGE AND SYLVIA ALMA R. W. D. L. ROMANO

DEPARTMENT OF HERPETOLOGY, INSTITUTO BUTANTAN, SÃO PAULO, BRAZIL

I. CLASSIFICATION

For practical reasons, the species of northern Mexico, although belonging to the Nearctic fauna, are included here.

The classification of neotropical poisonous snakes has undergone profound changes in the past 2 years. Accordingly, our objective has been to list all recognized species and subspecies with full understanding of their preliminary status.

No list of cited literature is given, since the text citations allow quick identification of the papers. Except for some changes in *Bothrops* and *Crotalus* species, this list is chiefly based on:

Gloyd, H. K. (1940). The rattlesnakes genera *Sistrurus* and *Crotalus*. *Chicago Acad. Sci.*

Hoge, A. R. (1966). Preliminary account on neotropical vipers. *Mem. Inst. Butantan (São Paulo)* **32**.

Klauber, L. M. (1956). Rattlesnakes **1** and **2**.

Roze, J. A. (1967). Checklist of New World venomous coral snakes. *Am. Museum Novitates* **2287**.

A. Family Elapidae

LEPTOMICRURUS Schmidt

Thread Coral Snakes
  Leptomicrurus Schmidt 1937 [Zool. Ser. Field Mus. Nat. Hist., Vol. **20**: 363]
  Genotype: *Elaps collaris* Schlegel

Leptomicrurus collaris (Schlegel)
  Elaps collaris Schlegel 1837 [Physion, Serp., Vol. 2: 448]
  Elaps gastrodelus Duméril 1854, Bibron et Duméral [Erp. Gen. Vol 7: 1212]—type locality, unknown.
  Hemibungarus collaris; Boulenger 1896 [Cat. Sn. Brit. Mus. Vol. 3: 393]
  Elaps collaris; Thompson 1913 [Notes Leyden Mus., Vol. 35: 171]
  Leptomicrurus collaris; Schmidt 1937 [Zool. Ser. Field Mus. Nat. Hist., Vol. 20: 363]

Type locality: Not given.
Range: Southeastern Venezuela, the Guianas, and Brazil (specimen from near Belem, Para).

Leptomicrurus narduccii (Jan)
  Elaps narduccii Jan 1893 [Arch. Zool. Anat. Fisiol., Vol. 2: 222]
  Elaps scutiventris Copo 1869 [Proc. Am. Phil. Soc., Vol. 11: 156]—type locality, Pebas, Ecuador.
  Elaps melanotus Peters 1881 [W., Sb. Ges. Naturf. Freunde Berlin, p. 51]—type locality, Sarayacu, Ecuador.
  Leptomicrurus narduccii, Schmidt 1937 [Zool. Ser. Field Mus. Nat. Hist., Vol. 20: 363]

Type locality: Bolivia and Peru.
Range: Eastern Peru, Ecuador, Colombia, Bolivia and Brazil (known from the State of Acre).

Leptomicrurus schmidti Hoge & Romano
Schmidt's Thread Coral Snake
  Leptomicrus schmidti Hoge & Romano 1966 (*error typographicus pro Leptomicrurus schmidti)*————[Mem. Inst. Butantan, Vol. 32: 1–9, Pl. 2, Fig. 2; Pl. 3, Fig. 2a; Pl. 4, Fig. 2b]

Type locality: Tapurucuara, Amazonas, Brazil.
Range: Known only from the type locality.

MICRUROIDES Schmidt

  Micruroides Schmidt 1928 [Bull. Antivenin Inst. Amer., Vol. 2: 63]
  Genotype: *Elaps euryxanthus* Kennicott

Micruroides euryxanthus euryxanthus (Kennicott)
Arizona Coral Snake
 Elaps euryxanthus Kennicott 1860 [Proc. Acad. Nat. Sci. Phil., Vol. 12: 337]
 Elaps euryxanthus; Boulenger 1896, [Cat. Sn. Brit. Mus., Vol. 3: 415]
 Micruroides euryxanthus; Schmidt 1928 [Bull Antivenin Inst. Amer., Vol. 2, No. 3: 63]
 Micruroides euryxanthus euryxanthus; Zweifel & Norris 1955 [Amer. Midl. Nat., Vol. 54: 246; Table 1]

Type locality: Restricted (Schmidt, 1953)—Guaymas, Sonora, Mexico.
Range: U.S.A., central Arizona, from the edge of the plateau region south and eastward to southwestern New Mexico and extreme western Texas. Mexico, northwestern Chihuahua and northern Sonora, including Tiburon Island; Intergrades with *M. e. australis* known from central west Sonora.

Micruroides euryxanthus australis Zweifel & Norris
Southern Arizona Coral Snake
 Micruroides euryxanthus australis Zweifel & Norris 1955 [Amer. Midl. Nat., Vol. 54: 230; Figs. 1–3, 1 Pl.]

Type locality: Guirocoba, Sonora, Mexico.
Range: Known from the type locality and from Batopilar, Chihuahua, Mexico (this specimen, reported by Cope, presumably belongs to this subspecies /Zweifel & Norris, *l. c.*/).

Micruroides euryxanthus neglectus Roze
 Micruroides euryxanthus neglectus Roze 1967 [Amer. Mus. Novit. No. 2287: 4; Fig. 1]

Type locality: Sixteen and three-tenths miles north-northwest of Mazatlan, Sinaloa, Mexico.
Range: North of Mazatlan, Sinaloa, Mexico.

MICRURUS Wagler

True Coral Snakes
Micrurus Wagler 1824, [*In* Spix, Serp. Brazil., p. 48]
 Genotype: *Micrurus spixii* Wagler

Micrurus albicinctus Amaral
White-ringed Coral Snake
 Micrurus albicinctus Amaral 1926 [Comm. Linh. Telegr. Mato Grosso ao Amazonas, Publ. 84, Annex 5, p. 48]
 Micrurus waehnerorum Meise 1938 [Zool. Anz., Vol. 123: 20]—type locality, São Paulo de Olivença, Brazil.
 Micrurus albicinctus; Roze 1967 [Amer. Mus. Novit., No. 2287: 5]

Type locality: Northern and central Matto Grosso, Brazil.
Range: Known only from the type locality and from São Paulo de Olivença, Amazonas, Brazil.

Micrurus alleni alleni Schmidt
Allen's Coral Snake
  Micrurus nigrocinctus alleni Schmidt 1936 [Zool. Ser. Field Mus. Nat. Hist., Vol. 20: 209; Fig. 25]
  Micrurus alleni alleni; Taylor 1951, Kansas Univ. Sci. Bull., Vol. 34: 172]
  Micrurus alleni richardi Taylor 1951, Kansas Univ. Sci. Bull. Vol. 34: 169; Pl. 23]—type locality, Los Diamantes, 2 km southeast of Guapiles, Costa Rica.

Type locality: Rio Mico, 7 miles above Rama, Siquia, District, Nicaragua.
Range: Eastern Nicaragua and Costa Rica to northwestern Panama.

Micrurus alleni vatesi Dunn
Yate's Coral Snake
  Micrurus nigrocinctus vatesi Dunn 1942, [Notul. Nat. Acad. Phil., No. 108: 8]
  Micrurus alleni vatesi; Roze 1967 [Amer. Mus. Novit., No. 2287: 6]

Type locality: Farm Two, Chiriqui Land Company, near Puerto Armuelles, Chiriqui, Panama.
Range: Pacific slopes of southeastern Costa Rica and southwestern Panama.

Micrurus ancoralis ancoralis (Jan)
Anchor Coral Snake
  Elaps marcgravii var. *ancoralis* Jan 1872 [*In* Jan & Sordelli, Icon. Gen. Ophid., Vol. 3: Pl. 4, Fig. 2]
  Elaps ancoralis; Boulenger 1896 [Cat. Sn. Brit. Mus., Vol. 3: 432]
  Micrurus ancoralis; Amaral 1925 [Proc. U.S. Nat. Mus., Vol. 67, No. 24: 19]
  Micrurus ancoralis ancoralis; Schmidt 1936 [Zool. Ser. Field Mus. Nat. Hist., Vol. 20: 197]

Type locality: Ecuador.
Range: Pacific drainage of Ecuador.

Micrurus ancoralis jani Schmidt
Jan's Coral Snake
  Micrurus ancoralis jani Schmidt 1936 [Zool. Ser. Field Mus. Nat. Hist., Vol. 20: 197]

Type locality: Andagoya, Chocó, Colombia.
Range: Chocó region from Colombia to Panamá.

Micrurus annellatus annellatus (W. Peters)
Lowland Annelated Coral Snake
  Elaps annellatus, Peters, W. 1871 [Mber. Dtsch. Akad. Wiss. Berlin, p. 402]
  Elaps annellatus; Boulenger 1896 [Cat. Sn. Brit. Mus. Vol. 3: 418]
  Micrurus annellatus; Amaral 1930 (*partim*) [Mem. Inst. Butantan, Vol. 4: 228]
  Micrurus annellatus annellatus; Schmidt 1954 [Fieldiana, Zool., Vol. 36, No. 30: 321–323; Fig. 62]

Type locality: Pozuzu, Peru.
Range: Lowlands of northeastern Peru and southeastern Ecuador.

Micrurus annellatus balzani (Boulenger)
Balzan's Coral Snake
  Elaps balzani Boulenger 1898 [Ann. Mus. Stor. Nat. Genova, Vol. 19, No. 2: 130]
  Elaps regularis; Boulenger 1902 [Ann. Mag. Nat. Hist., Vol. 9, No. 7: 402] —type locality, Chulumani, Bolivia.
  Micrurus corallinus corallinus; Amaral 1930 (partim) [Mem. Inst. Butantan, Vol. 4: 228]
  Micrurus balzani; Schmidt 1936 [Zool. Ser. Field Mus. Nat. Hist., Vol. 20: 192]

Type locality: Yungas, Bolivia.
Range: Amazonian drainage of Bolivia.

Micrurus annellatus bolivianus Roze
Bolivian Ringed Coral Snake
  Micrurus annellatus bolivianus Roze 1967 [Amer. Mus. Novit., No. 2287: 7; Fig. 2]

Type locality: Charobamba River, about 50 km northeast of Zudanez, Chuquisaca, Bolivia.
Range: Known only from the type locality and from Yungas de Cochabamba, Bolivia.

Micrurus bernadi (Cope)
Bernard's Coral Snake
  Elaps bernadi; Cope 1886 [*In* Perez, Proc. U.S. Nat. Mus., Vol. 9: 190 (n.rn.)]
  Elaps bernadi; Cope 1887 [Bull. U.S. Nat. Mus., No. 32: 87]
  Micrurus bernadi; Schmidt 1933 [Zool. Ser. Field Mus. Nat. Hist., Vol. 20: 40]

Type locality: Zacualtipan, Hidalgo, Mexico.
Range: Western Hidalgo and northern Puebla, Mexico.

Micrurus bocourti bocourti (Jan)
Bocourt's Coral Snake
  Elaps bocourti Jan 1872 [*In* Jan & Sordelli, Icon. Gen. Ophid., Vol., 3;
    Pl. 6, Fig. 2]
  Elaps corallinus; Boulenger 1896 (*partim*) [Cat. Sn. Brit. Mus., Vol. 3:
    420–421]
  Micrurus ecuadorianus Schmidt 1936 [Zool. Ser. Field Mus. Nat. Hist.,
    Vol. 20: 196]—type locality, Rio Daule, western Ecuador.
  Micrurus ecuadorianus ecuadorianus; Peters, J. A. 1960 [Bull. Mus. Comp.
    Zool. Harv., Vol. 122, No. 9: 530]
  Micrurus bocourti bocourti; Roze 1967 [Amer. Mus. Novit., No. 2287: 8]

Type locality: Restricted (Roze, 1967)—Rio Daule, Guayas, Ecuador.
Range: Pacific coast of Ecuador.

Micrurus annelatus montananus Schmidt
Mountain Ringed Coral Snake
  Micrurus annellatus montanus Schmidt 1954 [Fieldiana, Zool., Vol. 34,
    No. 30: 322–324]

Type locality: Camp 4, about 10 km north of Santo Domingo Mine, Puno,
  Peru; altitude, 2,000 m.
Range: Southeastern Peru to central Bolivia.

Micrurus averyi Schmidt
Avery's Coral Snake
  Micrurus averyi Schmidt 1939 [Zool. Ser. Field Mus. Nat. Hist., Vol. 24,
    No. 6: 45–47]

Type locality: Courantyne District, near Brazilian border, latitude 1° 40′ N,
  longitude 58° W, Guyana (formerly British Guiana).
Range: Known only from the type locality.

Micrurus bocourti sangilensis Niceforo Maria
San Gil Coral Snakes
  Micrurus ecuadorianus sangilensis Niceforo Maria 1942 [Rev. Acad.
    Colomb., Vol. 5, No. 17: 98; Pls. 3–10]

Type locality: (Schmidt, 1955, p. 344)—San Gil, Santander, Colombia.
Range: Northern Colombia, between the Cordillera Central and the Cordillera
  Oriental.

Micrurus bogerti Roze
Bogert's Coral Snake
  Micrurus bogerti Roze 1967 [Amer. Mus. Novit., No. 2287: 9; Fig. 3]

Type locality: Tangola-Tangola (Tangolunda), east of Puerto Angel, Oaxaca, Mexico.

Range: Extreme southern Pacific coast of Oaxaca, Mexico.

Micrurus browni browni Schmidt & Smith
Brown's Coral Snake
  Micrurus browni Schmidt & Smith 1943 [Zool. Ser. Field Mus. Nat. Hist., Vol. 29: 29]
  Micrurus nigrocinctus browni; Mittleman & Smith 1949 [Trans. Kansas Acad. Sci., Vol. 52: 88]
  Micrurus browni browni; Roze 1967 [Amer. Mus. Novit., No. 2287: 11]

Type locality: Chilpancingo, Guerrero, Mexico.

Range: Sierra Madre del Sur of central Guerrero and Oaxaca to Chiapas, Mexico, and the mountains of western Guatemala.

Micrurus browni importunus Roze
Antigua Basin Coral Snake
  Micrurus browni importunus Roze 1967 [Amer. Mus. Novit., No. 2287: 11; Fig. 4]

Type locality: Dueras, about 25 km west-southwest of Guatemala City in the Antigua Basin, Sacatepequez, Guatemala.

Range: Known only from the type locality.

Micrurus browni taylori Schmidt & Smith
Taylor's Coral Snake
  Micrurus nuchalis taylori Schmidt & Smith 1943 [Zool. Ser. Field Mus. Nat. Hist., Vol. 29: 30]
  Micrurus browni taylori; Roze 1967 [Amer. Mus. Novit., No. 2287: 12]

Type locality: Acapulco, Guerrero, Mexico.

Range: Acapulco and coastal regions of Guerrero, Mexico. Intergrades with *M. b. browni* between Acapulco and Chilpancingo.

Micrurus carinicaudus carinicaudus Schmidt
Rough-tailed Coral Snake
  Micrurus carinicauda Schmidt 1936 [Zool. Ser. Field Mus. Nat. Hist., Vol. 20: 194]
  Micrurus carinicaudus carinicaudus; Schmidt 1955 [Fieldiana, Zool., Vol. 34, No. 34: 343]

Type locality: Orope, Zulia, Venezuela.

Range: Northwestern Venezuela to North of Santander, Colombia.

Micrurus carinicaudus antioquiensis Schmidt
Antioquian Coral Snake

Micrurus antioquiensis Schmidt 1936 [Zool. Ser. Field Mus. Nat. Hist., Vol. 20: 195]
Micrurus carinicaudus antioquiensis; Schmidt 1955 [Fieldiana, Zool., Vol. 34, No. 34: 343]

Type locality: Santa Rita, north of Medellin, Antioquia, Colombia.
Range: Restricted to the Cauca Valley, Colombia. Intergrades with *M. c. carinicaudus* in central Magdalena Valley, Colombia, and with *M. c. transandinus* in northern Ecuador.

Micrurus carinicaudus colombianus (Griffin)
Santa Marta Rough-tailed Coral Snake
  Elaps colombianus Griffin 1916 [Mem. Carnegie Mus., Vol. 7, No. 3: 216]
  Micrurus carinicaudus colombianus; Roze 1967 [Amer. Mus. Novit., No. 2287: 13]

Type locality: Minca, Colombia.
Range: Santa Marta region, northern Colombia.

Micrurus carinicaudus dumerilii; (Jan)
Dumeril's Coral Snake
  Elaps dumerilii Jan 1858 [Rov. Mag. Zool., Vol 10, No. 2: 522]
  Elaps dumerilii; Boulenger 1896 [Cat. Sn. Brit. Mus., Vol. 3: 419]
  Micrurus carinicauda dumerilii; Roze 1967 [Amer. Mus. Novit., No. 2287: 13]

Type locality: Cartagena, Colombia.
Range: Cartagena to Santa Marta and southward to Nontander Santander, Colombia.

Micrurus carinicaudus transandinus Schmidt
Transandian Coral Snake
  Micrurus transandinus Schmidt 1936 [Zool. Ser. Field Mus. Nat. Hist., Vol. 20: 195]
  Micrurus carinicaudus transandinus; Schmidt 1955 [Fieldiana, Zool., Vol. 34, No. 34: 343–344]

Type locality: Andagoya, Choco, Colombia.
Range: The Choco region, Colombia, west of the Andes from the Gulf of Uraba to Ecuador.

Micrurus clarki Schmidt
Clark's Coral Snake
  Micrurus clarki; Schmidt 1936 [Zool. Ser. Field Mus. Nat. Hist., Vol. 20: 211–212]

Type locality: Yavisa, Darien, Panama.

Range: Eastern Costa Rica, Panama to western Colombia.

Micrurus corallinus Merrem

 Elaps corallinus Merrem 1820 [Nova Acta Acad. Leop. Carol., Vol. 10, No. 1: 108; Pl. 4]

 Coluber corallinus Raddi 1820 [Mem. Soc. Nat. Sci. Modena, p. 336]—type locality, Rio de Janeiro

 Elaps corallinus; Boulenger 1896 *(partim)* [Cat. Sn. Brit. Mus. Vol. 3: 420–421]

 Micrurus corallinus; Amaral 1925 [Proc. U.S. Nat. Mus., Vol. 67 No. 24: 20]

Type locality: Parahyba, Rio Frio, Rio de Janeiro, Brazil.

Range: Eastern portion of Brazil from Espirito Santo to Santa Catarina, westward to Mato Grosso, and southward to northern Argentina.

Micrurus decoratus (Jan)

Decorated Coral Snake

 Elaps decoratus Jan 1858 [Rev. Mag. Zool., Vol. 10, No. 2: 525]

 Elaps decoratus; Boulenger 1896 [Cat. Sn. Brit. Mus., Vol. 3: 419]

 Elaps fischeri Amaral 1921 [Anex. Mem. Inst. Butantan, Vol. 1, No. 1: 59; Pl. 2, Figs. 1–5]—type locality, Serra Bocaina, São Paulo, Brazil.

 Elaps ezequïeli Lutz & Mello 1922 [Mem. Inst. Oswaldo Cruz, Vol. 15: 235, Pl. 31]—type locality, Caxambu, Minas Gerais, Brazil.

 Micrurus decoratus; Amaral 1926 [Rev. Mus. Paul., Vol. 14: 32]

Type locality: (in error)—"Mexico."

Range: Rio de Janeiro to Santa Catarina, Brazil. A single specimen, without other data, is known from Rio Grande do Sul.

Micrurus diastema diastema (Duméril, Bibron & Duméril)

 Elaps diastema Duméril 1854, Bibron & Duméril *(partim)* [Erp. Gen., Vol. 7, No. 2: 1222]

 Elaps epistema Duméril 1854, Bibron & Duméril [Erp. Gen., Vol. 7, No. 2: 1222]—type locality, Mexico.

 Elaps corallinus var. *crebipunctatus* Peters, W. 1869 [Mber. Dtsch. Akad. Wiss. Berlin, p. 877]—type locality, Matamoras, Puebla, Mexico.

 Micrurus affinis affinis; (*non* Jan) Schmidt 1933 [Zool. Ser. Field Mus. Nat. Hist. Vol. 20: 36]

 Micrurus diastema diastema; Roze 1967 [Amer. Mus. Novit., No. 2287: 14]

Type locality: Restricted (Schmidt, 1933)—Colima, Mexico. According to Rose *(l. c.)*, Schmidt's restriction was based on misinformation; consequently, the type locality for *M. d. diastema* should be designated as Mexico.

Range: Central Vera Cruz and eastern Puebla, Mexico.

Micrurus diastema affinis (Jan)
  Elaps affinis Jan 1858 [Rev. Mag. Zool., Vol. 10, No. 2: 525]
  Elaps fulvius; Boulenger 1986 *(partim)* [Cat. Sn. Brit. Mus., Vol. 3:
    422–424]
  Micrurus affinis affinis; Schmidt 1933 *(partim)* [Zool. Ser. Field Mus. Nat.
    Hist., Vol. 20: 36]
  Micrurus diastema affinis; Roze 1967 [Amer. Mus. Novit., No. 2287: 14]

Type locality: Mexico.
Range: Northern Oaxaca, Mexico.

Micurrus diastema aglaeope (Cope)
  Elaps aglaeope Cope, 1859 [Proc. Acad. Nat. Sci. Phil., p. 344]
  Elaps fulvius; Boulenger 1896 *(partim)* [Cat. Sn. Brit. Mus. Vol. 3: 422–424]
  Micrurus affinis aglaeope; Schmidt 1936 [Zool. Ser. Field Mus. Nat. Hist.
    Vol. 20: 214]
  Micrurus diastema aglaeope; Roze 1967 [Amer. Mus. Novit., No. 2287: 15]

Type locality: Honduras.
Range: Mountains of northwestern Honduras.

Micrurus diastema alienus (Werner)
  Elaps alienus Werner 1903 [Zool. Anz., Vol. 26: 240]
  Micrurus affinis mavensis Schmidt 1933 [Zool. Ser. Field Mus. Nat. Hist.,
    Vol. 20: 37]—type locality: Chichen Itza, Yucatan, Mexico.
  Micrurus diastema alienus; Roze 1967 [Amer. Mus. Novit., No. 2287: 15]

Type locality: Restricted (Roze, *l. c.*)—Chichen Itza, Yucatan, Mexico.
Range: Yucatan Peninsula (State of Yucatan and Quintana Roo Territory),
  Mexico.

Micrurus diastema apiatus (Jan)
  Elaps apiatus Jan 1858 [Rev. Mag. Zool., Vol 10, No. 2: 522]
  Elaps fulvius; Boulenger 1896 *(partim)* [Cat. Sn. Brit. Mus., Vol. 3:
    422–424]
  Micrurus affinis apiatus Schmidt 1936 [Zool. Ser. Field Mus. Nat. Hist.,
    Vol. 20: 37]
  Micrurus diastema apiatus; Roze 1967 [Amer. Mus. Novit., No. 2287: 15]

Type locality: Vera Cruz, Mexico.
Range: Moderate elevations of the Caribbean slopes from Chiapas, Mexico,
  through central Guatemala, and probably Tabasco and southern
  Vera Cruz, also.

Micrurus diastema macdougalli Roze
MacDougall's Coral Snake
  Micrurus diastema macdougalli Roze 1967 [Amer. Mus. Novit., No. 2287: 15]

Type locality: El Modelo, Rio Chalchijapa and Rio del Corte, Oaxaca, Mexico.
Range: Known only in the type locality.

Micrurus diastema sapperi (Werner)
  Elaps fulvius var. *sapperi* Werner 1903 [Zool. Anz., Vol. 26: 350]
  Elaps guatemalensis Ahl 1927 [Zool. Anz., Vol. 70: 251]—type locality, Guatemala.
  Micrurus affinis stantoni Schmidt 1933 [Zool, Ser. Field Mus. Nat. Hist., Vol. 20: 36]—type locality, Belize, British Honduras.
  Micrurus affinis alienus; (*non* Werner) Schmidt, 1936 [Zool. Ser. Field Mus. Nat. Hist., Vol. 20: 212]
  Micrurus diastema sapperi; Roze 1967 [Amer. Mus. Novit., No. 2287: 17]

Type locality: Guatemala.
Range: Southern Quintana Roo Territory, Campeche, Tabasco to southern Vera Cruz, Mexico, and lowlands of northern Guatemala and British Honduras.

Micrurus dissoleucus dissoleucus (Cope)
  Elaps dissoleucus Cope 1859 [Proc. Acad. Nat. Sci. Phil., Vol. 11: 345]
  Elaps dissoleucus; Boulenger 1896 [Cat. Sn. Brit. Mus., Vol. 3: 422]
  Micrurus dissoleucus; Amaral 1925 [Proc. U.S. Nat. Mus., Vol. 67, No. 24: 18]
  Micrurus dissoleucus dissoleucus; Schmidt 1936 [Zool. Ser. Field Mus. Nat. Hist., Vol. 20: 202]

Type locality: Restricted (Roze, 1955)—Maracaibo, Venezuela.
Range: Northeast Colombia to northeast Venezuela.

Micrurus dissoleucus dunni Barbour
Dunns Coral Snake
  Micrurus dunni Barbour 1923 [Occ. Pap. Mus. Zool. Univ. Mich., No. 129: 15]
  Micrurus dissoleucus dunni; Schmidt 1936 [Zool. Ser. Field Mus. Nat. Hist., Vol. 20: 202]

Type locality: Ancon, Canal Zone, Panama.
Range: From the Canal Zone to eastern Panama.

Micrurus dissoleucus melanogenys (Cope)
  Elaps melanogenys Cope 1860 [Proc. Acad. Nat. Sci. Phil., Vol. 12: 72]
  Micrurus dissoleucus melanogenys; Schmidt 1936 [Zool. Ser. Field Mus. Nat. Hist., Vol. 20: 203]

Type locality: Restricted (Schmidt, 1955)—Santa Marta region in northeastern Colombia.
Range: Santa Marta region, Colombia.

Micrurus dissoleucus nigrirostris Schmidt
  Elaps gravenhorstii Jan 1852 [Rev. Mag. Zool., Vol. 10, No. 2: 523]—type locality (in error) "Brazil."
  Micrurus dissoleucus nigrirostris Schmidt 1955 [Fieldiana, Zool., Vol. 34, No. 34: 355]

Type locality: Barranquilla, Colombia.
Range: Lower Magdalena region, Colombia.

Micrurus distans distans (Kennicott)
  Elaps distans Kennicott 1869 [Proc. Acad. Nat. Sci. Phil., Vol. 12, 338]
  Elaps fulvius; Boulenger 1896 *(partim)* [Cat Sn. Brit. Mus., Vol. 3: 423]
  Micrurus diastema distans; Schmidt 1933 *(partim)* [Zool. Ser. Field Mus. Nat. Hist., Vol. 20: 39]
  Micrurus distans distans; Zweifel 1959 [Amer. Mus. Novit., No. 1953: 7]

Type locality: Batesogachic, Chihuahua, Mexico.
Range: Southwestern Chihuahua and southern Sonora southward through Sinaloa to northwestern Nayarit, Mexico.

Micrurus distans michoacanensis (Dugès)
Michoacan Coral Snake
  Elaps distans var. *michoacanensis* Dugès 1891 [Naturaleza, Vol. 1, No. 2: 487; Pl. 32]
  Elaps fulvius; Boulenger 1896 *(partim)* [Cat. Sn. Brit. Mus., Vol. 3: 422–426]
  Micrurus diastema michoacanensis; Schmidt & Smith 1943 [Zool. Ser. Field Mus. Nat. Hist., Vol. 29, No. 2: 28–29]
  Micrurus distans michoacanensis; Zweifel 1959 [Amer. Mus. Novit., No. 1953: 9]

Type locality: Michoacán, Mexico.
Range: The Balsas River basin, Michoacán and Guerrero, Mexico.

Micrurus distans oliveri Roze
Oliver's Coral Snake

Micrurus diastema diastema; (*non* Duméril, Bibron & Duméril) Schmidt & Smith 1943 [Zool. Ser. Field. Mus. Nat. Hist., Vol. 29, No. 2: 28]
Micrurus distans oliveri Roze 1967 [Amer. Mus. Novit., No. 2287: 18; Fig. 6]

Type locality: Periquillo, Colima, Mexico.
Range: Colima, Mexico.

Micrurus distans zweifeli Roze
Zweifel's Coral Snake
  Micrurus diastema distans; (*non* Kennicott) Schmidt & Smith 1943 [Zool. Ser. Field Mus. Nat. Hist., Vol. 29, No. 2: 28]
  Micrurus diastema proximans Smith & Chrapliwy 1958 *(partim)* [Herpetologica, Vol. 13, No. 4: 270]
  Micrurus distans zweifeli Roze 1967 [Amer. Mus. Novit., No. 2287: 21; Fig. 7]

Type locality: ——— Santa Maria, Nayarit, Mexico.
Range: Nayarit and Jalisco, Mexico; intergrades with *M. d. distans* in central Nayarit.

Micrurus elegans elegans (Jan)
Elegant Coral Snake
  Elaps elegans Jan 1858 [Rev. Mag. Zool., Vol. 10, No. 2: 524]
  Elaps elegans; Boulenger 1896 [Cat. Sn. Brit. Mus., Vol. 3: 418]
  Micrurus elegans Amaral 1930 [Mem. Inst. Butantan, Vol. 4: 229]
  Micrurus elegans elegans; Schmidt 1933 [Zool. Ser. Field Mus. Nat. Hist., Vol. 20: 32]

Type locality: Mexico.
Range: Central Vera Cruz and eastern Oaxaca to western Tabasco, Mexico.

Micrurus elegans veraepacis Schmidt
  Micrurus elegans veraepacis Schmidt 1933 [Zool. Ser. Field Mus. Nat. Hist., Vol. 20: 32]

Type locality: Campur, Alta Verapaz, Guatemala.
Range: Southern Tabasco and Chiapas, Mexico, to Alta Verapaz, Guatemala.

Micrurus ephippifer (Cope)
Elaps ephippifer Cope 1886 [Proc. Am. Phil. Soc., Vol. 23: 281]
  Elaps fulvius; Boulenger 1896 *(partim)* [Cat. Sn. Brit. Mus., Vol. 3: 423]
  Micrurus ephippifer; Schmidt 1933 [Zool. Ser. Field Mus. Nat. Hist., Vol. 20: 38]

Type locality: Pacific side of the Isthmus of Tehuantepec, Oaxaca, Mexico.
Range: Sierra Madre, del Sur, Oaxaca, to the Isthmus of Tehuantepec, Oaxaca, Mexico.

Micrurus filiformis filiformis (Günther)
Thread-form Coral Snake
  Elaps filiformis Günther 1859 [Proc. Zool. Soc. London, p. 86, Pl. 183]
  Elaps filiformis; Boulenger 1896 [Cat. Sn. Brit. Mus., Vol. 3: 430]
  Micrurus filiformis; Amaral 1925 [Proc. U.S. Nat. Mus., Vol. 67, No. 24: 19]
  Micrurus filiformis filiformis; Roze 1967 [Amer. Mus. Novit., No. 2287: 22]

Type locality: Pará, Brazil.
Range: Northern Amazon region, Brazil, southern Colombia to northern Peru.

Micrurus filiformis subtilis Roze
Slender Coral Snake
  Micrurus filiformis subtilis Roze 1967 [Amer. Mus. Novit., No. 2287: 22; Fig. 8]

Type locality: Carurú, Rio Uaupés, Colombia-Brazil boundary.
Range: Colombia-Brazil border region.

Micrurus fitzingeri (Jan)
Fitzinger's Coral Snake
  Elaps fitzingeri Jan 1858 [Rev. Mag. Zool., Vol. 10, No. 2: 521]
  Elaps fulvius; Boulenger 1896 *(partim)* [Cat. Sn. Brit. Mus., Vol. 3: 422]
  Micrurus fitzingeri; Schmidt 1933 [Zool. Ser. Field Mus. Nat. Hist., Vol. 20: 38]
  Micrurus fitzingeri fitzingeri; Brown & Smith 1942 [Proc. Biol. Soc. Wash., Vol. 55: 63]
  Micrurus fitzingeri; Roze 1967 [Amer. Mus. Novit., No. 2287: 24]

Type locality: Mexico.
Range: Distrito Federal and Morelos, Mexico.

Micrurus frontalis frontalis (Duméril, Bibron & Duméril)
  Elaps frontalis Duméril 1854 Bibron & Duméril [Erp. Gen., Vol. 7, No. 2: 1223]
  Elaps baliocoryphus Cope 1859 [Proc. Acad. Nat. Sci. Phil., Vol. 11: 346]
    —type locality, Buenos Aires.
  Elaps frontalis; Boulenger 1896 *(partim)* [Cat. Sn. Brit. Mus., Vol. 3: 427]
  Micrurus frontalis; Amaral 1925 [Proc. U.S. Nat. Mus., Vol. 67, No. 24: 19]

Micrurus frontalis frontalis; Schmidt 1936 [Zool. Ser. Field Mus. Nat. Hist., Vol. 20: 199]

Micrurus lemniscatus frontalis; Amaral 1944 *(partim)* [Pap. Avul. Depto. Zool. São Paulo, Vol. 5, No. 11: 92]

Micrurus frontalis frontalis; Roze 1967 [Amer. Mus. Novit., No. 2287: 24]

Type locality: Corrientes and Missiones, Argentina.
Range: Southern Brazil, southern Paraguay, and adjacent Argentina.

Micrurus frontalis altirostris (Cope)
  Elaps altirostris Cope 1859 [Proc. Acad. Nat. Sci. Phil., Vol. 11: 345]
  Elaps heterochilus Mocquard 1887 *(partim)* [Bull, Soc. Philom., Ser. 7, Vol. 11: 39]—type locality, Brazil.
  Elaps frontalis; Boulenger 1896 *(partim)* [Cat. Sn. Brit. Mus., Vol. 3: 427]
  Micrurus frontalis altirostris; Schmidt 1936 [Zool. Ser. Field Mus. Nat. Hist., Vol. 20: 199]
  Micrurus lemniscatus multicinctus Amaral 1944 *(partim)* [Pap. Avul. Depto. Zool. São Paulo, Vol. 5, No. 11: 91]—type locality, Teixeira Soares, Parana, Brazil.
  Micrurus frontalis altirostris; Roze 1967 [Amer. Mus. Novit., No. 2287: 25]

Type locality: South America.
Range: Southern Brazil, Uruguay, and northeastern Argentina; intergrades with *M. f. frontalis* in Parana and Santa Catarina, Brazil. The type specimens of *M. heterochilus* and *M. l. multicinctus* are obviously intergrades.

Micrurus frontalis brasiliensis Roze
  Micrurus frontalis brasiliensis Roze 1967 [Amer. Mus. Novit., No. 2287: 25; Fig. 9]

Type locality: Barreiras, Bahia, Brazil.
Range: The states of Bahia, Minas Gerais, and Goias, Brazil.

Micrurus frontalis pyrrhocryptus (Cope)
  Elaps pyrrhocryptus Cope 1862 [Proc. Acad. Nat. Sci. Phil., Vol. 14: 347]
  Elaps simonsii Boulenger 1902 [Ann. Mag. Nat. Hist., Ser. 7, Vol. 9: 338]—type locality, Cruz del Eje, Cordoba, Argentina.
  Micrurus pyrrhocryptus; Schmidt 1936 [Zool. Ser. Field Mus. Nat. Hist., Vol. 20, No. 27: 199]
  Micrurus lemniscatus frontalis; Amaral 1944 *(partim)* [Pap. Avul. Depto. Zool. São Paulo, Vol. 5, No. 11: 92]
  Micrurus frontalis pyrrhocryptus; Shreve 1953 [Breviora, No. 16: 5]
  Micrurus tricolor Hoge 1956 [Mem. Inst. Butantan, Vol. 27: 67; Figs. 1–6]

Type locality: Garandazal, Mato Grosso, Brazil.

Micrurus pyrrhocryptus; Hoge & Lancini 1959 [Mem. Inst. Butantan, Vol. 29: 12]

Micrurus frontalis pyrrhocryptus; Roze 1967 [Amer. Mus. Novit., No. 2287: 26]

Type locality: Vermejo River, Argentine Choco.

Range: Southwestern Mato Grosso, Brazil, western and southwestern Bolivia, adjacent Paraguay, southward to Mendoza and Santa Fe, Argentina, east of the Andes.

Micrurus fulvius maculatus Roze
Spotted Coral Snake
  Micrurus fulvius maculatus Roze 1967 [Amer. Mus. Novit., No. 2287: 27; Fig. 10]

Type locality: Tampico, Tamaulipas, Mexico.

Range: Known only from the type locality; intergrades with *M. f. microgalbineus* north and west of the Tanpico region.

Micrurus fulvius microgalbineus Brown & Smith
  Micrurus fitzingeri microgalbineus Brown & Smith 1942 [Proc. Biol. Soc. Wash., Vol. 55: 63]
  Micrurus fitzingeri fitzingeri; (*non* Jan) Brown & Smith 1942 [Proc. Biol. Soc. Wash., Vol. 55: 64]
  Micrurus fulvius microgalbineus; Roze 1967 [Amer. Mus. Novit., No. 2287: 29]

Type locality: Seven kilometers south of Antigo Morelos, Tamaulipas, Mexico.

Range: Tamaulipas and San Luis Potosi, Mexico.

Micrurus fulvius tenere (Baird & Girard)
Texas Coral Snake
  Elaps tenere Baird & Girard 1853 [Cat. Sn. Am. Rept., Vol. 1: 22 and 156]
  Elaps tristis Baird & Girard 1853 [Cat. Sn. Am. Rept., Vol. 1: 23]—type locality, Kemper County, Mississippi; Rio Grande, west of San Antonio, Texas.
  Micrurus fulvius tenere; Schmidt 1933 [Zool. Ser. Field Mus. Nat. Hist., Vol. 20: 40]

Type locality: Restricted (Schmidt, 1953)—New Braunfels, Texas, U.S.A.

Range: West of the Mississippi River from Louisiana, Arkansas, and Texas, U.S.A., to northern Coahuila, Nuevo Leon, and Tamaulipas, Mexico.

Micrurus hemprichii hemprichii (Jan)
Hemprich's Coral Snake
  Elaps hemprichii Jan 1858 [Rev. Mag. Zool., Vol. 10, No. 2: 523]
  Elaps hemprichii; Boulenger 1896 [Cat. Sn. Brit. Mus., Vol. 3: 421]
  Micrurus hemprichii; Amaral 1925 [Proc. U.S. Nat. Mus., Vol. 67, No. 24:
      17]
  Micrurus hemprichii hemprichii; Schmidt 1953 [Fieldiana, Zool., Vol. 34,
      No. 30: 166; Fig. 31]

Type locality: *emendata* (Hoge & Lancini, 1962)—Venezuela.
Range: Eastern Colombia, the Guianas, and Venezuela.

Micrurus hemprichii ortoni Schmidt
Orton's Coral Snake
  Micrurus hemprichii ortoni Schmidt 1953 [Fieldiana, Zool., Vol. 34, No.
      30: 166]

Type locality: Pebas, Peru.
Range: Amazon Basin in Ecuador, Peru, and Bolivia. Two specimens are
    known from Pará, Brazil. This subspecies is not known from the
    central Amazon Basin.

Micrurus hippocrepis (W. Peters)
  Elaps hippocrepis Peters, W. 1862 [Mber. Dtsch. Akad. Wiss. Berlin, for
      1861, 925]
  Micrurus affinis hippocrepis; Schmidt 1936 *(partim)* [Zool. Ser. Field Mus.
      Nat. Hist., Vol. 20: 214]
  Micrurus hippocrepis; Roze 1967 [Amer. Mus. Novit., No. 2287: 29]

Type locality: Santo Tomás (—Puerto Matias de Galvez), Guatemala.
Range: Stann Creek region, eastern British Honduras, south to Puerto
    Matias Galvez, Guatemala.

Micrurus ibiboboca (Merrem)
Ibiboboca
  Elaps ibiboboca Merrem 1820 [Tent. Syst. Amph., p. 142]
  Elaps marcgravii Wied 1820 [Nova Act Acad. Leop. Carol., Vol. 10: 109]
      —type locality, Brazil.
  Elaps marcgravii Boulenger 1896 [Cat. Sn. Brit. Mus., Vol. 3: 428]
  Micrurus ibiboboca; Amaral 1925 [Rev. Mus. Paul., Vol. 15: 7]

Type locality: Brazil.
Range: Northeastern Brazil.

Micrurus isozonus (Cope)
  Elaps isozonus Cope 1860 [Proc. Acad. Nat. Sci. Phil., Vol. 12: 73]

Elaps omissus Boulenger 1920 [Ann. Mag. Nat. Hist., Ser. 9, Vol. 6: 109]
—type locality, Venezuela.
Micrurus isozonus; Schmidt 1936 [Zool. Ser. Field Mus. Nat. Hist., Vol.
20: 198]

Type locality: Restricted (Roze, 1955)—Caracas, Venezuela.
Range: Northern Venezuela.

Micrurus langsdorffi langsdorffi Wagler
Micrurus langsdorffi Wagler 1824 [*In* Spix, Serp. Brazil., p. 10; Pl. 2]
Elaps inperator Cope 1868 [Proc. Acad. Nat. Sci. Phil., Vol. 20: 110]—
type locality, Napo and Marañon, Peru.
Elaps batesi Günther 1868 [Ann. Mag. Nat. Hist., Ser. 4, Vol. 1: 428;
Pl. 17-D]—type locality, Pebas, northern Peru.
Elaps langsdorffi; Boulenger 1896 [Cat. Sn. Brit. Mus., Vol. 3: 416]
Micrurus minosus; Amaral 1935 [Mem. Inst. Butantan, Vol. 9: 221; Fig.
6]—type locality, Rio Putumayo, Colombia.
Micrurus langsdorffi; Schmidt 1936 *(partim)* [Zool. Ser. Field Mus. Nat.
Hist., Vol. 20: 191]
Micrurus ornatissimus; *(non* Jan) Schmidt 1955 [Fieldiana, Zool., Vol. 34,
No. 34: 345]
Micrurus langsdorffi; Peters, J. 1960 *(partim)* [Bull. Mus. Comp. Zool.
Harv., Vol. 122, No. 9: 531]
Micrurus langsdorffi langsdorffi; Roze 1967 [Amer. Mus. Novit., No. 2287:
30]

Type locality: Rio Japura, Amazonas, Brazil.
Range: Upper Amazon region of southern Colombia, northwestern Brazil,
northern Peru, and adjacent Ecuador.

Micrurus langsdorffi ornatissimus (Jan)
Elaps ornatissimus Jan 1858 [Rev. Mag. Zool., Ser. 2, Vol. 10: 521]
Elaps buckleyi Boulenger 1896 [Cat. Sn. Brit. Mus., Vol. 3: 416; Pl. 22,
Fig. 1]—type locality, Canelos, Ecuador, and Para, Brazil.
Micrurus ornatissimus; Schmidt 1936 [Zool. Ser. Field Mus. Nat. Hist.,
Vol. 20, No. 19: 191]
Micrurus ornatissimus; Peters, J. 1960 [Bull. Mus. Comp. Zool. Harv.,
Vol. 122, No. 9: 532]
Micrurus langsdorffi ornatissimus; Roze 1967 [Amer. Mus. Novit., No.
2287: 30]

Type locality: Mexico (in error).
Range: Amazonian slopes of the Andes in eastern Ecuador and northern
Peru.

Micrurus laticollaris laticollaris (W. Peters)
  Elaps marcgravi var. *laticollaris*; Peters, W. 1870 [Mber. Dtsch. Akad. Wiss. Berlin, for 1868, p. 877]
  Micrurus laticollaris; Schmidt 1933 [Zool. Ser. Field Mus. Nat. Hist., Vol. 20: 39]

Type locality: restricted (Smith & Taylor, 1950)—Izucar de Matamoros, Puebla, Mexico.
Range: Balsas River Basin in Michoacan, Guerrero, Puebla, and Morelos, Mexico.

Micrurus laticollaris maculirostris Roze
  Micrurus laticollaris maculirostris Roze 1967 [Amer. Mus. Novit., No. 2287: 31]

Type locality: Vicinity of Colima, Colima, Mexico.
Range: Colima and Jalisco, Mexico.

Micrurus latifasciatus Schmidt
  Micrurus latifasciatus Schmidt 1933 [Zool. Ser. Field Mus. Nat. Hist., Vol. 20: 35]

Type locality: Finca El Cipres, Volcan Zunil, Suchitepequez, Guatemala.
Range: Moderate elevations of the Pacific slope of southern Chiapas, Mexico, south to western Guatemala.

Micrurus lemniscatus lemniscatus (Linnaeus)
  Coluber lemniscatus Linnaeus 1758 [Syst, Nat., 10th ed., Vol. 1: 224]
  Elaps lemniscatus; Boulenger 1896 (partim) [Cat. Sn. Brit. Mus., Vol. 3: 430]
  Micrurus lemniscatus; Beebe 1919 [Zoologica, Vol. 2: 216]
  Micrurus lemniscatus lemniscatus; Amaral 1944 [Pap. Avul. Depto. Zool. São Paulo, Vol. 5, No. 11: 89]

Type locality: Designated (Schmidt & Walker, 1143)—Belem, Para, Brazil. Roze (1967, p. 32) questions the validity of this selection.
Range: The Guayanas and Amapa in Brazil.

Micrurus lemniscatus carvalhoi Roze
  Micrurus lemniscatus carvalhoi Roze 1967 [Amer. Mus. Novit., No. 2287: 33; Fig. 11]

Type locality: Catanduva, São Paulo, Brazil.
Range: Parana, São Paulo, Minas Gerais, Mato Grosso, Goias, Pernambuco, Bahia, and Rio Grande do Norte, Brazil.

Micrurus lemniscatus diutius Burger
  Micrurus lemniscatus diutius Burger 1955 [Bol. Mus. Cien. Nat. Caracas, Vol. 1, No. 1: 8]

Type locality: Tunapuna, Trinidad.

Range: Tunapuna, Trinidad, adjacent coast of Venezuela and Guyana, and central parts of Surinam and French Guiana.

Micrurus lemniscatus frontifasciatus (Werner)
   Elaps frontifasciatus; Werner 1927 [Sitzber. Akad. Wiss. Vienna, Abt, 1, Vol. 135: 250]
   Micrurus lemniscatus frontifasciatus; Roze 1967 [Amer. Mus. Novit., No. 2287: 34]

Type locality: Bolivia.

Range: Eastern slopes of the Bolivian Andes.

Micrurus lemniscatus helleri Schmidt & Schmidt
Heller's Coral Snake
   Micrurus helleri Schmidt & Schmidt 1925 [Zool. Ser. Field Mus. Nat. Hist., Vol. 12: 129]
   Micrurus lemniscatus helleri; Roze 1967 [Amer. Mus. Novit., No. 2287: 35]

Type locality: Pozuzo, Huanuco, Peru.

Range: Amazon region to the foothills of the Andes from northern Brazil, southern Venezuela, and Colombia, to Ecuador, Peru, and Bolivia.

Micrurus limbatus Fraser
   Micrurus limbatus Fraser 1964 [Copeia, No. 3: 570; Fig. 1]

Type locality: Southern slope of Volcan San Martin, 7 airline miles north of San Andrés Tuxtla, Vera Cruz, Mexico.

Range: Tuxtla region of southern Vera Cruz, Mexico.

Micrurus margaritiferus Roze
   Micrurus margaritiferus Roze 1967 [Amer. Mus. Novit., No. 2287: 35; Fig. 12]

Type locality: Boca Rio Santiago-Rio Marañon. Peru.

Range: Known only from the type locality.

Micrurus mertensi Schmidt
Merten's Coral Snake
   Micrurus mertensi Schmidt 1936 [Zool. Ser. Field Mus. Nat. Hist., Vol. 20: 192]

Type locality: Pascamayo, Peru.

Range: Coastal region of northwestern and central Peru to southeastern Ecuador.

Micrurus mipartitus mipartitus (Duméril, Bibron & Duméril)

Elaps mipartitus Duméril, Bibron & Duméril 1854 [Erp. Gen., Vol. 7: 1220]
Elaps mipartitus; Boulenger 1896 [Cat. Sn. Brit. Mus., Vol. 3: 431]
Elaps aequicinctus Werner 1903 [Zool. Anz., Vol. 26, No. 693: 243]—type locality, unknown; supposedly Venezuela or Ecuador.
Micrurus mipartitus mipartitus; Schmidt 1955 [Fieldiana, Zool., Vol. 34, No. 34: 341]

Type locality: Rio Sucio or Senio (=Sinu?), Colombia.
Range: Darien region of eastern Panama to the Pacific side of Colombia, west of the Andes.

Micrurus mipartitus anomalus (Boulenger)
Elaps anomalus Boulenger 1896 [Cat. Sn. Brit. Mus., Vol. 3: 417; Pl. 22, Fig. 2]
Micrurus mipartitus; (*non* Duméril, Bibron & Duméril) Ruthven 1922 [Misc. Publ. Mus. Zool. Univ. Mich., No. 68]
Micrurus mipartitus anomalus; Roze 1967 [Amer. Mus. Novit., No. 2287: 37]

Type locality: Colombia.
Range: Santa Marta Mountains and Cordillera Oriental, east of Magdalena River, Colombia, and the Andes in western Venezuela.

Micrurus mipartitus hertwigi (Werner)
Elaps hertwigi Werner 1897 [Sitzber. Akad. Wiss. Munich, Vol. 27: 354]
Micrurus mipartitus multifasciatus; (*non* Jan) Taylor 1956 [Kansas Univ. Sci. Bull., Vol. 34, No. 1: 158]
Micrurus mipartitus hertwigi; Roze 1967 [Amer. Mus. Novit., No. 2287: 37]

Type locality: Central America.
Range: Atlantic slopes of Nicaragua, Costa Rica to northwestern Panama.

Micrurus mipartitus multifasciatus (Jan)
Elaps multifasciatus Jan 1858 [Rev. Mag. Zool., Vol. 10, No. 2: 521]
Elaps mipartitus; Boulenger 1896 (*partim*) [Cat. Sn. Brit. Mus., No. 3: 431]
Micrurus mipartitus multifasciatus; Roze 1955 (*partim*) [Acta Biol. Ven., Vol. 1, No. 17: 467]
Micrurus mipartitus multifasciatus; Roze 1967 [Amer. Mus. Novit., No. 2287: 37]

Type locality: Central America.
Range: Central Panama including the Canal Zone.

Micrurus mipartitus semipartitus (Jan)

Elaps semipartitus Jan 1858 (*ex errore pro mipartitus Duméril, Bibron &
Duméril*) [Rev. Mag. Zool., Vol. 10, No. 2: 516]
Micrurus mipartitus semipartitus; Roze 1955 [Acta Biol. Ven., Vol. 1, No.
17: 465

Type locality: Cayenne (in error); *emendata* (Roze, 1955)—Caracas, Venezuela.
Range: Cordillera de la Costa, Venezuela.

Micrurus nigrocinctus nigrocinctus (Girard)
Elaps nigrocinctus Girard 1854 [Proc. Acad. Nat. Sci. Phil., Vol. 7: 226]
Elaps fulvius; Boulenger 1896 (*partim*) [Cat. Sn. Brit. Mus., Vol. 3: 422]
Micrurus nigrocinctus; Schmidt 1928 (*partim*) [Bull. Antivenin Inst. Amer.,
Vol. 2: 64]
Micrurus nigrocinctus nigrocinctus; Schmidt 1933 (*partim*) [Zool. Ser. Field
Mus. Nat. Hist., Vol. 20: 33]

Type locality: Taboga Island, Bay of Panama.
Range: Pacific coast of southeastern Costa Rica, Panama, including the
Canal Zone, to adjacent Colombia.

Micrurus nigrocinctus babaspul Roze
Babaspul
Micrurus nigrocinctus babaspul Roze 1967 [Amer. Mus. Novit., No. 2287:
38; Fig. 13]

Type locality: Little Hill, Great Corn Island (Isla del Maiz Grande), about
55 km east-northeast of Bluefields, Nicaragua.
Range: Great Corn Island, Nicaragua.

Micrurus nigrocinctus coibensis Schmidt
Coiban Coral Snake
Micrurus nigrocinctus coibensis; Schmidt 1936 [Zool. Ser. Field Mus. Nat.
Hist., Vol. 20: 208]

Type locality: Coiba Island, Panama.
Range: Known only from Colbay Islands.

Micrurus nigrocinctus divaricatus (Hallowell)
Elaps divaricatus Hallowell 1855 [Journ. Acad. Nat. Sci. Phil., Vol. 3,
No. 2: 36]
Elaps fulvius; Boulenger 1896 (*partim*) [Cat. Sn. Brit. Mus., Vol. 3: 422–423]
Micrurus nigrocinctus divaricatus; Schmidt 1933 [Zool. Ser. Field Mus. Nat.
Hist., Vol. 20: 33]

Type locality: Honduras.
Range: Northern and central Honduras.

Micrurus nigrocinctus melanocephalus (Hallowell)
  Elaps melanocephalus Hallowell 1860 [Proc. Acad. Nat. Sci. Phil., Vol. 12: 266]
  Micrurus pachecoi Taylor 1951 [Kansas Univ. Sci. Bull., Vol. 34: 165]— type locality: Guanacaste, Costa Rica.
  Micrurus nigrocinctus melanocephalus; Roze 1967 [Amer. Mus. Novit., No. 2287: 39]

Type locality: Ometepec, Nicaragua.
Range: Pacific slopes of Nicaragua and southwestern Costa Rica.

Micrurus nigrocinctus mosquitensis Schmidt
  Micrurus nigrocinctus mosquitensis Schmidt 1933 [Zool. Ser. Field Mus. Nat. Hist., Vol. 20: 33]

Type locality: Limón, Costa Rica.
Range: Caribbean slopes of eastern and southern Nicaragua through Costa Rica and northwestern Panama.

Micrurus nigrocinctus zunilensis Schmidt
  Micrurus nigrocinctus zunilensis Schmidt 1932 [Proc. Calif. Acad. Sci., Vol. 20, No. 4: 266]
  Micrurus nigrocinctus wagneri; Mertens 1941 [Senckenbergiana, Vol. 23: 216]
    —type locality, Finca Germania, Sierra Madre, Chiapas, Mexico; altitude, 400–1,300 m.
  Micrurus nigrocinctus ovoandensis; Schmidt & Smith 1943 [Zool. Ser. Field Mus. Nat. Hist., Vol. 29: 26]—type locality, Salto de Agua (Monte Ovando), Chiapas, Mexico.

Type locality: Finco El Ciprés, Volcan Zunil, Suchitepeques, Guatemala.
Range: Pacific slopes from Chiapas, Mexico, Guatemala, El Salvador, and southern Honduras.

Micrurus nuchalis Schmidt
  Micrurus nuchalis Schmidt 1933 [Zool. Ser. Field Mus. Nat. Hist., Vol. 20: 35]
  Micrurus nuchalis nuchalis; Schmidt & Smith 1943 [Zool. Ser. Field Mus. Nat. Hist., Vol. 29: 30]
  Micrurus nuchalis; Roze 1967 [Amer. Mus. Novit., No. 2287: 40]

Type locality: Tapanatepec, Oaxaca, Mexico.
Range: Pacific side of the Isthmus of Tehuantepec, Mexico.

Micrurus peruvianus Schmidt
Peruvian Coral Snake

Micrurus peruvianus Schmidt 1936 [Zool. Ser. Field Mus. Nat. Hist., Vol. 20: 193]

Type locality: Perico, Cajamarca, Peru.
Range: Northwestern Andes of Peru.

Micrurus proximans (Smith & Chrapliwy)
    Micrurus diastema proximans Smith & Chrapliwy 1958 [Herpetologica, Vol. 13: 270]
    Micrurus nigrocinctus browni; Zweifel 1959 [Amer. Mus. Novit., No. 1953: 7]
    Micrurus proximans; Roze 1967 [Amer. Mus. Novit., No. 2287: 40]

Type locality: Four miles north of San Blas, Nayarit, Mexico.
Range: Nayarit, Mexico.

Micrurus psyches psyches (Daudin)
    Vipera psyches Daudin 1803 [Hist. Nat. Rept., Vol. 8: 320; Pl. 100: 1]
    Elaps psyches; Boulenger 1896 [Cat. Sn. Brit. Mus., Vol. 3: 426]
    Micrurus psyches; Beebe 1919 [Zoologica, Vol. 2: 216]
    Micrurus psyches psyches; Roze 1967 [Amer. Mus. Novit., No. 2287: 40]

Type locality: Surinam.
Range: Surinam, Guyana, and French Guiana, southern Venezuela, and extreme southern Colombia.

Micrurus psyches circinalis (Duméril, Bibron & Duméril)
Circled Coral Snake
    Elaps circinalis Duméril Bibron & Duméril 1854 [Erp. Gen., Vol. 7, No. 2: 1210]
    Elaps riisei; Jan 1858 [Rev. Mag. Zool., Ser. 2, Vol. 10: 525]—type locality, Ile Saint Thomas, Antilles.
    Elaps corallinus; Boulenger 1896 (*partim*) [Cat. Sn. Brit. Mus., Vol. 3: 420]
    Micrurus psyches riisei; Amaral 1931 [Bull. Antivenin Inst. Amer., Vol. 4: 89]
    Micrurus circinalis; Schmidt 1936 [Zool. Ser. Field Mus. Nat. Hist., Vol. 20: 192]
    Micrurus psyches circinalis; Roze 1967 [Amer. Mus. Novit., No. 2287: 40]

Type locality: (in error)—"Martinique" (=Trinidad).
Range: Trinidad Island and adjacent mainland of Venezuela.

Micrurus psyches medemi Roze
Medem's Coral Snake
    Micrurus psyches medemi Roze 1967 [Amer. Mus. Novit., No. 2287: 41; Fig. 14]

Type locality: Villavicencio, Meta, Colombia.
Range: Known only from the type locality.

Micrurus putomayensis Lancini
Putomayan Coral Snake
    Micrurus schmidti Lancini 1962 (previously occupied by *Micrurus schmidti* Dunn, 1940) [Publ. Ocas. Mus. Cien. Nat., Caracas, Zool., Vol. 2: 1; Fig. 1]
    Micrurus putomayensis; Lancini 1962 (*nomen novum pro Micrurus schmidti Lancini, 1962*) [Publ. Ocas. Mus. Cien. Nat., Caracas, Zool., Vol. 3: 1]

Type locality: Puerto Socorro, Putomayo River, Departamento de Loreto, Peru.
Range: Known only from the type locality.

Micrurus ruatanus (Günther)
Ruatan Coral Snake
    Elaps ruatanus Günther 1895 [Biol. Cent. Amer. Rept. Batr., p. 185; Pl. 57b]
    Elaps fulvius; Boulenger 1896 (*partim*) [Cat. Sn. Brit. Mus., Vol. 3: 422]
    Micrurus ruatanus; Schmidt 1933 [Zool. Ser. Field Mus. Nat. Hist., Vol. 20: 34]

Type locality: Ruatan (=Roatan) Island, Honduras.
Range: Roatan Island and adjacent mainland, Honduras.

Micrurus spixii spixii Wagler
Giant Coral Snake
    Micrurus spixii Wagler 1824 [*In* Spix, Serp. Brazil., p. 48; Pl. 18]
    Elaps spixii; Boulenger 1896 (*partim*) [Cat. Sn. Brit. Mus., Vol. 3: 427]
    Elaps ehrhardti Müller 1926 [Zool. Anz., Vols. 7/8: 198]—type locality, Manacapuru on the Solimoes, Amazonas, Brazil.
    Micrurus spixii spixii; Schmidt 1953 [Fieldiana, Zool., Vol. 34, No. 14: 175]

Type locality: Solimoes River, Brazil.
Range: Middle Amazon region, Amazonas, Brazil.

Micrurus spixii martiusi Schmidt
Martius Coral Snake
    Micrurus spixii martiusi Schmidt 1953 [Fieldiana, Zool., Vol. 34, No. 14: 175; Figs. 33 and 34b]

Type locality: Santarem, Pará, Brazil.
Range: Lower Amazon region of Para and northeastern Mato Grosso, Brazil.

Micrurus spixii obscurus (Jan)
  Elaps corallinus var. *obscura* Jan 1872 [*In* Jan & Sordelli, Icon. Gen.
    Ophid., Vol. 3; Pl. 6: 3]
  Elaps heterozonus Peters, W. 1881 [Sitzber. Ges. Naturf. Freunde Berlin,
    p. 52]—type locality, Sarayacu, Ecuador.
  Elaps heterozonus; Boulenger 1896 [Cat. Sn. Brit. Mus., Vol. 3: 417]
  Micrurus spixii obscura; Schmidt & Walker 1943 [Zool. Ser. Field Mus.
    Nat. Hist., Vol. 24, No. 26: 294]

Type locality: (in error)—"Lima," designated (Schmidt, 1953)—Iquitos, Peru.
Range: Along Andean tributaries of Amazon from Colombia to southern
  Peru and east to Amazonas, Venezuela.

Micrurus spixii princeps (Boulenger)
  Elaps princeps Boulenger 1905 [Ann. Mag. Nat. Hist., Vol. 15, No. 7: 456]
  Micrurus spixii princeps; Schmidt 1953 [Fieldiana, Zool., Vol. 34, No. 14:
    175; Fig. 34d]

Type locality: Province Sara, Santa Cruz de la Sierra, Bolivia.
Range: Central and northwestern Bolivia.

Micrurus spurrelli (Boulenger)
  Elaps spurrelli Boulenger 1914 [Proc. Zool. Soc. London, p. 817]
  Micrurus nicefori; Schmidt 1955 [Fieldiana, Zool., Vol. 34, No. 34: 346;
    Pl. 65]—type locality, Villavicencio, (Cundinamarca), Colombia.

Type locality: Penna Lisa, Condoto River, Colombia.
Range: Western and central Colombia.

Micrurus steindachneri steindachneri (Werner)
Steindachner's Coral Snake
  Elaps steindachneri Werner 1901 [Verh. Zool. Bot. Gesl. Vienna, Vol. 51:
    599]
  Elaps fasslii Werner 1926 [Sitzber. Akad. Wiss. Vienna, Abt. 1, Vol. 135:
    249]—type locality, Colombia.
  Micrurus langsdorffii; Schmidt 1936 (*partim*) [Zool. Ser. Field Mus. Nat.
    Hist., Vol. 20: 191]
  Micrurus steindachneri steindachneri; Roze 1967 [Amer. Mus. Novit., No.
    2287: 43]

Type locality: Ecuador.
Range: Eastern slopes of the Andes in the Macas-Mendez region, Ecuador.

Micrurus steindachneri orcesi Roze
Orces' Coral Snake
  Micrurus steindachneri orcesi Roze 1967 [Amer. Mus. Novit., No. 2287:
    43; Fig. 15]

Type locality: Meta Trail, Baños, Ecuador; altitude, 1,200 m.
Range: Higher elevations of upper Rio Pastaza valley, from 1,000 to 1,800 m.

Micrurus steindachneri petersi Roze
Peters' Coral Snake
  Micrurus steindachneri petersi Roze 1967 [Amer. Mus. Novit., No. 2287: 45; Fig. 16]

Type locality: One mile south of Plan de Milagro on the trail to Pan de Azucar, Morona-Santiago Province, Ecuador.
Range: Known only from the type locality.

Micrurus stewarti Barbour & Amaral
Stewart's Coral Snake
  Micrurus stewarti Barbour & Amaral 1928 [Bull. Antivenin Inst. Amer., Vol. 1: 100]
  Micrurus schmidt; Dunn 1940 [Proc. Acad. Nat. Sci. Phil., Vol. 92: 119; Pl. 2]—type locality, Valle de Anton, 50 miles west of the Canal Zone, Panama.
  Micrurus stewarti stewarti; Smith 1958 [Herpetologica, Vol. 14: 224]
  Micrurus stewarti; Roze 1967 [Amer. Mus. Novit., No. 2287: 47]

Type locality: Sierra de la Bruja, Panama.
Range: Intermediate elevations east and west of the Canal Zone, Panama.

Micrurus stuarti Roze
Stuart's Coral Snake
  Micrurus wagneri; (*non* Mertens) Stuart 1963 [Misc. Publ. Mus. Zool. Univ. Mich., No. 122: 127]
  Micrurus stuarti Roze 1967 [Amer. Mus. Novit., No. 2287: 47]

Type locality: Finca La Paz, San Marcos, Guatemala; altitude, 1,345 m.
Range: Known from San Marcos and Suchitepequez, Guatemala.

Micrurus surinamensis surinamensis (Cuvier)
Surinam Coral Snake
  Elaps surinamensis Cuvier 1817 [Regne Anim., 1st ed., Vol. 2: 84]
  Elaps surinamensis; Boulenger 1896 [Cat. Sn. Brit. Mus., Vol. 3: 414]
  Micrurus surinamensis; Beebe 1919 [Zoologica, Vol. 2: 216]
  Micrurus surinamensis surinamensis; Schmidt 1952 [Fieldiana, Zool., Vol. 34, No. 4: 29]

Type locality: Surinam.
Range: The Guianas, the Amazon basins of Brazil, Ecuador, Colombia, and Bolivia, and the Oriental region of Venezuela.

Micrurus surinamensis nattereri Schmidt
Natterer's Coral Snake

Micrurus surinamensis nattereri Schmidt 1952 [Fieldiana, Zool., Vol. 34, No. 4: 27]

Type locality: Between Guaramoca and San Fernando, Venezuela; (corrected by Hoge & Lancini, 1962)—between Guaramaco and San Fernando de Atabapo, upper Orinoco, Ter. Fed. Amazonas and State Bolivar, Venezuela.

Range: Upper Orinoco, Venezuela, and upper Rio Negro, Brazil.

Micrurus tschudii tschudii (Jan)
Tschudi's Coral Snake
  Elaps tschudii Jan 1858 [Rev. Mag. Zool., Vol. 10, No. 2: 524]
  Elaps tschudii; Boulenger 1896 [Cat. Sn. Brit. Mus., Vol. 3: 422]
  Micrurus tschudii; Schmidt & Schmidt 1925 [Zool. Ser. Field Mus. Nat. Hist., Vol. 12, No. 10: 132; Pl. 12]
  Micrurus tschudii tschudii; Schmidt 1936 [Zool. Ser. Field Mus. Nat. Hist., Vol. 20: 202]

Type locality: Peru.
Range: Pacific coast of Peru from Departamento de Libertad probably to northwestern Bolivia.

Micrurus tschudii olssoni Schmidt & Schmidt
Olsson's Coral Snake
  Micrurus olssoni Schmidt & Schmidt 1925 [Zool. Ser. Field Mus. Nat. Hist., Vol. 12, No. 10: 130; Pl. 11]
  Micrurus tschudii olssoni; Schmidt 1936 [Zool. Ser. Field Mus. Nat. Hist., Vol. 20: 202]

Type locality: Negritos, Piura, Peru.
Range: Desert coastal region of northwestern Peru and southwestern Ecuador Intergrades with *M. t. tschudii* in Cajamarca, Peru.

B. Family Hydrophidae

PELAMIS Daudin

Sea Snakes
Pelamis platurus (Linnaeus)
  Anguis platura Linnaeus 1766 [Syst. Nat. 12th ed., Vol. 1: 301]
  Pelamis platurus; Gray 1825 [Ann. Phil. London Nem Ser., Vol. 10: 207]
  Hydrus platurus; Boulenger 1896 [Cat. Sn. Brit. Mus., Vol. 3: 267; pl. 19]
  Pelamis platurus; Klemmer 1966 [Rez. Giftschl. d. Erde, p. 359]

Type locality: Not given.

Range: Tropical regions of Indian and Pacific Ocean, westwards till the east coast of Africa, eastwards to the west coast of America, from California to Peruvian coasts.

C. Family Viperidae

Subfamily *Crotalinae*

AGKISTRODON Beauvois

Agkistrodon bilineatus bilineatus Günther
  Ancistrodon bilineatus Günther 1863 [Ann. Mag. Nat. Hist., Vol. 12, No. 3: 364]
  Ancistrodon bilineatus; Boulenger 1896 [Cat. Sn. Brit. Mus., Vol. 3: 521]
  Agkistrodon bilineatus bilineatus; Burger & Robertson 1951 [Kansas Univ. Sci. Bull., Vol. 34: 214; Pl. 25: 3]

Type locality: Pacific coast of Guatemala.
Range: Coastal region from southern Mexico to Nicaragua.

Agkistrodon bilineatus Taylori Burger & Robertson
Taylor's Mocassin
  Agkistrodon bilineatus taylori Burger & Robertson 1951 [Kansas Univ. Sci. Bull., Vol. 34: 213]

Type locality: Twenty-one kilometers north of Villagrán, Tamaulipas, Mexico.
Range: Mexico, Tamaulipas, and Nuevo Leon.

Agkistrodon contortrix pictigaster Gloyd et Connant
Trans-Pecos Copperhead
  Agkistrodon mokeson Pictigaster Gloyd et Connant 1943 [Bull. Chicago Acad. Sci., Vol. 7: 56; Fig. 10)
  Agkistrodon contortrix pictigaster; Klauber 1948 [Copeia, p. 8]

Type locality: Maple Canyon (Chisos mountains), Brewster County, Texas, U.S.A.
Range: Known from Trans-Pecos, Texas, U.S.A. Possible in the adjacent Mexican Territory.

BOTHROPS Wagler

Type species *B. megaera* Wagler—*B. leucurus* Wagler
American Lance-Headed Pit Vipers
Bothrops leucurus Wagler
  Bothrops megaera Wagler 1824 (homonym of *Megaera* Shaw—*Bothrops lanceolatus* (Lacépède) [*In* Spix Serp. Brazil.. Sp. Nov, p. 50; Pl. XIX]—type locality, City of Bahia (Salvador), Brazil.

Bothrops leucurus Wagler 1824 [*In* Spix, Serp. Brazil., Sp. Nov. p. 57; Pl. XXII, Fig. 2]—type locality, Province Bahia, Brazil.

Bothrops megaera; Hoge 1966 [Mem. Inst. Butantan, Vol. 32: 110]

Type locality: Bahia, actually Salvador, Brazil.

Bothrops albocarinatus Shreve
  Bothrops albocarinatus Shreve 1934 [Occ. Pap. Boston Soc. Nat. Hist., Vol. 8: 130]

Type locality: Pastaza River, between Canelos and Marañon River, Ecuador.
Range: Pastaza River, drainage.

Bothrops alternatus Duméril, Bibron et Duméril
The Urutu Pit Viper
  Bothrops alternatus Duméril, Bibron et Duméril 1854 [Erp. Gen., Vol. 7, No. 2; Atlas Pl. 82 (lower), Fig. 1a]
  Lachesis alternatus; Boulenger 1896 [Cat. Sn. Brit. Mus., Vol. 3: 543]
  Lachesis inaequalis Magalhaes [Mem. Inst. Oswaldo Cruz, Vol. 18, No. 1: 153; Pl. 7–12]

Type locality: South America, Argentine, and Paraguay.
Range: Argentine: provinces of Buenos Aires, Entre Rios, Corrientes, Missiones, Santa Fe, Cordoba, Santiago del Estero, Chaco, Tucuman, San Luiz, La Pamapa, Rio Negro. Uruguay. Brazil: State of Rio Grande do Sul, Santa Catarina (except coastal region), Parana and São Paulo (except coastal regions), Minas Gerais (only in the broadleaved forest), Mato Grosso (only along the Paraná River and the extreme south-center of the state. Paraguay (southern parts only.)

Bothrops alticolus Parker
  Bothrops alticola Parker 1934 [Ann. Mag. Nat. Hist., Vol. 14, No. 10: 272]

Type locality: Five kilometers east of Loja, 9,200 ft. Ecuador.
Range: Known only from type locality.

Bothrops ammodytoides Leybold
Patagonian Pit Viper
  Bothrops ammodytoides Leybold 1873 [Excurs. Pamp. Argent., p. 80]
  Rhinocerophis nasus Garman 1881 [Bull. Mus. Comp. Zool. Harv., Vol. 8: 85]
  Bothrops nasus Berg 1884 [Acta Acad. Cordoba, Vol. 5: 96]
  Bothrops patagonicus Müller 1885 [Verh. Nat. Ges. Basel, Vol. 7: 697]
  Bothrops burmeisteri Koslowsky 1895 [Rev. Mus. La Plata, Vol. 6: 369; Pl. 4]
  Lachesis ammodytoides; Boulenger 1896 [Cat. Sn. Brit. Mus., Vol. 3: 543]

Type locality: Northern Argentina.

Range: Argentina: Province of Buenos Aires (southern), Chubut, Cordoba, Mendonza, Neuquen, Rioja, San Juan, San Luiz, Santa Cruz, Tucuman (only in the mountains).

Bothrops andianus Amaral
Andian Pit Viper
  Bothrops andiana Amaral 1923 [Proc. New Engl. Zool. Club, Vol. 8: 103]

Type locality: Machu Pichu, Department of Cuzco, 9,000–10,000 ft. Peru.
Range: Known from Department of Cuzco, Peru.

Bothrops asper Garman
  Trigonocephalus asper var. n. c. *lanceolati* Garman 1883 [Bull. Mus. Comp. Zool. Harv., Vol. 8: 124]

  Lachesis atrox; Boulenger 1896 (*partim*) [Cat. Sn. Brit. Mus., Vol. 3: 537]
  Bothrops atrox asper; Morfin 1918 [Informe rendido por la Comission Geografica exploradora de Quintana Roo al C. Secretario de Fomento: 1–57, 1–10 (not seen)]
  Bothrops asper; Hoge 1966 [Mem. Inst. Butantan, Vol. 32: 113; Pl. VI]

Type locality: Obispo, Isthmus of Darien, Panama.
Range: Mexico, St. Campèche, Chiapas, Oaxaca, San Luiz Potosí, Tabasco, Vera Cruz and Yucatán. In the low and moderate elevations of Guatemala, Costa Rica, Honduras and Panamá. In south America along the Pacific coast from Panamenian border through Colombia southward to Guayaquil, Ecuador, and Gorgona Island.

Bothrops atrox Linnaeus
  Coluber atrox Linnaeus 1758 [Syst. Nat., 10th ed., Vol. 1: 222]
  Bothrops furia Wagler 1824 [*In* Spix, Serp. Brazil., Sp. Nov. p. 52]
  Bothrops taeniatus Wagler 1824 [*In* Spix, Serp. Brazil., Sp. Nov. p. 55; Pl. XXI]
  Lachesis atrox; Boulenger 1896 (*partim*) [Cat. Sn. Brit. Mus., Vol. 3: 537]
  Bothrops atrox; Hoge 1966 [Mem. Inst. Butantan, Vol. 32: 113; Pl. V, Figs. 1, 1a, and 1b]

Type locality: Restricted to Surinam.
Range: The equatorial forests of Colombia, Venezuela, the Guianas, Brazil, Peru, Ecuador, and Bolivia.

Bothrops barbouri (Dunn)
Barbours Pit Viper
  Lachesis barbouri Dunn 1919 [Proc. Biol. Sci. Wash., Vol. 32: 213]
  Bothrops barbouri; Amaral 1930 [Mem. Inst. Butantan, Vol. 4: 232]
  Agkistrodon browni Shreve 1938 [Copeia, No. 1]

Type locality: Omilteme, Guerrero, Mexico.
Range: Sierra Madre del Sur, Guerrero, Mexico.

Bothrops barnetti Parker
Barnett's Pit Viper
  Bothrops barnetti Parker 1938 [Ann. Mag. Hist. Nat., Vol. 2, No. 11: 447]

Type locality: Between Lobitos and Talara, northern Peru.
Range: Northern Peru.

Bothrops bicolor Bocourt
  Bothrops bicolor Bocourt 1868 [Ann. Sci. Nat. Zool. Paleont., Vol. 10, No. 5: 202]
  Bothrops (Bothriechis) bernouillii Müller 1878 [Verh. Nat. Ges. Basel, Vol. 6: 399]
  Lachesis bicolor; Boulenger 1896 [Cat. Sn. Brit. Mus., Vol. 3: 566]

Type locality: Saint Augustin de Sololá, Guatemala.
Range: Pacific foothills of Guatemala and Mexico (Chicharas and Mount Ovando in extreme south of Chiapas).

Bothrops bilineatus bilineatus (Wied)
  Cophias bilineatus Wied 1821 [Reise Brazil, Vol. 2: 339]
  Trigonocephalus bilineatus; Schinz 1822 [Cuv. Thier., Vol. II: 143]
  Cophias bilineatus; Wied 1824 [Abbild. Naturg. Brazil, Pls. 5 and 6]
  Cophias bilineatus; Wied 1824 [*In* Isis v. Oken, p. 446]
  Cophias bilineatus; Wied 1825 [Beitr. Nat. Brazil, Vol. 1: 483]
  Bothrops . . . species . . . Cophias bilineatus Neuw.; Wagler 1830 [Syst. Amph., p. 174]
  Trigonocephalus (Bothrops) arboreus Cope 1869 [Proc. Amer. Phil. Soc., Vol. 9: 157]
  Lachesis bilineatus; Boulenger 1896 (*partim*) [Cat. Sn. Brit. Mus.]
  Bothrops bilineatus bilineatus; Hoge 1966 [Mem. Inst. Butantan, Vol. 32: 114; Pl. I, Fig. 1]

Type locality: "Villa Viçosa" on Peruhybe River (now Marobá State of Bahia, Brazil.
Range: The equatorial forests of Venezuela and the Guianas. In Brazil Territorio Federal do Amapa, and an isolated population in the tropical forests of the Atlantic slope from the State of Bahia to Rio de Janeiro.

Bothrops bilineatus smaragdinus Hoge
  Bothrops bilineatus smaragdinus; Hoge 1966 [Mem. Inst. Butantan, Vol. 32: 114; Pl. I, Fig. 2a and 2b]

FIG. 1. Distribution of *Bothrops alternatus*.

Type locality: Upper Purús, State of Amazonas, Brazil.
Range: Known from Ecuador, Peru, Bolivia, and Brazil, probably also in Colombia.

Bothrops brazili Hoge
Brazil's Pit Viper
   Bothrops neglecta Amaral 1923 (*partim paratype*) [Proc. New Engl. Zool. Club, Vol. 8: 99]
   Bothrops brazili Hoge 1953 [Mem Inst. Butantan, Vol. 25: 15–21]
   Bothrops brazili; Hoge 1966 [Mem. Inst. Butantan, Vol. 32: 115]

Type locality: Tome Assú on Acará Mirim River, State of Pará, Brazil.
Range: The equatorial forest, known from Venezuela, the Guianas, Colombia and Brazil, States of Para, Amazonas and extreme north Mato Grosso (Map. 1 in Brazil), Map 12 (Brazil only).
Bothrops castelnaudi Duméril, Bibron et Duméril
Castelnaud's Pit Viper

Bothrops castelnaudii Duméril 1853 [Mem. Acad. Sci., Vol. 23 (p. 139 of reprint) (no description)]

Bothrops castelnaudi Duméril 1854 [Bibron et Duméril, Erp. Gen., Vol. 7, No. 2: 1511]

Bothropsis quadricarinatus; Peter 1861 [Mber, Berlin Acad., p. 359]

Thanatophis montanus Posada-Arango 1889 [Bull. Soc. Zool. France, p. 244]

Lachesis castelnaudi; Boulenger 1896 [Cat. Sn. Brit. Mus., Vol. 3: 544]

Type locality: Unknown (Guichenot in Castelnau gives) "Province du Goyas," now State of Goias, Brazil.

Range: Equatorial forests of Brazil, Ecuador, and Peru. This species will be divided in subspecies for one of which the name *B. quadricarinatus* Peters is available.

Bothrops caribbaeus (Garman)
Caribbean Pit Viper

Trigonocephalus caribbaeus Garman 1887 [Proc. Amer. Phil. Soc., Vol. 24: 285]

Lachesis lanceolatus; Boulenger 1896 (*partim*) [Cat. Sn. Brit. Mus., Vol. 3: 535]

Bothrops caribbaeus; Lazell 1964 [Bull. Mus. Comp. Zool. Harv., Vol. 132, No. 3: 250]

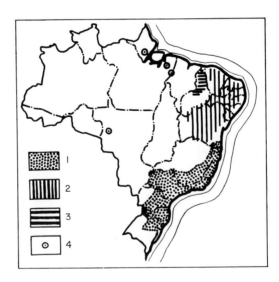

Fig. 2. Distribution of *Bothrops jararaca* (1), *B. erythromelas* (2), *B. iglesiasi* (3), and *B. brazili* (4).

Type locality: Sainte Lucia Island—restricted (Lazell l. c. : 251) to Grande Anse, Sainte Lucia.

Range: Coastal lowlands of northern part, except extreme north and southern parts of Sainte Lucia Island.

Bothrops colombiensis (Hallowell)
> *Trigonocephalus colombiensis* Hallowell 1845 [Proc. Acad. Sci. Phil., Vol. 2: 241–247]
> *Bothrops neuwiedii Venezuelenzi* Briceno Rossi 1934 [Bol. Min. Salubrid. Agricola, Vol. 2, No. 15: 46]
> *Bothrops jararaca*; Briceno Rossi 1933 [Bol. Min. Salubrid. Agricola, Vol. 2, No. 15: 48]
> *Bothrops amarali* Briceno Rossi 1933 [Bol. Min. Salubrid. Agricola, Vol. 2, No. 15: 53]
> *Bothrops colombiensis*; Hoge 1966 [Mem. Inst. Butantan, Vol. 34: 164]

Type locality: "Republic of Colombia within 200 miles of Caracas.

Range: Northern and northwestern Venezuela and northeastern Colombia.

Bothrops cotiara (Gomes)
Cotiara
> *Lachesis cotiara* Gomes 1913 [Ann. Paul. Med. Cirurg. São Paulo, Vol. 1, No. 3: 65]
> *Bothrops cotiara* Amaral 1925 [Contr. Harv. Inst. Trop. Biol. Med., Vol. 2: 53]

Type locality: Marechal Mallet, Parana, Brazil.

Range: The Araucaria forests of Argentine (Missiones) and Brazil (states of Rio Grande do Sul, Santa Catarina, and Parana, also known from two localities in southeastern São Paulo, near the border of Parana), Map 3.

Bothrops dunni (Hartweg and Oliver)
Dunn's Pit Viper
> *Trimeresurus dunni* Hartweg and Oliver 1938 [Occ. Pap. Mus. Zool. Univ. Mich., No. 390: 6]
> *Bothrops dunni* Smith and Taylor 1945 [Bull. U.S. Nat. Mus., Vol. 187: 181]

Type locality: Vicinity of Village of Tehuantepec, Oaxaca, Mexico.
Range: Mexico, the Pacific slopes of the State of Oaxaca.

Bothrops erythromelas Amaral
> *Bothrops erythromelas* Amaral 1923 [Proc. New Engl. Zool. Club., Vol. 8: 96]

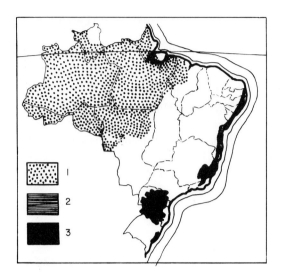

FIG. 3. Distribution of *Lachesis* (1), (2), and *Bothrops cotiara* (3).

Type locality: Near Joazeiro, State of Bahia, Brazil.

Range: Known from Brazil (the caatinga vegetation of States Ceará and Bahia), possibly also the other states with same vegetation (Map 2).

Bothrops fonsecai Hoge et Belluomini
Fonseca's Pit Viper
  Bothrops fonsecai Hoge et Belluomini 1959 [Mem. Inst. Butantan, Vol. 28: 195]

Type locality: Santo Antonio do Capivary, State of Rio de Janeiro, Brazil.
Range: Brazil: northeast São Paulo, south of Rio de Janeiro and extreme south of Minas Gerais.

Bothrops godmanni (Günther)
Godmann's Pit Viper
  Bothriechis godmanni Günther 1863 [Ann. Mag. Nat. Hist., Vol. 12, No. 3: 364]
  Bothrops brammianus Bocourt 1868 [Ann. Sci. Nat. Zool. Paleont., Vol. 10, No. 5: 201]
  Bothrops (Bothriopsis) godmanni; Müller 1878 [Verh. Nat. Ges. Basel., Vol. 6: 402]
  Bothriechis scutigera Fisher 1880 [Arch. Nat., p. 218]
  Bothriechis triangulifera Fisher 1883 [Oster. Prg. Akad. Gymm. Hamburg, p. 13]
  Lachesis godmanni; Boulenger 1896 [Cat. Sn. Brit. Mus., Vol. 3: 545]

Bothrops godmanni; Barbour & Loveridge 1929 [Bull. Antivenin Inst. Amer., Vol. 3, No. 1: 3]

Type locality: Dueñas and other parts of tableland of Guatemala.
Range: From Mexico (Chiapas) along the moderate to high elevation to Panama. Known on Mexico only from a few localities.

Bothrops hyoprorus Amaral
    Bothrops hyoprora Amaral 1935 [Mem. Inst. Butantan, Vol. 9: 222]

Type locality: La Pedrera, Colombia.
Range: Known from a few localities from the equatorial forests of Colombia, Ecuador, Peru, and western Brazil.

Bothrops iglesiasi Amaral
Iglesias' Pit Viper
    Bothrops iglesiasi Amaral 1923 [Proc. New Engl. Zool. Club, Vol. 8: 97]

Type locality: Near Fazenda Grande on the Right riverside of the Gurgueia River, State of Piauí, Brazil.
Range: Known only from northern Piauí, Brazil. (Map 2).

Bothrops insularis Amaral
Jararaca Ilhôa (Island's Jararaca)
    Lachesis insularis Amaral 1921 [Anex. Mem. Inst. Butantan, Sec. Ofiol., Vol. 1, No. 1: 18]
    Bothrops insularis; Amaral 1930 [Mem. Inst. Butantan, Vol. 4: 114]

Type locality: Island "Queimada Grande" on the coast of the State of São Paulo, Brazil.
Range: Known only from type locality.

Bothrops itapetiningae (Boulenger)
Cotiarinha
    Lachesis itapetiningae Boulenger 1907 [Ann. Mag. Nat. Hist., Vol. 20, No. 7: 338]
    Lachesis neuwiedii itapetiningae; Ihering 1910 *(partim)* [Rev. Mus. Paul., Vol. 8: 360]
    Bothrops itapetiningae; Amaral 1930 [Mem. Inst. Butantan, Vol. 4: 235]

Type locality: Itapetininga, State of São Paulo, Brazil.
Range: Brazil, from northeastern Paraná, through the State of São Paulo and Minas Gerais, northward to Brasilia, Distrito Federal; also known from one locality on State of Mato Grosso (there is a specimen from Rio Grande do Sul but this occurrence must be confirmed) (Map 4).

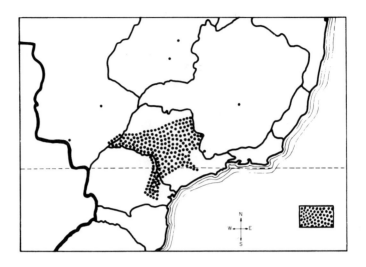

FIG. 4. Distribution of *Bothrops itapetiningae.*

Bothrops jararaca (Wied)
Jararaca

> *Cophias jararaca* (in text) *Cophias atrox "pullus"* (on plate) non *Cophias jararaca* Merrem 1822 nom.nov. pro *Coluber jauanus* Gmelin [iconotype in Seba I, Pl. LXX, 12]—type locality, Java "in error." —*Crotalus durissus* subsp. (pos. C. d. cascavella Wagler 1824) Wied [Abbild. Nat. Brazil, Lief. 7]

> *Cophias atrox . . . jararaca*; Wied 1824 [*In* Isis v. Oken, Vol. 14, No. 9: 987]

> *Cophias jararaca*; Wied 1824 [Abbild. Nat. Brazil, Lief, 8]

> *Cophias jajaraca*; (misspelling of jararaca) Wied 1824 [*In* Isis v. Oken, Vol. 14, No. 10: 1103]

> *Cophias jararaca*; Wied 1825 [Beitr. Nat. Brazil, Vol. I: 470]

> *Bothrops . . .* species . . . *Cophias jararaca* Neuw., has syn. of *Coluber Ambiguus* Gmelin, Wagler 1830 [Syst. Amph., p. 174]

> *Lachesis lanceolatus*; Boulenger 1896 *(partim)* [Cat. Sn. Brit. Mus., Vol. 3: 53]

Type locality: Espirito Santo, Brazil.

Range: Northern Argentina (Missiones); Paraguay and Brazil, states of Rio Grande do Sul, Sta. Catarina, Paraná, São Paulo, extreme eastern Mato Grosso, Rio de Janeiro, Espirito Santo and southern Bahia, also known from the broad-leaved forests of the State of Minas Gerais (Map 2) (Brazil only).

Bothrops jararacussu Lacerda
Jararacussu
  Bothrops jararacussu Lacerda 1884 [Lec. Ven. Serp. Bresil, p. 8]
  Lachesis lanceolatus; Boulenger 1896 *(partim)* [Cat. Sn. Brit. Mus., Vol.
    3: 535]
Type locality: Province of Rio de Janeiro, Brazil.
Range: Northeastern Argentine, Brazil, states of Sta. Catarina, Paraná, Mato
    Grosso, São Paulo, southern Minas Gerais, Rio de Janeiro, Espírito
    Santo and extreme southern Bahia. Paraguay and extreme southern
    Bolivia.

Bothrops lanceolatus (Lacepede)
The Fer de Lance or Martinican Pit Viper
  Vipera coerulescens Laurenti 1768 [Syn Rept., p. 19]—type locality,
    Martinique.
  C(oluber) glaucus Gmelin 1788 [Caroli Linnei S. Nat., 13th ed., Vol. 1:
    1092 (based on *Vipera coerulescens* Laurenti)]
  C(oluber) Lanceolatus La Cépède 1789 [Hist. Nat. Serp., Vol. II: 80
    and 121, Pl. V, Fig. 1]—type locality, Martinique.
  Coluber hastatus Suckow 1798 [Naturg. Thiere, Vol. III: 239 (based on
    La Cepede's *Lanceolatus*)]
  Coluber megaera Shaw 1802 [Gen. Zool., Vol. 3, No. 2: 406 (based on La
    Cepede's *Lanceolatus*)]
  Trigonocephalus lanceolatus; Oppel 1811 [Ord. Rept., p. 66]
  C. (ophias) lanceolatus; Merrem 1820 [Tent. Syst. Amph., p. 155]
  Bothrops lanceolatus; Wagler 1830 [Syst. Amph., p. 174]
  Lachesis lanceolatus; Boulenger 1896 *(partim)* Martinique.
  Bothrops lanceolatus; Hoge 1952 [Mem. Inst. Butantan, Vol. 24, No. 2:
    231–236]
  Bothrops lanceolatus; Lazell 1964 [Bull. Mus. Comp. Zool. Harv., Vol. 132,
    No. 3: 254]

Type locality: Martinique.
Range: Island of Martinique, in the wet regions, two disconnected populas
    tions, one in the highlands above Fort de France and northward in
    the mountains to the Mount Pelée massif and also along the coastal
    wet regions; the other one is confined to the southern highlands from
    Morne Serpent and Morne Vauclin to the hills between Trois-Ilets
    and Les-Anses D'Arlets.

Bothrops lansbergii lansbergii Schlegel
Lansberg's Pit Viper
  Trigonocephalus lansbergii Schlegel 1841 [Mag. Zool. Rept., Nos. 1–3]

Bothrops lansbergii; Jan 1863 [Elenco Sist. Ofid., p. 127]
Lachesis lansbergii; Boulenger 1896 *(partim)* [Cat. Sn. Brit. Mus., Vol. 3:
546]

Type locality: Turbaco, Colombia.
Range: Colombia, the arid and semiarid region of the Costa del Caribe and
Baja Magdalena.

Bothrops lansbergii annectens (Schmidt)
Trimeresurus lansbergii annectens Schmidt 1936 [Proc. Biol. Soc. Wash.,
Vol. 9: 50]

Type locality: Subirana-Tal, Yoro, Honduras.
Range: Honduras.

Bothrops lansbergii janisrozei (nomen novum pro *B. lansbergii venezuelensis*
Roze)
Roze's Pit Viper
Bothrops lansbergii venezuelensis Roze 1959 [Amer. Mus. New York,
No. 1934: 11 (Homonym of *Bothrops venezuelensis* Montilla, 1952)]

Type locality: Caripito, Monagas, Venezuela.
Range: Northern Venezuela, from State of Monaga to State of Zulia (Vide
Roze).

Bothrops lateralis (Peters)
Bothriechis latralis Peters 1863 [Mn. Akad. Wiss. Berlin, p. 674]
Bothrops (Bothriechis) lateralis; Müller 1878 [Verh. Nat. Ges. Basel,
Vol. 6: 401]
Lachesis lateralis; Boulenger 1896 [Cat. Sn. Brit. Mus., Vol. 3: 566]

Type locality: Veragua and Volcan Barbo, Costa Rica.
Range: Costa Rica and Panama.

Bothrops lichenosus Roze
Bothrops lichenosa Roze 1958 [Acta Biol. Venez., Vol. 2: 308]

Type locality: Chimanta-Tepui, Estado Bolivar, Venezuela.
Range: Known only from type locality.

Bothrops lojanus Parker
Lojan Pit Viper
Bothrops lojana Parker 1930 [Ann. Mag. Nat. Hist., Vol. 5, No. 10: 568]

Type locality: Loja, Ecuador.
Range: Only known from the vicinity of type locality.

Bothrops marajoensis Hoge
Bothrops marajoensis Hoge 1966 [Mem. Inst. Butantan, Vol. 32: 123]

Type locality: Severino, Island of Marajó, State of Pará, Brazil.
Range: The savannah of Marajó.

Bothrops medusa (Sternfeld)
    Lachesis medusa Sternfeld 1920 [Senckenbergiana, Vol. 2: 180]
    Bothrops medusa; Amaral 1930 [Mem. Inst. Butantan, Vol. 4, No. 1929: 23b]

Type locality: Caracas, Venezuela.
Range: Central region of "Cordillera de la Costa" from Caracas to Valencia 1,400 m–2,000 m.

Bothrops melanurus (Müller)
Black-tailed Pit Viper
    Trimeresurus melanurus Müller 1924 [Mitt. Zool. Mus. Berlin, Vol. 11: 92]
    Bothrops melanura; Amaral 1930 [Mem. Inst. Butantan, Vol. 4: 236]
    Trimeresurus garciai Smith 1940 [Proc. Biol. Soc. Wash., Vol. 53: 62]

Type locality: Mexico.
Range: Mexico. Desert region of southern Puebla and probably northern Oaxaca.

Bothrops microphthalmus microphthalmus Cope
    Bothrops microphthalmus Cope 1876 [Journ. Acad. Art. Sci. Phil., Vol. 8, No. 2: 182]
    Lachesis microphthalmus; Boulenger 1896 [Cat. Sn. Brit. Mus., Vol. 3: 540]
    Lachesis pleuroxanthus Boulenger 1912 [Am. Mag. Nat. Hist., Vol. 10, No. 8: 423]
    Bothrops microphthalma microphthalma; Peters 1960 [Bull. Mus. Comp. Zool. Harv., Vol. 122, No. 9: 510]

Type locality: Between Balsas Puerto and Moyabamba, Peru.
Range: Amazonian equatorial forests of Ecuador, Peru, known from Bolivia by a single specimen.

Bothrops microphthalmus colombianus Rendahl & Vestergren
    Bothrops microphthalmus colombianus Rendahl et Vestergren 1940 [Ark. Zool., Vol. 33A: 15]

Type locality: La Costa, Cauca, Colombia.
Range: Colombia.

Bothrops moojeni Hoge
    Bothrops moojeni Hoge 1966 [Mem. Inst. Butantan, Vol. 32: 126; Pl. IV]

Type locality: Brasilia, Distrito Federal, Brazil.
Range: The savannah of Central Brazil, southward to the State of Paraná.

Bothrops nasutus Bocourt
Hog-nosed Pit Viper
  Bothrops nasutus Bocourt 1868 [Ann. Sci. Nat. Paris, Vol. 10, No. 5: 202]
  Bothriopsis proboscideus Cope 1876 [Journ. Acad. Nat. Sci. Phil., Vol. 8,
    No. 2: 150]
  Lachesis brachystoma; Boulenger 1896 [Cat. Sn. Brit. Mus., Vol. 3: 547]

Type locality: Panzos, Rio Polochic, Guatemala.
Range: Mexico (Vera Cruz), southward through Guatemala, Costa Rica,
  Panama, Colombia, and Ecuador.

Bothrops neuwiedi neuwiedi Wagler
  Bothrops neuwiedi Wagler 1824 [*In* Spix, Serp. Brazil, Sp. Nov., p. 56]
  Lachesis neuwiedi; Boulenger 1896 *(partim)* [Cat. Sn. Brit. Mus., Vol. 3:
    542]
  Bothrops neuwiedii neuwiedii; Amaral 1925 [Contr. Harv. Inst. Trop. Biol.
    Med., Vol. 2: 57]

Type locality: Province of Bahia, Brazil.
Range: Southern Bahia, Brazil.

Bothrops neuwiedi bolivianus Amaral
  Bothrops neuwiedii boliviana Amaral 1927 [Bull. Antivenin, Inst. Amer.,
    Vol. 1: 6]

Type locality: Buenavista, Province Sara, Department of Santa Cruz de la
  Sierra, Bolivia.
Range: Known from Bolivia, Department of Santa Cruz de la Sierra, provinces of Chiquitos, Ibañez, Ichilo, Velasco, Department of Cochabamba. Brazil: State of Mato Grosso, extreme west.

Bothrops neuwiedi goyazensis Amaral
  Bothrops neuwiedi goyazensis; Amaral 1925 [Contr. Harv. Inst. Trop. Biol.
    Med., Vol. 2: 58; Table XIV: 3; Table XV: 3]

Type locality: Ypamery, Goias, Brazil.
Range: State of Goias, Brazil.

Bothrops neuwiedi lutzi (Miranda-Ribeiro)
Lutz's Pit Viper
  Lachesis lutzi Miranda-Ribeiro 1915 [Arch, Mus. Nac. Rio de Janeiro,
    Vol. 17: 4]
  Bothrops neuwiedii bahiensis Amaral 1925 [Contr. Harv. Inst. Trop. Biol.
    Med., Vol. 2: 57]
  Bothrops neuwiedii lutzi; Amaral 1930 [Mem. Inst. Butantan, Vol. 4: 238]

Type locality: Sao Francisco River, Bahia, Brazil.
Range: Dry regions of State of Bahia, Brazil.

Bothrops neuwiedi mattogrossensis Amaral
  Bothrops neuwiedii mattogrossensis Amaral 1925 [Contr. Harv. Inst. Trop.
  Biol. Med., Vol. 2: 60; Table 14: 6; Table 16: 6]

Type locality: Miranda, State of Mato Grosso, Brazil.
Range: Southern Mato Grosso, Brazil.

Bothrops neuwiedi meridionalis Müller
  Bothrops atrox meridionalis Müller 1885 [Verh. Nat. Ges. Basel, Vol. 7:
  699]
  Lachesis neuwiedii; Boulenger 1896 *(partim)* [Cat. Sn. Brit. Mus., Vol. 3:
  542]
  Bothrops neuwiedii fluminensis Amaral 1932 [Mem. Inst. Butantan, Vol.
  7: 97]
  Bothrops neuwiedii meridionalis; Hoge 1966 [Mem. Inst. Butantan, Vol.
  32]

Type locality: Andaray, State of Rio de Janeiro, Brazil.
Range: State of Rio de Janeiro, Guanabara, and Espirito Santo.

Bothrops neuwiedi paranaensis Amaral
  Bothrops neuwiedi paranaensis Amaral 1925 [Contr. Harv. Inst. Trop.
  Biol. Med., Vol. 2: 61; Pl. 14: 7; Pl. 16: 7]

Type locality: Castro, Paraná, Brazil.
Range: State of Paraná, Brazil.

Bothrops neuwiedi pauloensis Amaral
  Bothrops neuwiedii pauloensis; Amaral 1925 [Contr. Harv. Inst. Trop.
  Biol. Med., Vol. 2: 59]

Type locality: Leme, São Paulo, Brazil.
Range: Southern parts of State of São Paulo, Brazil.

Bothrops neuwiedi piauhyensis Amaral
  Bothrops neuwiedii piauhiense Gomes 1916 [*In* Neiva et Penna . . . (no
  diagnosis)]
  Bothrops neuwiedii piauhyensis Amaral 1925 [Contr. Harv. Inst. Trop.
  Biol. Med., Vol. 2: 58]

Type locality: Regeneração, Piauí.
Range: State of Piauhy, Pernambuco, Ceará, and southern Maranhão,
Brazil.

Bothrops neuwiedi pubescens (Cope)
 Trigonocephalus (Bothrops) pubescens Cope 1869 [Amer. Phil. Soc. Phil.,
  Vol. 11:57]
 Lachesis neuwiedii; Boulenger 1896 [Cat. Sn. Brit. Mus., Vol. 3: 542]
 Bothrops neuwiedii riograndensis Amaral 1925 [Contr. Harv. Inst. Trop.
  Biol. Med., Vol. 2: 61]
 Bothrops neuwiedi pubescens; Hoge 1959 [Mem. Inst. Butantan, Vol. 28: 84]

Type locality: Rio Grande do Sul, Brazil.
Range: State of Rio Grande do Sul, Brazil.

Bothrops neuwiedi urutu Lacerda
 Bothrops urutu; Lacerda 1884 [Leç. Ven. Serp. Brésil, No. 11]
 Lachesis neuwiedii; Boulenger 1896 *(partim)* [Cat. Sn. Brit. Mus., Vol. 3:
  542]
 Bothrops neuwiedi urutu; Amaral 1937 [Mem. Inst. Butantan, Vol. 10]

Type locality: "Province de Minas Gerais," now State of Minas Gerais.
Range: Southern parts of Minas Gerais and northern State of São Paulo,
  Brazil.

Bothrops nigroviridis nigroviridis (Peters)
 Bothriechis nigroviridis Peters 1859 [Mber. Dtsch. Akad. Wiss. Berlin,
  p. 278]
 Bothrops (Bothriechis) nigroviridis; F. Müller 1878 [Verh. Nat. Ges. Basel,
  Vol. 6: 401]
 Lachesis nigroviridis nigroviridis; Barbour et Loveridge 1896 [Bull. Anti-
  venin, Inst. Amer., Vol. 3: 2]

Type locality: Vulcan Barbo, Costa Rica.
Range: From Costa Rica to Panama.

Bothrops nigroviridis aurifer (Salvin)
 Thamnocenchis aurifer Salvin 1860 [Proc. Zool. Soc. London, Vol. 1860:
  459]
 Bothrops aurifer; F. Müller 1878 [Verh, Nat. Ges. Basel. Vol. 6: 401]
 Lachesis aurifer; Boulenger 1896 [Cat. Sn. Brit. Mus., Vol. 3: 568]
 Bothrops nigroviridis aurifera; Barbour et Loveridge 1929 [Bull. Antivenin,
  Inst. Amer., Vol. 3, No. 1: 1–3]

Type locality: Cobán, Alto Verapaz, Guatemala.
Range: Moderate and intermediate elevations of the Caribbean versant from
  Chiapas, Mexico throughout Guatemala.

Bothrops nigroviridis marchi Barbour & Loveridge
 Bothrops nigroviridis marchi Barbour & Loveridge 1929 [Bull. Antivenin,
  Inst. Amer., Vol. 3, No. 1; Pl. 1]

Type locality: Gold mines near Quimistan (Santa Barbara), Honduras.
Range: Honduras.

Bothrops nummifer nummifer (Rüppel)
  Atropos nummifer Rüppel 1845 [Ver. Mus. Senckenberg, Vol. 3: 313]
  Bothrops nummifer; Jan 1863 [Elenco Syst., p. 126]
  Lachesis nummifer; Boulenger 1896 *(partim)* [Cat. Sn. Brit. Mus., Vol. 3: 540]
  Bothrops nummifer veraecrucis Burger 1950 [Bull. Chicago Acad. Sci., Vol. 9, No. 3: 65]

Type locality: Restricted (Burger 1950, *l. c.*) to: Teapa, Tabasco; Mexico.
Range: The dry tropical shrub vegetation on the southeastern edge of the Mexican plateau from São Luiz do Potosi, southward to Oaxaca.

Bothrops nummifer mexicanus (Duméril, Bibron & Duméril)
  Atropos mexicanus; Duméril 1854 Bibron & Dumeril [Erp. Gen., Vol. 7, No. 2: 1521]
  Bothriechis nummifer var. *notata* Fischer 1880 [Arch. Nat., Vol. 46: 222]
  Bothrops mexicanus; F. Müller 1882 [Verh. Nat. Ges. Basel, Vol. 7: 154]
  Lachesis nummifer; Boulenger 1896 *(partim)* [Cat. Sn. Brit. Mus., Vol. 3: 544]
  Bothrops nummifer mexicanus; Mertens 1952 [Abh. Senckenb. Naturf. Ges., Vol. 487: 79]

Type locality: Cobán (Alta) Verapaz, Guatemala.
Range: The Caribbean versant (low to intermediate elevations) from extreme southern Mexico to Panama.

Bothrops nummifer occidduus Hoge 1966 (nomen novum pro *B. affinis* Bocourt, 868)
  Bothrops affinis Bocourt 1868 [Ann. Sci. Nat., Vol. 10, No. 5: 10–201 (homonym of *Bothrops affinis* Gray 1849 = *Bothrops atrox* Linnaeus, 1758)]
  Bothrops nummifer affinis; Stuart 1963 [Misc. Publ. Mus. Zool. Univ. Mich., No. 122: 130]
  Bothrops nummifer occidduus; Hoge 1966 [Mem. Inst. Butantan, Vol. 32: 130]

Type locality: San Augustine, on the west (south) slope of the mountains, Guatemala; 610 m.
Range: The deciduous moist monsoon forest on the low to moderate elevations along the Pacific slopes in El Salvador (possible in Eastern Chiapas, Mexico into El Salvador).

Bothrops oligolepis (Werner)

Lachesis bilineatus var. *oligolepis* Werner 1901 [Abh. Ber. Mus. Dresden, Vol. 9, No. 2: 13]
Lachesis chloromelas Boulenger 1912 [Ann. Mag. Nat. Hist., Vol. 10, No. 8: 423]

Type locality: Bolivia.
Range: Perú, Bolivia.

Bothrops ophryomegas Bocourt
  Bothrops ophryomegas; Bocourt 1868 [Ann. Sci. Nat. Zool. Paleont., Vol. 10, No. 5: 201]
  Lachesis lansbergi; Boulenger 1896 *(partim)* [Cat. Sn. Brit. Mus., Vol. 3: No. 546]

Type locality: Occidental slopes of Escuntla range, Guatemala.
Range: Pacific versant of Central America from western Guatemala to Panama.

Bothrops peruvianus (Boulenger)
  Lachesis peruvianus Boulenger 1903 [Ann. Mag. Nat. Hist., Vol. 12, No. 7: 354]
  Bothrops peruvianus; Amaral 1930 (1929) [Mem. Inst. Butantan, Vol. 4: 240]

Type locality: La Oroya, Carabaya, southeastern Peru.
Range: Southeastern Peru.

Bothrops picadoi (Dunn)
  Trimeresurus nummifer picadoi Dunn 1939 [Proc. Biol. Soc. Wash., Vol. 52: 165]
  Bothrops picadoi; Smith & Taylor 1945 [Bull. U.S. Nat. Mus. Wash., Vol. 187: 183]

Type locality: La Palma, Costa Rica, 4,500 ft.
Range: Central plateau of Cośta Rica and surrounding mountains.

Bothrops pictus (Tschudi)
  Lachesis picta Tschudi 1845 [Faun. Per. Herp., p. 61]
  Bothrops pictus; Jan 1863 [Elenco Syst. Ofid., p. 126]
  Lachesis pictus; Boulenger 1896 [Cat. Sn. Brit. Mus., Vol. 3: 540]

Type locality: "The high mountains of Peru."
Range: Coastal region of Peru.

Bothrops pirajai Amaral
Piraja's Pit Viper—Jararacussu
  Bothrops pirajai Amaral 1923 [Proc. New Engl. Zool. Club, Vol. 8: 99]

Bothrops neglecta Amaral 1923 *(partim)* [Proc. New Engl. Zool. Club, Vol. 8: 100]
Bothrops pirajai; Hoge 1966 [Mem. Inst. Butantan, Vol. 34: 132]

Type locality: Ilheos, State of Bahia, Brazil.
Range: Known only from southern Bahia.

Bothrops pradoi (Hoge)
Trimeresurus pradoi Hoge 1948 [Mem. Inst. Butantan, Vol. 20: 193–202]
Bothrops atrox; Amaral 1955 [Mem. Inst. Butantan, Vol. 26: 215–220]
Bothrops pradoi; Hoge 1966 [Mem. Inst. Butantan, Vol. 32: 132]

Type locality: Pau Gigante, State Espirito Santo, Brazil.
Range: Known from type locality northward to southern Bahia.

Bothrops pulcher (Peters)
Trigonocephalus pulcher Peters 1862 [Mber. Dtsch. Akad. Wiss. Berlin, p. 672]
Lachesis pulcher; Boulenger 1896 [Cat. Sn. Brit. Mus., Vol. 3: 539]
Bothrops pulchra; Amaral 1930 [Mem. Inst. Butantan, Vol. 4: 240]

Type locality: Quito, Equador.
Range: Equatorial forests of Equador and Peru.

Bothrops punctatus (Garcia)
Lachesis punctata Garcia 1896 [Los Ofid. Ven. del Cauca, Cali Colombia, No. 31]
Lachesis monticelli Peracca 1910 [Amer. Mus. Napoli, Vol. 3, No. 12: 1–3]
Bothrops leptura Amaral 1923 [Proc. New Engl. Zool. Club, Vol. 8: 102]
Bothrops punctatus; Dunn 1944 [Caldasia, Vol. 3, No. 12: 215]

Type locality: "Las montanhas del Dagua." Colombia.
Range: From Darien, Panama, Colombia, to Chocó, Equador.

Bothrops roedingeri Mertens
Roedinger's Pit Viper
Bothrops roedingeri Mertens 1942 [Beitr. Fauna, Perus, Vol. 11: 284]

Type locality: "Hacienda Huayri," southern Peru.
Range: The desert region along the Pacific coast of Peru.

Bothrops sanctaecrucis Hoge
Bothrops sanctaecrucis; Hoge 1966 [Mem. Inst. Butantan, Vol. 32: 133; Pl. 9]

Type locality: Oromomo, Rio Secure, upper Beni, Bolivia.
Range: Known from Bolivia.

Bothrops schlegelli (Berthold)
Schlegel's Pit Viper
Trigonocephalus schlegelli Berthold 1946 [Nachr. Univ. Ges. Wiss. Güttingen, p. 147]
B. (othrops) schlegeli; Jan 1863 [Elenc. Syst. Ofid., p. 127]
Lachesis schlegelii; Boulenger 1896 [Cat. Sn. Brit. Mus., Vol. 3: 567]

Type locality: Popayon, restricted to Popayon, Colombia (Dunn & Stuart, 1954).
Range: From southern Mexico to Equador and Venezuela.

Bothrops sphenophrys Smith
Bothrops sphenophrys Smith 1960 [Trans. Kansas Acad. Sci., Vol. 62: 267]

Type locality: La Soledad, Oaxaca, Mexico.
Range: Known from South Oaxaca, Mexico.

Bothrops supraciliaris Taylor
Bothrops schlegelii supraciliaris Taylor 1954 [Kansas Univ. Sci. Bull., Vol. 36, No. 2, 11: 791]
Bothrops supraciliaris; Stuart 1963 [Misc. Publ. Mus. Zool. Univ. Michigan, No. 122: 131]

Type locality: Mountains near San Isidoro del General, San José Province, Costa Rica.
Range: Known only from type specimens and a specimen without locality (banana shipping) in MCZ.

Bothrops undulatus (Jan)
Trigonocephalus (Atropos) undulatus Jan 1859 [Rev. Mag. Zool., p. 157]
Bothrops undulatus; Günther 1895 [Biol. Centr. Amer. Rept. Amph., p. 187]
Lachesis undulatus; Boulenger 1896 [Cat. Sn. Brit. Mus., Vol. 3: 565]

Type locality: Mexico.
Range: High elevations central Vera Cruz, southward through central Oaxaca, and northward in central Guerrero in the Sierra Madre del Sur (known from Omilteme and Chilpaziungo Guerrero); Oaxaca, Oaxaca; Adopan and Orizaba; Vera Cruz.

Bothrops venezuelensis Sandner Montilla
Bothrops venezuelensis; sp. no., Sandner Montilla 1952 [Mon. Cien. Inst. Ter. Exp. Lab. "Veros" Ltda., No. 21: 4 (not *Bothrops neuwiedii venezuelenzi* Rossi 1933=*Bothrops colombiensis*]

Bothrops venezuelae; sp. no., Sandner Montilla 1961 [Mon. Cienc. Centr. Ocas. Mus. Nat. La Salle, Caracas, Zool., No. 30: 3]
Bothrops pifanoi; sp. no., Sandner Montilla *l. c.*

Type locality: "Boca de Tigre," Serrania de El Avila, Distrito Federal al Norte de Caracas.

Range: Northern and central part of Venezuela, Avila mountains, the western mountains of Los Tigres, the forests of Rancho Grande and Fil Miranda, from the State of Aragua and finally the forests of Gualapo (vide Sandner Montilla).

Bothrops xantogrammus (Cope)
   Trigonocephalus xantogrammus Cope 1868 [Proc. Acad. Nat. Sci. Phil., Vol. 10: 110]
   Lachesis xantogramma; Amaral 1896 [Mem. Inst. Butantan, Vol. 4: 241]

Type locality: Pallatanga, Ecuador.
Range: Apparently to be found in the highlands of Ecuador and Colombia.

Bothrops yucatannicus (Smith)
Yucatan Pit Viper
   Trimeresurus yucatannicus Smith 1941 [Zoologica, Vol. 26: 62]
   Bothrops yucatannicus; Smith 1944 [Carnegie Mus., Vol. 30: 92]

Type locality: Chichzen Itzá, Yucatan, Mexico.
Range: Northern and northeastern Yucatán Peninsula, Mexico. Known only by a few specimens from several localities in Yucatán.

CROTALUS Linnaeus

Crotalus [Crotalus] Linnaeus*
Rattlesnakes
Crotalus atrox Baird & Girard
Western Diamond Rattlesnake
   Crotalus atrox Baird & Girard 1853 [Cat. North. Amer. Rept., No. 1: 5]
   Caudisona atrox var. *atrox*; Kennicott 1861 [Proc. Acad. Nat. Sci. Phil., Vol. 13: 206]
   Caudisona atrox sonoraensis Kennicott 1861 [Proc. Acad. Nat. Sci. Phil., Vol. 13: 206]
   Crotalus confluentus; Boulenger 1896 *(partim)* [Cat. Sn. Brit. Mus., Vol. 3: 576]
   Crotalus atrox; Klauber 1956 [Rattlesnakes, Vol. 1: 29]

* Subgenera are not repeated in citations.

Type locality: Indianola (Calhoun County), Texas, U.S.A.

Range: U.S.A. from Arkansas and Oklahoma south to Mexico, Sonora, Chihuahua, Durango, Coahiula, Nuevo Leon, Tamaulipas, San Luiz de Potosí and northern Vera Cruz, isolated colonies in central Vera Cruz and southern Oaxaca probably present in northern Zacatecas and possibly Hidalgo, also Tiburon, Turna on San Pedro Matir Islands (Map 5).

Crotalus basiliscus basiliscus (Cope)
Mexican West Coast Rattlesnake

    Caudisona basilisca Cope 1864 [Proc. Acad. Nat. Sci. Phil., Vol. 16, No. 3: 166]

    Crotalus basiliscus Cope 1875 [*In* Yarrow, Surv. W. of 100th Merid. (Wheeler) Vol. 5, No. 4: 532]

    Crotalus terrificus; Boulenger 1896 *(partim)* [Cat. Sn. Brit. Mus., Vol. 3: 573]

    Crotalus basiliscus basiliscus; Gloyd 1948 [Nat. Hist. Misc., No. 17: 1]

    Crotalus basiliscus basiliscus; Klauber 1956 [Rattlesnakes, Vol. 1: 30]

Type locality: Near Colima, Mexico restricted (Smith & Taylor, 1950) to Colima, Mexico.

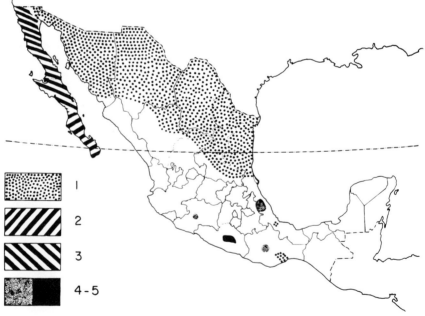

Fig. 5. Distribution of *Crotalus atrox* (1), *Crotalus ruber lucasensis* (2), *Crotalus ruber ruber* (3), *Crotalus intermedius intermedius* (4), *Crotalus intermedius omiltemanus* (5).

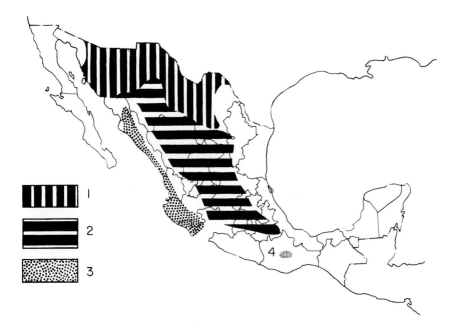

FIG. 6. Distribution of *Crotalus mollosus mollosus* (1), *Crotalus mollosus nigrescens* (2), *Crotalus basiliscus basiliscus* (3), and *Crotalus basiliscus oaxacus* (4).

Range: Mexico, from extreme southern Sonora along west coast of Mexico through Sinaloa, Jalisco, Colima, and western Michoacan, and Nayarit (Map 6).

Crotalus basiliscus oaxacus Gloyd
Oaxacan Rattlesnake
    Crotalus basiliscus oaxacus Gloyd 1948 [Nat. Hist. Misc., No. 17: 1]
    Crotalus basiliscus oaxacus; Klauber 1956 [Rattlesnakes, Vol. 1: 30]

Type locality: Oaxaca, Oaxaca, Mexico.
Range: Known from type locality and Chilpanzingo Region (Map 6).

Crotalus catalinensis Cliff
Santa Catalina Island Rattlesnake
    Crotalus catalinensis Cliff 1954 [Trans. San Diego Soc. Nat. Hist., Vol. 12, No. 5: 80]
    Crotalus catalinensis; Klauber 1956 [Rattlesnakes, Vol. 1: 30]

Type locality: Santa Catalina Island, Gulf of California, Mexico.
Range: Known only from type locality (Map 7).

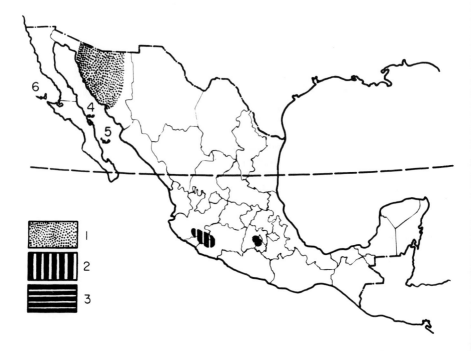

FIG. 7. Distribution of *Crotalus tigris* (1), *Crotalus pusilus* (2), *Crotalus transversus* (3), *Crotalus tortuguensis* (4), *Crotalus catalinensis* (5), *Crotalus exsul* (6).

Crotalus cerastes cercobombus Savage & Cliff
Sonoran Desert Sidewinder
> *Crotalus cerastes cercobombus* Savage & Cliff 1953 [Nat. Hist. Misc., No. 119: 2]
> *Crotalus cerastes cercobombus*; Klauber 1956 [Rattlesnakes, Vol. 1: 31]

Type locality: Near Gila Band, Maricopa County, Arizona, U.S.A.
Range: U.S.A., Arizona, Maricopa, Mexico, northwestern Sonora (except parts of the state north and west of Bahia Adair) Tiburon Island (Map 8)

Crotalus cerastes laterorepens Klauber
Colorado Desert Sidewinder
> *Crotalus cerastes laterorepens* Klauber 1944 (*partim*) [Trans. San Diego Soc. Nat. Hist., Vol. 10, No. 8: 94]
> *Crotalus cerastes laterorepens*; Klauber 1956 [Rattlesnakes, Vol. 1: 31]

Type locality: The Narrows, San Diego County, California, U.S.A.

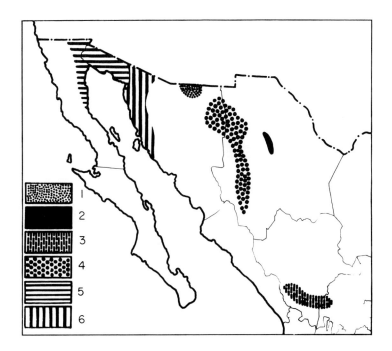

FIG. 8. Distribution of *Crotalus willardi willardi* (1), *C. willardi amabilis* (2), *C. willardi meridionalis* (3), *C. willardi silus* (4), *C. cerastes laterorepens* (5), and *C. cerastes cercobombus* (6).

Range: U.S.A., the desert areas of central and eastern Riverside, northeastern San Diego and Imperial County in California, western Yuma County, Arizona; Northeastern Baja California, Mexico, Sonora (Map 8)

Crotalus durissus durissus Linnaeus

Central American Rattlesnake

   Crotalus durissus Linnaeus 1758 [Syst. Nat., 10th ed., Vol. 1: 214]

   Crotalus simus Latreille 1802 [*In* Sonnini et Latreille, Hist. Nat. Rept., Vol. 3: 202; Vol. 4: 323]

   Crotalus terrificus; Boulenger 1896 (*partim*) [Cat. Sn. Brit. Mus., Vol. 3: 573]

   Crotalus durissus durissus; Klauber 1936 (*partim*) [Occ. Pap. San Diego Soc. Nat. Hist., No. 2]

   Crotalus durissus durissus; Klauber 1956 [Rattlesnakes, Vol. 1: 31–32]

   Crotalus durissus durissus; Hoge 1966 [Mem. Inst. Butantan, Vol. 32]

Type locality: North America. Restricted (Taylor & Smith, 1950) to Jalapa, Vera Cruz, Mexico.

Range: Mexico, central Vera Cruz, southeastern Oaxaca, Tabasco and Chiapas. Central and southern Guatemala, western and southern Honduras, El Salvador, southwestern Nicaragua. Northwestern and central Costa Rica.

Crotalus durissus cascavella (Wagler, 1824)

> *Crotalus cascavella* Wagler 1824 [*In* Spix. Serp. Brazil, Spec. Nov., p. 60] —type locality, "in campis provinciae Bahiae" (Caatinga of interior of Bahia), Brazil. Type specimen, none designated. [Neotype Hoge, 1966, Mina Caraiba, Bahia (This locality is near Spix itinerary]
>
> *Crotalus terrificus* var. *collirhombeatus* Amaral 1925 [Rev. Mus. Paul., Vol. 15: 90]
>
> *Crotalus durissus cascavella*; Hoge 1966 [Mem. Inst. Butantan, Vol. 32]

Type locality: Mina Caraiba, State of Bahia, Brazil.

Range: Brazil; The dry Caatinga regions of states of Maranhão, Ceará, Piauhy, Pernambuco, Alagoas, Rio Grande do Norte and Bahia, possibly extreme northeastern Minas Gerais (Map 9).

Crotalus durissus collilineatus Amaral

> *Crotalus terrificus* var. *collilineatus* Amaral 1926 (*partim*) [Rev. Mus. Paul., Vol. 15: 90]
>
> *Crotalus durissus terrificus*; Klauber 1956 (*partim*) [Rattlesnakes, Vol. 1: 33]
>
> *Crotalus durissus collilineatus*; Hoge 1966 [Mem. Inst. Butantan, Vol. 32: 139–142]

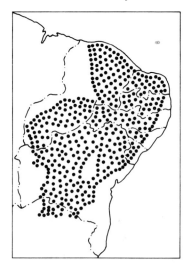

Fig. 9. Distribution of *Crotalus durissus cascavela.*

FIG. 10. Distribution of *Crotalus durissus*.

Type locality: Central, southeastern and southern Brazil, Argentina, Paraguay and probably Bolivia. Restricted (Hoge, 1966) to State of Mato Grosso, Brazil.

Range: Southeastern State of Mato Grosso, states of Goias, Federal District; Minas Gerais and northeastern São Paulo. Intergradation with *c. d. terrificus*, over a large area in the State of São Paulo.

Crotalus durissus culminatus Klauber
Northwestern Neotropical Rattlesnake
    Crotalus durissus culminatus Klauber 1952 [Bull. Zool. Soc. San Diego, Vol. 26: 65]

Type locality: Hacienda el Sabino, near Uruapan, Michoacan, Mexico.
Range: Southwestern Michoacan, southern and western Morilos Guerrero and southwestern Oaxaca, Mexico. Possibly western Puebla and Distrito Federal.

Crotalus durissus cumanensis Humbold
    Crotalus cumanensis Humbold 1833 [*In* Humbold et Bonpland, Rec. Obs. Zool. Anat. Comp., Vol. 2: 6]

Crotalus loeflingi Humbold [*In* Humbold Bonpland, Rec. Obs. Zool. Anat. Comp., Vol. 2: 6]
  Crotalus terrificus; Boulenger 1896 (*partim*) [Cat. Sn. Brit. Mus., Vol. 3: 373]
  Crotalus durissus terrificus; Klauber 1956 (*partim*) [Rattlesnakes, Vol. 1: 32]
  Crotalus durissus cumanensis; Hoge 1966 [Mem. Inst. Butantan, Vol. 32: 142]

Type locality: Cumaná, Venezuela.
Range: Known from about ten specimens all from or near Cumaná, Venezuela, except the high mountains, the savannah of Monagas and the equatorial forests of the Delta, Bolivar and Amazonas. Colombia in the extreme northeast.

Crotalus durissus dryinus Linnaeus
  Crotalus dryinus Linnaeus 1758 [Syst. Nat., 10th ed., Vol. 1: 214]
  Crotalus terrificus; Boulenger 1896 [Cat. Sn. Brit. Mus., Vol. 3: 573]
  Crotalus dryinus; Klauber 1956 (as senior, but rejected synonym for the South American Rattler)
  Crotalus durissus dryinus; Hoge 1966 [Mem. Inst. Butantan, Vol. 32:142; Pl. XIV]

Type locality: South America, restricted by Hoge (1966) to Paramaribo, Surinam.)
Range: The Guianas.

Crotalus durissus marajoensis Hoge
  Crotalus durissus marajoensis Hoge 1966 [Mem. Inst. Butantan, Vol. 32: 143; Pl. XV]

Type locality: Tuyuyu, Ilha do Marajó, State of Pará, Brazil.
Range: The Savannah of Marajó, Brazil.

Crotalus durissus ruruima Hoge
  Crotalus durissus ruruima Hoge 1966 [Mem. Inst. Butantan, Vol. 32: 145; Pl. XVI]

Type locality: Paulo Camp, Mt. Roraima, Venezuela.
Range: Known from Venezuelan slopes from Mt. Roraima, and Cariman Paru, Amazonas. Brazil, a single specimen from Territorio Federal of Roraima.

Crotalus durissus terrificus (Laurenti)
  Caudisona terrifica Laurenti 1768 [Syn. Rept., p. 93]
  Crotalus terrificus; Boulenger 1896 (*partim*) [Cat. Sn. Brit. Mus., Vol. 3: 573]

Crotalus terrificus collilineatus Amaral 1926 (*partim*) [Rev. Mus. Paul., Vol. 15: 90]
Crotalus durissus terrificus; Klauber 1936 (*partim*) [Rattlesnakes, Vol. 1: 32]
Crotalus durissus terrificus; Hoge 1966 [Mem. Inst. Butantan, Vol. 34: 147; Pl. XVII]

Type locality: Julio de Castillo, Municipio Taquari, State Rio Grande do Sul, Brazil (by indication of neotype Hoge, 1966).
Range: Extreme southern South America, southern Brazil, Uruguay, Argentina, Paraguay and Bolivia.

Crotalus durissus totonacus Gloyd & Kauffeld
Totonacan Rattlesnake
  Crotalus totonacus Gloyd & Kauffeld 1940 [Bull. Chicago Acad. Sci., Vol. 6, No. 2: 12]
  Crotalus durissus totonacus; Smith & Taylor 1945 [Bull. U.S. Mus., pp. 187–190]

Type locality: Panaco Island, about 75 miles south of Tampico, Vera Cruz, Mexico, 12 miles inland from Cabo Rojo.
Range: Southern Tamaulipas, southeastern San Luiz Potosí, and northern Vera Cruz.

Crotalus durissus tzabcan Klauber
Yucatan Neotropical Rattlesnake
  Crotalus durissus tzabcan Klauber 1952 [Bull. Zool. Soc. San Diego, No. 26: 71]

Type locality: Kantunil, Yucatan, Mexico.
Range: From Yucatan south into northern El Petén, Guatemala, and British Honduras.

Crotalus durissus unicolor Van Lidth de Geude
Aruba Island Rattlesnake
  Crotalus horridus var. *unicolor* Van Lidth de Geude 1887 [Notes Leyden Mus., Vol. 2, No. 8: 133]
  Crotalus terrificus; Boulenger 1896 (*partim*) [Cat. Sn. Brit. Mus., Vol. 3: 573]
  Crotalus durissus unicolor; Brongersma 1940 (Studies of Fauma of Curaçâo, Aruba, Bonaire and Venezuelan Islands).
  Crotalus unicolor; Klauber 1956 [Rattlesnakes, Vol. 1: 44]
  Crotalus durissus unicolor; Hoge 1966 [Mem. Inst. Butantan, Vol. 32: 148]

Type locality: Aruba Island, Netherlands West Indies.

Range: Known from Aruba Island only. A specimen described as *Crotalus pulvis* by Ditmars from Managua, Nicaragua is probably based on a specimen with erroneous locality.

Crotalus enyo enyo (Cope)
Lower California Rattlesnake
  Caudisona enyo Cope 1861 [Proc. Acad. Sci. Phil., Vol. 13: 208]
  Crotalus enyo; Cope 1875 [Bull. U.S. Mus., Vol. 1: 33]
  Crotalus enyo enyo; Lowe & Norris 1954 [Trans. San Diego Soc. Nat. Hist., Vol. 12, No. 4: 52]
  Crotalus enyo enyo; Klauber 1956 [Rattlesnakes, Vol. 1: 34]

Type locality: Lower California.
Range: Mexico, Baja California, from the vicinity of El Marmol (lat. 30°N) South to Cape San Lucas and the Islands Magdalena, Santa Margarida Espirito Santo, Partida, San Francisco, and Carmen (Map 9).

Crotalus enyo cerralvensis Cliff
Cerralvo Island Snake
  Crotalus enyo cerralvensis Cliff 1854 [Trans. San Diego Soc. Nat. Hist., Vol. 12, No. 5: 82]
  Crotalus enyo cerralvensis; Klauber 1956 [Rattlesnakes, Vol. 1: 34]

Type locality: Mexico, Island of Cerralvo, Gulf of California.
Range: Cerralvo Island (Map 11).

Crotalus enyo furvus Lowe & Norris
Rosario Rattlesnakes
  Crotalus enyo furvus Lowe & Norris 1954 [Trans. San Diego Soc. Nat. Hist., Vol. 12, No. 4: 52]
  Crotalus enyo furvus; Klauber 1956 [Rattlesnakes, Vol. 1: 34]

Type locality: 10.9 miles north of El Rosario, Baja California del Norte, Mexico.
Range: Mexico, Baja California del Norte, on west coast from San Telmo River to El Rosario (Map 11).

Crotalus exsul Garman
Cedros Island Diamond Rattlesnake
  Crotalus exsul Garman 1883 [Mem. Mus. Comp. Zool. Harv., Vol. 8, No. 3: 114]
  Crotalus exsul Klauber 1956 [Rattlesnakes, Vol. 1: 34]

Type locality: Cedros Island (or Cerros), Pacific Coast of Baja California, Mexico.
Range: Mexico, Cedros (or Cerros) Island, Pacific Coast of Baja California (Map 7).

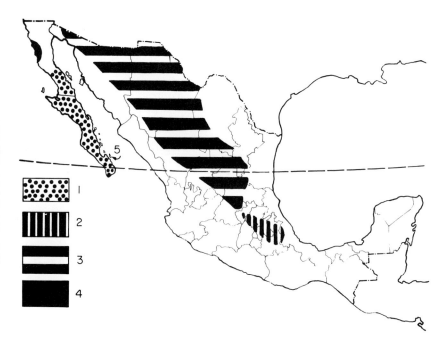

FIG. 11. Distribution of *Crotalus enyo enyo* (1), *C. scutalatus salvini* (2), *C. scutulatus* (3), *C. furvus* (4), and *C. cerralvensis* (5).

Crotalus intermedius intermedius Troschel
Totalcan Small-Headed Rattlesnake

> *Crotalus intermedius*, Troschel 1865 [*In* Müller, Reisen Ver. Staaten, Canada, Mexico, Vol. 3: 613]
>
> *Crotalus triseriatus*; Boulenger 1896 [*partim*) [Cat. Sn. Brit. Mus., Vol. 3: 581]
>
> *Crotalus triseriatus anahuacus*, Gloyd 1940 (*partim*) [Chicago Sci. Soc. Publ., No. 4: 91]
>
> *Crotalus triseriatus gloydi* Taylor 1841 [Kansas Univ. Sci. Bull., Vol. 27, No. 1: 130]
>
> *Crotalus gloydi gloydi*; Smith 1946 [Kansas Univ. Sci. Bull., Vol. 31, No. 1: 73]
>
> *Crotalus intermedius intermedius*; Klauber 1952 [Bull. Zool. Soc. San Diego, No. 26: 9]
>
> *Crotalus intermedius intermedius*; Klauber 1956 [Rattlesnakes, Vol. 1: 36]

Type locality: Mexico.

Range: Imperfectly known from Michoacan, northeastern Puebla, west central Vera Cruz, and central Oaxaca (Map 5).

Crotalus intermedius omiltemanus Günther
Omilteman Small-Headed Rattlesnake
   Crotalus omiltemanus Günther 1895 [Biol. Centr. Amer., Rept. Batr., p. 192]
   Crotalus triseriatus; Boulenger 1896 (*partim*) [Cat. Sn. Brit. Mus., Vol. 3: 381]

   Crotalus triseriatus omiltemanus; Klauber 1938 [Copeia, No. 4: 196]
   Crotalus intermedius omiltamanus; Klauber 1952 [Bull. Zool. Soc. San Diego, No. 26: 14]
   Crotalus intemedius omiltemanus; Klauber 1956 [Rattlesnakes, Vol. 1: 36]

Type locality: Omilteme Guerero, Mexico.
Range: Mexico, Central Guerero (Map 5).

Crotalus Lannomi Tanner
   Crotalus lannomi Tanner 1966 [Herpetologica, Vol. 22, No. 4: 298]

Type locality: 1.8 miles west of the pass Puerto Los Mazos or 22 miles west by road from Tuxcacuisco, a branch of the Rio Armenria on Mexican Highway n.80, Jalisco, Mexico.

Range: Known from type locality only.

Crotalus lepidus lepidus (Kennicott)
Mottled Rattlesnake
   Caudisona lepida Kennicott 1861 [Proc. Acad. Nat. Sci. Phil., Vol. 13: 206]
   Crotalus lepidus; Cope 1883 (*partim*) [Proc. Acad. Nat. Sci. Phil., Vol. 35: 13]
   Crotalus (tigris) palmeri Garman 1887 [Bull. Exxes Inst., Vol. 19: 124]
   Crotalus lepidus lepidus; Gloyd 1936 [Occ. Pap. Mus. Zool. Univ. Mich., No. 337: 2]
   Crotalus lepidus lepidus; Klauber 1956 [Rattlesnake, Vol. 1: 37]

Type locality: Presidio del Norte and Eagle pass Texas, Restricted [Smith & Taylor, 1950]: Presidio del Norte, Presidio County, Texas, U.S.A.
Range: U.S.A., southern New Mexico, southeastern Texas. Mexico, northeastern Mexico, Cahuila Nuevo León, Western San Luiz Potosí and southeastern Zacatecas (intergradation with *c. l. klauberi*) (Map 12).

Crotalus lepidus klauberi Gloyd
Banded Rock Rattlesnake
   Crotalus lepidus klauberi Gloyd 1936 [Occ. Pap. Mus. Zool. Univ. Mich., No. 337: 2]
   Crotalus semicornutus Taylor 1944 [Kansas Univ. Sci. Bull., Vol. 30, No. 1: 52]
   Crotalus lepidus klauberi; Klauber 1956 [Rattlesnakes, Vol. 1: 37]

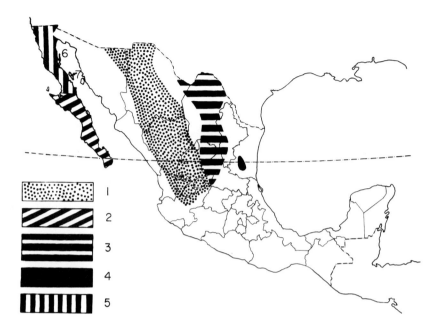

FIG. 12. Distribution of *C. lepidus klauberi* (1), *C. mitchelli* (2), *C. l. lepidus* (3), *C. l. morulus* (4), *C. pyrrhus* (5), *C. mitchelli muertensis* (6), *C. mitchelli angelicus* (7).

Type locality: Carr Canyon, Huachuca Mountains, Cochise County, Arizona, U.S.A.

Range: U.S.A., southeastern Arizona, southwestern New Mexico, the El Paso area in Texas, north central Mexico, including the Santa Rita Huachuca, Draggon Dez Cabezas and Chiricahua Mountains of southeastern Arizona. Mexico (mountains): northeastern Sonora, southeastern Sinalora, Chihuahua, Durango, Nayarit (in this area *Crotalus lepidus klauberi* are different and may justify the recognition of another subspecies (vide Klauber 1956) (Map 12).

Crotalus lepidus morulus Klauber
Taumalipan Rock Rattlesnake
  Crotalus lepidus morulus Klauber 1952 [Bull. Zool. Soc. San Diego, No. 26: 52]
  Crotalus lepidus morulus; Klauber 1956 [Rattlesnakes, Vol. 1: 37]

Type locality: 10 miles northwest of Gomez Farias on the trail to la Voya de Salas, Tamaulipas, Mexico.
Range: Mexico, mountains northwest of Gomez Farias, and near Chinas, Tamaulipas, Mexico (Map 12).

Crotalus mitchellii mitchellii (Cope)
San Lucas Speckled Rattlesnake
  Caudiona mitchellii Cope 1861 [Proc. Acad. Nat. Sci. Phil., Vol. 13: 293]
  Crotalus mitchellii; Cope 1875 [*In* Yarrow, Surv. W of 100th Merid.
    (Wheeler), Vol. 5, No. 4: 535]
  Crotalus mitchellii mitchellii; Stejneger 1895 (*partim*) [Stejneger, Rept.
    U.S. Nat. Mus., p. 454]
  Crotalus mitchellii mitchellii Klauber 1936 [Trans. San Diego Soc. Nat.
    Hist., Vol. 8, No. 19: 154]
  Crotalus mitchellii mitchellii; Klauber 1956 [Rattlesnakes, Vol. 1: 38]

Type locality: Cape San Lucas, Baja California, Mexico.
Range: Mexico, District del Sur of Baja California, Islands of Cerralvo,
  Espiritu Santo, San Jose, Santa Cruz, Carmen (Gulf Coast and
  Margarita Pacific coast. Intergrades with *C. m. pyrrhus* (Map 12).

Crotalus mitchellii angelicus Klauber
  Crotalus mitchellii angelicus Klauber 1964 [Trans. San Diego Soc. Nat.
    Hist., Vol. 13, No. 5: 75–80]

Type locality: Isla Angel de la Guardia, Mexico.
Range: Mexico, Isla Angel de la Guardia (Map 12).

Crotalus mitchellii muertensis Klauber
El Muerto Island Speckled Rattlesnakes
  Crotalus mitchellii muertensis; Klauber 1949 [Trans. San Diego Soc. Nat.
    Hist., Vol. 11, No. 6: 97]
  Crotalus mitchellii muertensis; Klauber 1957 [Bull. Zool. Soc. San Diego,
    No. 26: 123]

Type locality: El Muerto Island, San Luzi Group, Gulf of California, coast of
  Baja California, Mexico.
Range: Known only from type locality (Map 12).

Crotalus mitchellii pyrrhus Cope
Southwestern Speckled Rattlesnake
  Caudisona pyrrhus Cope 1866 [Proc. Acad. Nat. Sci. Phil., Vol. 18: 308]
  Crotalus pyrrhus; Cope 1875 [*In* Yarrow, Surv. W of 100th Merid.(Wheeler)
    Vol. 5, No. 5: 535]
  Crotalus mitchellii pyrrhus; Stejneger 1895 (*partim*) [Rept. U.S. Nat. Mus.
    p. 456]
  Crotalus godmani; Schmidt 1922 [Bull. Amer. Mus. Nat. Hist., Vol. 46,
    No. 11: 70]
  Crotalus mitchellii pyrrhus; Klauber 1936 [Trans. San Diego Soc. Nat.
    Hist.]
  Crotalus mitchellii pyrrhus; Klauber 1958 [Rattlesnakes, Vol. 1]

Type locality: "Not stated" (stated by Klauber 1956) as a "Canyon Prieto, Yavapac County, Arizona," U.S.A.

Range: U.S.A., Southern California, southern Nevada, western Arizona, northwestern Sonora, Mexico, northern Baja California (Map 12).

Crotalus molossus molossus Baird & Girard
Northern Black-Tailed Rattlesnake
   Crotalus molossus Baird & Girard 1893 [Cat. Sn. Amer. Rept., No. 1: 10]
   Crotalus ornatus; Hallowell 1854 [Proc. Acad. Nat. Sci. Phil., Vol. 7: 192]
   Crotalus molossus molossus; Gloyd 1936 [Occ. Pap. Mus. Zool. Univ. Mich., No. 325: 2]
   Crotalus molossus molossus; Klauber 1956 [Rattlesnakes, Vol. 1: 39]

Type locality: Fort Webster, Santa Rita del Cobre, Grand County, New Mexico, U.S.A.

Range: U.S.A. from central Texas to western Arizona, Mexico, northern Mexico (Map 7).

Crotalus molossus estebanensis Klauber
San Esteban Island Rattlesnake
   Crotalus molossus estebanensis Klauber 1949 [Trans. San Diego Soc. Nat. Hist., Vol. 11, No. 6: 104]

Type locality: San Esteban Island, Gulf of California, Mexico.
Range: Mexico, only on San Esteban Island, Gulf of California.

Crotalus molossus nigrescens Gloyd
Mexican Black-Tailed Rattlesnake
   Crotalus molossus nigrescens; Gloyd 1936 [Occ. Pap. Mus. Zool. Univ. Mich., No. 325: 2]
   Crotalus molossus nigrescens; Klauber 1956 [Rattlesnakes, Vol. 1: 39–40]

Type locality: 4 miles west of La Colorada, Zacatecas, Mexico.
Range: Mexico, tableland from southern Sonora, southwestern Chihuahua and southern Coahiula to Oaxaca and Vera Cruz. Probably also in Eastern Jalisco, Aguas Calientes and Morelos (Map 6).

Crotalus polystictus Cope
Mexican Lace-Headed Rattlesnake
   Caudisona polysticta Cope 1865 [Proc. Acad. Nat. Hist. Phil., Vol. 17: 191]
   Crotalus polystictus; Cope 1875 [*In* Yarrow, Surv. W. of 100th Merid. (Wheeler), Vol. 5, No. 4: 533]
   Crotalus jimenezi Dugés 1877 [Naturaleza, Vol. 4: 23]
   Crotalus polystictus; Klauber 1896 [Rattlesnakes, Vol. 1]

Type locality: Restricted [Taylor & Smith 1950] Tupataro, Guanaputo, Mexico.

Range: Mexico, tableland of Central Mexico from southern Zacatecas to central Vera Cruz including eastern Jalisco, Guanajuato, Michoacan, Distrito Federal Morelos and central Vera Cruz, probably also in the State of Aguas Calientes, Gueretaro, Hidalgo, Mexico, Tlaxcala, Puebla, and Oaxaca (Map 13) .

Crotalus pricei pricei Van Denburgh
Arizona Twin-Spotted Rattlesnake
  Crotalus pricei Van Denburgh 1895 [Proc. Calif. Acad. Sci., Vol. 5, No. 2: 856]
  Crotalus triseriatus pricei; Klauber 1931 [*In* Githens and George, Bull. Antivenin Inst. Amer., Vol. 5, No. 2: 33]
  Crotalus pricei pricei; Smith 1946 [Kansas Univ. Sci. Bull., Vol. 31, No. 1: 79]
  Crotalus pricei pricei; Klauber 1956 [Rattlesnakes, Vol. 1: 40]

Type locality: Huachuca Mountain, Cochise County, Arizona, U.S.A.
Range: U.S.A., southeastern mountains of Arizona. Mexico, northwestern mountains. Including the Pinaleo, Santa Rita, Huachuca and Chiricahua mountains in Arizona, and the Sierra Tarahumare and Sierra Madre in eastern Sonora, western Chihuahua, and Durango, probably

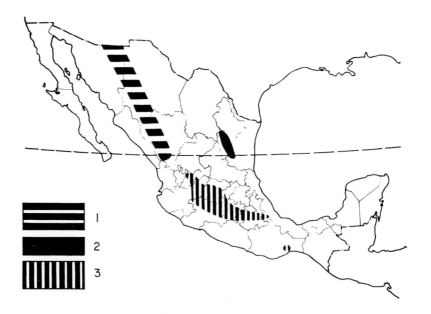

FIG. 13. Distribution of *Crotalus pricei pricei* (1), *C. miquihuanus* (2), and *C. polystictus* (3).

also in the mountains of eastern Sinalon and northern Nayarit (Map 13).

Crotalus pricei miquihuanus Gloyd
Miquihuanan Twin-Spotted Rattlesnakes
> *Crotalus triseriatus triseriatus*; Klauber 1936 (*partim*) [Trans. San Diego Sci. Hist. Nat., Vol. 8, No. 20: 208]
> *Crotalus triseriatus miquihuanus* Gloyd 1940 [Chicago Acad. Sci. Spec. Publ., Vol. 4: 102]
> *Crotalus pricei miquihuanus*; Smith 1946 [Kansas Univ. Sci. Bull., Vol. 31, No. 1: 79]
> *Crotalus pricei miquihuanus*; Klauber 1956 [Rattlesnakes, Vol. 1]

Type locality: Cerro Potosí, near Galeana, Nuevo Leon, Mexico.
Range: Mexico, southeastern Nuevo Leon and southwestern Tamaulipas, and extreme southeastern Coahuila (Map 13).

Crotalus pusillus Klauber
Tancitaron Dusky Rattlesnake
> *Crotalus pusillus* Klauber 1952 [Bull. Zool. Soc. San Diego Soc. Nat. Hist., Vol. 8, No. 20: 247]
> *Crotalus pusillus*; Klauber 1956 [Rattlesnakes, Vol. 1: 41]

Type locality: Tancitaro Michoacan, alt. 5,000 ft., Mexico.
Range: Mexico, mountains of western Michoacan and southern Jalisco (Map 7).

Crotalus ruber ruber Cope
Red Diamond Rattlesnakes
> *Crotalus adamanteus ruber* Cope 1892 [Proc. U.S. Nat. Mus., Vol. 14: 690]
> *Crotalus atrox elegans* Schmidt 1922 [Bull. Am. Mus. Nat. Hist., Vol. 46, No. 11: 699]
> *Crotalus ruber ruber*, Klauber 1949 [Trans. San Diego Soc. Nat. Hist., Vol. 11, No. 5: 59]
> *Crotalus ruber ruber*; Klauber 1956 [Rattlesnakes, Vol. 1: 41]

Type locality: Not designated. Restricted [Smith & Taylor 1959] to Dulzura, San Diego County, California, U.S.A.
Range: U.S.A., southwest California to Mexico, northern Baja California (Map 5).

Crotalus ruber lucasensis Van Denburgh
San Lucas Diamond Rattlesnake
> *Crotalus lucasensis* Van Denburgh 1920 [Proc. Calif. Acad. Sci., Ser. 4, Vol. 10, No. 2: 29]

Crotalus ruber lucasensis; Klauber 1949 [Trans. San Diego Soc. Nat. Hist., Vol. 11, No. 5: 59]

Crotalus ruber lucasensis; Klauber 1956 [Rattlesnakes, Vol. 1: 42]

Type locality: Agua Caliente, Cape region of Baja California, peninsula from Loreto to Cape San Lucas, and the islands of Santa Margarita and San José (Map 5).

Crotalus scutulatus scutulatus (Kennicott)
Mojave Rattlesnake
  Caudisona scutulata Kennicott 1861 [Proc. Acad. Nat. Sci. Phil., Vol. 13: 207]
  Crotalus scutulatus; Cope 1875 [*In* Yarrow, Surv. W. of 100th Merid. (Wheeler), Vol. 5, No. 4: 533]
  Crotalus scutulatus; Boulenger 1896 (*partim*) [Cat. Sn. Brit. Mus., Vol. 3: 575]
  Crotalus confluentus kellyi Amaral 1929 [Bull. Antivenin Inst. Amer., Vol. 2, No. 4: 91]
  Crotalus scutulatus scutulatus; Gloyd 1940 [Chicago Acad. Sci. Spec. Publ., No. 4: 200]
  Crotalus scutulatus scutulatus; Klauber 1956 [Rattlesnakes, Vol. 1: 42]

Type locality: Designated [Smith & Taylor 1950] Wickenburg, Maricopa, County, Arizona, U.S.A.

Range: U.S.A., Mexico, from the Mojave Desert, California, southeast to south central Mexico (Map 11).

Crotalus scutulatus salvini Günther
Huamanthan Rattlesnake
  Crotalus salvini Günther 1895 [Biol. Centr. Amer. Rept. Batr., p. 193]
  Crotalus scutulatus; Boulenger 1896 (*partim*) [Cat. Sn. Brit. Mus., Vol. 3: 575]
  Crotalus scutulatus salvini; Gloyd 1940 [Chicago Acad. Sci. Spec. Publ., No. 4: 201]
  Crotalus scutulatus salvini; Klauber 1956 [Rattlesnakes, Vol. 1: 201]

Type locality: Huamantla [Tlaxcala] Mexico, alt. 8,000 ft.

Range: Mexico, Tlaxcala, Pueblo, Grerétaro and West Central Vera Cruz probably present in Hidalgo (Map 11).

Crotalus stejnegeri Dunn
Long Tailed Rattlesnake
  Crotalus stejnegeri Dunn 1919 [Proc. Biol. Soc. Wash., Vol. 32: 214]
  Crotalus stejnegeri; Klauber 1956 [Rattlesnakes, Vol. 1: 43]

Type locality: Plumosos=Plomosas, Sinaloa, Mexico.
Range: Mexico. The mountains of southwest and western Durango.

Crotalus tigris Kennicott
Tiger Rattlesnakes
  Crotalus tigris Kennicott 1859 [*In* Baird, Rept. Boundary U.S. Mex.
    Boundary Surv., Vol 2: 14]
  Crotalus tigris; Boulenger 1896 [Cat. Sn. Brit. Mus., Vol. 3: 580]
  Crotalus tigris; Klauber 1956 [Rattlesnakes, Vol. 1: 43]

Type locality: Sierra Verde, and Pozo Verde (on the Sonoran side of the
  U.S.A. near Sarbe, Sonora), Mexico.
Range: U.S.A., the rocky desert foothills of south central Arizona. Mexico,
  northern Sonora (Map 7).

Crotalus tortuguensis Van Denburg & Slevin
Tortuga Island Diamond Rattlesnakes
  Crotalus tortuguensis Van Denburg and Slevin 1921 [Proc. Calif. Acad.
    Sci., Vol. 11, No. 4: 398]
  Crotalus atrox tortuguensis; Stejneger & Barbour 1933
  Crotalus tortuguensis; Klauber 1956 [Rattlesnakes, Vol. 1: 43]

Type locality: Tortuga Island, Gulf of California, Mexico.
Range: Only known from Tortuga Island, Mexico (Map 7).

Crotalus transversus Taylor
Cross-Banded Mountain Rattlesnakes
  Crotalus transversus Taylor 1944 [Kansas Univ. Sci. Bull., Vol. 30, No. 1:
    47]
  Crotalus transversus, Klauber 1956 [Rattlesnakes, No. 1: 43]

Type locality: Near Tres Marias (Tres Cumbre) about 55 km southwest of
  Mexico, D. F. Mexico.
Range: Mexico, Distrito Federal and northwestern Morelos (Map 7).

Crotalus triseriatus triseriatus (Wagler)
Central Plateau Dusky Rattlesnakes
  Urosophis triseriatus Wagler 1830 [Nat. Syst. Amph., p. 176]
  Crotalus triseriatus Gray 1831 [Syn. Spec. Clas. Rept. In Griffith, Animal
    Kingdom by Cuvier]
  Crotalus lugubris Jan 1859 [Rev. Mag. Zool., Vol. 10: 153 and 1956]
  Crotalus pallidus Günther 1895 [Biol. Centr. Amer. Rept. Batr., p. 193
    (*partim*)]
  Crotalus triseriatus; Boulenger 1896 [Cat. Sn. Brit. Mus., Vol. 3: 581]

Crotalus triseriatus triseriatus Klauber 1931 (*partim*) [*In* Githens and
George, Bull. Antivenin Inst. Amer., Vol. 5, No. 2: 33]
Crotalus triseriatus anahuacus Gloyd 1940 [Chicago Acad. Sci. Spec. Publ.,
No. 4: 91]
Crotalus triseriatus triseriatus, Klauber 1952 [Rattlesnakes, Vol. 1: 44]
Crotalus triseriatus triseriatus; Klauber 1956 [Rattlesnakes, Vol. 1: 44]

Type locality: Restricted (Smith & Taylor 1950) Alvarez, San Luiz Potosí,
Mexico.

Range: Mexico, Nayarit, west central Vera Cruz, Jalisco, Michoacan,
Distrito Federal, Morelos, Pueblo, and west central Vera Cruz,
probably in Tlaxcala.

Crotalus triseriatus aquilus Klauber
Queretaran Dusky Rattlesnakes
Crotalus triseriatus aquilus Klauber 1952 [Bull. Zool. Soc. San Diego,
No. 26: 24]
Crotalus triseriatus aquilus; Klauber 1956 [Rattlesnakes, Vol. 1: 44]

Type locality: Near Alvarez, San Luiz Potosí, Mexico.

Range: Mexico, southern San Luiz Potosí, Guanajuato, northeast Michoacan,
Queretaro and Hidalgo, probably in northwestern Vera Cruz.

Crotalus vegrandis Klauber
Uracoan Rattlesnake
Crotalus vegrandis Klauber 1941 [Trans. San Diego Soc. Nat. Hist., Vol.
9, No. 30: 331]
Crotalus durissus vegrandis; Klauber 1956 [Rattlesnakes, Vol. 1: 34]
Crotalus vegrandis; Hoge 1966(1955) [Mem. Inst. Butantan, Vol. 34, 149;
Pl. 18]

Type locality: Maturin Savannah, near Uracoa, distr. Sotillo, State of
Monagas, Venezuela.

Range: Known from several localities in states of Monagas and Anzoateguy,
Venezuela.

Crotalus viridis viridis (Rafinesque)
Prairie Rattlesnakes
Crotalus viridis: Rafinesque 1818 [Amer. Mon. Mag. Crit. Rev., Vol. 4, No.
1: 41]
Crotalus confluentus; Say 1823 [*In* Long's Exped. from Pittsburgh to
Rocky Mountains]
Crotalus lecontei; Hallowell 1852 [Proc. Acad. Nat. Sci. Phil., Vol. 6: 100]
Crotalus confluentus var. *pulverulentus*; Cope 1883 [Proc. Acad. Nat. Sci.
Phil., Vol. 35: 11]

Crotalus confluentus; Boulenger 1897 (*partim*) [Cat. Sn. Brit. Mus., Vol. 3: 576]

Crotalus viridis viridis; Klauber 1936 [Trans. San Diego Soc. Nat. Hist., Vol. 8, No. 2: 191]

Crotalus viridis viridis; Klauber 1956 [Rattlesnakes, Vol. 1: 45]

Type locality: Restricted (Smith & Taylor 1950), Cross Boyd County, Nebraska, U.S.A.

Range: North America, the Great Plains from 96°W to the Rocky Mountains and from southern Canada to Mexico, extreme northern Sonora, northern Chihuahua, and northern Coahuila.

Crotalus viridis caliginis Klauber
Coronado Island Rattlesnake
  Crotalus confluentus; Boulenger 1896 (*partim*) [Cat. Sn. Brit. Mus., Vol. 3: 576]
  Crotalus viridis caliginis; Klauber 1949 [Trans. San Diego Soc. Nat. Hist., Vol. 11, No. 6: 90]
  Crotalus viridis caliginis; Klauber 1956 [Rattlesnakes, Vol. 1: 45–46]

Type locality: South Coronado Island, northwest coast of Baja California, Mexico.

Range: Known from type locality.

Crotalus viridis helleri Meek
Southern Pacific Rattlesnake
  Crotalus helleri Meek 1905 [Field. Cat. Mus. Zool. Sci., Vol. 3, No. 1: 7]
  Crotalus viridis helleri; Klauber 1949 [Trans. San Diego Soc. Nat. Hist., Vol. 11, No. 6: 77]
  Crotalus viridis helleri; Klauber 1956 [Rattlesnakes, Vol. 1: 46–47]

Type locality: San José, California, Baja California, Mexico.
Range: U.S.A. southwest California, Mexico.

Crotalus willardi willardi Meek
Arizona Ridge-Nosed Rattlesnake
  Crotalus willardi Meek 1905 [Field. Columbian Misc. Publ. 104, Zool. Ser., Vol. 7, No. 1: 18]
  Crotalus willardi willardi; Klauber 1956 [Rattlesnakes, Vol. 1: 48]

Type locality: (Corrected Swarth 1921) above Hamburg, Middle branch of Ramsey Canyon. Huachuca Mountains alt. above 7,000 ft. Cochise County Arizona, U.S.A.

Range: U.S.A. the Huachuca and Santa Rita Mountains, southern Arizona, Mexico, Sierra de Ojos and Sierra Azul, northern Sonora.

Crotalus willardi amabilis Anderson
  Crotalus willardi amabilis; Anderson 1962 [Copeia, No. 1: 160–163]

Type locality: Arroyo Mesteño 8,500 ft Sierra del Mido, Chihuahua, Mexico
Range: Known only from the Sierra del Mido proper, but may also occur
    in the high county of the Sierra Santa Clara and Cerro Campañas
    which are included in the Sierra del Mido complex.

Crotalus willardi meridionalis Klauber
Southern Ridge-Nosed Rattlesnake
  Crotalus willardi meridionalis; Klauber 1949 [Trans. San Diego Soc. Nat.
    Hist., Vol. 11, No. 8: 131]
  Crotalus willardi silus; Klauber 1956.
Chihuahua Ridge-Nosed Rattlesnake
  Crotalus willardi silus; Klauber 1949 [Trans. San Diego Soc. Nat. Hist.,
    Vol. 11, No. 8: 128]
  Crotalus willardi silus; Klauber 1956 [Rattlesnakes, Vol. 1: 49]

Type locality: On the Rio Cavilán, 7 miles southwest of Pacheco, Chihuahua,
    Mexico, alt. 6,200 ft.
Range: Mexico, northeastern Sonora and western Chihuahua.

CROTALUS (Sistrurus) Garman

Crotalus (Sistrurus) catenatus edwardsi Baird & Girard
Edward's Massasauga
  Crotalophorus edwardsi; Baird & Girard 1853.
  Crotalus catenatus; Boulenger 1896 *(partim)* [Cat. Sn. Brit. Mus., Vol. 3:
    571]
  Sistrurus catenatus edwardsii; Gloyd 1955 [Bull. Chicago Acad. Sci.,
    Vol. 10, No. 6: 83]

Type locality: Tamaulipas, Mexico.
Range: U.S.A. extreme southern and trans-Pecos Texas, Colorado, New
    Mexico, and Arizona. Mexico, extreme northern Tamaulipas.

Crotalus (Sistrurus) ravus (Cope)
Mexican Pigmy Rattlesnake
  Crotalus ravus; Cope 1865 [Proc. Acad. Nat. Sci. Phil., Vol. 17, No. 4: 191]
  Sistrurus ravus; Boulenger 1896 [Cat. Sn. Brit. Mus., Vol. 3: 571]
  Sistrurus ravus; Klauber 1956 [Rattlesnakes, Vol. 1: 51]

Type locality: Tableland of Mexico, restricted (Smith & Taylor) Tolalco,
    Vera Cruz, Mexico.
Range: Mexico, eastern and south-central. State of Mexico, Distrito Federal
    Morelos, Tlaxcala, Puebla, west-central Vera Cruz and east-central
    Oaxaca.

LACHESIS Daudin

Lachesis muta muta (Linnaeus)
Bushmaster, Surucucu (Brazil)
  Crotalus mutus Linnaeus 1766 [Syst. Nat. 12th ed., p. 373]
  Lachesis mutus; Daudin 1803 [Hist. Nat. Rept., Vol. 5: 351]
  Lachesis muta; Boulenger 1896 [Cat. Sn. Brit. Mus., Vol. 3:]
  Lachesis muta muta; Taylor 1951 (as cons. of the use of *L. stenophrys* Cope
    as a subspecies of *muta*); [Kansas Univ. Sci. Bull., Vol. 34, No. 1:
    184]
  Lachesis muta muta; Hoge 1966 [Mem. Inst. Butantan, Vol. 32: 161]

Type locality: Surinam.
Range: the equatorial forests of Brazil, Guianas, Venezuela, Trinidad, Bolivia,
  Peru, Ecuador, and Colombia. [Map 3 (1)].

Lachesis muta stenophrys Cope
  Lachesis stenophrys Cope 1876 (1875) [Journ. Acad. Nat. Sci. Phil.,
    Vol. 8, No. 2: 152]
  Lachesis muta stenophrys; Taylor 1951 [Kansas Univ. Sci. Bull., Vol. 24,
    No. 1: 184]

Type locality: Sipurio, Costa Rica.
Range: The forests of Costa Rica and Panama (the intergradation zone with
  L. muta, either in southern Panama or northern Colombia is not
  known).

Lachesis muta noctivaga Hoge
  Lachesis muta noctivaga Hoge 1966 [Mem. Inst. Butantan, Vol. 32: 162]

Type locality: Vitoria, Espirito Santo, Brazil.
Range: The forests along the Atlantic, from the State of Alagoas, to Rio de
  Janeiro. All specimens are from the coastal forest, and in the gallery
  forests of Rio Doce. [Map 3 (2)].

D. Faunal Lists

ARGENTINA

Micrurus corallinus
Micrurus frontalis frontalis
Micrurus frontalis altirostris
Micrurus frontalis pyrrhocryptus
Bothrops alternatus
Bothrops ammodytoides
Bothrops cotiara

Bothrops jararaca
Bothrops jararacussu
Bothrops neuwiedi diporus
Crotalus durissus collilineatus
Crotalus durissus terrificus

BOLIVIA

Leptomicrurus narduccii
Micrurus annellatus balzani
Micrurus annellatus bolivianus
Micrurus annellatus montanus
Micrurus frontalis pyrrhocryptus
Micrurus lemniscatus frontifasciatus
Micrurus lemniscatus helleri
Micrurus spixii princeps
Micrurus surinamensis surinamensis
Micrurus tschudii tschudii
Bothrops atrox
Bothrops bilineatus smaragdinus
Bothrops brazili
Bothrops castelnaudi
Bothrops hyoprorus
Bothrops jararacussu
Bothrops microphthalmus microphthalmus
Bothrops neuwiedi bolivianus
Bothrops oligolepis
Bothrops sanctaecrucis
Crotalus durissus collilineatus
Crotalus durissus terrificus
Lachesis muta muta

BRAZIL

Leptomicrurus narduccii
Leptomicrurus collaris
Leptomicrurus schmidti
Micrurus albicinctus
Micrurus corallinus
Micrurus decoratus
Micrurus filiformis filiformis
Micrurus filiformis subtilis
Micrurus frontalis frontalis

Micrurus frontalis altirostris
Micrurus frontalis brasiliensis
Micrurus frontalis pyrrhocryptus
Micrurus hemprichii ortoni
Micrurus ibiboboca
Micrurus langdorffi langsdorffi
Micrurus lemniscatus carvalhoi
Micrurus lemniscatus lemniscatus
Micrurus lemniscatus helleri
Micrurus spixii spixii
Micrurus spixii martiusi
Micrurus surinamensis surinamensis
Micrurus surinamensis nattereri
Bothrops alternatus
Bothrops atrox
Bothrops bilineatus bilineatus
Bothrops bilineatus smaragdinus
Bothrops brazili
Bothrops castelnaudi
Bothrops cotiara
Bothrops erythromelas
Bothrops fonsecai
Bothrops hyoprorus
Bothrops iglesiasi
Bothrops insularis
Bothrops itapetiningae
Bothrops jararaca
Bothrops jararacussu
Bothrops marajoensis
Bothrops moojeni
Bothrops neuwiedi neuwiedi
Bothrops neuwiedi bolivianus
Bothrops neuwiedi diporus
Bothrops neuwiedi goyazensis
Bothrops neuwiedi lutzi
Bothrops neuwiedi mattogrossensis
Bothrops neuwiedi meridionalis
Bothrops neuwiedi paranaensis
Bothrops neuwiedi pauloensis
Bothrops neuwiedi piauhyensis
Bothrops neuwiedi pubescens
Bothrops neuwiedi urutu

Bothrops pirajai
Crotalus durissus cascavella
Crotalus durissus collilineatus
Crotalus durissus marajoensis
Crotalus durissus ruruima
Crotalus durissus terrificus
Lachesis muta muta
Lachesis muta noctivaga

BRITISH HONDURAS

Micrurus diastema sapperi
Micrurus hippocrepis
Micrurus nigrcinctus divaricatus
Bothrops asper
Bothrops schlegeli
Crotalus durissus tzabcan

CARIBBEAN ISLANDS

Martinique

Bothrops lanceolatus

Saint Lucie

Bothrops caribaeus

Netherland West Indies

Crotalus durissus unicolor

COLOMBIA

Leptomicrus narduccii
Micrurus ancoralis jani
Micrurus bocourti sangilensis
Micrurus carinicauda carinicauda
Micrurus carinicauda antioquensis
Micrurus carinicauda colombianus
Micrurus carinicauda dumerilii
Micrurus carinicauda transandinus
Micrurus clarki
Micrurus dissoleucus dissoleucus
Micrurus dissoleucus melanogenys

Micrurus dissoleucus nigrirostris
Micrurus filiformis filiformis
Micrurus filiformis subtilis
Micrurus hemprichii hemprichii
Micrurus hemprichii ortoni
Micrurus isozonus
Micrurus langsdorffi langsdorffi
Micrurus lemniscatus helleri
Micrurus mipartitus mipartitus
Micrurus mipartitus anomalus
Micrurus mipartitus decussatus
Micrurus nigrocinctus nigrocinctus
Micrurus psyches psyches
Micrurus psyches medemi
Micrurus spixii obscurus
Micrurus spurrelli
Micrurus surinamensis surinamensis
Pelamis platurus
Bothrops atrox
Bothrops asper
Bothrops bilineatus smaragdinus
Bothrops brazili
Bothrops castelnaudi
Bothrops hyoprora
Bothrops lansbergii lansbergii
Bothrops nasutus
Bothrops microphthalmus colombianus
Bothrops punctatus
Bothrops schlegelli
Bothrops xantogrammus
Crotalus cumanensis
Lachesis muta muta

COSTA RICA

Micrurus alleni alleni
Micrurus alleni yatesi
Micrurus clarki
Micrurus mipartitus hertwegi
Micrurus nigrocinctus nigrocinctus
Micrurus nigrocinctus melanocephalus
Micrurus nigrocinctus mosquitensis
Pelamis platurus

Bolthrops asper
Bothrops godmanni
Bothrops lateralis
Bothrops nasutus
Bothrops nigroviridis nigroviridis
Bothrops neuwiedi mexicanus
Bothrops ophryomegas
Bothrops picadoi
Bothrops schlegeli
Bothrops supraciliaris
Crotalus durissus durissus
Lachesis muta stenophrys

ECUADOR

Leptomicrurus narduccii
Micrurus ancoralis ancoralis
Micrurus annellatus annellatus
Micrurus bocourti bocourti
Micrurus carinicaudus transandinus
Micrurus hemprichi ortoni
Micrurus langsdorffi langsdorffi
Micrurus langsdorffi ornatissimus
Micrurus lemniscatus helleri
Micrurus mertensi
Micrurus mipartitus decussatus
Micrurus spixii obscurus
Micrurus steindachneri steindachneri
Micrurus steindachneri orcesi
Micrurus steindachneri petersi
Micrurus surinamensis surinamensis
Micrurus tschudii olssoni
Pelamis platurus
Bothrops albocarinatus
Bothrops alticolus
Bothrops atrox
Bothrops asper
Bothrops bilineatus smaragdinus
Bothrops castelnaudi
Bothrops hyoprorus
Bothrops lojanus

Bothrops nasutus
Bothrops microphthalmus microphthalmus
Bothrops pulcher
Bothrops punctatus
Bothrops schlegeli
Bothrops xantogrammus
Lachesis muta muta

EL SALVADOR

Micrurus nigrocinctus zunilensis
Pelamis platurus
Bothrops asper
Bothrops godmanni
Bothrops nummifer occiduus
Bothrops ophryomegas
Bothrops schlegeli
Crotalus durissus durissus

GUATEMALA

Micrurus browni browni
Micrurus browni importunus
Micrurus diastema apiatus
Micrurus diastema sapperi
Micrurus elegans veraepacis
Micrurus hyppocrepis
Micrurus latifasciatus
Micrurus nigrocinctus zunilensis
Micrurus stuarti
Pelamis platurus
Bothrops bicolor
Bothrops godmanni
Bothrops nasutus
Bothrops nigroviridis aurifer
Bothrops nummifer mexicanus
Bothrops nummifer occiduus
Bothrops ophryomegas
Bothrops schlegeli
Crotalus durissus durissus
Crotalus durissus tzabcan

THE GUIANAS

Leptomicrurus collaris
Micrurus averyi
Micrurus hemprichii hemprichii
Micrurus lemniscatus lemniscatus
Micrurus lemniscatus diutius
Micrurus psyches psyches
Micrurus surinamensis surinamensis
Bothrops atrox
Bothrops bilineatus bilineatus
Bothrops brazili
Crotalus durissus dryinus
Crotalus durissus cumanensis
Lachesis muta muta

HONDURAS

Micrurus diastema aglaeope
Micrurus nigrocinctus divaricatus
Micrurus nigrocinctus zunilensis
Micrurus ruatanus
Bothrops asper
Bothrops godmanni
Bothrops langsbergii annectens
Bothrops nasutus
Bothrops nigroviridis marchi
Bothrops nummifer mexicanus
Bothrops schlegeli
Crotalus durissus durissus

MEXICO

Micruroides euryxanthus euryxanthus
Micruroides euryxanthus australis
Micruroides euryxanthus neglectus
Micrurus bernardi
Micrurus bogerti
Micrurus browni browni
Micrurus browni taylori
Micrurus diastema diastema
Micrurus diastema affinis
Micrurus diastema alienus

Micrurus diastema apiatus
Micrurus diastema macdougalli
Micrurus diastema sapperi
Micrurus distans distans
Micrurus distans michoacanensis
Micrurus distans oliveri
Micrurus distans zweifeli
Micrurus elegans elegans
Micrurus elegans veraepacis
Micrurus ephippifer
Micrurus fitzingeri
Micrurus fulvius maculatus
Micrurus fulvius microgalbineus
Micrurus fulvius tenere
Micrurus laticollaris laticollaris
Micrurus laticollaris maculirostris
Micrurus latifasciatus
Micrurus limbatus
Micrurus nigrocinctus zunilensis
Micrurus nuchalis
Micrurus proximans
Pelamis platurus
Agkistrodon bilineatus bilineatus
Agkistrodon bilineatus taylori
Agkistrodon contortrix pictigaster
Bothrops asper
Bothrops barbouri
Bothrops bicolor
Bothrops dunni
Bothrops melanurus
Bothrops nasutus
Bothrops nigroviridis aurifer
Bothrops nummifer nummifer
Bothrops nummifer mexicanus
Bothrops nummifer occiduus
Bothrops schlegeli
Bothrops sphenophrys
Bothrops undulatus
Bothrops yucatanicus
Crotalus atrox
Crotalus basiliscus basiliscus
Crotalus basiliscus oaxacus

Crotalus catalinensis
Crotalus cerastes cercobombus
Crotalus cerastes laterorepens
Crotalus durissus durissus
Crotalus durissus culminatus
Crotalus durissus totonacus
Crotalus durissus tzabcan
Crotalus enyo enyo
Crotalus enyo cerralvensis
Crotalus enyo furvus
Crotalus exsul
Crotalus intermedius intermedius
Crotalus intermedius omiltemanus
Crotalus lannomi
Crotalus lepidus lepidus
Crotalus lepidus klauberi
Crotalus lepidus morulus
Crotalus mitchellii mitchellii
Crotalus mitchellii angelicus
Crotalus mitchellii muertensis
Crotalus mitchellii pyrrhus
Crotalus mollossus molossus
Crotalus molossus estebanensis
Crotalus molossus nigrescens
Crotalus polystictus
Crotalus pricei pricei
Crotalus pricei miquihuanus
Crotalus pussillus
Crotalus ruber ruber
Crotalus lucasensis
Crotalus scutelatus scutelatus
Crotalus scutelatus salvini
Crotalus stejnegeri
Crotalus tigris
Crotalus tortugensis
Crotalus transversus
Crotalus triseriatus triseriatus
Crotalus triseriatus aquilus
Crotalus viridis viridis
Crotalus viridis caliginis
Crotalus viridis helleri
Crotalus willardi willardi

Crotalus willardi amabilis
Crotalus willardi meridionalis
Crotalus (Sistrurus) catenatus edwardsi
Crotalus (Sistrurus) ravus

NICARAGUA

Micrurus alleni alleni
Micrurus mipartitus hertwegi
Micrurus nigrocinctus babaspul (Corn Island)
Micrurus nigrocinctus melanocephalus
Micrurus nigrocinctus mosquitensis
Pelamis platurus
Bothrops asper
Bothrops godmanni
Bothrops nasutus
Bothrops nummifer mexicanus
Bothrops ophryomegas
Bothrops schlegeli
Crotalus durissus durissus

PANAMA (incl. Canal Zone)

Micrurus alleni alleni
Micrurus alleni yatesi
Micrurus ancoralis jani
Micrurus clarki
Micrurus dissoleucus dunni
Micrurus mipartitus mipartitus
Micrurus mipartitus hertwegi
Micrurus mipartitus multifasciatus
Micrurus nigrocinctus nigrocinctus
Micrurus nigrocinctus coibensis (Coiba Island)
Micrurus nigrocinctus mosquitensis
Micrurus stewarti
Pelamis platurus
Bothrops asper
Bothrops godmanni
Bothrops lateralis
Bothrops nasutus
Bothrops nigroviridis nigroviridis
Bothrops nummifer mexicanus
Bothrops ophryomegas

Bothrops punctatus
Bothrops schlegeli
Lachesis muta stenophrys

PARAGUAY

Micrurus frontalis frontalis
Micrurus frontalis pyrrhocryptus
Bothrops alternatus
Bothrops jararaca
Bothrops jararacussu
Bothrops neuwiedi diporus
Crotalus curissus collilineatus
Crotalus durissus terrificus

PERU

Leptomicrurus narduccii
Micrurus annellatus annellatus
Micrurus annellatus montanus
Micrurus filiformis filiformis
Micrurus hemprichii ortoni
Micrurus langsdorffi langsdorffi
Micrurus langsdorffi ornatissimus
Micrurus lemniscatus helleri
Micrurus margaritiferus
Micrurus mertensi
Micrurus mipartitus decussatus
Micrurus peruvianus
Micrurus putumayensis
Micrurus spixii obscurus
Micrurus surinamensis surinamensis
Micrurus tschudii tschudii
Micrurus tschudii olssoni
Pelamis platurus
Bothrops andianus
Bothrops atrox
Bothrops barnetti
Bothrops bilineatus smaragdinus
Bothrops brazili
Bothrops castelnaudi
Bothrops hyoprorus
Bothrops microphthalmus microphthalmus

Bothrops oligolepis
Bothrops peruvianus
Bothrops pictus
Bothrops pulcher
Bothrops roedingeri
Crotalus durissus subsp.
Lachesis muta muta

URUGUAY

Micrurus corallinus
Micrurus frontalis altirostris
Bothrops alternatus
Bothrops neuwiedii pubescens
Crotalis durissus terrificus

VENEZUELA

Leptomicrurus collaris
Micrurus carinicauda carinicauda
Micrurus dissoleucus dissoleucus
Micrurus hemprichii hemprichii
Micrurus isozonus
Micrurus lemniscatus diutius
Micrurus lemniscatus helleri
Micrurus mipartitus anomalus
Micrurus mipartitus semipartitus
Micrurus psyches psyches
Micrurus psyches circinalis
Micrurus spixii obscurus
Micrurus surinamensis nattereri
Bothrops atrox
Bothrops bilineatus bilineatus
Bothrops brazili
Bothrops colombiensis
Bothrops hyoprorus
Bothrops langsbergii janisrozei
Bothrops lichenosus
Bothrops medusa
Bothrops schlegelii
Bothrops venezuelensis
Crotalus cumanensis
Crotalus vegrandis
Lachesis muta muta

Chapter 31

Lethal Doses of Some Snake Venoms

D. DE KLOBUSITZKY

STATE UNIVERSITY OF SÃO PAULO, NATIONAL ACADEMY OF MEDICINE
OF BRAZIL, SÃO PAULO, BRAZIL

I. INTRODUCTION

Paracelsus' (1589) aphorism, well known to pharmacologists: "The dose determines whether a substance is a venom or not," must be completed by adding: "for a definite test object" in order to give it general validity. The justification of a formulation completed in this way ("The dose and the object of test determine whether a substance is a venom or not") is perhaps most evident when the saying is applied to animal venoms. The *Vipera berus* introduces about 20 mg of venom into the body of its victim in one bite. 2.25 mg (about $\frac{1}{8}$ of the amount corresponding to one bite) kills a 1-kg rabbit, but to kill a 1-kg shrew or a 1-kg viper, 192 and 300 mg of venom, respectively, (amounts corresponding to 9.5 and 15 bites) are needed. These data show that the mentioned gland secretion is a strong venom for the rabbit, a very weak one for the shrew and a harmless substance for the viper.

The example quoted shows very clearly that in the determination of the toxicity of a snake venom, since it can only be established by animal experiments, the choice of the animal species to be used for the tests is a very important factor. The way the venom is introduced into the test animal must be considered as a further factor. Besides these circumstances, which decisively

295

influence the result, it is necessary to take also into account certain practical considerations in the choice of the determination method (the limit of error wanted, the price and the possibility of obtaining the test animal, the time within which the result is obtained, the comparability of one's own results with those of others, the purpose of the determination and other similar points).

As the toxic and pharmacologic action of venom from snakes of the same species—just like the composition of other secretions of biological origin— shows provable fluctuations (among other things, it may be influenced by the season, the state of nutrition, age), it is useless to strive for a greater precision when the question is to establish the toxicity of the venom of a species (or subspecies) in regard to an animal species or to titrate the fixation power of an antiserum. A precision of a less than 10% error limit cannot be attained with the methods generally used.

II. THEORETICAL BASES OF THE DOSAGE

A. Conception and Various Definitions and Interpretations of the Mortal Doses

That amount of an unquestionably identified venom, dried so as to attain the weight constant that kills the test animal with the indicated administration method, is considered the mortal dose (LD). As a basic condition, it is required that death occur after the appearance of the symptoms of poisoning which are characteristic for the toxin of the species (subspecies) involved.

In toxicology it is, however, important to establish the smallest venom amount just sufficient to cause death. This amount, with reference also to the method of administration used, is designated as the *minimum lethal dose* (MLD) or *Dosis lethalis minima* (DLM) and it is expressed by the animal species or by a weight unit of this animal. Many authors, especially those who use white mice, give as toxic value the amount that kills 50% of the animals inoculated with the venom and designate it as MLD_{50}. They reckon the value on the basis of mathematical considerations. Some of them apply Reed and Muench's formula (1938), others that of Lichtfield and Wilcoxon (1949) and the ascertainment of the standard error of the MLD_{50} can be calculated according to Pizzi (1950). For a change, others give the mean neurotoxic value, meaning the smallest amount that kills 2 out of 3 animals within 24 hours.

Calmette (1894), the founder of serum therapy against snake bites, used rabbits and defined the MLD per kilogram of animal concerned. The Butantan Institute uses the determination method introduced by Brazil (1907). It consists in injecting at most a 2-ml volume intravenously into pigeons weighing 300–320 gm. In the case of the *Bothrops* or *Micrurus* species, a time of 20 minutes was fixed, for *Crotalus durissus* one of 2 hours. To the definition of Brazil's MLD belongs also the indication of period of time before death.

Ghosh and De (1938) also used pigeons. They fixed the weight of the animal at 300–310 gm and chose intramuscular injection as the method of administration. Their definition of the toxic value is based on these requirements.

In Russia, pigeons are used too. Apparently they do not inject the venom intravenously, for Nemtsov (1956) has found the MLD of the venom of the *Vipera lebetina* to be 13 mg/kg.

Kellaway (1929a,b,c,d) has established his known dosages on rabbits, guinea pigs, rats, and white mice to which he administered the venom subcutaneously or intravenously and calculated the MLD values per kilogram of body weight. Michael and Jung (1936), Wieland and Koncz (1936), Slotta and Szyszka (1937), as well as Schöttler (1938) used white mice and gave them the venom subcutaneously. With the exception of Wieland and Koncz, who defined the toxic value per animal, the other authors reckoned it per gram of body weight. Slotta and Szyszka determined as time 8–22 hours, Schöttler fixed 7 days as a limit—certainly a too long period for a neurotoxin death.

B. Technique and Its Relation to the Purpose of the Dosage

As methods of administering the venom, we use subcutaneous, intramuscular, intraperitoneal, and intravenous injections as well as the instillation into the outer opening of the nostrils. The results obtained by means of the different forms of injection give different values for otherwise identical test conditions. It is generally admitted that the intravenous MLD doses are lower than subcutaneous or intramuscular. This, however, corresponds to the facts only with some limitations, and that rule only can be considered generally valid that states that after an intravenous injection the space of time between the administering and the appearance of the first symptoms of poisoning is shorter than after the two other ways of administration, in the cases of doses around the MLD. On the contrary, when the results obtained with many toxins by the two most frequently used ways of administering, that is, when subcutaneous and intravenous injections are compared, one finds that in the relationship between the values found not only does there not exist a generally valid rule, but there are cases in which the subcutaneous MLD is smaller than the intravenous MLD. From Kellaway's following test results (1929a,b,c), this appears very clearly (Table I).

A similar discrepancy from the above-mentioned generally accepted rule has, besides, been observed, as far as I know, in the following cases: the venoms of *Pseudechis porphyriacus* and of *Crotalus adamanteus*, dosed on

TABLE I

MLD WITH SUBCUTANEOUS AND INTRAVENOUS APPLICATION OF THE VENOMS OF THREE
AUSTRALIAN SNAKES

Venom	Rabbits		Guinea pigs		Rats		White mice	
	sc	iv	sc	iv	sc	iv	sc	iv
Acanthopis antarcticus	0.14	0.12	0.15	0.06	0.5	0.2	0.7	0.4
Denisonia superba	0.7	0.3	0.06	0.015	1.4	0.9	1.2	0.5
Notechis scutatus	0.002	0.045[a]	0.007	0.02	0.7[a]	0.4	0.04[a]	0.25

[a] Result is only approximate owing to the small number of observations.

rabbits, show for both injection methods (sc and iv) the same MLD, namely 0.7 mg (see Klobusitzky, 1941) and 0.25 mg (Camus *et al.,* 1916) respectively. *Naja flava* venom behaves similarly as in pigeons in that its MLD was found to be 0.1 mg per animal (Klobusitzky, 1941). Césari and Boquet (1936) have established that the venom of *Naja naja* behaves similarly in guinea pigs; its MLD, determined by both injection methods, is 0.45 mg per kg of body weight. Boquet (1943) has found the same in a test of the venom of *Dendraspis viridis* (MLD—sc: 0.285; iv, 0.28 mg/kg guinea pig). The venom of *Pseudechis*, besides the particularity described, occupies in still another respect a special position. Its toxic action (when given subcutaneously) is the same for rats, guinea pigs, and rabbits, that is, again 0.7 mg/kg (Klobusitzky, 1941).

The choice of the injection method must first of all depend on the purpose of the dosage. If the mortal dose in case of a bite of man has to be determined, it is better to inject the venom subcutaneously or intramuscularly. The fangs of some snakes are not long enough to penetrate into the muscles of the leg [according to the statistics of the Butantan Institute, by far most bites—76.2% —in human beings occur under the knee (da Fonseca, 1949)], so that, in our opinion, for these species, the logically choice of method is subcutaneous injection. On the contrary, in the case of species which have especially long fangs, so that they introduce the main amount of their toxin in the muscle layer (*Bothrops atrox, B. jararacussu, Crotalus adamanteus, Dendraspis angusticeps, Naja haie, N. naja, N. nigricollis, Ophiophagus hannah* to mention only the most common ones), we consider the dosage by intramuscular injection to be more appropriate.

If the dosage is undertaken to determine the fixation power of an antitoxin, than the injection method has unquestionably to be the intravenous one, as the neutralization of the toxin occurs in the blood. When, however, white mice are used for this dosing, it is important to carry it out on animals of the same sex for it can be proved that the females are more resistant than the males (Dossena, 1949; Schöttler, 1952). To be complete it should also be mentioned here that Roger *et al.* (1961) have attempted to establish the value of the snake venom antitoxins by other methods, but they came to the conclusion that, for the time being, only animal tests can be used for that purpose.

If with the dosing purely scientific ends are pursued, then the way of injecting or, in general, the mode of introducing the venom is chosen according to the object of the research (intraperitoneal, intralumbar, intraventricular, intranerval, intraganglionic, etc., injections, painting of the organ, the resorption conditions of which, in relation to a venom, must be investigated).

The error limit of the dosage of the neurotoxic activity, performed by injection, is in general 15–20%; when not very active venoms are involved, it can be reduced to 10% through certain precautions. When white mice are used and the venom is administered subcutaneously, it is enough to choose animals of the same weight (± 2 gm) and to maintain a constant environmental temperature to bring the error limit down to 10% (Micheael and Jung, 1936). Should the toxin be very active for the test animal, the determination of the MLD will then show a bigger error. We see this in the dosing of the toxin of the *Crotalus d. terrificus* on pigeons by iv injection. The values obtained vary between 0.0015 and 0.002 gm per animal; the error limit is 33%. This proves to be still clearer in these tests of Schöttler (1951b) in which he dosed the toxicity of two different samples from the *Vipera aspis* and the *Sepedon haemachates* on white mice through venom solutions injected subcutaneously. The differences were for both venom kinds 100% (0.001 mg and 0.002 mg/gm mouse).

Schöttler (1951a) has been able to establish the toxicity of the venoms of eleven of the snake species most frequently found in Brazil (*Crotalus d. terrificus, Micrurus frontalis, Bothrops jararaca, B. jararacussu, B. cotiara, B. insularis, B. neuwiedi, B. alternata, B. itapeteningae, B. atrox, L. muta*) and those of ten not South American species (*C. t. basiliscus, C. atrox, C. horridus, Agkistrodon halys, A. piscivorus, A. mokasen, A. blomhoffi, Sistrurus catenatus, Trimeresurus flavoviridis, T. anamallensis*) by means of the abovementioned technique with an error limit of about 3%. This precision was reached because he always determined the highest dose not lethal, the MLD, and that minimal dose that killed all animals (on the average 81 per sample) of a test series. The absolute values obtained (for instance, for the *B. alternatus*: 0.01284 mg per 1 gm mouse) can only be considered valid for the examined sample of venom, as he himself has also stressed.

A low error limit (5%) has also been obtained by Ohsaka *et al.* (1960) in the dosing of the toxicity of the venom of the *Trimeserurus flavoviridis* (the MLD_{50} was 0.061 mg $\pm$ 0.003 mg/animal, the standard error of the MLD_{50}

being calculated according to Pizzi, 1950). The authors injected 14–17 gm white mice intravenously and fixed 4 days as observation period. The toxicity fluctuation in one and the same species has been studied in detail by Mitsuhashi *et al.* (1959) in eight samples of the venom of the *Trimeresurus flavoviridis* on white mice through intravenous or intraperitoneal injections. The MLD values found showed indisputably how deceptive the attempt is to obtain a greater precision in the toxicity determination. The MLD per animal values were:

dried pool collected 1942	0.1 mg
dried pool collected 1952	0.125 mg
fresh venom collected 1958	0.057 mg
fresh venom collected 1958	0.075 mg

In the toxicity determination it is very important to know the age of the venom involved, as many snake venoms lose much of their activity with time. Klobusitzky (1935) has mentioned that the venom of *Bothrops jararaca* is very unstable. He found that the MLD of the fresh venom of this viper species administered intravenously was 0.02–0.04 mg per pigeon; venoms that had been kept over 10 years in a closed glass jar in a dark place showed values of 0.26–0.27 mg per pigeon. Macht (1937), who has examined the venoms of many *Crotalus* species for their toxicity depending on time, ascertained that the venom of the *Crotalus d. terrificus,* kept dry and protected from light, did not suffer any significant change in its toxic activity. Schöttler (1951b) as well as Ramsey and Gennaro (1959) have recently determined the toxicity decrease of different snake venoms, depending on time.

Schöttler—who confirms the findings of previous authors—found that the nonhemorrhagic (pure neurotoxic) venoms had lost little or nothing of their original toxicity in 8 or 9 years of storage in glass tubes in the dark. The

TABLE II

REDUCTION OF TOXIC VALUE OF THE *Bothrops atrox* VENOM, PRESERVED AT ROOM TEMPERATURE UNDER AIR AND UNDER CO_2 ATMOSPHERE[a]

Date of dosage	Sample I preserved under		Sample II preserved under	
	Air	CO_2	Air	CO_2
8/2/1961	0.072 mg	0.055 mg	0.80 mg	0.40 mg
14/8/1961	0.13 mg	0.07 mg	1.00 mg	0.60 mg
16/8/1962	0.125 mg	0.085 mg	1.5 mg	0.60 mg
18/2/1963	0.20 mg	0.105 mg	2.0 mg	0.80 mg

[a] Initial toxic value (MLD) determined 23/8/1960 on pigeons via iv injections: sample I, 0.036; sample II, 0.28 mg/pigeon.

TABLE III

DOSAGE OF MLD OF THE VENOM OF *Acanthopis antarcticus*[a]

	Rabbit				Guinea pig						
	sc		iv		sc					iv	
Total of animals used	22		22		93					22	
mg/kg of venom injected	0.14	0.13	0.12	0.10	0.15	0.14	0.13	0.12	0.10	0.06	0.04
Number of animals injected	4	4	7		23	10	24	22	14	6	2
Number of animals dead	4	3	7		23	9	24	15	6	6	—
Number of animals survived	—	1	—		—	1	—	7	8	—	2

	Rat									Mouse							
	sc				iv					sc				iv			
Total of animals used	27				47					49				110			
mg/kg of venom injected	0.5	0.4	0.35	0.3	0.2	0.18	0.16	0.14	0.12	0.7	0.6	0.5	0.4	0.4	0.3	0.25	0.2
Number of animals injected	4	9	6	2	12	12	10	2	2	10	10	9	10	20	30	30	30
Number of animals dead	4	3	5	—	12	10	8	—	2	10	9	6	5	20	21	15	12
Number of animals survived	—	6	1	2	—	2	2	2	—	—	1	3	5	—	9	15	18

[a] From Kellaway (1929a).

hemorrhagic venoms kept under the same conditions had lost 41–56% of their power in 9 years and 84% in 13 years (viper) and 78% (*Bitis*) and only 14% (*Agkistrodon*) in 8 years.

Ramsey and Gennaro had determined the toxicity decrease on samples of the venoms of *Agkistrodon piscivorus* and of *Crotalus adamanteus* in 20–22 gm white mice by interperitoneal administration. The animals were observed for 72 hours. Their results are as follows:

Agkistrodon piscivorus
1-year-old MLD   0.004 mg/animal   12-years old   0.004 mg/animal
1-year-old LD_{100} 0.005 mg/animal   12-years old   0.006 mg/animal
Crotalus adamanteus
1-year-old MLD 0.003 mg/animal   9,12,13-years old 0.004 mg/animal
1-year-old LD_{100} 0.004 mg/animal   9,12,13-years old 0.006 mg/animal

The mode of preservation apparently plays a part to be considered in all venoms, even those that are sensitive to time. Schöttler (1951b) has reported that the *Agkistrodon* venom and the toxin of the *Crotalus d. terrificus*, when stored in a frequently opened and reevacuated dessicator lost approximately 70 and 36%, respectively, of their activity within 1 year.

The decrease of the toxic action is certainly tied to an oxidation process of the neurotoxins. However, unfinished tests of Klobusitzky have shown that the *Bothrops atrox* venom kept under air in ampoules, closed by melting, loses much more of its activity than the one kept under CO_2 (Table II).

The subcutaneous, intramuscular, or intraperitoneal administration of the venom to be given certainly does not cause any difficulties. Intravenous injection of pigeons or other birds requires some practice. In general, the wing veins near the wing shoulder articulation are used (preferably a distal vein before the anastomosis). The external vein is mostly covered by tendon. If an injection is given into an area where the sinew presses against the vein, no bleeding occurs after the needle is removed. A syringe with an excentrically placed needle base makes the introduction of the needle in the vein much easier.

TABLE IV

MLD OF THE VENOM OF *Vipera russelli* ON PIGEONS, 305–340 GM BY IV INJECTION[a]

0.026 mg/kg: without sympt.	0.033 mg/kg: without sympt.
0.034 mg/kg: without sympt.	0.036 mg/kg.:
	death − 4 minutes = MLD
0.036 mg/kg: death − 2 minutes	0.074 mg/kg: death − 3 minutes

[a] From Klobusitzky (1961).

TABLE V

MLD OF THE VENOM OF *Lachesis muta*[a]

	im		intralumbar		iv		MLD		
	Quantity	Symptoms	Quantity	Symptoms	Quantity	Symptoms	Route	Quantity	Symptoms
Dog (mg/kg)	10.3	Death, 1 hour 40 minutes	0.96	Death, 55 minutes	0.72	Death, 5 hours 36 minutes	im	2–3	Death, 10–30 hours
	3.1	Death, 36 hours			0.67	Death, 1 hour 30 minutes	*i. lumbar*	0.5	Death, 40–60 minutes
	1.8	Serious symptoms			0.3	Death, 6 hours	iv	0.3–0.5	Death, 2–5 hours
Rabbit (mg/kg)	8	Death, 20–24 hours			4	Death, 5–6 hours			
Guinea pig (mg/kg)	14	Death, 12 hours			3.1	Death, 5 hours 30 minutes		13–16.5	Death, 10–12 hours
Pigeon (per animal)	0.5	Death, 28 minutes	0.01	With symptoms			im	0.30	Death, 30–40 minutes
			0.025	Slight symptoms			iv	0.07	Death, 15–30 minutes
			0.05	Death, 27 minutes					
			0.07	Death, 14 minutes					
			0.10	Death, 11 minutes					
			1.0	Death, 4 minutes					

[a] From Vellard (1948).

REFERENCES

Boquet, P. (1943). *Bull. Soc. Pathol. Exotique*, **36**, 189.

Brazil, V. (1907). *Rev. Med. S. Paulo* **10**, 196.

Calmette, A. (1894). *Ann. Inst. Pasteur* **8**, 275.

Camus, J., Césari, E., and Jouan, C. (1916). *Ann. Inst. Pasteur* **30**, 180.

Césari, A., and Boquet, P. (1936). *Ann. Inst. Pasteur* **56**, 511.

Christensen, P. A. (1955). "South African Snake Venoms and Antivenoms." Johannesburg.

da Fonseca, F. (1949). "Animais peçonhentos." Inst. Butantan, São Paulo.

Dossena, P. (1949). *Acta trop.* **6**, 263.

Ghosh, B. N., and De, S. S. (1938). *Indian J. Med. Res.* **25**, 779.

Kellaway, C. H. (1929a). *Med. J. Austr.* **1**, 764.

Kellaway, C. H. (1929b). *Med. J. Austr.* **1**, 358.

Kellaway, C. H. (1929c). *Med. J. Austr.* **1**, 348.

Kellaway, C. H. (1929d). *Med. J. Austr.* **1**, 372.

Klobusitzky, D. von (1935). *Arch Exptl. Pathol. Pharmakol.* **179**, 204.

Klobusitzky, D. von (1941). *Ergeb. Hyg. Bakteriol., Immunitaetsforsch. Expertl. Therap.* **24**, 226.

Klobusitzky, D. von (1961). *Naturwissenschaflen* **48**, 407.

Lichtfield, J. T. and Wilcoxon, F. J. (1949). *J. Pharmacol. Exptl. Therap.* **96**, 99.

Macht, D. J. (1937). *Proc. Soc. Exptl. Biol. Med.* **36**, 499.

Michael, F., and Jung, F. (1936). *Z. Physiol. Chem.* **239**, 217.

Mitsuhashi, S., Maeno, H., Kawakami, M., Hashimoto, H., Sawai, Y., Miyazaki, S., Makino, M., Kobayashi, M., Okonogi, T., and Yamaguchi, K. (1959). *Japan. J. Microbiol.* **3**, 95.

Nemtsov, A. V. (1956). *C.R. Acad. Sci. U.R.S.S.* **107** (3), 485.

Ohsaka, A., Ikezawa, H., Kondo, H., Kondo, S., and Uchida, N. (1960). *Brit. J. Exptl. Pathol.* **41**, 478.

Paracelsus, T. (1589). "De ente veneni," Collected Publ. Vol. 1. Huser, Basel.

Pizzi, M. (1950). *Human Biol.* **22**, 152.

Ramsey, H. W., and Gennaro, J. F. Jr., (1959). *Am. J. Trop. Med.* **8**, 522.

Reed, L. J., and Muench, H. (1938). *Am. J. Hyg.* **27**, 493.

Roger, F., Roger, A., Lamy, R., and Rouyer, M. (1961). *Ann. Inst. Pasteur* **101**, 389.

Schöttler, W. H. A. (1938). *Z. Hyg. Infektionskrankh.* **120**, 408.

Schöttler, W. H. A. (1951a). *J. Trop. Med.* **31**, 489.

Schöttler, W. H. A. (1951b). *J. Immunol.* **67**, 299.

Schöttler, W. H. A. (1952). *Bull. Osaka Med. School* **5**, 293.

Slotta, C. H., and Szyszka, G. (1937). *Mem. Inst. Butantan (São Paulo)* **11**, 109.

Vellard, J. (1948). "El veneno de *Lachesis muta* (L.)." Univ. Nacl. Mayor San Marcos, Lima, Peru.

Wieland, H., and Konz, W. (1936). *Sitzber math.-nat. Abt. bayr.* p. 177.

CHEMISTRY AND PHARMACOLOGY OF THE VENOMS OF BOTHROPS AND LACHESIS

Chapter 32

Chemistry and Pharmacology of the Venoms of Bothrops and Lachesis

E. KAISER AND H. MICHL

DEPARTMENT OF MEDICAL CHEMISTRY, AND DEPARTMENT OF CHEMISTRY,
UNIVERSITY OF VIENNA, VIENNA, AUSTRIA

Several very venomous snakes belong to the genus *Bothrops*, with the chief biotope being in South and Central America. Table I shows the most important snakes. The genus *Lachesis* seems to be represented by only one species, *L. mutus*, the "bushmaster."

In southern Brazil, Rio Grande do Sul, Santa Catarina, Paraná, São Paulo, Mato Grosso, Goiás and Rio de Janeiro, in Paraguay, Uruguay, and Argentina *Bothrops jararaca* is the most frequent venomous snake, followed by *Crotalus durissus terrificus*. Both constitute 84% of all venomous snakes sent to the Instituto Butantan. The quantity of the other *Bothrops* species and venomous coral snakes received by Butantan in the last 45 years is insignificant. Only four or six bushmasters have been received during this time (Bücherl, 1963).

In an investigation of the 730 human accidents caused by poisonous snakes in 7 years, *Bothrops jararaca* is first with 87.4%, followed by *Crotalus durissus terrificus* with only 10%; *B. jararacussu* was responsible for 1.2% of

TABLE I

VENOMOUS SNAKES OF CENTRAL AND SOUTH AMERICA

Species	Distribution	Length
Bothrops		
B. alternatus Duméril, Bibron	Argentine, Uruguay, Paraguay, southern Brazil	to 140 cm
B. ammodytoides Leybold	Argentine, Uruguay, southern Brazil	to 50 cm
B. atrox L.	Central America, Antilles, tropical South America	to 250 cm
B. cotiara Gomes	Brazil	to 100 cm
B. erythromelas Amaral	Brazil	to 125 cm
B. jararaca Wied	Argentine, Uruguay, Paraguay, Brazil	to 150 cm
B. jararacussu Lacerda	Argentine, Uruguay, Paraguay, Brazil	to 220 cm
B. neuwiedi Wagler	Argentine, Paraguay, Bolivia, Brazil	to 130 cm
B. nummifer Rüppell	Mexico, Central America to Panama	to 90 cm
Lachesis		
L. mutus L.	Tropical South America, Central America, Trinidad	to 400 cm

the bites; the Brazilian coral snakes, 1.1 %, and *B. alternatus*, 0.3 % (Rosenfeld, 1963).

I. QUANTITIES OF VENOMS AND TOXICITY

In Table II the average and the maximum venom quantities and the toxicity of several *Bothrops* venoms can be seen.

The comparison of the average and the maximum venom quantities with the LD_{100} gives a good idea on the potential danger of a snake bite in South and Central America. The bite of *B. jararacussu* seems to be very dangerous because this snake has a large quantity of venom; it is followed by *B. insularis* from the Queimada Island near Santos, *B. jararaca*, *B. alternatus*, *B. cotiara*, *B. atrox*, *B. neuwiedi*, and others. The toxicity of the venoms of South and Central American arboreal snakes seem to be higher than that of *B. jararaca*.

II. SYMPTOMATOLOGY

The symptomatology after venom injection is more or less the same for all the *Bothrops* snakes, the intensity of effects being in direct relation to the amount of injected venom and the location of the bite. Edema and hemorrhages may develop around the affected region. A few days later necrosis may occur necessitating amputation in severe local intoxications. Intensive serum treatment soon after the accident may generally avoid amputation. Cicatrices

TABLE II

VENOM QUANTITIES AND TOXICITY OF SEVERAL *Bothrops* SPECIES AND OF *L. mutus* [a]

Species	Quantity of venom		Toxicity
	Average	Maximum	
B. alternatus	70 mg (1)	435 mg (1)	LD_{100} mouse sc 25 μg/gm (1 and 3)
B. ammodytoides	50 mg (2)		Toxic dose: Rabbit sc 7 μg/gm (2); Mouse sc 3 μg/gm (2)
B. atrox	72 mg (1)	310 mg (1)	LD_{100} mouse sc 31 μg/gm (1 and 3); MLD mouse sc 22 μg/gm (3); LD_{100} mouse sc 15 μg/gm (1 and 3)
B. cotiara	37 mg (1)		
B. erythromelas	20–30 mg (4)		
B. jararaca	36 mg (1)	185 mg (1)	LD_{100} mouse sc 7 μg/gm (1 and 3)
B. jararacussu	151 mg (1)	1,000 mg (1)	LD_{100} mouse sc 13 μg/gm (1 and 3)
B. neuwiedi	32 mg (1)	120 mg (1)	LD_{100} mouse sc 14 μg/gm (1 and 3)
B. nummifer	120 mg (5)	280 mg (5)	
L. mutus	200–300 mg (6)	500 mg (6)	Toxic doses: Rabbits sc 4 μg/gm (6); Pigeon im 300 μg/350 gm

[a] Numbers in parentheses refer to the following references:

(1) Bücherl (1963)
(2) Houssay and Negrete (1923)
(3) Schöttler (1951)
(4) Vellard (1938)
(5) Jimenez-Porras (1964b)
(6) Vellard (1948)

may be eliminated by later surgery. The possibility of local infections always exists. When the venom quantity injected by the bite is high, one may observe hemorrhages from the mucous membranes of the nose, gums, from the margins of the fingernails, from the skin of the head, even from the stomach, with hematemesis, melena, and hematuria. The blood pressure is generally normal, low pressure being present only concomitantly with larger wounds, perhaps due to liberation of histamine. The pulse also may be normal, tachycardia rarely being present. Fever generally accompanies acute intoxication as well as the larger necroses. Blood coagulation is affected only in severe cases, when the blood will become uncoagulable 15 to 30 minutes after the bite. Necrose is often accompanied by leukocytosis or neutrophilia. Immediately after the bite the number of erythrocytes and the erythrocytic volume are generally increased, but anemia may be present later (Rosenfeld, 1963).

The human mortality rate, according to do Amaral (1930), is about 20% when not treated with antiserum; according to Rosenfeld (1963) it is only 8% (not including the arboreal *Bothrops* snakes). When serum treatment is started immediately, the human mortality is very low and occurs only in patients who have received the serum 6–10 hours after the bite (1.1%). Death is due to collapse of peripheral circulation or cerebral hemorrhage.

The pathological alterations not only show local changes at the site of the bite but extensive damage to the circulatory system. Rotter (1938) pointed out significant lesions on the walls of the small arterial vessels with a coagulation necrosis allied with leukocytosis of the necrotized intima. Similar changes may occur concomitantly in the precapillaries. MacClure (1935) found diffuse glomerulonephritis (*B. jararacussu*), and Azevedo and Castro-Teixeira (1938) necrosis of the cortex of the kidneys allied with uremia. Edema of the brain, and hemorrhages, are relatively frequent in severe cases (Jutzy *et al.*, 1953; Rotter, 1938).

III. PHARMACOLOGICAL EFFECTS

All the *Bothrops* venoms, except that of *B. cotiara* (Kelen *et al.*, 1960–1962) show necrotizing and coagulating effects. A very powerful local action has been demonstrated in the venom of *B. alternatus* and *B. neuwiedi* (Houssay, 1923), while *B. erythromelas* (Vellard, 1938) has a relatively slow necrotizing action. Necrosis is due to the proteolytic activity of these venoms. Following slow intravenous injection into dogs (Rosenfeld, 1963), defibrinogenation (fibrinolysis) of the blood occurs. Consequently a definite retardation of clotting is frequently observed, or even a complete abolition of clotting, probably due to fibrinogenopenia. Small amounts of venoms may cause at first acceleration of clotting, followed later by retardation; large amounts may induce death in a few minutes with massive intravasal clotting. The bothropic venoms may occasionally cause a lowering of blood pressure (Amuchastegui, 1940; Houssay and Negrete, 1923; Vellard and Huidobro, 1941); not only the liberation of histamine but also the formation of brady-kinin may be partly responsible (Kaiser and Raab, 1965). The effect of all *Bothrops* venoms on the nervous system is generally minimal, except for a more or less significant neurotropic action of the *B. erythromelas* venom (Vellard, 1938).

The proteolytic and the necrotic effects of the venom of *Lachesis mutus* are much less severe than that of the *Bothrops* venoms, but the neurotoxic action of the former seems to be much higher (Vellard, 1948).

IV. ENZYMATIC ACTIVITIES

Crude nonfractioned venom samples of *Bothrops* species and of *L. mutus* show a large range of enzymatic effects (Kaiser and Michl, 1958) as pointed out in Table III.

Several other factors may exist that will influence blood coagulation, e.g. accelerating clotting (thrombin- or thromboplastinlike activity) or fibrinolysis (Table IV).

V. FRACTIONATION OF VENOMS

A. *Bothrops atrox*

Fresh, undesiccated venom of *B. atrox* from the Atlantic coast of Costa Rica gives after electrophoresis in starch gel (pH 8.6) eight anodic fractions (A_1–A_8) and six cathodic fractions (K_1–K_6). Phospholipase A has been demonstrated in both of the most rapid anodic zones A_7 and A_8. The fraction A_6 has L-amino acid oxidase activity. The phosphoesterase activity was found in the fractions K_1 and K_2. The other zones show pro-teases, partly with thrombinlike effects. The percentages of these fractions do not seem to be uniform in all venom samples. In *B. atrox* venoms of the Pacific coast of Costa Rica, however, some of the anodic fractions are missing. The electrophoretic pattern from the same animal, milked at different times, have been relatively similar (Jimenez-Porras, 1964a).

The venom of *B. atrox* is highly suitable for the preparation of phospho-esterases. The venom solution is precipitated with acetone (42% v/v) at

TABLE III
ENZYMATIC ACTIVITIES IN CRUDE VENOMS OF *Bothrops* AND *Lachesis*[a]

Venoms	L-Amino acid oxidase	Hyaluronidase	Ribonuclease	Deoxyribonuclease	Phospholipase A	Phospholipase C	Protease	ATPase	Chondroitinase
B. alternatus	(1, 16)	(2)	(3)	(4)	(5)	(5)	(6, 17)	—	—
B. ammodytoides	(16)				(7)		(7)		
B. atrox	(1)	(8)	(3)	(4)	(19)		(17)	(14)	
B. cotiara	(1)	(9)			(8)		(9, 10, 17)	(8)	
B. jararaca	(16)	(9, 11, 12)	(3)	(4)	(19)		(17)	(14)	(13)
B. jararacussu	(1, 16)		(3)	(4)	(19)		(17)		
B. neuwiedi	(1, 16)	(15)	(3)	(4)	(19)		(17)	(14)	
B. nummifer	(18)		(18)	(18)	(18)		(18)	(18)	
L. mutus							(17, 20)		

[a] Numbers in parentheses refer to the following references:

(1) Zeller (1948)
(2) Favilli (1940a)
(3) Taborda *et al.* (1952a)
(4) Taborda *et al.* (1952b)
(5) Vidal-Breard and Elias (1950)
(6) Croxatto *et al.* (1943)
(7) Houssay and Negrete (1923)
(8) Kaiser (1960)
(9) Kaiser (1953)
(10) Janszky (1950a)
(11) Favilli (1940b)
(12) Eichbaum (1947)
(13) Bergamini (1953)
(14) Zeller (1951)
(15) Jaques (1955)
(16) Barrio (1954)
(17) Janszky (1951)
(18) Jimenez-Porras (1964b)
(19) Kaiser and Michl (1958)
(20) Deutsch and Diniz (1955)

TABLE IV
BLOOD-CLOTTING ACTIVITY OF SOME *Bothrops* VENOMS[a]

Venom	Coagulant activity	Fibrinolysis
B. alternatus	+ (1, 2, 3); thrombinlike (4, 5)	+ (5)
B. atrox	Thrombinlike (4, 6, 7); thromboplastinlike (8)	+ (5)
B. cotiara	Thrombinlike (4)	
B. erythromelas	+ (9)	
B. jararaca	Thrombinlike (10)	+ (5)
B. jararacussu	Thrombinlike (4, 5)	+ (5)
B. neuwiedi	Thrombinlike (5)	+ (5)

[a] Numbers in parentheses refer to the following references:

(1) Hanut (1938)
(2) Croxatto and Sainz (1944)
(3) Ahuja *et al.* (1947)
(4) Janszky (1950a)
(5) Didisheim and Lewis (1956)
(6) Janszky (1949)
(7) Fayos *et al.* (1960)
(8) Janszky (1950b)
(9) Vellard (1938)
(10) von Klobusitzky and König (1936)

−20°C according to Koerner and Sinsheimer (1957). 5′-Nucleotidase activity is found in the precipitate, phosphodiesterase and a nonspecific phosphomonoesterase in the supernatant. To obtain further purification of the 5′-nucleotidase, precipitation was repeated yielding 9-fold purification. The precipitate was fractionated with ammonium sulfate; a product with an activity of 57 times of the raw venom was precipitated at 0.65–0.70 saturation. Further steps of purification were chromatography on Sephadex G-100 and DEAE-cellulose (Tris buffer pH 7.5, elution steps 0.01 M, 0.05 M, 0.2 M). The final enzyme was 1000-fold more active than the starting material. Venom 5′-nucleotidase hydrolyzes only 5′-mononucleotides. It is ineffective against ribose-5′-phosphate, *p*-nitrophenylphosphate, ATP, mononucleoside-3′,5′-diphosphate, or higher oligonucleotides. The rate of hydrolysis of the 5′-monoribonucleotides or deoxyribonucleotides depends on the base: 5′-adenosine monophosphate is hydrolyzed twice as easily as 5′-guanosine monophosphate. The pH optimum is 9.0. EDTA ($10^{-4}M$) inhibits markedly: $10^{-3}M$ Mn^{++} and Co^{++} ions act as activators (Sulkowski *et al.*, 1963).

The nonspecific phosphomonoesterase of the supernatant has been purified 200-fold by chromatography on CM-cellulose at pH 6.0 (Tris buffer, 0.05 M, 0.2 M, 0.5 M) followed by treatment with calcium phosphate gel (phosphomonoesterase in the supernatant), chromatography on Sephadex G-100 and finally on DEAE-cellulose (pH 8.0, Tris buffer, 0.05–0.95 M gradient). The pH optimum is 9.5; Ca^{++} and Mg^{++} ions are activating. 5′-Adenosine monophosphate, ribose-5-phosphate, ATP, *p*-nitrophenylphosphate and flavine mononucleotide are hydrolyzed. On the other hand pyrophosphate and the cyclic uridine-2′,3′-phosphate are not hydrolyzed (Sulkowski *et al.*, 1963).

For the preparation of the phosphodiesterase, the supernatant obtained by acetone precipitation of the venom according to Koerner and Sinsheimer (1957), has to be fractionated with alcohol. The fractions precipitating with 33–66% ethanol are used for further purification. This includes continuous electrophoresis at pH 8.9 and chromatography on CM- and DEAE-cellulose (Williams *et al.*, 1961; Björk, 1963). The enzyme, purified approximately 250-fold, has a similar substrate specificity as the enzyme from *Crotalus adamanteus* (Razzell and Khorana, 1959). Studies of the action of the diesterase on deoxyribooligonucleotides and ribonucleotides bearing 3′-hydroxy end groups have shown that the hydrolysis starts from the tail end of the nucleic acid and results in the successive release of nucleoside-5′-phosphate units.

B. *Bothrops jararaca*

Electrophoresis at pH 8.7 on filter paper yields six components with anodic migration. The material with enzymatic activity is localized in three sections of the paper strip. In the sections with the slowest migration part of the ATPase and the deoxyribonuclease activity has been found. This fraction could be separated at pH 6.3 into three components; the middle fraction has shown enzymatic activities. The fraction with moderate mobility at pH 8.7 includes ATPase and phospholipase A activity, coagulating factors and a highly toxic principle. The fraction with the highest mobility at pH 8.7 contains L-amino acid oxidase and a proteolytic enzyme (Michl, 1954).

The venom of *Bothrops jararaca,* like all other venoms of the genus *Bothrops*, shows a multiplicity of proteolytic and esterolytic activities. It is still extremely difficult to isolate the different enzymes. Von Klobusitzky and König (1936) made an attempt, more than 30 years ago, to separate a thrombinlike principle from the other toxic substances by precipitating *B. jararaca* venom with ammonium sulfate and lead acetate. The isolated substance has been assayed therapeutically. Drugs like "reptilase," "bothropase," "hemobotrase," etc. are still used for the treatment of hemorrhages of different origin. Habermann (1958) partially denatured the venom with 0.05 N HCl, and fractionated with acetone and electrophoresis at pH 8.6. The fractions obtained were 12 times more effective (thrombin activity) than the crude venom and 6 to 7 times less toxic. They show a weak phospholipase activity, but a strong action on gelatin. Another attempt for purification of a thrombinlike principle has been made by precipitation with ammonium sulfate (0.60–0.65 saturation) and starch column electrophoresis (Henriques *et al.*, 1960b) or by chromatography on DEAE-cellulose at pH 6.5 by stepwise elution with cacodylate buffer (0.05 M, 0.1 M, 0.2 M, 0.4 M) (Fichman and Henriques, 1962). This blood-clotting enzyme is inhibited by Tris- and phosphate-ions. It is, however, not affected by Cd^{++} ions which inhibit the other proteolytic enzymes of *B. jararaca* venom (Henriques *et al.*, 1960b). The pH optimum is about 7.0. The protease hydrolyzes benzoylarginineamide 3 times more rapidly than the crude venom; casein, however, is not hydrolyzed (Fichman and Henriques, 1962).

Banerjee *et al.* (1960) have succeeded in obtaining from *B. jararaca* venom a clotting principle 40 times more active than the crude venom. The venom was fractionated with ammonium sulfate, partially heat denatured and treated with calium phosphate gel. The clotting activity could be activated with Ca^{++} ions and blocked by diisopropylfluorophosphate. Similar to thrombin it hydrolyzes tosylarginine methyl ester, but does not split hemoglobin. In respect to inhibitors the clotting principle shows fundamental differences to thrombin. It also seems that the clot produced by thrombin or by "reptilase" may be chemically different. Clegg and Bailey (1962) have split both kinds of clots with sulfite and studied electrophoretically the three SO_3H-peptides; the T-chain was not influenced by thrombin or "reptilase," however, the B-chain was altered by thrombin but not by "reptilase." The A-chain was changed by both agents but in a different way.

B. jararaca venom contains at least two more proteolytic enzymes. One of them may be precipitated with ammonium sulfate at a saturation of 0.50–0.55. It has a strong action on casein ("caseinase activity") but fails to hydrolyze benzoyl-arginineamide. The pH optimum of the caseinase is 8.8. It is inhibited by Co^{++} and Cd^{++} ions, by cyanide, EDTA and cysteine (Henriques *et al.*, 1958, 1966).

Another enzyme, called "protease A" is precipitated by ammonium sulfate at a saturation of 0.7–0.8. This enzyme is relatively heat-stable and has a pH optimum of 8.8. When purified it hydrolyzes tosylarginine methyl ester and benzoyl-arginineamide (Mandelbaum and Henriques, 1964; Henriques *et al.*, 1966).

Rocha e Silva *et al.* (1949) were the first to demonstrate a bradykinin-forming factor in *B. jararaca* venom. Bradykinin is a polypeptide, which may be formed by trypsin ("T-bradykinin") or by the venom of *Bothrops jararaca* ("B-bradykinin") from bradykininogen. It was demonstrated later that other venoms of the genus *Bothrops* and even of other genera may produce bradykinin (Deutsch and Diniz, 1955). Bradykininogen is found in Cohn fraction IV/4 or IV/6b (van Arman, 1955; Gaddum, 1955). Proteases of the venoms are responsible for bradykinin formation and destruction after prolonged incubation (cf. Table V).

By treating the venom of *B. jararaca* with ammonium sulfate it was possible to demonstrate a bradykinin-forming as well as a bradykinin-inactivating fraction (Holtz and Raudonat, 1956). The bradykinin-forming principle in the

TABLE V

PROTEOLYTIC AND ESTEROLYTIC ACTIVITIES; BRADYKININ FORMATION AND DESTRUCTION BY SOME *Bothrops* VENOMS IN RELATION TO TRYPSIN[a]

| | Activities | | Bradykinin formation | Bradykinin destruction |
	Proteolytic (hemoglobin)	Esterolytic (BAEE)		
Trypsin	100	100	100	100
B. alternatus	0.25	0.71	9.60	250
B. atrox	0.25	6.80	71.10	350
B. cotiara	0.34	14.10	48.00	664
B. jararaca	0.51	6.60	330.70	514
B. jararacussu	0.72	6.80	—	107
B. neuwiedi	1.30	3.90	17.30	28.60

[a] After Deutsch and Diniz (1955).

venoms is partly dialyzable; the dialyzates do not contain proteolytic, coagulating, or tryptic activities (benzoylarginineamide), but could hydrolyze benzoylarginine methyl ester (esterase activity) (Holtz *et al.*, 1960). The bradykinin-forming factor is independent of protease A (Henriques *et al.*, 1960a). A fraction, obtained after saturation (0.7–0.8) with ammonium sulfate, contains the bradykinin-forming principle, and is 8 times more active than the crude venom. The B-bradykinin from snake venoms seems to be identical with the T-bradykinin and could be isolated (Zuber and Jaques, 1960). Ferreira (1965) and Ferreira and Rocha e Silva (1963) have pointed out in recent years that the venom of *B. jararaca* contains another factor which increases the pharmacological effects of bradykinin. This principle is partly soluble in water, thermostable, dialyzable, but could not be as yet isolated in pure form.

C. Bothrops jararacussu

The electrophoresis of the venom at pH 8.6 (0.1 barbital-citrate buffer) has demonstrated nine anodic fractions (Goncalves and Vieira, 1950). A basic fraction has been obtained by chromatography on Amberlite IRC-50 (0.2 M phosphate buffer, pH 7.3). This fraction produced in white mice symptoms similar to those of crotamine (Goncalves, 1956).

D. Bothrops neuwiedi

It has been possible to separate a phospholipase A and a protease by paper electrophoresis (0.1 barbital-acetate buffer, pH 8.6) of the venom (Grassmann and Hannig, 1954). Phospholipase A was purified by a procedure involving heat-treatment at pH 3.0, gel filtration (Sephadex G-25) and chromatography on DEAE-cellulose (Vidal *et al.*, 1966).

E. Bothrops nummifer

Jimenez-Porras (1964b) has studied venom samples of several subspecies of *B. nummifer* with starch gel electrophoresis at pH 8.9. It was possible to demonstrate four cathodic fractions (K_1–K_4) and six fractions migrating to the anode (A_1–A_6) but with interspecific variations. The fraction with the highest anodic mobility (A_6) showed phospholipase A and thrombin-like activities. Fraction A_3 has L-amino acid oxidase activity. Near the start (A_1 and K_1) very strong protease activities could be found. In the venoms of snake specimens captured along the Atlantic coast of Costa Rica the toxic principles were located near the start (A_1 and K_1). In the venoms of animals collected near the Pacific coast of the same country, the most toxic fractions have been found in K_3 and K_4. The same fractions also show protease and phosphatase activities.

F. *Lachesis mutus*

It seems that no fractionations of the venom of *L. mutus* have been performed. Some data on the unfractionated venom have been presented in Tables II and III.

REFERENCES

Ahuja, M. L., Veeraraghavan, N., and Menon, I. G. K. (1947). *Indian J. Med. Res.* **35**, 227.
Amuchastegui, S. R. (1940). *Compt. Rend. Soc. Biol.* **133**, 317.
Azevedo, A., and Castro-Teixeira, J. (1938). *Mem. Inst. Oswaldo Cruz.* **33**, 23.
Banerjee, R., Devi, A., and Sarkar, N. (1960). *Thromb. Diath. Haemorrhag.* **5**, 296.
Barrio, A. (1954). *Ciencia Invest. (Buenos Aires)* **8**, 36.
Bergamini, C. (1953). *Giorn. Biochim.* **2**, 320.
Björk, W. (1963). *J. Biol. Chem.* **238**, 2487.
Bücherl, W. (1963). *In* "Die Giftschlangen der Erde," p. 67. Behringw. Mitt. N. G. Elwert, Marburg.
Clegg, J. B., and Bailey, K. (1962). *Biochim. Biophys. Acta* **63**, 525.
Croxatto, H., and Sainz, N. (1944). *Bol. Soc. Biol. Santiago Chile* **1**, 115.
Croxatto, H., Marsano, A., and Croxatto, R. (1943). *Anales. Soc. Biol. Bogota* **1**, 148.
Deutsch, H. F., and Diniz, C. R. (1955). *J. Biol. Chem.* **216**, 17.
Didisheim, P., and Lewis, J. H. (1956). *Proc. Soc. Exptl. Biol. Med.* **93**, 10.
do Amaral, A. (1930). *Mem. Inst. Butantan (São Paulo)* **5**, 193.
Eichbaum, F. W. (1947). *Mem. Inst. Butantan (São Paulo)* **20**, 95.
Favilli, G. (1940a). *Boll. Ist. Sieroterap. Milan.* **19**, 481.
Favilli, G. (1940b). *Nature* **145**, 866.
Fayos, J. S., Outeirino, J., Paniagua, G., Serrano, J., and Martinez, M. T. (1960). *Rev. Clin. Espan.* **77**, 215.
Ferreira, S. H. (1965). *Brit. J. Pharmacol.* **24**, 163.
Ferreira, S. H., and Rocha e Silva, M. (1963). *Ciencia Cult. (São Paulo)* **15**, 276.
Fichman, M., and Henriques, O. B. (1962). *Arch. Biochem. Biophys.* **98**, 95.
Gaddum, J. H., ed. (1955). "Polypeptides Which Stimulate Plain Muscle." Livingstone, Edinburgh and London.
Goncalves, J. M. (1956). *In* "Venoms," Publ. No. 44, p. 261. Am. Assoc. Advance Sci., Washington, D.C.
Goncalves, J. M., and Vieira, L. B. (1950). *Anais. Acad. Brasil. Cien.* **22**, 141.
Grassmann, W., and Hannig, K. (1954). *Z. Physiol. Chem.* **296**, 30.
Habermann, E. (1958). *Arch. Exptl. Pathol. Pharmakol.* **234**, 291.
Hanut, C. J. (1938). *Arch. Intern. Physiol.* **47**, 377.
Henriques, O. B., Fichman, M., and Beraldo, W. T. (1960a). *Nature* **187**, 414.
Henriques, O. B., Fichman, M., and Henriques, S. B. (1960b). *Biochem. J.* **75**, 551.
Henriques, O. B., Lavras, A. A. C., Fichman, M., Mandelbaum, F. R., and Henriques, S. B. (1958). *Biochem. J.* **68**, 597.
Henriques, O. B., Mandelbaum, F. R., and Henriques, S. B. (1966). *Mem. Inst. Butantan (São Paulo)* **33**, 359.
Holtz, P., and Raudonat, H. W. (1956). *Arch. Exptl. Pathol. Pharmakol.* **229**, 113.
Holtz, P., Raudonat, H. W., and Contzen, C. (1960). *Arch. Exptl. Pathol. Pharmakol.* **239**, 54.
Houssay, B. A. (1923). *Compt. Rend. Soc. Biol.* **89**, 55.

Houssay, B. A., and Negrete, J. (1923). *Compt. Rend. Soc. Biol.* **89**, 751.

Janszky, B. (1949). *Science* **110**, 307.

Janszky, B. (1950a). *Arch. Biochem.* **28**, 139.

Janszky, B. (1950b). *Nature* **165**, 246.

Jaques, R. (1955). *Helv. Physiol. Pharmacol. Acta* **13**, 113.

Jimenez-Porras, J. M. (1964a). *Toxicon* **2**, 155.

Jimenez-Porras, J. M. (1964b). *Toxicon* **2**, 187.

Jutzy, D. A., Biber, S. H., Elton, N. W., and Lowry, E. C. (1953). *Amer. J. Trop. Med.* **2**, 129.

Kaiser, E. (1953). *Monatsh. Chem.* **84**, 482.

Kaiser, E. (1960). Unpublished experiences.

Kaiser, E., and Michl, H. (1958). "Die Biochemie der tierischen Gifte." F. Deutficke; Wien.

Kaiser, E., and Raab, W. (1965). *Z. Angew. Zool.* **52**, 1.

Kelen, E. M. A., Rosenfeld, G., and Nudel, F. (1960–1962). *Mem. Inst. Butantan (São Paulo)* **30**, 133.

Koerner, J. F., and Sinsheimer, R. L. (1957). *J. Biol. Chem.* **228**, 1049.

MacClure, E. (1935). *Bol. Secret. Geral Saude (Rio de Janeiro)* **1**, 35.

Mandelbaum, F. R., and Henriques, O. B. (1964). *Arch. Biochem. Biophys.* **104**, 369.

Michl, H. (1954). *Monatsh.* **85**, 1240.

Razzell, W. E., and Khorana, H. G. (1959). *J. Biol. Chem.* **234**, 114.

Rocha e Silva, M., Beraldo, W. T., and Rosenfeld, G. (1949). *Amer. J. Physiol.* **156**, 261.

Rosenfeld, G. (1963). In "Die Giftschlangen der Erde." p. 161. Behringwerk-Mitt., N. G. Elwert; Marburg.

Rotter, W. (1938). *Arch. Pathol. Anat.* **301**, 409.

Schöttler, W. H. A. (1951). *J. Trop. Med. Hyg.* **31**, 489.

Sulkowski, E., Björk, W., and Laskowski, M. (1963). *J. Biol. Chem.* **238**, 2477.

Taborda, A. R., Taborda, L. C., Williams, J. N. Jr., and Elvehjem, C. A. (1952a). *J. Biol. Chem.* **194**, 227.

Taborda, A. R., Taborda, L. C., Williams, J. N. Jr., and Elvehjem, C. A. (1952b). *J. Biol. Chem.* **195**, 207.

van Arman, C. G. (1955). *In* "Polypeptides Which Stimulate Plain Muscle" (J. H. Gaddum, ed.). Livingstone, Edinburgh and London.

Vellard, J. (1938). *Compt. Rend. Soc. Biol.* **127**, 38.

Vellard, J. (1948). *Publ. Museo. Hist. Nat.* (Lima) **1**, No. 1.

Vellard, J., and Huidobro, F. (1941). *Rev. Soc. Arg. Biol.* **17**, 477.

Vidal, J. C., Badano, B. N., Stoppani, A. O. M., and Boveris, A. (1966). *Mem. Inst. Butantan (São Paulo)* **33**, 913.

Vidal-Breard, J. J., and Elias, V. E. (1950). *Arch. Farm. Bioquim. Tucuman* **5**, 77.

von Klobusitzky, D., and König P. (1936). *Arch. Exptl. Pathol. Pharmakol.* **181**, 387.

Williams, E. J., Sung, S. S., and Laskowski, M. (1961). *J. Biol. Chem.* **236**, 1130.

Zeller, E. A. (1948). *Advan. Enzymol.* **8**, 459.

Zeller, E. A. (1951). *Helv. Chim. Acta* **33**, 821.

Zuber, H., and Jaques, R. (1960). *Helv. Chim. Acta* **43**, 1128.

Chapter 33

Intermediate Nephron Nephrosis in Human and Experimental Crotalic Poisoning

MOACYR DE F. AMORIM

DEPARTMENT OF PATHOLOGICAL ANATOMY AND PHYSIOPATHOLOGY,
ESCOLA PAULISTA DE MEDICINA, SÃO PAULO, BRAZIL

I. INTRODUCTION

Snake venoms are extremely complex mixtures composed chiefly of proteins. Their composition changes from species to species, depending on habitat and physical conditions (Porges, 1953; Rosenfeld, 1965; Rosenfeld *et al.,* 1968). Therefore, the lesions produced by snakebite are variable. The type of lesion seen varies with the organ or tissue affected as well as the species in which envenomation has occurred.

Although there have been only a few studies of the pathological anatomy of snakebite, many other aspects have been reported over a period of years. In earlier papers, in collaboration with Mello (Amorim and Mello, 1952, 1954) and Mello and Saliba (Amorim *et al.,* 1951), we analyzed the

319

material available in the literature. We will not repeat these findings, but refer the reader to these papers for more detailed bibliographical data.

However, we will review some of the principal facts here. In a case of human bothropic poisoning, embolisms were found in the capillaries of the lung (Mallory, 1926). We were able to reproduce these lesions experimentally (Amorim *et al.*, 1951). In another case, MacClure (1935) observed acute diffuse glomerulonephritis, after *Bothrops jararacussú* poisoning and Pena de Azevedo and Teixeira (1938) observed symmetric necrosis of the renal cortex due to poisoning by *Bothrops jararaca*. Rotter (1937), in an interesting paper based on autopsy of 3 fatal cases of poisoning by *Bothrops atrox* in Costa Rica, studied lesions of the vascular and central nervous system; he reported no other injury directly related to the snake venom.

The crotalic poisoning bibliography is still poorer, referring only to lesions induced by North American pit vipers. Most authors have described vascular and hemorrhagic damage to various organs from envenomation by these snakes (Mitchell, 1860, 1868; Mitchell and Reichert, 1886; Taube and Essex, 1937). Pearce (1909) was the only one to describe exudative glomerular lesions, which he called experimental vascular nephritis, with leukocyte and fibrin infiltration and fibrinoid degeneration of the capillary tuft; no tubular degenerative lesions were found. Pearce worked on rabbits, using the venom of North American *Crotalus adamanteus*. In one experiment he injected 0.5 mg of dry venom in 1 cc fluid intravenously, every 2 or 3 days. In other experiments he injected one dose of 1 or 2 mg of dry venom. In these he obtained hematuria or hemoglobinuria with albuminuria in 24 hours, sacrificing the animals when they clearly presented hematuria or albuminuria. In 21 rabbits, only 14 presented renal lesions and 7 other animals, which had received only 1 injection or several small doses, did not present lesions of any organ. In the first group he described hemorrhagic glomerular lesions or glomerular exudative lesions (leukocyte and fibrin infiltration), with fibrinoid degeneration of the capillary tuft. He inferred that the venom acted selectively on the glomerular epithelium.

It is remarkable that Pearce found no tubular epithelial lesions in most of his rabbits. In some of them he did find a typical cloudy swelling and, rarely, vacuolar degeneration. He found no nuclear necrosis or cellular destruction. Pearce described casts of exudative origin, with "products of hemolysis" not of tubular origin, which he classified as characteristic of an experimental vascular nephritis, undescribed until then.

Fidler *et al.* (1940) injected 7 to 10 mg of dry venom from *Crotalus atrox* (rattlesnake)/kg body weight into 9 monkeys (*Macaca mulatta*). Death occurred within 5 hours. They did not observe hematuria. In 3 of the animals they observed a fatty degeneration of the renal tubules. They also remarked upon the absence of hemorrhagic or exudative glomerular lesions.

Lesions produced by South American crotalic venom have been discussed only briefly by some authors (for more bibliographical data, see general reviews from da Fonseca, 1949; Schoettler, 1951; Rosenfeld, 1960; Rosenfeld *et al.*, 1968, and more detailed experimental work in our previous paper on the subject; Amorim *et al.*, 1951, 1954a,b, 1966). However, in 1952–54, while studying 3 cases of fatal snakebite in humans, we have observed lesions in the kidney, which presented a hemoglobinuric necrotizing nephrosis, identical to the so-called crush syndrome, recently called by some authors inferior nephron nephrosis. We prefer to call it intermediate nephron nephrosis for reasons we will explain later.

We have produced, in the dog, renal lesions absolutely identical to those found in man (Amorim *et al.*, 1954a,b, 1966).

Hadler and Vital Brazil (1966), using crotoxin instead of crude venom in dogs, reported that the kidney lesion "was not exactly the same as that which Amorim and Mello described in human cases of *C. d. terrificus* snakebite." They found that "the furthermost segment of the renal tubule [involved] was the proximal one." According to them, the lesions were purely degenerative, that is, a "microvacuolar degeneration" was seen. They said nothing about the nature of this degeneration—whether it was simple hydropic degeneration, true fat degeneration, or degenerative steatosis of the proximal segments of the tubules.

Rodrigues Lima (1956), however, found only moderate degenerative lesions at the proximal segments while the more severe lesions which were also true "necrotic lesions," were more frequently found in the distal segments; this seems to confirm our earlier findings.

Raab and Kaiser (1966) adopted the term "intermediate nephron nephrosis" when the nephrosis is produced directly by the action of the toxin on the tubular cells. They reserve the term "hemoglobinuric/myoglobinuric nephrosis" for cases in which the lesion is secondary to an extrarenal cellular degeneration involving erythrocytes or muscle cells.

Because of these differing opinions and the growing interest in this form of nephrosis in modern renal pathology, we will review our findings and discuss some of the related questions of pathology and nomenclature which have not appeared in the literature. We also wish to evaluate crotalic poison, with respect to its selective action on various parts of the renal tubules. Our work covers both human and experimental renal lesions.

II. RENAL LESIONS IN HUMAN CROTALIC POISONING

As mentioned above (Amorim and Mello, 1952, 1954), the lesions found by us in 3 cases of individuals bitten by *Crotalus d. terrificus*, at autopsy, correspond exactly to those occurring in hemoglobinuric nephrosis, also called inferior nephron nephrosis, which we prefer to call intermediate nephron nephrosis.

In our 3 cases, the individuals were white male farm workers, 58, 30, and 12 years of age, respectively. After snakebite, which occurred in the foot in the first 2 cases, the patients presented visual disturbances, dizziness, oliguria, or total anuria in one case, and "hemorrhagic aspect" of the urine. All 3 cases were treated with anticrotalic serum. The first was also treated with an artificial kidney. In this case, death occurred on the fifth day; the patient had not urinated for 3 days of his hospitalization at the Instituto Butantan. The second case presented intense hemoglobinuria, without hematuria, and died after 3 days. The third presented uremia, acidosis, albuminuria, hematuria, and hemoglobinuria with oliguria followed by anuria; he died within 7 days.

On necropsy, the kidneys were enlarged; in case 1, they were of a pale pink color, opaque with a firm consistency. In case 3, the cortical and medullary zones were very distinct; the cortex was 7 to 8 mm. thick and of a pale yellow color; the medulla was brown. In the histopathological examination, we found intense hyperemia of the renal cortex and of the glomeruli, in 2 cases and ischemia in the third. The renal cortex of case 3 was hyperemic, with interstitial hemorrhages in the boundary zone. There were no inflammatory lesions or hemorrhagic areas in the glomeruli.

The proximal convoluted tubules showed cloudy swelling, with an intact brush border. Certain cells were in mitosis, one, in the convoluted tubules (in case 1), being atypical.

In the ascending limbs of Henle's loop, as well as in the second convoluted tubules, and principally in the intermediate zone, the lumina was frequently obliterated by casts which stained intensely with eosin. These sometimes were homogeneous or hyaline and formed by irregular blocks or fragments. The cells were degenerated or desquamated, with pycnotic and karyolytic nuclei. Some cells showed a high degree of vacuolization or granulohyaline degeneration or they were necrotic (Fig. 1).

Many collecting tubules in the medullary zone were cellularly intact, but with lumina filled with casts identical to those described above. By the Lepehne method the casts showed a dark brown color with benzidine and were positive for hemoglobin, even after formalin fixation, in case 1.

By the Mallory-Masson method, the hemoglobin casts stained bright red, pink, or dark carmine (in the last 2 cases). They were stained blue by the Weigert method for fibrin and gold-yellow or light orange by the Van-Gieson method.

In cases 1 and 3, we noticed, in addition to severe degenerative tubular lesions, the presence of inflammatory lesions. These were focally distributed in both the cortex and the boundary zone. It was remarkable that many tubules contained neutrophilic granulocytes mixed with hemoglobin casts

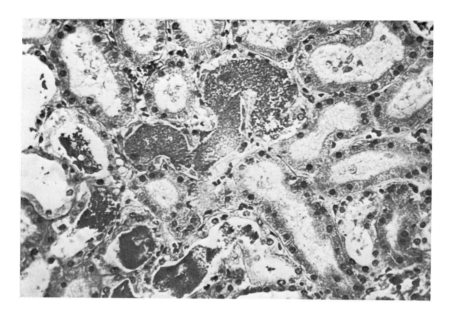

FIG. 1. Human kidney (H.E.). A portion of the distal convoluted tubule is shown, with vacuolized cells. Some cells are partly eroded away; and the nuclei appear pycnotic or karyolytic. Granular hemoglobin casts are visible. The proximal convoluted tubules with a well-preserved brush border are also shown.

with edematous peritubular connective tissue. At some points, in the boundary zone, there was a definite proliferation of histiocytic macrophages associated with lymphocytes and plasma cells.

In the third case, this proliferation of histiocytic macrophages at the boundary zone sometimes constituted granulomata the size of Aschoff nodes. Identical granulomas were also described by Zollinger(1952). In the medullary zone, the absence of inflammatory foci around the collecting tubules presenting casts is remarkable. Since the cells of these tubules were free from degenerative or necrotic lesions, we are led to believe that the inflammatory lesions of the intermediate and cortical zones are due to degenerative or necrotic lesions, following the purely nephrotic character of the syndrome. Also the medullary zone in these cases appears to be "clean," in contrast to other zones of the kidney, only presenting casts in the lumina of the tubules. The interstitial tissue showed no inflammatory infiltration.

Having thus observed that the lesions produced by crotalic poisoning in man correspond exactly to those seen in hemoglobinuric nephrosis or intermediate nephron nephrosis, which are identical to those seen in the crush syndrome, we attempted to reproduce this nephrosis experimentally, using crotalic venom.

III. RENAL LESIONS IN EXPERIMENTAL
CROTALIC POISONING

We began, with Mello and Saliba (Amorim *et al.*, 1954), a series of experiments to induce a nephrosis identical to that observed in crotalic poisoning in man. When death occurs in a few hours of poisoning, no histological lesions are to be found (Bell, 1950; and others); we sought, therefore, to reproduce conditions identical to those which occur in human poisoning, employing sublethal doses and injecting, in some of the animals, doses of anticrotalic serum calculated to be sufficient to neutralize the poison.

One female dog and 14 male dogs, weighing from 6 to 14 kg, were injected (subcutaneously) with 0.5 to 2 mg of dry crotalic venom/kg body weight; in 3 dogs the doses were fractionated over 2 or 3 injections. In 7 animals we injected from 9 to 30 cc antivenin in from 1 to 3 doses; these doses were believed sufficient to neutralize the venom.

Observations were made for 7 to 9 days. Two animals recovered partially and 4 completely. In 9 animals, death occurred spontaneously from the poisoning itself, and their survival times were as follows: 7 days and 20 hours, 2 days (2 dogs), 17 hours, 7 hours, 9 hours (3 dogs), and 11 hours.

In 2 cases the experiment was continued 9 days; however, since the animals recovered completely, they were not autopsied. The other 2 dogs, that also recovered completely, were sacrificed by ether and autopsied 7 and 8 days after injection. Two dogs, in partial recovery, were sacrificed by ether after 8 and 9 days of observation.

Only 1 dog presented anuria: death was at $11\frac{1}{2}$ hours and there was no urine left, even at the moment of autopsy, to test with benzidine.

Oliguria was found in 7 animals and was very intense in 2 cases; the other 8 animals urinated normally, including the 4 animals which recovered.

No erythrocytes, at least in appreciable quantity, were found in the urine; this was confirmed histopathologically for no red blood cells were found at any time in the sections examined.

The Lepehne test for free hemoglobin in urine was positive in 12 of 14 cases examined; no test could be run on the dog which presented complete anuria. Of the 2 dogs without hemoglobinuria, one presented oliguria and lived 7 hours; the other lived 9 hours, with no oliguria.

The dogs, which presented hemoglobinuria early in the experiment, gave negative Lepehne reactions at autopsy, or in the last few days before autopsy. They were supposed in clinical recovery.

The 2 surviving dogs which were not autopsied recovered completely. The dog which survived 8 days, and which was then autopsied, did not present any lesions of interest in the histological sections.

A. Histopathology of Experimentally Induced Nephrosis

The kidney lesions induced by injection of crotalic venom in the dog are identical to those found in man. Briefly, the following points are observed: (1) No glomerular lesions are seen. At times, Bowman's capsule contains albumin. (2) A slight hyperemia is seen in the labyrinth zone of the kidney; a more acute hyperemia occurs in the straight vessels of the boundary zone of the medulla. The cortex is not ischemic. (3) No important lesions are seen in the convoluted tubules, which show, at times, vacuolar degeneration and cloudy swelling; they may contain albumin. (4) Pigmented casts of hemoglobin-derived substances are seen principally in the ascending limb of Henle's loop and the distal convoluted tubules (that is to say, in that part of the nephron classically known as the intermediate section of Schweigger-Seidel), as well as in the collecting tubules of the cortex and medulla. In some cases, however, lesions were observed extending from the "cortex corticis," under the capsule, to the renal papillary zone. (5) In the intermediate section of the tubule, that is to say, the ascending limb of Henle's loop and the distal convoluted tubule, the lesions are more severe, showing not only cloudy swelling and vacuolar and lipoid degeneration, but also pycnotic and karyolytic nuclei; necrosis and desquamation occur in large sections of these segments of the nephron, with loss of tubular epithelium. Here, the hemoglobin casts are in contact with the connective tissue of the kidney. This was constantly observed, in the cortex, at the level where the intermediate segment, turning from the renal pyramids, joins the glomerulus. (6) The most severe and numerous lesions, however, occur in the "juxtamedullary" or "boundary zone" of the kidney, and in the transition zone between the convoluted tubules and the medulla, as has been observed by various authors, among them Bywaters and Dible (1942), Ogilvie (1951), Zollinger (1952), and Amorim and Mello (1952, 1954). In man, the ascending limb of Henle's loop or the so-called intermediate segment of the nephron is mainly affected. (7) In some animals, there occurs, in the "juxtamedullary zone" of the cortex and in the adjacent medulla, small inflammatory secondary foci characterized, at times, by neutrophilic granulocytes, and, at times, by plasma cells; these surround those tubules in which the epithelial cells have disappeared, and are also in contact with the casts. We did not see, in our experimental animals, the granulomas described by us in man (Amorim and Mello, 1952, 1954), the chronic sclerotic lesions described by Ogilvie (1951) and Zollinger (1952), or venous thrombosis as described by these authors, Bywaters and Dible (1942), and others. Neither did we see tubulovenous aneurisms, which were described in the excellent Zollinger monograph (1952).

We conclude that, in our experimental animals, the presence of hemoglobin in the urine was the most characteristic clinical sign, being almost constantly present (12 out of 14 cases). Therefore we agree with Mallory and Allen and prefer to call the clinical syndrome hemoglobinuric nephrosis.

It is remarkable that hemoglobinuria and the occurrence of hemic casts cannot be correlated with the anuria or oliguria seen clinically. In the only case which presented evident anuria, where the animal lived 11½ hours, no hemic casts were found. However, of the 3 animals with pronounced hemoglobinuria and with numerable casts in the sections the 2 first cases were oliguric and urinated an average of once a day; one lived 7 days and the other 9 hours. The third died after 2 days, but urinated normally, in spite of an intense hemoglobinuria, from the first day, to the moment of death; autopsy showed the presence of many hemic casts (Fig. 2). The dog with a small number of hemic casts, presented hemoglobinuria and a slight oliguria, urinating 3 times in the 9 hours it lived after the injection of crotalic venom. We were therefore unable to correlate oliguria and hemoglobinuria and the occurrence of hemic casts in the sections.

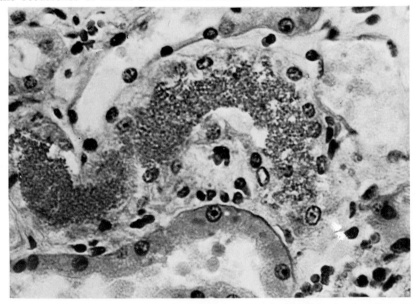

Fig. 2. Dog kidney (Formalin, H.E. 200×) as seen in crotalic poisoning. Granulous hemoglobin casts are seen in the distal convoluted tubule. Several cells are tumefact and vacuolized; many nuclei are pycnotic or karyolytic. A typical proximal convoluted tubule is shown above.

IV. DISCUSSION AND GENERAL CONSIDERATIONS

These observations on hemoglobinuric nephrosis in man and in experimental animals, lead us to a further discussion of some questions involving the anatomy, nomenclature, and physiology of the syndrome.

A. Experimental Induction and Pathogenesis of Hemoglobinuric Nephrosis

As we have seen, the experimental induction of hemoglobinuric nephrosis presents great difficulties, and various authors have called attention to this point.

Youile *et al.* (1945) refer to "the mechanism involved in the production of this type of renal failure which has remained a mystery; despite the efforts of many investigators it would appear to be due, at least in part, to the difficulties encountered in attempting to produce a similar lesion in the kidneys of experimental animals." This is perhaps the reason why so few investigators have attempted to produce experimentally the clinical and pathological lesions characteristically seen in man (Fig. 3).

Yorke and Nauss, in 1911, produced hemoglobinuria in rabbits only when the animals were fed a dry diet, completely without green vegetables, before hemoglobin was injected.

Baker and Dodds (1925) elaborated a method to prepare strong solutions of hemoglobin, which they injected intravenously in rabbits, and arrived at the conclusion that there were 2 factors necessary to produce a hemoglobinuria: (1) an acid reaction to convert the pigment to metahemoglobin. (2) The presence of a certain amount of inorganic salts. If the urine is above pH 6, hemoglobin will be excreted as oxyhemoglobin; the urine will be red,

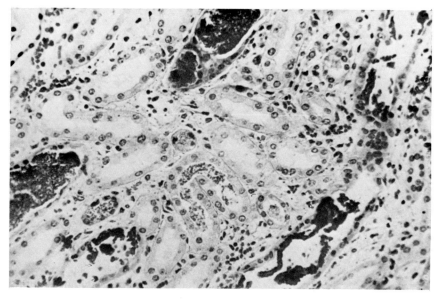

FIG. 3. Human kidney (H.E.). Cortex—several proximal convoluted tubules are seen. Among these, several intermediate segments are distinguished by their flat cells containing hemoglobin casts.

but there is no kidney damage. If, however, the pH is less than 6, and there is a sufficient concentration of NaCl (about 1%), heme will precipitate in the tubules. Hemoglobin, then, will pass through Bowman's capsule in the diluted filtrate. When the hemoglobin reaches the tubules, reabsorption and concentration occur with an increase in urine acidity. They conclude by advising the production of an alkaline diuresis for the treatment of such conditions.

Youile et al. (1945) could not confirm the results of Baker and Dodds, Gowin, Bywaters, Anderson, and Bing, regarding the role of pH in the production of hemoglobinuria. In attempting to produce this nephrosis, Youile et al. employed 2 synergic factors: hemoglobin injection and ligation of the renal artery, for less than an hour, followed by sodium tartrate injection.

Crotalic venom seemed to us to be an easy and convenient method to obtain this form of nephrosis, since, working only with this agent, we succeeded in the induction of this nephrosis in a significant number of cases. This does not mean, however, that, by means of this poison, we induced hemoglobinuric nephrosis with only 1 chemical agent or substance, for this would contradict the previously cited experimental facts. Crotalic poison is, as we said, a complex mixture of proteins; its chemical composition is known through the research work of Slotta and Szyska (1937), Rocha e Silva et al. (1949), Gonçalves (1956), Rosenfeld (1965), Devi (1968), Vital Brazil et al. (1966), and many other authors. It is possible that its effect could be explained by the synergic action of the various compounds contained in this complex mixture of crotoxin, neurotoxin, hemolysin, crotamine, enzyme, etc. One active agent, certainly, would be its hemolytic fraction, or hemolysin, which is known to produce at least some of the conditions necessary for the induction of the hemoglobinuric syndrome, as seen in crotalic poisoning (da Fonseca, 1949; Schoettler, 1951; Rosenfeld, 1968; Devi, 1968; and others). However, hemoglobinemia or hemolysis alone, will not produce hemoglobinuria, as demonstrated by Baker and Dodds (1925); according to these workers, 2 or more factors are necessary to precipitate hemoglobin in the urine. According to Youile et al. (1945), precipitation of hemoglobin in the renal tubules does not depend exclusively upon some functional abnormality of the nephrons. Precipitation was obtained only when moderate lesions had been produced in the tubules by other means (ligating the renal artery for less than an hour, or by injecting a specific poison such as sodium tartrate).

Lalich (1954) concludes, from his research in rabbits, that associated factors may be more significant in the production of hemoglobinuric nephrosis, than hemoglobinemia per se. Among these factors he mentions: previous lesions, injection of histamine or arsenic, starvation, restricted water intake, etc. Bull and Dible (1953) also conclude that: "tubular damage may be produced 2 ways: first, by the action of a poison and second, by ischemia. Not infrequently both factors operate" (p. 283).

In the cases we observed, we never found ischemia; on the contrary, there generally occurred a glomerular and cortico-medullary hyperemia. Thus, our ·observations seem to uphold the conclusion that the crotalic poison may contain a possible nephrotoxic fraction, capable of injuring the renal parenchyma, perhaps selectively, on the level of the intermediate segment. The poison could produce hemolysis, through the action of a hemolytic agent, with the release of hemoglobin in the blood, this being a requisite condition for hemoglobin precipitation in the glomerular filtrate tubular preurine. The nephrotoxic fraction of the poison would induce tubular lesions; thus, the 2 processes would act synergically to produce hemoglobinuria and the associated nephrotic syndrome.

B. Macroscopic Features of the Kidney in Hemoglobinuric Nephrosis

Very few macroscopic changes are seen in the kidney in this disorder, and this is an important feature in its differential diagnosis and pathogenesis. For a thorough review of the changes which have been observed, we refer the reader to the excellent monograph of Zollinger (1952).

From our observation of human cases, as well as experimental ones, we conclude that the macroscopic changes in the kidney are generally insignificant, when they do occur, in hemoglobinuric nephrosis. We have observed a few exceptions, as in a case of eclampsia, where a large number of casts had turned the medulla greenish brown in color. In this case it was possible to make a tentative diagnosis based on a macroscopic examination, which was confirmed later by the histological examination.

In most cases, however, the kidney appears normal, though increased in size and weight. The external surface is smooth with no change in color, but sectioning shows a reddish brown, or dark brown medulla, in contrast to the lighter cortex. Only once did we see a really pale cortex, in case 3 of our paper with Mello (Amorim and Mello, 1954). This was the case of a 12-year-old boy, who had been bitten by a rattlesnake 7 days before. In this case we did observe, also histologically, cortical ischemia. In case 1 of the cited paper, at autopsy of a 58-year-old patient bitten by a rattlesnake 5 days previously, the cortex was pale pink, mottled darker pink. Hyperemia and vasodilatation were observed throughout the cortex, including the glomeruli, on histopathological examination. In case 2, the patient, about 30 years old, died 3 days after being bitten by *Crotalus d. terrificus*. Autopsy revealed many hyperemic glomeruli. Experimentally we did not observe any marked change; the medulla was only generally darker than the cortex.

Because of the barely recognizable macroscopic alterations of hemo-globinuric nephrosis (only slight increase in size and weight; Zollinger, 1952), this entity has frequently been confused in the literature with bilateral cortical necrosis of the kidney. This is perhaps due to the fact that degenerative phenomena are seen at the intermediate nephron level, with necrosis and desquamation of cells. However, the necrosis in hemoglobinuric nephrosis is slight and widely scattered. These necrotic areas are only seen microscopically and, often, only by the exhaustive examination of many sections. They are seldom seen macroscopically and are not easily recognizable microscopically. The same cannot be said, of course, of hemic casts, which, when numerous, are immediately apparent. In less severe cases, the casts may be hard to find, as noted above.

Very different indeed is the picture of the so-called symmetrical cortical necrosis, in which compact zones of necrosis are found. These zones may extend through almost the entire thickness of the cortex, and coagulation necrosis is easily recognized.

As an example of the problems of diagnosis, see Fig. 3 of Lucke's paper (1946), which, in our opinion, is absolutely characteristic, at least macroscopic-ally, of bilateral symmetrical cortical necrosis, in spite of the author's description of it as the "external appearance of the kidney in a representative case of lower nephron nephrosis." The picture clearly shows a kidney with a large pale-white band seizing the cortex. This is not a typical case of hemoglobin-uric nephrosis; it resembles, rather, symmetrical cortical necrosis. If further examination showed it to be hemoglobinuric nephrosis, and supporting data for this are not in the paper, then the coexistence of both lesions should have been suggested. The diagnosis is again in doubt in the paper of Pizarro (1950) who refers to both symmetrical cortical necrosis and hemoglobinuric nephro-sis (p. 744). Now, in obstetrical pathology, the differential diagnosis is im-portant. For example, in eclampsia, symmetrical necrosis of the cortex occurs in some cases and inferior nephron nephrosis in others.

In the recent work of Lalich (1954), in a photograph of a macroscopic kidney section (Fig. B), great necrotic areas are seen; these suggest anemic infarcts, an observation which is not consistent with the diagnosis of inter-mediate nephron nephrosis. Dible (1950, p. 301) maintains that: "severe tubular necrosis is in no way a stage in the development of cortical sym-metrical necrosis, which is essentially a vascular lesion."

As mentioned previously there is the possibility that the 2 types of lesions could coexist in the same case and even with anemic infarcts. However, we have not observed any such case, nor is such a case mentioned in the litera-ture, not even by Zollinger (1952) in 17,000 autopsies over a 10-year period at the Zürich Institute. The pathological features of one process have nothing in common with the other, since both syndromes are well distinguished and independent entities, involving different pathogenetic processes even when observed under identical etiological conditions.

C. Nomenclature and Synonymy

In 1950, Bell remarked that "the introduction of the term 'lower nephron nephrosis' has been a disservice to renal pathology in that it has added confusion instead of clarity." We cannot entirely agree with this assertion, for, in studying the literature, we see that the nomenclature of this syndrome has never been as simple as it seems to be. On the contrary, this syndrome has, over many years, been called by a great number of names. Zollinger (1952) has listed many of these synonyms in his excellent monograph. It is important that these be known, so as to avoid any confusion on the subject. From the bibliography available to us we have culled out the following synonyms for the syndrome.

1. Hemoglobinuric glomerulonephritis (Afanasjew, cited by Fahr, 1925).
2. Diffuse severe nephritis (Ponfick, cited by Fahr, 1925).
3. Post-crushing nephrosis ("Verschuettungs-Nephrose") (Borst, 1900; Hackradt, 1917; Bredauer, 1920; Minami, 1923).
4. Hemoglobinemic nephrosis (or "Nephrose durch Hemoglobinspeicherung") (Fahr, 1925).
5. Intolerance nephritis ("Nephrite d'intolérance") (Tzanck, 1935).
6. Crush syndrome (Bywaters and Dible, 1942; Ogilvie, 1951).
7. Laminar nephrosis (Russee, 1944).
8. Renal anoxic syndrome (Malgraith, 1944, 1949).
9. Hemoglobin nephrosis with interstitial exudative inflammation (Hemoglobinnephrose mit interstitieller exudativer Entzuendung) (Zollinger, 1945).
10. Lower nephron nephrosis (Lucke, 1946).
11. Hemoglobinuric nephrosis (Mallory, 1947; Allen, 1951; and others).
12. Toxic renal anoxia (Junet, 1949).
13. Distal tubular nephritis (Bergstrand and Bergstrand, 1949).
14. Erytrolytic nephrosis (Letterer and Masshoff, 1949).
15. Tubular nephritis (Hamburger, 1950).
16. Chromo-protein kidney (Chromo-protein-Niere, hemolysis kidney or crush-kidney) (Zollinger, 1952).
17. Acute tubular necrosis (Bull and Dible, 1953).

Bull and Dible (1953) also assert that lower nephron nephrosis is not a happily chosen term (p. 284) for the anatomical lesions, as well as the functional changes, are not limited to a portion of the nephron, even when the more pronounced histological changes are found in the distal portion. They propose, therefore, the term acute tubular necrosis (p. 284).

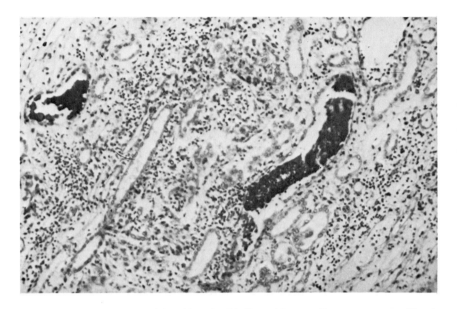

Fig. 4. Boundary zone of the kidney (H.E.). Several intermediate segments are dilated, with hemoglobin casts and a great number of neutrophilic granulocytes. These lesions resemble those of pyelonephritis.

Bell too affirms that "there are serious objections to the concept of lower nephrosis" and that "the advocates of 'lower nephron nephrosis' have ignored the lesions in the proximal tubules or have attempted to explain them as a result of obstruction of the distal segments."

Evidently, Lucke used lower nephron nephrosis to stress the point at which the lesion was typically more severe. We agree with this author as to the site of these lesions, but we disagree with him as to which name to use for that segment of the nephron, and, consequently, for the syndrome. Therefore we preferred, in our earlier papers (Amorim and Mello, 1952, 1954), to use the term intermediate nephron nephrosis, derived from the original name given to this segment of the renal tubule by Schweigger-Seidel and still used by various authors (Poirier and Charpy, 1925; Hovelacque and Turchini, 1938; Policard, 1950) as a synonym for distal or inferior segment of the nephron. We prefer to use, as a generic name, as "real" or "genuine," inferior nephron nephrosis, to mean that syndrome in which the lesions predominate along the collecting tubules, even in the medulla (Bergstrand and Bergstrand, 1949; White and Mori-Chavez, 1952; and others).

Today, we have discarded the old concept of the nephron as consisting of secretory and resorbing units, with the collecting tubule (medulla) acting only for the excretion of urine. Regarding this, Bell (1950) actually says:

"Resorption of water apparently takes place all along the tubule. The fact that uric acid precipitates out in the collecting tubules in the so-called uric acid infarcts of infants suggests that resorption may continue even into the collecting tubule." "Similar evidence that concentration continues into the collecting tubules is afforded by the precipitation of sulfonamide crystals in the collecting tubules and similar deposits of uric crystals in gout nephritis."

Dible (1950), also, corroborates this, asserting that "the specific action which many chemical poisons (e.g., uranium salts) have on the convoluted tubules is explicable on the ground that they only reach an effective concentration when water has been abstracted in this portion of the kidney" (p. 712). Now, if this is true for the proximal tubules, there is no reason to conclude otherwise for the collecting ducts. The selective precipitation of uric acid crystals in the collecting tubules has been known for a long time in classical pathology, and, more recently, the precipitation of sulfathiazole crystals (Horack, 1940; Bergstrand, 1944, 1947), as well as various other agents, has been observed (White and Mori-Chavez, 1952).

Bull and Dible (1953) mention that "in poisoning, the kidney is not infrequently, far more severely affected than other parts of the body. This may be the result of water reabsorption in the tubules during excretion of the poison, the concentration of the poison becoming higher in the tubules than anywhere else in the body, except possibly at the portal of entry" (p. 282).

Thus, from histophysiological, as well as physiopathological considerations, we prefer to call the referred to segment of the renal tubule as

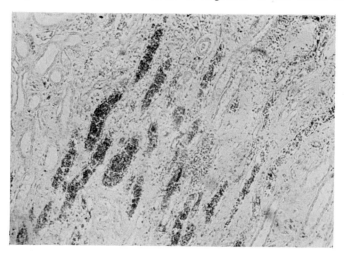

FIG. 5. Human kidney (Lepehne's method). Boundary zone—several collecting tubules and straight parts of the intermediate segment contain dark brown casts.

intermediate rather than distal or inferior, and to call the syndrome under discussion intermediate nephron nephrosis. As mentioned before, Raab and Kaiser (1966) prefer this designation.

D. A Histotopographical Classification of Nephrosis: Site of Crotalic Poison Action

Bull and Dibgle (1953) point out that "in tubular necrosis of all sorts there is a general disturbance of function affecting all parts of the tubule" (p. 286). Evidently, we can expect to see physiopathological changes throughout the nephron, including the glomeruli, if our methods are sensitive enough. Of those available, the physiological methods are more sensitive than the histopathological. However, it is important to remember that the more severe anatomical lesions are those seen by the pathologist. These lesions also indicate the areas in which the toxin has had its most pronounced effect, or, conversely, those tissues which are most sensitive to the toxin.

Bull and Dible (1953), as we have already mentioned, present the best arguments for a new nomenclature and classification of nephrosis, which could describe the pathological genesis and morphology of the lesions more precisely. However, these would not include the purely functional disturbances occurring throughout the nephron. These are of less importance, pathologically, not having reached what may be called here the *anatomopathological threshold.*

Thus, Dible (Bull and Dible, 1953), in his study of 62 cases of anuria, divides tubular anuria into 2 main groups (p. 291): (1) "those in which there is gross necrosis of tubular epithelium with maximum incidence of this in the first convoluted tubules." These include cases of poisoning by mercuric chloride, carbon tetrachloride, methyl bromide, and Nembutal. (2) "those in which there is selective damage to the nephron, the grosser visible changes being found more especially in the boundary zone, and in the second convoluted and collecting tubules." These include cases of incompatible blood transfusion, massive intravascular hemolysis, extensive muscle damage in the crush syndrome, or traumatic anuria; and others, included under the rubric of inferior nephron nephrosis, as are surgical shock, severe malaria, burns, septic abortion, reaction to sulfonamide, quinine poisoning, and hemolytic anemia.

In group (1), Dible describes a case of poisoning by Nembutal with a 48-hour anuria, in which there was "almost complete necrosis of the cells of the proximal convoluted tubule" "with striking change." By contrast, the thin portions of the loop and the ascending limb of Henle and the second convoluted tubules appear relatively normal though collapsed. "And these portions of the nephron show in strong relief in the cortex against the mass of necrotic proximal tubules" (p. 292).

In group (2), "selective epithelial necrosis with a maximum incidence in the second convoluted tubule is the outstanding lesion" (p. 293), constituting 40 of the 62 cases studied.

Dible stated also that the earliest detectable lesions were found only after anuria of 30- to 40-hour duration, confirming the experimental findings of Scarf and Keele (1943) in rabbits; these authors found ischemic lesions only after 2 days of renal ischemia.

Dible proceeds (pp. 295 and 296): "Our positive findings indicate a selective necrosis of tubules in the boundary zone of the cortex and evidence of stagnation in the second convoluted tubules of the nephron and beyond." "Lastly" (p. 298) it "is suggested that the main brunt of the lesion falls upon the intermediate or boundary zone of the organ." "The early necrosis

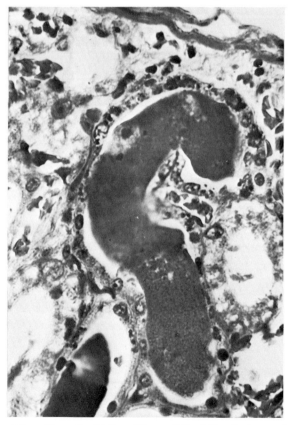

FIG. 6. Dog kidney (formalin, Mallory-Masson 200 ×). Two finely granular hemoglobin casts are seen in the secondary convoluted segments, the cubic or lower epithelium being fairly visible in the greater tubule. The proximal convoluted tubules show a microvacuolar hydropic degeneration.

of the descending limb of Henle, mentioned as the first visible change, is followed by a concentration of pathological alterations in this area" (p. 298); these include: inflammation, congestion of the straight vessels, thrombosis, swelling of the boundary zone with fibrosis, as well as edema of the papillary zone, and desquamation of the cells of the collecting tubules. These changes suggest "that a veritable pathological barrier is formed across the base of each renal pyramid at this level" (p. 298).

In Dible's cases, the kidneys were generally hyperemic, with the glomeruli presenting pronounced vascular congestion, as we too observed in our material. It is interesting that Dible does not comment upon the use of either the term lower nephron nephrosis (Lucke) or hemoglobinuric nephrosis (Mallory, Allen, and others).

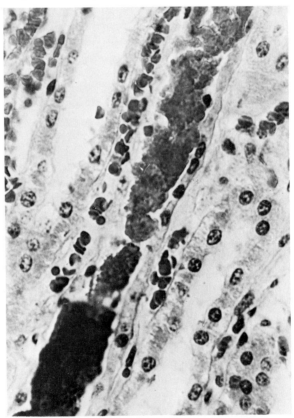

FIG. 7. Dog kidney (formalin, H.E. 200×). Radial zone of the cortex presents typical brown hemoglobin casts in the lower portion and in the descending straight part of the intermediate segment. The epithelium of the latter shows disappearance of some of its nuclei; others are pycnotic or karyolytic. Hyperemia of the straight capillaries is seen.

Ellis (1942), and Bull and Dible (1953) object to the general use of the term nephrosis. Objections have also been raised for the use of Munk's lipoidic nephrosis and Volhard or Fahr's pure or genuine nephrosis, since this syndrome has not been demonstrated as existing apart from Ellis' type 2 glomerulonephritis. The necrotic forms of nephrosis, however, are very common and are caused by a great number of pathogenic agents. The pathological features here, where the primary lesions are degenerative or parenchymatous, are quite distinct from the inflammatory lesions of nephritis. Only Hamburger (1950) interprets these necrotic lesions as tubular nephritis.

Dible's (see Bull and Dible, 1953) term, acute tubular nephrosis, could be used to include all forms of nephrosis, from those involving the proximal nephron to those of the intermediate and of the true inferior nephron. This would be a very general usage and would not apply to mild cases of nephrosis in which no necrosis occurs and recovery is complete.

Allen (1951) also classifies renal lesions histotopographically. They are subdivided by him as follows: Proximal nephron nephrosis due to osmotic changes and poisoning by mercury, potassium bichromate, glycol diethylene, etc.; proximal and distal nephron nephrosis (hemoglobinuric lower nephron

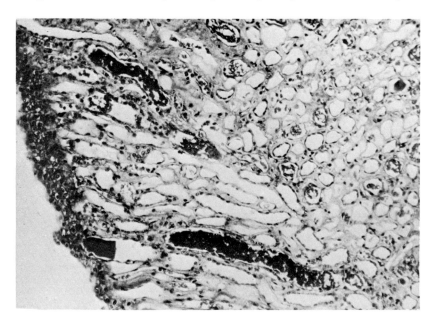

FIG. 8. Dog kidney (formalin, Mallory-Masson, 45 ×). Papillary zone of the kidney with the renal pelvis. Several papillary tubules have lumina filled with hemoglobin casts, some being sectioned longitudinally and others transversely.

nephrosis, cholemic nephrosis) due to Gierke's disease, poisoning by hydro-quinone-pyrogallic acid and the sulfonamides; distal nephron nephrosis due to myeloma, uric acid infarcts in gout, inclusion disease, and parenchymatous nephrocalcinosis. In the text, this author groups the 2 latter nephroses together. Moreover, he does not distinguish among the different segments of the nephron from which he draws his classification.

Referring to Lucke's inferior nephron nephrosis, which Allen prefers to call hemoglobinuric nephrosis, this author says (p. 256): "There are many 'nephroses' of the lower nephron including those due to myeloma, gout, sulfonamides, nephrocalcinosis (secondary to alcalosis, bony lesions, and hyperparathyroidism), uric acid, and hepatic diseases (cholemic nephrosis). Each of these conditions is a lower nephron nephrosis in a sense as complete as, or more definitive than, hemoglobinuric nephrosis."

Though Allen does consider the lesions of the medulla nephrotic, thus placing this syndrome with the inferior nephron nephroses, he also places hemoglobinuric nephrosis (Lucke's lower nephron nephrosis) with the inferior or distal nephron nephroses without specifying the segment of the nephron involved. Thus, hemoglobinuric nephrosis is first classed as proximal

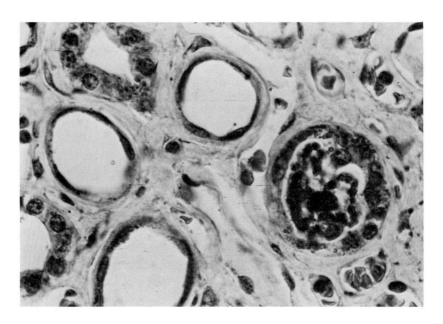

FIG. 9. Dog kidney (formalin, Scarlet R-H. 45×). Boundary zone of the kidney showing, in the center, 3 typical sections of the descending limbs of the Henle's loop (intermediate segment) with empty lumina. At one side, the ascending limb of the loop containing hemoglobin cast with fat degeneration of the cubic cells. Some small blood capillaries are filled with erythrocytes.

and distal nephron nephrosis (p. 207), and then (p. 256) as inferior or distal nephron nephrosis, which includes, here, lesions of the medullary segment of the nephron.

Following our own observations and those in the literature, we think it possible to classify nephrosis from a physiological and histotopographical point of view, according to the segment of the nephron in which the greatest damage occurs through the selective action of venoms and toxins. Such a classification might also lead to advances in the treatment of these lesions.

A brief scheme of our classification follows:

a. Proximal or superior nephron nephrosis (mainly cortical nephrosis). Selective damage to the proximal tubule by:

Nembutal overdose (Dible, see Bull and Dible, 1953)
Sucrose intoxication
Corrosive sulfamate poisoning
Potassium bichromate or potassium chloride poisoning
Glycol diethylene poisoning
Racemic tartaric acid poisoning

b. Intermediate nephron nephrosis (boundary zone nephrosis, or cortico-medullary nephrosis). Selective damage to the intermediate segment of Schweiger-Seidel by:

Severe or prolonged crushing
Blood transfusion reactions
Eclampsia, placentary uterine hemorrhage, toxemia of pregnancy
Uranium nitrate poisoning (Meessen, 1952)
Sulfur drug reactions
Shock associated with abdominal operations; surgical shock
Burns
Sunstroke
Severe malaria (falciparum)
Carbon tetrachloride poisoning (Lucke, 1946)
Carbon monoxide poisoning
Ferrol and other venom poisoning
Photographic developer poisoning
Rickettsiosis
Acute pancreatitis
Shock due to various causes
Hemolytic anemia
Hepatorenal syndrome
Poisoning by South American crotalic snakes

c. Distal or inferior nephron nephrosis (medullar nephrosis). Selective damage to collecting tubules of the medullary zone by:

Uric acid infarcts in gout
Vinylamine poisoning

Mono-*N*-methylamine experiments on mice (White and Mori-Chavez 1952)
Tetrahydroquinoline poisoning
Bromotylaminbromohydrate poisoning
Sulfonamide precipitation (Bell, 1950; Bergstrand and Bergstrand, 1949)
Multiple myeloma (Bence-Jones protein or serum globulin precipitation)
  d. Glomerular and glomerulotubular nephrosis (*Glomerulonephrosis* Randerath). Selective damage to the glomerular apparatus by:
Amyloidosis
Diabetes (Kimmelstiel-Wilson)

V. SUMMARY

The author has reviewed the bibliography on the pathological anatomy of snake venom poisoning referring chiefly to the works of Mallory, Mac-Clure, Rotter, Pena de Azevedo and Teixeira (bothropic poisoning) and to those of Pearce, Mitchell, Fidler, and others (North American crotalic poisoning). Snake poison is a complex mixture of proteins, which varies in composition from species to species and according to the habitat and physical condition of the snake. Lesions due to snake venom poisoning vary in site and severity; they also vary with the animal used for experiments or observation (for references, see Amorim *et al.*, 1954a,b, 1966; Rosenfeld *et al.*, (1968)).

In South American crotalic poisoning, renal lesions are similar to those seen in the so-called crush syndrome, *Verschuettungs-nephrose*, or nephrosis of the inferior or intermediate nephron, as the author demonstrated together with Mello (Amorim and Mello, 1952, 1954) in humans and later with Mello and Saliba (Amorim *et al.*, 1954 a,b, 1966) in experimental research in dogs; this has also been demonstrated by Raab and Kaiser (1966).

Because of the interest in this syndrome in modern renal pathology, the author has compared renal lesions in human crotalic poisoning and those produced experimentally in the dog, emphasizing the fact that all the authors have stressed the difficulty of obtaining experimental hemoglobinuric nephrosis (Youile *et al.*, 1945), there being always the necessity of a synergic action of 2 factors to produce it (Yorke and Nauss, 1911; Baker and Dodds, 1925; Lalich, 1954). Youile *et al.* (1945) ligated the renal artery or injected sodium tartrate before the administration of hemoglobin to produce this nephrosis.

Meanwhile, poisoning by South American crotalic venom proved to be a relatively easy way of reproducing this nephrosis (7 out of 15 animals in our experiments). This research seems to uphold the conclusion that South

American crotalic venom contains a nephrotoxic, as well as a hemolytic, fraction. The nephrotoxic fraction appears to act selectively on the intermediate segment of the renal tubule. It has not been isolated, nor is its nature known.

The author also describes certain correlated problems regarding the nomenclature and classification of the known nephroses, including those of descriptive pathology and differential diagnosis. In the nephrosis typically produced by South American crotalic venom, he observes that the kidney, macroscopically, shows little change except in size and weight. Microscopically, necrosis is seen at the level of the intermediate tubule, though it may not be easily demonstrable in a small sampling of sections.

New arguments are presented in favor of the designation "nephrosis of the intermediate nephron," based principally on the classic name of the involved segments of the nephron as given by Schweiger-Seidel and on a physiopathological, histotopographical approach. The author distinguishes in this way: (1) *Nephrosis of the proximal or superior nephron* (cortical nephrosis), induced by Nembutal (Dible), mercury, etc. (2) *Nephrosis of the intermediate nephron* (nephrosis of the boundary zone or corticomedullary nephrosis), induced by crotalic venom, incompatible blood transfusion, crushing, and a great number of different etiological conditions enumerated by Fahr, Lucke, Bell, Allen, Messen, Zollinger, and others. (3) *Nephrosis of the lower or distal nephron* (medullary nephrosis) induced by sulfathiazole (Bergstrand, 1944), uric acid (as in gout), Bence-Jones protein (as in multiple myeloma), etc.

A modification in the classic concept of the nephron is necessary, if we are to admit this classification. The nephron would be defined to include the medullary zone of the tubule, to which we, today, attribute functions of absorption and concentration of the urine (Bell, 1950), as demonstrated by precipitation of certain crystals in this zone.

REFERENCES

Allen, A. C. (1951). "The Kidney. Medical and Surgical Diseases." Grune & Stratton, New York.

Amorim, M. de F., and Mello, R. F. (1952). *Mem. Inst. Butantan (São Paulo)* 24, 281–316.

Amorim, M. de F., and Mello, R. F. (1954). *Am. J. Pathol.* 3, 479–499.

Amorim, M. de F., Mello, R. F., and Saliba, F. (1951). *Intern. Congr. Clin. Pathol.*, London, 1951.

Amorim, M. de F., Mello, R. F., and Saliba, F. (1954a). *Comun. Soc. Biol. (São Paulo)*.

Amorim, M. de F., Mello, R. F., and Saliba, F. (1954b). *Comun. Sess. Patol. Assoc. Paul. Med.*

Amorim, M. de F., Mello, R. F., and Saliba, F. (1966). *Frankfurter Z. Pathol.* 75, 87–99.

Baker, S. L., and Dodds, E. C. (1925). *Brit. J. Exptl. Pathol.* 6, 247–260.

Bell, E. T. (1950). "Renal Diseases," 2nd ed. Lea & Fibiger, Philadelphia, Pennsylvania.

Bergstrand, H. (1944). *Acta Med. Scand.* **118**, 97–113.

Bergstrand, H. (1947). *Acta Med. Scand.* Suppl. 196, 268–272.

Bergstrand, H., and Bergstrand, H. (1949). *Scand. J. Clin. & Lab. Invest.* **1**, No. 4, 334–338.

Borst, M. (1916–1917). *Beitr. Z. Pathol. Anat. Path.* **63**, 725–754 [cited by Fahr in "Handbuch der Speziellen pathologischen Anatomie und Histologie" (Hencke, F., and O. Lubarsh, eds). Springer, Berlin, 1925].

Bredauer, K. (1920). Inaugural Dissertation, Muenchen.

Buechner, F. (1950). "Allgemeine Pathologie." Urban & Schwarzenberg, Munich.

Bull, G. M., and Dible, J. H. (1953). *In* "Recent Advances in Pathology" (G. Hadfield, ed.), 6th ed., pp. 273–308. London.

Bywaters, E. G. L., and Dible, J. H. (1942). *J. Pathol. Bacteriol.* **54**, 111.

Celestino da Costa, A. (1949). "Tratado Elementar de Histologia e Anatomia Microscopica," 2nd ed., Vol. 2, Lisboa.

da Fonseca, F. (1949). "Animais peçonhentos." Inst. Butantan, São Paulo.

Devi, A. (1968). *In* "Venomous Animals and Their Venoms" (W. Bucherl *et al.*, eds.), Vol. I, pp. 119–165. Academic Press, New York.

Dible, J. H. (1950). "Pathology." 3rd ed. London.

Ellis, A. W. M. (1942). *Lancet* **1**, 1–7; 34–36; 72–76.

Essex, H. E., and Markowitz, J. (1930). *Am. J. Physiol.* **92**, 335.

Fahr, T. (1925). *In* "Handbuch der Speziellen pathologischen Anatomie und Histologie" (F. Hencke and O. Lubarsch, eds.), Vol. 6, Part 1, pp. 279–281. Springer, Berlin.

Fidler, H. K., Glasgow, R. D., and Carmichael, E. B. (1940). *Am. J. Pathol.* **16**, 355–364.

Gonçalves, J. M. (1956). *In* "Venoms," Publ. No. 44, p. 189. Am. Assoc. Advanc. Sci., Washington, D.C.

Hackradt, A. (1917). Inaugural Dissertation, Muenchen.

Hadler, W. A., and Vital Brazil, O. (1966). *Mem. Inst. Butantan (Sao Paulo)* **33**, No. 3, 1001–1008.

Hamburger, J. (1948). *In* "Les acquisitions medicales recentes," Vol. 1, p. 304. Editions Med., Flamarion, Paris.

Hamburger, J. (1950). *In* "Les acquisitions medicales recentes," Vol. 1. Editions Med., Flamarion, Paris.

Hawk, P. B., Oser, B. L., and Summerson, W. H. (1953). "Practical Physiological Chemistry," 12th ed., pp. 437–767. McGraw-Hill (Blakiston), New York.

Hovelacque, A., and Turchini, J. (1938). "Anatomie et histologie de l'appareil urinaire et de l'appareil genital de l'homme," pp. 209–211.

Huth, F., and MacClure, E. (1964). *Frankfurter Z. Pathol.* **74**, 91–108.

Kaiser, E. (1953). *Monatsh. Chemi.* **84**, 483–490.

Lalich, J. J. (1954). *A.M.A. Arch. Pathol.* **57**, 36–43.

Lison, L. (1966). "Histochemie et cytochemie animales." Gauthier-Villars, Paris.

Lucke, E. (1946). *Military Surgeon* **99**, 371–396.

MacClure, E. (1935). *Bol. Secret. Saude Assistencia (Rio de Janeiro)* **1**, 35–49.

MacClure, E. (1946). "Glomerulonefrite aguda difusa." Rio de Janeiro.

Mallory, F. B. (1926). "Cases of Snake-Bite Treated in Almirante Hospital, Panamá—1922–1926."

Mallory, F. B. (1947). *Am. J. Clin. Pathol.* **17**, 427–443.

Meessen, H. (1952). "Experimentelle Histopathologie." Thieme, Stuttgart.

Minami, S. (1923). *Arch. Pathol. Anat. Physiol.* **245**, 247–267.

Mitchell, S. W. (1868). *N. Y. Med. J.* **23**, January [cited by Pearce 1909].

Mitchell, S. W., and Reichert, E. T. (1886). *Smithsonian Contribution to Know* **26**, 647 (cited by Noguchi).

Noguchi, H. (1909). "Snake venoms—an investigation of venomous snakes with special reference to the phenomena of their venoms." Carnegie Inst. of Washington, Publ. 111.

Ogilvie, R. F. (1951). "Pathological Histology," 4th ed. Livingstone, Edinburgh and London.

Pearce, R. M. (1909). *J. Exptl. Med.* 11, 532–540.

Pena de Azevedo, A., and Teixeira, J. C. (1938). *Mem. Inst. Oswaldo Cruz* 33, 23–38.

Pizarro, J. J. (1950). *Rev. Ginecol. Obstet. (Rio de Janeiro)* 2, 743–766.

Poirier, P., and Charpy, A. (1925). "Traité d'Anatomie humaine," Part 5, pp. 83–85.

Policard, A. (1950). "Précie d'histologie physiologique," 5th ed. Doin, Paris.

Porges, N. (1953). *Science* 117, 47–51.

Raab, W., and Kaiser, E. (1966). *Mem. Inst. Butantan (Sao Paulo)* 33, No. 3, 1017–1020.

Rocha e Silva, M., Beraldo, W. T., and Rosenfeld, G. (1949). *Am. J. Physiol.* 156, 261.

Rodrigues Lima, J. P. (1956). Tese Dout., Sao Paulo.

Rosenfeld, G. (1951). *Proc. 3rd Intern. Congr. Hematol., Cambridge,* 1950. pp. 84–91. Grune & Stratton, New York.

Rosenfeld, G. (1965). *Pinheiros Terap.* 17, 84 and 315.

Rosenfeld, G., Nahas, L., and Kelen, E. M. A. (1968). *In* "Venomous Animals and their Venoms," (W. Bucherl *et al.*, eds.), Vol. I, pp. 229–273. Academic Press, New York.

Rotter, W. (1938). *Virchows Archiv. Pathol. Anat.* 301, 409–416.

Scarf, R. W., and Keele, C. A. (1943). *Brit. J. Exptl. Pathol.* 24, 147 (cited by Bull and Dible, 1953, p. 294).

Schoettler, W. B. A. (1951). *Am. J. Trop. Med.* 31, No. 4.

Slotta, K. H., and Szyska, G. (1937). *Mem. Inst. Butantan (Sao Paulo)* 11, 109.

Spielmeyer, W. (1922). "Histopathologie des Nervensystems." Springer, Berlin.

Taube, H. N., and Essex, H. W. (1937). *Arch. Pathol.* 24, 43–51.

Torgersen, O. (1949). *Proc. 6th Intern. Congr. Exptl Cytol., Stockholm,* 1947, pp. 415–422. Academic Press, New York.

Trueta, J., Daniel, P. M., Franklin, K. L., and Prichard, M. M. L. (1947). "Studies of the Renal Circulation," Vol. 1, Blackwell, Oxford.

Van Slyke, D. D. (1948). *Ann. Internal Med.* 28, 701–722.

Vital Brazil, O., Franceschi, J. P., and Waisbich, E. (1966). *Mem. Inst. Butantan (Sao Paulo)* 33, No. 3, 973–980.

Wajchenberg, B. L., Sesso, J., and Inagne, T. (1954). *Rev. Assoc. Med. Brasil* 1, 179.

White, J., and Mori-Chavez, P. (1952). *J. Natl. Cancer Inst.* 12, No. 4, 1–777.

Woods, W. W. (1946). *J. Pathol. Bacteriol.* 58, 767–773.

Yorke, W., and Nauss, R. W. (1911). *Ann. Trop. Med. Parasitol.* 5, 287.

Youile, C. L., Gold, M. A., and Hinds, E. G. (1945). *J. Exptl. Med.* 82, 361–374.

Zollinger, H. U. (1952). "Anurie bei Chromoproteinurie (Haemolyseniere, Crush-niere)." Thieme, Stuttgart.

Chapter 34

Symptomatology, Pathology, and Treatment of Snake Bites in South America

G. ROSENFELD*

INSTITUTO BUTANTAN, SÃO PAULO, BRAZIL

* Head of research with a grant from the National Council Research. Retired Head of the Department of Physiopathology and of the Hospital Vital Brazil of the Instituto Butantan.

I. INTRODUCTION

Since prehistoric times, man has always suffered from snakebite, and he has always been able to identify this disease by the objective relation of cause and effect. In our times, there still is a high incidence of snakebite in most areas of the world. Nevertheless study of the treatment of snakebite frequently is omitted in the curriculums of medical schools and in treatises on medicine and even on tropical medicine. On the other hand, rare diseases are sometimes extensively described. The ignorance of most physicians of the treatment of snakebite has been one of the causes of mortality.

According to Swaroop and Grab (1956), about 30,000 to 40,000 deaths caused by the bites of venomous snakes occur annually, not including the USSR, China, and Central Europe. In Brazil, according to da Fonseca (1949), the mortality rate is about 2.4%. This number permits the evaluation of approximately 1,250,000 to 1,665,000 accidents by venomous snakes in the world population excluding the above mentioned countries. This number probably is less than the actual one, since many unrecorded accidents occur. In the Hospital Vital Brazil of the Instituto Butantan in São Paulo, 15,709 patients bitten by different venomous animals came for medical treatment in a 12-year period (1954–1965). Among these, 1718 were caused by venomous and 1323 by nonvenomous snakes (see Table I).

Many victims are children under 14. In a partial statistical investigation by Braga (1964) among 1547 patients treated in 1964 in the Hospital Vital Brazil, 32.6% were children. They constitute 32.9% of the snake-bitten patients.

Table I is a record of data collected during 12 years (1954–1965) from patients treated in the Hospital Vital Brazil of the Instituto Butantan together with those of da Fonseca (1949) collected from reports received in a period of 44 years (1902–1945) from South and Central America.

Tables I and II indicate the mortality rate due to different venomous species and the respective death percentages.

Da Fonseca (1949) refers to a mortality of 2.43% in a total of 6601 treated cases, while we observed 1.75% in 1718 patients. This difference is an obvious consequence of more efficient serum therapy in the Hospital Vital Brazil. This percentage would be even smaller, reduced to 1.28%, if the

TABLE I

VENOMOUS SNAKE SPECIES AND FREQUENCY OF BITES

Species	Compiled data (1902–1945) by da Fonseca (1949)				Hospital Vital Brazil 12 years (1954–1965)			
	Number of cases	Percent	Deaths Number	Deaths Percent	Number of cases	Percent	Deaths Number	Deaths Percent
Crotalus durissus terrificus	738	11.18	90	12.2	143	8.32	17	11.89
Bothrops sp.	—	—	—	—	915	53.23	8	0.87
Bothrops alternatus	384	5.82	8	2.0	3	0.17	0	0
Bothrops atrox	83	1.26	1	1.2	—	—	—	—
Bothrops cotiara	96	1.45	1	1.0	1	0.058	0	0
Bothrops jararaca	3446	52.20	25	0.7	625	36.38	2	0.32
Bothrops jararacussu	657	9.95	11	1.6	14	0.82	1	7.14
Bothrops lansbergii	1	0.015	0	0	—	—	—	—
Bothrops neuwiedi	236	3.58	1	0.4	2	0.12	—	—
Bothrops neuwiedi paoloensis	—	—	—	—	1	0.058	—	—
Bothrops pradoi	—	—	—	—	1	0.058	—	—
Bothrops schlegelli	3	0.045	1	—	—	—	—	—
Lachesis muta muta	16	0.24	1	6.2	—	—	—	—
Micrurus corallinus	—	—	—	—	1	0.058	0	0
Micrurus frontalis	—	—	—	—	2	0.12	0	0
Micrurus lemniscatus	—	—	—	—	2	0.12	0	0
Micrurus sp.	15	0.23	0	0	8	0.47	2	25.00
Not specified	926	14.03	22	2.3	—	—	—	—
	6601		161	2.43	1718		30	1.75

TABLE II

MORTALITY IN PERCENTAGES CAUSED BY THE FOUR
GENERA OF VENOMOUS SNAKES[a]

Genus	da Fonseca	Hospital V Brazil	Combined
Crotalus	12.2	11.89	12.15
Bothrops	0.98	0.70	0.91
Lachesis	0.24	—	0.24
Micrurus	0	15.39	7.14
Not specified	2.3	—	2.3
	2.43	1.75	2.30

[a] 6601 cases (1902–1945) from data compiled by da Fonseca (1949) and 1718 cases observed in the Hospital Vital Brazil of the Instituto Butantan during 12 years (1954–1965).

cases arriving at the hospital *in extremis* were not included (8 of 30 fatal cases).

It is generally assumed that the most dangerous areas for snake bite are woods or fields. However, a survey of 563 cases of snake bite (Rosenfeld and Sawaya, 1957) showed that 18.83% occurred in woods or fields, 22.38% were incurred by agricultural workers, and 25.4% occurred around the farmhouse; 12.97% while fishing (Table III).

In 563 observations (Rosenfeld and Sawaya, 1957), the localization of bites was: foot, 60.39%; lower third of the leg, 11.01%; middle third, 3.9%; upper third, 1.42%; thighs, 1.59%; hands, 17.22%; wrist, 0.88%; forearm, 1.59%; arm, 0.88%; face, 0.17%; shoulder, 0.53%; back, 0.17%; and buttocks, 0.17%.

According to Rosenfeld and Sawaya, the highest frequency of bites in our country was observed during the dampest and warmest months. In relation to atmospheric humidity, half of the accidents happened when humidity was between 75% and 94%, with an accident frequency of 85% and 89%. In relation to temperature, lower incidence was observed while temperature was below 18°C. In relation to time, the highest frequency was between 7 and 12 A.M. even though bites can occur in any period of the day.

There was a need for establishing a symptomatologic clinical criterion that would permit diagnosis of the accident and consequent therapeutic handling and prognosis without identification of the snake. This was proposed some years ago (Rosenfeld, 1960, 1963a, 1965, 1966) because most physicians are unable to classify venomous snakes and it is rare for the patient or family to bring the snake that caused the accident.

TABLE III

CIRCUMSTANCES OF SNAKE BITES[a]

	Number	Percent	
House			
Indoors	12	2.13	
Surroundings	90	15.99	25.40
Kitchen and garden	41	7.28	
Plantations			
Eucalyptus	4	0.70	
Corn	10	1.78	22.38
Farming	69	12.26	
Cutting wood	43	7.64	
Miscellaneous			
Pasture, grass, field	55	9.77	
Fishing, moor, river	73	12.97	
Natural forest	51	9.06	
Road	19	3.37	41.92
Abandoned farm	12	2.13	
Snake handling	12	2.13	
Marsh	14	2.49	
Not specified	58	10.30	10.30
	563	100.00	

[a] 563 cases of snake bites (venomous and non-venomous) from 1954 to 1957 (Rosenfeld and Sawaya, 1957).

II. PATHOGENESIS

A. Definition of Venom

Venom is an immobilizing agent for prey seized by some animals for food; some venoms also have digestive properties. They contain proteinic and phosphatidic substances with enzymatic activity and various pharmacological properties.

B. Inoculation Route

Snake venoms do not enter the circulation through the skin or mucous membranes (Calmette, 1896) but are slowly absorbed by the serosa (Calmette, 1896). When inoculated subcutaneously or intramuscularly, venoms enter through the lymphatics into the circulatory system. Probably, some fractions also get into the cerebrospinal fluid by crossing the brain barrier.

The lethal dose of proteolytic and coagulant venoms varies with the inoculation route. Intravenously, it is smaller than subcutaneously, and paradoxically the intraarterial dose differs from the intravenous one. For the lethal effect after intraarterial injection, a fourfold venom dose is necessary (Rosenfeld and de Langlada, 1964c).

The amount of venom contained in the glands of different species of snakes varies. As observed experimentally in dogs, more venom can be inoculated when the bite occurs on the skin over a bone. Snakes never inject their entire gland contents in one bite. We observed one snake successively biting four dogs and killing them all; those bitten at a bony site died more rapidly.

There is also lack of correlation between the size of the snake and the severity of envenomation.

C. Composition and Properties

Venoms are viscous secretions, white, yellow, or orange colored, and have 25 to 40% crystallized residue after desiccation. Vacuum-dried venoms may present stronger specific activities than lyophilized ones. Kept in the dark in well-stoppered flasks at room temperature, venoms may remain active for 40 years, some properties lasting longer than others. Venom solutions are not stable even when kept at 4°C or deep-frozen at -20°C; there is a drastic decrease in the MLD or LD_{50}.

Venom properties are due to proteinic substances or phospholipids with strong enzymatic activity. Some venoms act on natural substrate causing the formation of lesions. Others act indirectly, modifying substrates, and the new components damage cells and tissues and cause serious physical disturbances by autopharmacological mechanisms.

The pharmacological properties experimentally observed *in vitro* and *in vivo* do not always simulate the symptomatology or physiopathological alterations observed clinically. One example is the hemolytic activity *in vitro* of *Bothrops jararaca* and *Crotalus durissus terrificus* venoms which are similar, while *in vivo* no evident hemolysis develops after a bite by *Bothrops jararaca*. A weak coagulant venom like *C. d. terrificus* may produce blood incoagulability *in vivo*; this appears frequently with very strong coagulant venoms like *B. jararaca*.

Barrio and Brazil (1951) described veratrinic activity in venoms of some *C. d. terrificus* snakes. Later, Moura Gonçalves (1951, 1956) isolated and purified the substance which he called crotamin, which differs from crotoxin isolated by Slotta and Fraenkel-Conrat (1938). Doses of 0.25 to 0.5 mg of venom rapidly produce a characteristic paralysis of the hind legs of a mouse weighing 20 gm: the legs remain stretched backwards. Patients bitten by rattlesnakes with crotamin in their venom have never shown such symptoms, probably because a proportional dose of 0.25–0.50 mg/20 gm could not be injected in man.

Some symptoms observed in man cannot be observed in animals. For instance, the neurotoxic symptoms produced by the *C.d. terrificus* venom, such as ophthalmoplegia and blepharoptosis, could not be seen in dogs, rabbits, bovids, and other animals.

Snake venoms are active allergens. Allergic reactions were observed in people who handle these substances (Mendes *et al.,* 1960).

III. PATHOLOGY

A. Classification of Snakes by the Venoms' Physiopathologic Activities

From their physiopathological activity, apart from the pharmacological properties which are not the subject of this chapter, all venoms can be classified into a few fundamental groups (Rosenfeld, 1965, 1966). Knowledge of such a classification enables a better understanding of lesions caused by different species or genera, and allows the physician to diagnose the kind of envenomation after an indirect classification of the snake in order to guide the treatment and prognosis of the case. Each physiopathologic activity is due to multiple factors, but each has a clinical course which could be called a syndrome.

The utilization of such a criterion in every country for all the snakes causing accidents would be of great help. It would provide the ability (Rosenfeld, 1960) (Table IV), to identify the genus or the species of the snake similar to what happens in Brazil (Rosenfeld, 1960), see Table IV.

TABLE IV

CLASSIFICATION OF SOUTH AMERICAN
SNAKE VENOMS BY PHYSIOPATHOLOGIC
ACTIVITY

Venoms	Snake genus
Proteolytic + Coagulant	*Bothrops–Lachesis*
Hemolytic + Neurotoxic	South American *Crotalus*
Neurotoxic	*Micrurus*

There are only four fundamental physiopathologic activities: *coagulant, proteolytic, hemolytic, neurotoxic*; there is a possibility of maybe an anti-coagulant activity.

According to genus or species, the fundamental physiopathologic activities may be found isolated or joined in different group formation or proportions. Species of the same genus do not always have same activities, like the venom of *Crotalus* snakes: one species in North America has proteolytic activity and none has neurotoxic activity; in the central southern portions of South America, these venoms are neurotoxic and not proteolytic. The same species may have a different proportion of components during different periods of life. The young *Bothrops jararaca* snake has weak proteolytic and more coagulant venom than the adult species (Rosenfeld *et al.,* 1959) which has strong proteolytic and coagulant venom. Minton (1967b) also observed activity variation in relation to the age of snakes.

Another difficulty in the understanding of the fundamental venom physiopathology is encountered when there is an attempt to study coagulant, proteolytic, neurotoxic, and hemolytic properties separately, and compare each one with the lethal activity. In fact, the lethal activity is a result of the whole complex contained in a venom. Some of these factors play a more important role than others in causing death. There is an integration of all activities.

B. Fundamental Physiopathologic Activities

1. Coagulant

There are venoms with direct coagulant activity (thrombin-like) and others with an indirect one (thromboplastin-like). These venoms may still have activity induced by one or more clotting factors (Rosenfeld *et al.,* 1968). However, all of them cause the same organic disturbances that can be summarized as follows: a brisk intravenous or intraarterial penetration into circulation starts massive intravascular clotting which causes death in a few minutes. This is rare and it may be easily observed by experimentation. In almost every case, clots form a barrier at the site of inoculation and block the diffusion of the venom. From this site there is a gradual penetration into circulation through the lymphatics (Fidler *et al.,* 1940), and thus, a gradual clotting of fibrinogen and fibrin microclot formation. These are retained in the different capillaries, especially in the lungs as stated by Amorim and co-workers (1951). If there is enough venom in the circulation (in the case of *Bothrops jararaca* venom, 0.1 mg/kg of body weight is sufficient), fibrinogen is completely clotted and blood becomes incoagulable by defibrination (Houssay and Sordelli, 1919–1921; Rosenfeld *et al.,* 1958b; Rosenfeld, 1964, 1968).

This is how a coagulant venom induces blood incoagulability, a simple but generally misunderstood mechanism. Thus, the protective action of previously injected heparin was described as a specific activity of this substance (Ahuja *et al.,* 1947). Heparin does not destroy venom, it only has an opposing pharmacodynamic activity. When given in sufficient quantity heparin can overcome the venom's coagulant activity (Rosenfeld and de Cillo, 1956; Rosenfeld and Kelen, 1963). Since the inverse mechanism is also true, the venom coagulant fraction can have therapeutic application as an antiheparinic agent (Rosenfeld et al., 1961).

This coagulant activity is caused by a proteinic fraction that was first isolated in an impure state (von Klobusitzky and König, 1936a,b). A pure substance was later obtained (Henriques et al., 1960) with a very low biologic activity. It has been used for some time in hemorrhagic processes and its application as a coagulant (Hanut, 1936; von Klobusitzky, 1938; Peck, 1932) or a hemostatic (Rosenfeld and de Cillo, 1958) is gradually increasing.

When clotting is induced in vivo by a coagulant venom, there is a shortening of the clotting time immediately followed by a fibrinolytic activity of the blood. This activity is not directly due to the venom but to activation of fibrinolysin by intravascular clotting (Rosenfeld, 1964, 1968).

Blood incoagulability by defibrination lasts as long as the venom enters the circulation, acting on fibrinogen which is being produced by the organism (Rosenfeld et al., 1958b). It may last for more than one day and will only stop after venom neutralization or exhaustion after being consumed by fixation. This incoagulable blood will not clot after the addition of thrombin and has thrombinic activity on normal citrated plasma; thus, the presence of venom is easily identified. This activity can be neutralized by the addition of specific antivenin (Rosenfeld, 1964, 1968).

No capillary lesions or hemorrhages are caused by the coagulant activity, despite the blood incoagulability by defibrination (Rosenfeld, 1964, 1968; Rosenfeld et al., 1958b).

In spite of regional edema and erythema, coagulant venoms do not induce local venous thrombosis as would be expected. This was experimentally observed by Bouabci (1964). Edema is not due to vein occlusion but to the proteolytic activity of the venom. This fact is confirmed by the clinical observation of patients bitten by young Bothrops jararaca snakes which have strong coagulant but less proteolytic venom than the adult snakes (Rosenfeld et al., 1959). In these cases there is little or no edema (Rosenfeld et al., 1957).

2. Proteolytic

The proteolytic activity is caused by a group of substances having this enzymatic activity, acting on different substrates. The degradating proteins produce some interesting substances from the pharmacologic point of view, as, for instance, bradykinin (Rocha e Silva et al., 1949). The aggression on proteins cause pain at the site of the bite with edema, erythema, ecchymosis, and phlyctena. These manifestations are directly proportional to the amount of injected venom and the diagnosis, the evaluation of envenomation's severity, and prognosis may be made from these. There is late destruction of all soft tissues and consequent necrotic formation which may necessitate amputation with large amounts of venom. There may be permanent scarring. The amount of venom of *B. jararaca* necessary to produce necrosis in the dog is 1 mg/1 kg; this is close to the lethal dose.

The enzymatic decomposition of proteins in tissues or in blood results in peptides, bradykinin, etc. These substances reach the circulatory system and cause proteinic shock like the peptonic one, and may induce a peripheral shock—the usual cause of death in this kind of envenomation.

Proteolytic activity can also damage the capillary walls, thus favoring hemorrhages in the form of petechia or blood suffusions in the skin and in various organs (Amorim *et al.,* 1951). This hemorrhagic activity is considered by Ohsaka and co-workers (1966) as a specific one, since they partially isolated the hemorrhagic from proteolytic fractions. In fact, one should consider that the proteolytic activity is a result of different enzymes, each one of which acts on a specific substrate. It is obvious that different results will be obtained if the activity is investigated on different substrates, such as casein and fibrin (Table V). The hemorrhagic activity of the fractions isolated by Ohsaka *et al.* (1966) is also due to proteolytic enzymes, and differs from those acting on casein. Thus, their biochemical assays will be different. Some of them will be more active on substances of the vascular endothelium, others on substances from the muscles, and so on. From the physiopathological point of view, only venoms with proteolytic activity *senso lato* have hemorrhagic activity.

The coagulant activity is also a proteolytic one from the biochemical point of view. It is not due to only one enzymatic substance since coagulant venoms act on this or that clotting factor (Rosenfeld *et al.,* 1968), but for all these venoms the physiopathologic picture is the same.

Coagulant venoms are nearly always proteolytic as well. This does not mean, however, that both activities are due to the same enzymes. There are strong coagulant venoms such as the one of the young *B. jararaca,* with small proteolytic activity on fibrin (Rosenfeld *et al.,* 1959), and strong fibrinolytic venoms, as that of *B. neuwiedi paoloensis,* with small coagulant activity. There is no direct proportion even between proteolytic activity on fibrin and on casein. Some venoms act more on the first substrate and less on the second one or vice versa (Table V).

As mentioned before with regard to coagulant activity, blood incoagulability does not disturb capillary hemostasis or cause hemorrhage. When present, these manifestations are due to capillary lesions caused by the action of some proteolysins, as demonstrated by Ohsaka and co-workers (1966). The incoagulable blood leaves the vessel through the damaged walls. That is why hemorrhages, epistaxis, hematemesis, hematuria, and focal visceral hemorrhages appear. These hemorrhages, primarily caused by the

TABLE V

COMPARISON BETWEEN PROTEOLYTIC ACTIVITY OF SOME SNAKE VENOMS ON CASEIN AND FIBRIN[a]

Venom	Casein (mg)	Fibrin (mg)	Index of activity on Casein	Index of activity on Fibrin
Bothrops jararacussu	2.60	0.082	2.36	0.21
Crotalus durissus durissus	2.50	0.192	2.27	0.49
Bothrops atrox	2.40	0.317	2.18	0.81
Bothrops neuwiedi	1.90	0.243	1.73	0.62
Trimeresurus flavoviridis	1.80	0.055	1.64	0.15
Bothrops neuwiedi paoloensis	1.70	0.298	1.55	0.76
Bothrops insularis	1.60	0.365	1.46	0.93
Agkistrodon piscivorus	1.50	0.267	1.36	0.68
Bothrops itapetiningae	1.40	0.251	1.27	0.64
Bothrops jararaca	1.10	0.392	1.00	1.00
Bothrops atrox asper	0.94	0.176	0.86	0.45
Lachesis muta muta	0.93	0.176	0.85	0.45
Bothrops jararaca (young snake)	0.78	0.196	0.71	0.50
Bothrops alternatus	0.54	0.094	0.49	0.24
Bothrops cotiara	0.50	0.074	0.45	0.19
Vipera lebetina	0.42	0.094	0.38	0.24
Vipera ammodytes montandoni	0.33	0.063	0.30	0.16
Bothrops fonsecai	0.30	0.081	0.27	0.20
Vipera russelli	0.07	0.055	0.06	0.14
Crotalus durissus terrificus (white venom)	0.00	0.034	0.00	0.09
Crotalus durissus terrificus (yellow venom)	0.00	0.068	0.00	0.17
Naja naja		0.066		0.17
Micrurus frontalis		0.086		0.22
Crotalus durissus terrificus (Marajó isle)		0.086		0.22

[a] Milligram of protein digested by 0.1 mg venom. *Bothrops jararaca* venom activity referred to as index 1.0. Data extracted from Rosenfeld *et al.* (1959) and Rzeppa and Rosenfeld (1966).

proteolytic activity, are aggravated by blood incoagulability and may rarely lead to death because of extreme hematuria, glomerular lesion, or permanent lesions in the central nervous system when the hemorrhagic focus destroys the local tissue (Teixeira, 1943). The neurologic symptomatology will indicate where the hemorrhagic focus occurred. Such focus can also destroy part of the renal tissue which will later calcify (Oram *et al.*, 1963).

3. Hemolytic

The hemolytic activity of snake venoms occurs indirectly. Venoms do not act on a suspension of washed erythrocytes unless plasma or serum is added.

The venom acts on serum lecithin and isolecithin is then formed. A detailed explanation of this is given in Volume I (Rosenfeld *et al.,* 1968).

However, it is necessary to stress that hemolysins are also multiple (Rosenfeld *et al.,* 1960–1962), and sometimes, extremely specific. This is how one venom may be hemolytic for erythrocytes of one or more species and not affect others. One example is the venom of *Micrurus frontalis.* Human and ox red blood cells are not hemolyzed *in vitro* by this venom, but monkey, dog, horse, sheep, rabbit, guinea pig, rat, and mouse erythrocytes are. *Bothrops cotiara* venom does not hemolyze *in vitro* human, monkey, ox, sheep, or rabbit erythrocytes, but hemolysis occurs with dog, horse, guinea pig, rat, and mouse erythrocytes (Kelen *et al.,* 1960–1962).

Strongly hemolytic venoms *in vitro* may have different characteristics *in vivo. Crotalus durissus terrificus* venom, *in vitro,* is somewhat more hemolytic than *Bothrops jararaca* on sheep erythrocytes (Rosenfeld *et al.,* 1960–1962). However, clinically envenomations by *C. d. terrificus* show intense and severe hemolysis; this is not the case with patients bitten by *Bothrops jararaca* where there is only a trace of damaged erythrocytes.

Hemolytic activity is expressed by intense hemoglobinuria, which actually is a methemoglobinuria (Rosenfeld, 1963b), its severity being proportional to the envenomation's intensity. An intermediate nephrotic syndrome develops simultaneously with lesions of the renal tubules (Amorim and de Mello, 1952) similar to what occurs after incompatible blood transfusions or in the crush syndrome. Anuria follows methemoglobinuria and if severe, it may be very dangerous. Moderate anuria is usually followed by a diuretic crisis after the fourth or fifth day. If, at this stage, the total or relative anuria continues (about 100 to 200 ml urine in 24 hours), renal dialysis should be done.

Most deaths are generally caused by this renal syndrome. They occur after the seventh to the fifteenth day and are due to secondary endogenous intoxication.

The renal syndrome probably is not only the direct consequence of hemolysis; there must be some hemolytic component that acts on the cells of the renal tubules. This mechanism is somewhat elucidated (Lima, 1966).

4. Neurotoxic

Snake venom neurotoxins act on the central nervous system. The symptomatology authorizes this statement in the absence of experimental demonstration of their presence in the nervous system. The clinical picture indicates that there is some action on the nuclei of mesencephalonic cranial nerves which are nearer to the surface and thus more easily reached by the neurotoxins. This suggests that neurotoxins probably cross the blood-brain barrier. It may also be assumed that the neurotoxins do not cause organic damage, since all specific symptoms disappear after cure. Neurotoxins probably act by destroying some chemical mediation substances that are necessary for the effectiveness of nerve cell activity. Regression takes some days, probably the time required for regeneration of the destroyed substance. Neurotoxins are antigenic because no neurologic manifestations are observed in cases in which antivenin therapy is given immediately after the bite. A specific antivenin is able to neutralize the neurotoxin, but not to cure the functional disturbances.

There must be several neurotoxic substances since venom from *Crotalus durissus terrificus* produces some neurological symptoms and *Micrurus frontalis* and *M. corallinus* (coral snakes) venom causes much more severe symptoms. It is possible that the coral snake venom has several neurotoxic substances and *C. d. terrificus* only one.

About 30 to 60 minutes after the bite a flaccid paralysis occurs in the motor muscles controlled by the third pair of cranial nerves. This results in symmetric partial external ophthalmoplegia and ptosis of the upper eyelids. If the envenomation is more severe the fourth and sixth cranial nerves are affected, then the eyeballs are completely paralyzed. Probably, the vestibular division of the eighth cranial nerve is also affected but not the cochlear portion since equilibrium is disturbed but not audition.

At first there is a loss of accommodation of the crystalline lens, distant vision is impaired more than close vision, indicating a flaccid paralysis of the ciliary muscles, or a partial internal ophthalmoplegia. Pupillary reflexes to light remain normal. In severe cases the reflexes are disturbed, and if mydriasis appears it is a fatal prognostic sign.

Briefly, there is a loss of visual acuity, turbid sight and diplopia, besides the palpebral ptosis which covers the pupil and impairs sight. These symptoms led to the false concept that blindness was caused by neurotoxic venoms.

Neurotoxic facies (Rosenfeld, 1960, 1963a,b, 1966) is a pathognomonic sign of the bite by a snake with neurotoxic venom. This "neurotoxic facies" is easily demonstrable when the patient is asked to look at an object: he contracts the frontal muscle in a vain effort to raise the upper eyelids, the eyes can be opened slightly, the head tilts backwards and the lower part of the face is relaxed, as with an air of superiority (Figs. 7, 9). If the patient is asked to follow an object he will turn his head because of the difficulty or the inability to move the eyeballs.

With venoms having stronger neurotoxic activity (*Micrurus*, Elapidae) the symptoms are more severe. In addition to the symptoms described for crotalic envenomation, when III, IV, VI and part of the VIII cranial nerves are affected, two other cranial nerves are disturbed: the fifth, since there is pain at the mandibular angle and thick saliva, the ninth, as swallowing and speaking are handicapped; and the tenth, since dorsal and testicular pains are experienced. Finally, there may be respiratory paralysis of central origin; this is the cause of death by venoms with neurotoxic activity. Actually, the respiratory centers are located around the IX and X cranial nerves.

Another argument favoring the hypothesis of central action in the nuclei of the cranial nerves is the symmetry of neurologic symptoms, showing a flaccid paralysis of motor nerves, and pain with respect to the sensory nerves.

At the site of the bite, neurotoxic venoms induce pain and burning which disappears in a short time, and then there is only pain when the area is palpated, followed by paresthesia which spreads to the whole distal region of the extremity with cutaneous anesthesia. In envenomations by *Micrurus* snakes containing powerful neurotoxic venom, paresthesia spreads more extensively and includes the proximal region up to the point of attachment of the bitten extremity.

IV. SYMPTOMATOLOGY

A. General Information

Rapid penetration of small amounts of venom into the circulation causes neutrophilic leukocytosis. Larger doses cause neutropenia and thrombocytopenia within a few minutes, this is followed in one hour by neutrophilic leukocytosis and thrombocytemia, which are in inverse relation to the initial depression. Leukocytosis and neutrophilia are directly proportional to the severity of envenomation and may be used for its evaluation (Rosenfeld, 1950).

Another alteration is a decreased sedimentation rate, which may reach zero during the first minutes after the penetration of the venom into the circulation (Rosenfeld, 1951) even if shock is not clinically apparent.

Hemolytic venoms *in vitro,* even those unable to cause hemolysis as evidenced by methemoglobinuria, induce an increase of the erythrocyte mean volume and there is a change of the cellular osmotic tension. The increased volume is due to liquid penetration into the cells (Rosenfeld, 1951). The loss of liquid from the circulation thus may be great if one considers the large number of existing erythrocytes. The same alterations appear in shock not induced by venoms (Rosenfeld *et al.,* 1958a). During the envenomation the mean diameter of the erythrocytes decreases due to a tendency toward sphericity. The cells resistance to hemolysis, as analyzed in relation to different concentrations of saline solutions, is decreased in high saline concentrations when the envenomation was induced by small doses of hemolytic venoms or by hemolytic venoms causing no clinical symptoms (Rosenfeld, 1951). However, almost paradoxically, this resistance is increased in less concentrated saline solutions. One explanation is that venoms increase the cell fragility to hemolysis. The more fragile ones are destroyed and disappear from the circulation, and only the resistant undamaged cells remain.

Anemia follows this action on the erythrocytes, even when hemorrhage is absent. It is more severe in envenomations by venoms with hemolytic activity *in vivo* and very weak when this activity is only observed *in vitro*.

B. Proteolytic and Coagulant Venoms

I. Local Symptoms

Severe pain follows the bite and spreads throughout the region. Erythematous edema and ecchymosis appear. Hemorrhagic phlyctena may or may not be present. This reaction may extend to the point of attachment of the bitten extremity and ecchymoses may be seen on the vessels paths. *Bothrops jararacussu* venom, as well as others, causes superficial capillary hemorrhages and generates petecchiae at different sites of the body, even at the hair roots and nail borders. Ecchymoses can be seen along the paths of the vessels of the arm. Pink alternated spots may be seen on the bitten extremity, similar to the cadaveric hypostasis, nevertheless, it is not an indication of irreversible lesions as is necrosis (Figs. 1–5).

Necrosis may be seen at the site of the bite or adjacent areas, with destruction of soft tissues. Sometimes, a fluctuation may be felt on palpation after several days, even when the skin does not show any prominent superficial

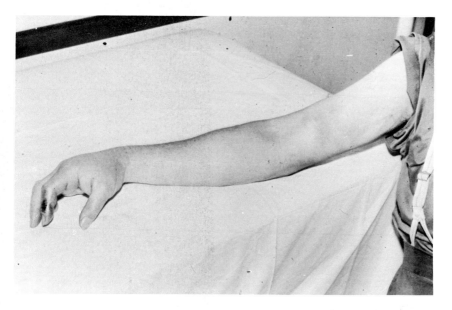

FIG. 1. Proteolytic and coagulant venom of *Bothrops jararaca*. Mild case, number 11639. Shown 17 hours after the bite. Note edema and erythema; cured without necrosis.

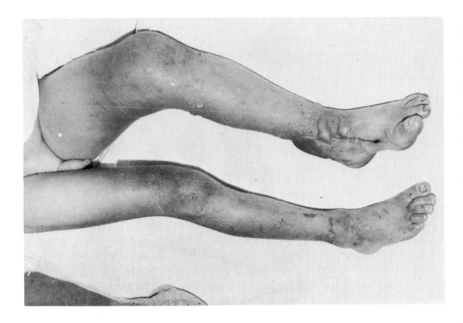

Fig. 2. Proteolytic and coagulant venom of *Bothrops jararaca*. Serious case, number 11635. Photographed 24 hours after the bite. Note phlyctenae; cured without necrosis.

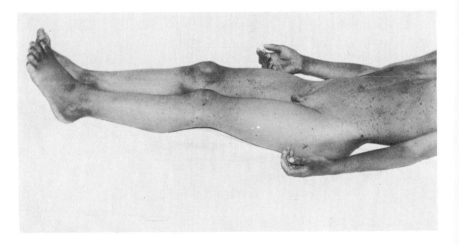

Fig. 3. Proteolytic and coagulant venom of *Bothrops jararacussu*. Serious case, number 10186. Shown third day after the bite. Note petechiae and echymosis; cured without necrosis.

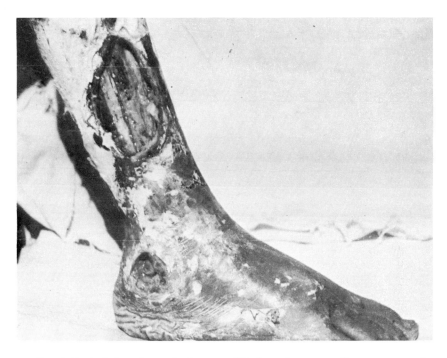

FIG. 4. Proteolytic and coagulant venom of *Bothrops jararacussu*. Severe case, number 2645. Shown several days after the bite. Necrosis; cured.

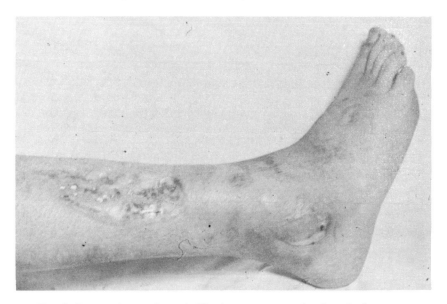

FIG. 5. Same patient as shown in Fig. 4 seen some months after plastic surgery.

alterations. This indicates subcutaneous necrosis with secondary infection and should be drained.

All of these disturbances are directly proportional to the amount of injected venom and provide a rapid evaluation of the severity of the bite.

2. General Symptoms

Dizziness, nausea, uneasiness, bilious or bloody vomitus, hematuria and melena, epistaxis or gingival hemorrhage are all generally found. Arterial pressure is normal, hypotension indicates the beginning of shock. The pulse is usually normal; it becomes rapid and threadlike in more severe cases or when infection and necrosis appear. The temperature is generally normal; if it drops it indicates shock, and if elevated it indicates infection or necrosis.

Venous blood remains incoagulable by defibrination; nevertheless, the capillary hemostasis is not modificated (Rosenfeld *et al.,* 1958b; Rosenfeld, 1968). Blood coagulability one hour after the bite indicates the inoculation of a small amount of venom. If no coagulability is found at this time, more than 0.1 mg/kg (for *B. jararaca* venom) was injected (Rosenfeld *et al.,* 1957). Hematologic alterations have been described.

C. Hemolytic and Neurotoxic Venoms

I. Local Symptoms

Strong pain is felt immediately after the bite; this is followed by local or sometimes regional paresthesia with sensation of formication and anesthesia. Edema is rare and depends on the part of the body that was affected. There is no erythema. Frequently, no lesion is apparent at the site of the bite (Fig. 6).

2. General Symptoms

Visual disturbances due to external and partially internal ophthalmoplegia, as well as blepharoptosis, appear about one hour after the bite. Direct and consensual pupillary reflexes to light are undisturbed. During this period, the neurotoxic facies (Rosenfeld, 1960, 1963a,b, 1966) occurs and it affords a positive diagnosis as well as an evaluation of the intensity of envenomation (Fig. 7). There are also generalized muscular pains or only those in the back of the neck. There is loss of equilibrium.

In very serious cases, due to large amounts of venom, there is agitation, bilious vomitus, and delayed pupillary reflexes. In severe envenomation, there is also prostration and coma. The observation of mydriasis and a slight depression of the anterior eye chamber is a fatal prognostic sign.

During the first hours, urine is clear and excreted in large volume; this

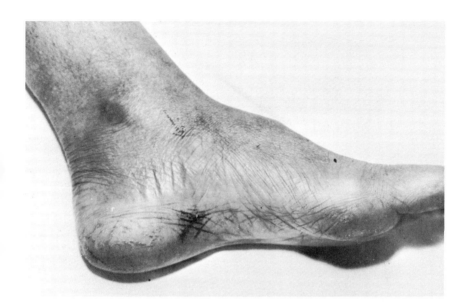

FIG. 6. Hemolytic and neurotoxic venom of *Crotalus durissus terrificus*. Serious case, number 14388. Photographed 24 hours after the bite. Note that the site of the bite shows no reaction or sign.

is followed by methemoglobinuria, which is characterized by its reddish brown color and translucence. Progressive anuria, which may be total, develops later. The occurrence of a diuretic crisis after the fourth or fifth day is good prognostically. Since the intensity of these symptoms is proportional to the envenomation, the absence of or mild methemoglobinuria indicates a mild case; methemoglobinuria without anuria is the next degree of intensity.

The temperature is normal. Arterial blood pressure is normal also, but frequently it is somewhat increased. Pulse may be normal or accelerated and full. In cases of shock, manifestations are those usually found in these shock conditions.

In some very serious cases blood may be incoagulable. Usually, clotting time is normal, even in fatal cases. A mild anemia may follow after a few days and is proportional to the intensity of hemolysis. No other indirect effects of hemolysis, such as subicterus and hyperbilirubinemia, are observed.

Except for those cases in which massive envenomation occurred causing agitation, prostration, and coma within a few hours, and death in the first to the third day, death is usually a consequence of endogenous intoxication due to the intermediate nephrotic syndrome caused by the venom. After one week

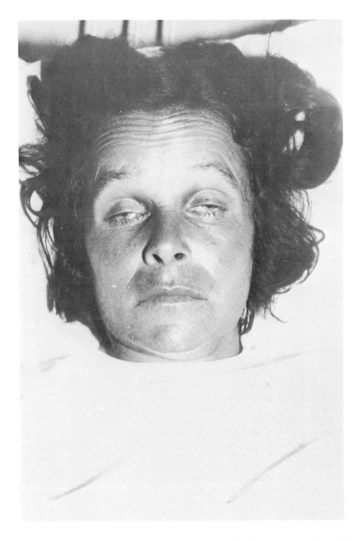

FIG. 7. Hemolytic and neurotoxic venom of *Crotalus durissus terrificus*. Serious case number 14388. Seen 24 hours after the bite. Note neurotoxic facies; cured.

it will cause death which was incorrectly described as secondary envenomation or late death by secondary liberation of venom.

Neurologic symptoms will not appear if serum therapy is given properly and in time. However, if the treatment starts after the manifestation of these symptoms, the symptoms remain for several days and gradually disappear. A motor difficulty of the eyes may remain for 15 days, and will only be observed if the eyes should suddenly move laterally.

D. Neurotoxic Venoms

I. Local Symptoms

Pain is felt immediately after the bite with variable intensity, depending on the snake species. There is an absence of edema or any local reaction (Fig. 8). Progressive paresthesia toward the proximal direction of the extremity, with drowsiness occurs. The site of the bite feels anesthetized and sometimes painful on palpation.

2. General Symptoms

Ophthalmoplegia and neurotoxic facies (Fig. 9) identical to those described for hemolytic and neurotoxic venoms are found. However, they appear more rapidly and with greater intensity. There is thick salivation and difficulty in swallowing and articulation of words as well as torpor and loss of equilibrium. Patients complain of dorsal or lumbar pains, as well as lower maxilar, testicular, and meatus urinarius. Urine is normal with normal excretion. The blood does not show any sign of hemolysis or of any clotting disturbance.

Neurologic signs of intoxication of the central nervous system are gradually accentuated until the death of the patient by respiratory paralysis of central origin. The course of fatal cases caused by these venoms is very rapid. The patient dies within 4 to 6 hours. Necroscopy of one case showed pulmonary edema (Machado, 1968).

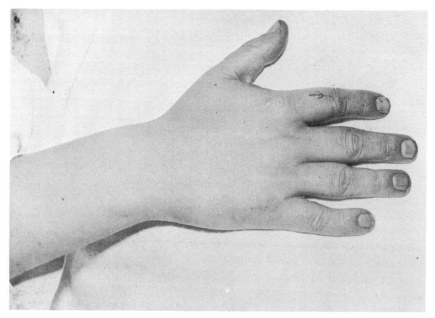

FIG. 8. Neurotoxic venom of *Micrurus frontalis*. Serious case, number 8265. Shown 24 hours after the bite. Note site of the bite with no reaction; cured.

Clear or bilious vomitus, agitation and prostration followed by coma may be observed in some cases.

Treatment as well as regression of neurologic symptoms is similar to that of hemolytic and neurotoxic venoms.

TABLE VI

DIFFERENTIAL DIAGNOSIS FOR ENVENOMATIONS BY SNAKES[a]

Venom type	Local reaction	Neurotoxic facies	Methemo-globinuria	Hematuria	Incoagulable blood
Proteolytic + Coagulant	+ +	−	−	+	+ +
Hemolytic + Neurotoxic	−	+	+ +	−	−(+)
Neurotoxic	−	+ +	−	−	−

[a] No symptoms (−); evident symptoms (+); pronounced symptoms (+ +).

V. EVALUATION OF ENVENOMATION SEVERITY AND PROGNOSIS

A. Proteolytic and Coagulant Venoms

In *mild* cases there is a small local reaction at the site of the bite. After 2 hours, venous blood is coagulable (Fig. 1).

In *serious* cases the blood becomes incoagulable 1 to 2 hours after the bite. There is local and regional reaction with edema, ecchymosis, and pink spots. Arterial blood pressure is normal. There may be some risk to life and local necrosis (Figs. 2, 3).

In *severe* cases the blood is incoagulable. There is a strong local reaction and phlyctenae may appear after 24 hours. Hematuria, vomiting and bloody evacuation, epistaxis, and gingival hemorrhages occur. There are generalized pains and prostration. It is very dangerous when shock occurs within the first three days. Tendency to shock is recidivist and ends in death.

Intensive renal hemorrhage and destructive or fatal kidney lesions may be observed in some cases. Others may show hemorrhages in the nervous system causing irreparable (Teixeira, 1943) or fatal lesions. In rare cases with severe renal hemorrhage which may cause anuria due to tissue destruction, a cicatricial renal calcification may result (Oram et al., 1963).

The most frequent complication is the necrosis which occurs in 15.7%

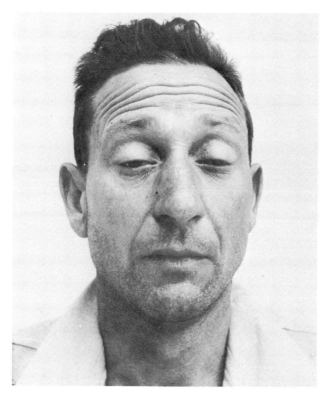

FIG. 9. Neurotoxic venom of *Micrurus corallinus*. Serious case, number 8343. Photographed 17 hours after the bite. Note neurotoxic facies; cured.

of the cases. There is also formation of subcutaneous abscesses in 2.6% of the patients (Rosenfeld and Takeuti, 1968). Necroses are established within the first days and may leave permanent scars that will have cosmetic and functional effects. Another difficulty is the distal edema of the necrotic area which may last for months due to the destruction of blood vessels (Figs. 4, 5).

There are no late consequences. All of them develop within the first days. Other sequelae are absent.

Mortality among untreated patients is about 8%; among those treated properly it is 0.7%, and usually these deaths are due to belated treatment. When serum therapy is given within 3 hours after the bite, death and necrosis can be avoided.

B. Hemolytic and Neurotoxic Venoms

With *mild* cases no neurotoxic facies is found after 3 hours. Urine is normal or may show moderate methemoglobinuria without anuria. There is no risk to life.

In *serious* cases, one finds neurotoxic facies (Fig. 7). There is moderate methemoglobinuria without anuria. Pupillary reflexes are normal. There is partial external and internal ophthalmoplegia. There is some risk to life with death after the seventh day.

Neurotoxic facies is also found in *severe* cases. The equilibrium is disturbed. Methemoglobinuria and anuria are found. There are delayed pupillary reflexes. The patient complains of generalized pains. There is agitation or prostration. Pupillary reflexes are suppressed; mydriasis and incoagulable blood indicate a fatal prognosis.

On recovery, all renal and neurologic symptoms disappear and no deficiency or late sequelae remain.

The mortality among the untreated patients is about 72% with bites by *Crotalus durissus terrificus* of South America. Among the treated cases it drops to 11.89%, many of them receiving antivenin more than 6 hours after the bite. However, there is a small number of cases that cannot be saved by antivenin administered even within 3 hours. Patients treated immediately do not show any symptomatology of envenomation.

C. Neurotoxic Venoms

In *mild* cases there is only local paresthesia with no neurotoxic facies and no risk to life.

Neurotoxic facies (Fig. 9) and paresthesia over the bitten extremity occurs with *serious* cases. There is partial external and internal ophthalmoplegia. The equilibrium is disturbed, but there is no risk to life.

With *severe* cases there is neurotoxic facies and extensive paresthesia. Total external ophthalmoplegia and delayed pupillary reflexes are found. There is thick salivation with difficulty in swallowing and speaking. Prostration occurs. Mydriasis and unconsciousness indicate fatal prognosis.

Patients who recover do not retain any of the effects of the neurotoxic symptoms or late sequelae.

Insufficient data about treated patients do not allow the statistical evaluation of mortality after the bite by *Micrurus*. The cause of death is respiratory paralysis. Immediate treatment prevents the entire symptomatology. Fatal cases run their course in less than 6 hours.

VI. TREATMENT

A. General Measures

1. Immediate

If possible, the snake should be captured for identification. Suction should be applied immediately to the site of the bite if it is bleeding. If there

TABLE VII

EVALUATION OF SEVERITY OF SNAKEBITE

Venom type	Mild cases	Serious cases	Severe cases
Proteolytic + Coagulant	Small local reaction, coagulable blood	Evident local reaction, incoagulable blood	Intensive local reaction, incoagulable blood, hematuria, generalized pain, prostration
Hemolytic + Neurotoxic	Absent neurotoxic facies normal urine	Neurotoxic facies, methemoglobinuria	Neurotoxic facies, methemoglobinuria generalized pain, agitation or torpor
Neurotoxic	Absent neurotoxic facies, local paresthesia	Neurotoxic facies, paresthesia over the entire extremity	Neurotoxic facies, thick salivation, difficulty in swallowing and speaking, prostration, unconsciousness

is no bleeding, punctures, 3 to 5 mm deep, should be made around the site with a needle, pin, or thorn without considering possibility of infection. The depth of the punctures depends on the affected site. Punctures should be made with quick brisk movements. With the needle held firmly, this can be done without the risk of deeper penetration. Suction should be applied with the mouth or a sucking cup if handy. There is no danger since the venom does not penetrate intact skin or mucosa. Dental caries would not cause a problem since Obiglio (1930) demonstrated experimentally that the venom introduced in a dental cavity obtained by perforation up to the pulp was not absorbed even though other substances were. Ingestion of small amounts of venom is also harmless, even when the venom is neurotoxic (Calmette, 1896; P. Russell, 1796).

Aspiration of blood, serous fluid, and venom is only beneficial when started within 15 minutes after the bite (Gennaro et al., 1961). Aspiration never removes all of the venom, but only a part of it which will decrease the intensity of envenomation. It may also avoid death if the amount of venom injected was close to the minimal lethal dose (MLD).

Depending on the site of the bite the application of a tourniquet may be an auxiliary procedure during blood aspiration. It must be such as to allow arterial circulation and render venous circulation difficult. This may be checked by noting the turgidity and cyanosis of the distal part of the extremity. More lymph and blood will exude because of the tourniquet thus removing more of the venom. We observed that aspiration under this manner actually removes part of the venom.

Thirty minutes after the bite, aspiration and ligature are practically useless.

The tourniquet should never be applied in order to keep the venom out of the circulation. We have seen many cases of envenomations by neurotoxic venoms where a ligature was employed showing typical symptoms due to circulating toxins. This was described by Calmette (1896) and P. Russell (1796). Also in envenomations by proteolytic and coagulant venom, some of the patients who arrived with a ligature showed blood incoagulability and stronger local reaction. If a patient comes with a ligature, it should be removed only after the beginning of antivenin and shock preventive therapy. Removal must not be done at once and arterial blood pressure should be controlled by a sphygmomanometer as this may cause shock if not handled carefully. The partial retention of circulation by a well placed tourniquet results in a protein degradation and its brisk removal will cause shock regardless of the venom activity.

In 1928, Jackson and Harrison demonstrated the advantages of incision and suction in dogs injected with four MLD. In 1956, Parrish demonstrated beneficial results with excision and suction in dogs injected with six MLD of *Crotalus atrox* venom. We do not advise excision or incision. The vessels are cut and bleed severely and no venom may be found. The venom is in the lymph and tissular fluid and these must be aspirated.

2. Movement

Immobility of the patient has been advised in order to avoid the activation of circulation which would favor venom dissemination. This does not seem to be a reasonable procedure, since if the venom is neurotoxic it diffuses rapidly by itself, also absorption of proteolytic and coagulant venoms is slow, due to the local reaction produced which keeps the venom at the site of the bite, and the longer it remains there the greater the risk of necrosis; also, as mentioned previously, venoms are absorbed into the circulation by the lymphatics and not by the bloodstream.

After suction of venom, the patient must seek specific antivenin treatment quickly.

While waiting for medical treatment, the patient should be given sweet or warm beverages as well as psychological support. Even brave patients, bitten more than once, are frightened and lipothymia induced by fear or insufficient nutrition may favor and aggravate the venom-induced shock. Fontana (1781) stated that faint-hearted patients die more easily from snake bites than the courageous ones. Alcoholic drinks are frequently counterindicated since they stimulate peripheral circulation and may favor more rapid diffusion of the venom. We do not agree with this; if stimulating nonalcoholic drinks are not available, one may give small doses of alcohol in order to obtain the same psychological effect. The beneficial action of alcohol was already described when no other specific therapywas known (Oliveras, 1858).

B. Systemic Secondary Therapy

I. Antihistaminics

Antihistaminics are always given while antivenin therapy is being prepared. They will act on an eventual liberation of histamine by the venom (Trethewie, 1939) and will avoid or decrease an eventual serum reaction. They also relax the patient. In serious and severe cases, 50 mg of Promethazine are given intramuscularly, in mild cases, orally, one to three tablets the first day (Promethazine, 25–75 mg).

2. Respiratory Analeptics

These are usually given simultaneously with antihistamine preparations. Intramuscularly, in the serious or severe cases, and orally, some drops in water with some sugar, in the mild cases, Nikethamide, 1.5 ml, 25%.

Since the patient usually feels pain or torpor, the general condition will be improved after the relief of respiratory conditions.

3. Antibiotics

Antibiotics are indicated in all cases and they are absolutely necessary in envenomations by proteolytic venoms. The application follows the antivenin therapy. The choice of the antibiotic is individual. We always indicate sulfanilamide, mainly the new derivatives, because of the easy application—two or one daily dose—since they remain a long time in the circulation. Besides acting on a large number of pyogenic germs, these substances act also on the gas gangrene bacteria. They can also cross the blood-brain barrier and reach the cerebrospinal fluid. Since antibiotics have been administered before the infection develops, results have been very good. Long acting penicilin in a dose of 1,200,000 units should be prescribed to the ambulatory cases, when one cannot be sure that the patient will continue to take antibiotics.

4. Neuroleptics

Phenothiazine derivatives are given intramuscularly when the patient vomits or shows signs of agitation or convulsions, or when the intradermal reaction for serum hypersensitivity is positive.

5. Sympathomimetics

An intramuscular 2-mg dose of adrenalin in oil is given before antivenin therapy if the intradermal reaction is positive; it is given immediately after the neuroleptics. A 1-mg dose in aqueous solution is given intramuscularly at the first sign of anaphylactic shock. Intravenous injection is necessary if the shock has advanced to facial cyanosis.

If the shock is induced by envenomation, aqueous adrenalin is injected intramuscularly in addition to the other therapy. In remission of shock or when shock may be expected, adrenalin in oil is injected intramuscularly, and immediately followed by an intravenous injection of 500 ml of glucose–sodium chloride solution with 4 mg of noradrenalin added. This should be given at the rate of 10 to 20 drops per minute while the arterial blood pressure is maintained at a minimum of 80 mm Hg.

6. Cardiotonics

One vial of strophantin (Cardiovitol or Ouabain), intramuscularly or intravenously, according to the severity of the case, when there is noticeable and persistent tachycardia.

7. Sedatives and Hypnotics

If the patient has difficulty sleeping because of pain, barbiturates may be given orally. If this is not sufficient to obtain rest, or in cases of intensive agitation, one vial (6 mg) of morphine or its derivatives may be injected intramuscularly.

8. Analgesics and Antipyretics

Dipirone (pyrazolone derivative) is the drug of choice and may be given by tablet or injection, according to the symptoms.

Strong local pain without apparent lesion can be treated with analgesic ointments such as butesin picrate or nupercain. Unfortunately, vaso-dilating ointments which were very useful are no longer available. Ointments with heparin derivatives, favoring more rapid reabsorption, can also be used.

C. Local Secondary Therapy

When there is no open wound at the site of the bite, or the accident was caused by a snake with neurotoxic or hemolytic venom, cleaning with Dakin's or other antiseptic is sufficient. Treatment of the wound with an antibiotic ointment is also recommended; it is mandatory when the wound is caused by proteolytic and coagulant venoms, and should be frequently observed for superficial necrotic formation or a deep necrosis without initial skin reaction.

A necrotic formation is treated with the usual methods since there is no specific treatment.

D. Symptomatic Therapy

The presence of severe tachycardia and tachypnea, as well as tendency toward shock, frequently necessitates oxygen therapy. In envenomations of neurotoxic venom, when death is due to respiratory paralysis of central origin, artificial respiratory apparatus should be utilized as well as an artificial lung. There are necessary since there is no organic lesion, but rather a functional lesion of the nervous system, which will slowly disappear after a few days.

To counteract the intermediate nephrotic syndrome, all the usual procedures should be employed, such as renal dialysis. One should only stress that excessive intervention in these cases is harmful. The clinical symptomatology must indicate the treatment to be followed. Food should be supplied if the patient is hungry. Close observation of time and volume of urine as well as the fluid intake should be instituted.

The intravenous administration of 10 gm mannitol is very useful as a protective measure against renal lesions.

In a series of 1562 cases of bites by *Bothrops jararaca* (coagulant and proteolytic venom) only a minimal number of patients needed amputation. These cases reached the hospital with dry, dead extremities, months after the bite. In other cases it was not necessary.

E. Specific Therapy—Antivenin Therapy

I. General Considerations

Antivenins are able to neutralize specifically the venoms used for their preparation. This classic definition is one of the contributions of Brazil (1901). Recently crossed neutralization by unspecific antivenins has been described for some of the venom's activities such as the coagulant (Rosenfeld and Kelen, 1966), or the proteolytic (Rzeppa and Rosenfeld, 1966). They observed it between *Bothrops, Crotalus,* and *Lachesis* venoms and their antivenins but not with *Micrurus* venom and antivenin. They related some very interesting peculiarities: an homologous antivenin is sometimes less effective than a heterologous one; the paraspecific activity of an antivenin is sometimes stronger than its specific activity; some venoms are very antigenic and may induce antibodies against a large number of species and genera. The same observations were made in relation to crossed neutralization of the venom's lethal activity (Rosenfeld, 1950, 1951; Keegan *et al.,* 1961, 1962, 1964; Minton, 1967a; Schöttler, 1951) or of the changes of red blood cells shape by venoms (Balozet, 1966).

These facts prove the necessity of more extensive investigation on some points previously considered settled. The clinical practical applications are great, in cases when no specific antivenin is available. This may bring about the utilization of venoms having larger antigenic spectrum for antivenin preparation. In Brazil, for instance, *Bothrops cotiara, B. jararaca, Crotalus durissus terrificus,* and *Lachesis muta muta* venoms are indicated for animal immunization. It seems also that only polyvalent, instead of monovalent antivenins should be prepared.

Other alterations of the now existing norms would be advisable and should be introduced in Institutions where antivenins are prepared. One of them is the indication on vials of neutralizing capacity as units. "Unit" would be the ability to neutralize one mg venom (Rosenfeld *et al.,* 1962). Another one is the production of more concentrated antivenins. If not possible, antivenins should be lyophilized and one vial filled with such volume as to contain a minimum of 50 units. The reason for this is the complete uselessness of an insufficient antivenin therapy. There is no use injecting 30 units if the snake inoculated 100 mg venom and its lethal dose is 50 mg. Obviously, death will be the consequence of the free 70 mg venom, giving the wrong impression of inefficient serum therapy when, in fact, it was insufficient. One of the reasons in applying small doses is the fear of injecting a considerable number of vials, necessary for the concentration. There is also the lack of knowledge about the venom amount that snakes of different species are able to inoculate. In Brazil, Rosenfeld and Belluomini (1960) (Table VIII) made an experimental research on this subject since up to this time, only the average amount of

TABLE VIII

AMOUNT OF SNAKE VENOM OBTAINED ON INITIAL EXTRACTION[a]

Number of snakes	Snake genus	Milligrams of venom				
		1–50	51–100	101–200	201	Maximum
162	*Crotalus durissus terrificus*	77.0	17.6	4.2	0.6	220
78	*Bothrops alternatus*	43.4	21.5	18.7	14.9	380
52	*Bothrops atrox*	34.3	40.1	20.9	3.8	300
58	*Bothrops cotiara*	42.7	51.5	5.1	—	120
16	*Bothrops fonsecai*	43.7	41.4	18.7	—	120
33	*Bothrops insularis*	53.8	29.2	20.3	—	200
100	*Bothrops jararaca*	67.0	28.0	5.0	—	160
11	*Bothrops jararacussu*	18.0	9.0	18.0	54.0	830
45	*Bothrops neuwiedi*	86.5	13.2	—	—	100
9	*Lachesis muta muta*					1760[b]
45	*Micrurus corallinus*					6[b]
201	*Micrurus frontalis*					53[b]
8	*Micrurus lemniscatus*					8[b]
1	*Micrurus multicinctus*					6[b]

[a] From Rosenfeld and Belluomini (1960).
[b] Data obtained from the Laboratory of Antivenoms (I. Butantan).

different species was known. For other countries data are lacking or are partial (Gennaro, 1963).

There is another notion often forgotten, fundamental to antivenin therapy: that antivenins neutralize venoms but do not cure lesions. Consequently they must be injected as soon as possible, and one should not wait for the symptoms to appear; there may be permanent lesions. There is a difference between serum therapy of envenomations by venomous animals and of infectious processes. In the latter, germs gradually produce toxins and in the first, there is inoculation of the venom at once. Thus, antivenin must be given immediately in one dose and not divided. The correct dose to be injected will be indicated by the knowledge of the amount of venom each snake species can inject.

The injection route is very important for successful treatment. Subcutaneously or intramuscularly, antivenin takes about 4 hours to penetrate the circulation where it remains for 1 week. Practically, hyaluronidase does not favor absorption (Banks *et al.,* 1948) even though some authors describe it as a useful agent for this purpose (Boquet *et al.,* 1952). In cases of rapidly diffusing venoms such as the hemolytic and the neurotoxic, death may occur within 4 hours. The time elapsed from the bite to the antivenin injection, added to the period necessary for its absorption may be enough for establishing irrecoverable lesions; this can occur in 6 hours with *Crotalus durissus terrificus* of southern and central South America.

In conclusion, the factors for successful antivenin therapy may be classified as follows:

1. Specificity of antivenin
2. Early administration
3. Sufficient dose
4. Whole dose at once
5. Proper injection route

2. Dosage

Since the indication of dosage in number of vials is not satisfactory, they will be referred to as "units." Enough vials must be used for the recommended doses.

Mild cases: 50 units subcutaneously.

Serious cases: 100 units, 50 units subcutaneously followed immediately by 50 units intravenously.

Severe cases: 150 to 300 units. More when the accident was caused by a snake able to inoculate larger venom amounts. Fifty units to be given subcutaneously and the remaining units intravenously.

If there is no doubt about a poisonous snake bite, the patient should be treated as a serious case, without waiting for the symptoms. Thus, lesions may be avoided

Outdated preparations should not be discarded. If they are kept away from

TABLE IX

SNAKE ANTIVENINS OF SOUTH AMERICA

Genus of snake	Antivenin	Units/Vial 10 ml	Manufacturer[a]	Purification
Bothrops	Antibothropic polyvalent	25	Instituto Butantan	Enzymatic
	Antibothropic	25	Instituto Pinheiros Instituto Vital Brazil	Enzymatic
			Instituto Nacional de Microbiologia	Enzymatic
			Laboratorio Behrens	Partial
Crotalus (South and Central America)	Anticrotalic	10 10	Instituto Butantan Instituto Pinheiros Instituto Vital Brazil	Enzymatic Enzymatic
			Instituto Nacional de Microbiologia	Enzymatic
			Laboratorio Behrens	Partial
Bothrops and *Crotalus* (South and Central America)	Antiophidic	20 *Bothrops* 10 *Crotalus*	Instituto Butantan	Enzymatic
		16 *Bothrops* 10 *Crotalus*	Instituto Pinheiros	Enzymatic
			Instituto Vital Brazil	
			Instituto Nacional de Microbiologia	Enzymatic
			Laboratorio Behrens	Partial
			Instituto Nacional de Salud	Ammonium sulfate precipitation. Lyophilized
Lachesis	Antilachetic	20	Instituto Butantan	Enzymatic
Micrurus	Antielapidic	10	Instituto Butantan	Enzymatic

[a] Instituto Butantan, Caixa Postal 65, São Paulo, Brazil; Instituto Pinheiros, Caixa Postal 951, São Paulo, Brazil; Instituto Vital Brazil, Caixa Postal 28, Niteroi, Brazil; Instituto Nacional de Microbiologia, Velez Sarsfield 563, Buenos Aires, Argentina, Laboratorio Behrens, Apartado 62, Caracas, Venezuela; Instituto Nacional de Salud, Calle 57 no. 8–35 Bogotá, Colombia.

heat and light, their activity decreases slightly and they may still be usable after 15 or 20 years. These can be considered as half as potent as when new. The anti-*Bothrops* fraction is more stable than the anti-*Crotalus* one. A precipitate does not mean they have deteriorated; the supernatant may be used, leaving the precipitate which is inactive.

3. Complications and Precautions with Antivenin Therapy

Before serum therapy is initiated, the patient must be asked about any allergic reactions he might have had. Intradermal testing should be done. The information of an early serum treatment proved not to be very valuable since some of the patients had negative skin tests and no reaction to the therapy, while other patients never having taken any serum before developed a positive reaction and even shock during the treatment. Similar observations have been mentioned by others (Tateno *et al.*, 1964).

If there is a positive intradermal reaction, 2 mg of adrenalin in oil should be injected intramuscularly after the usual previous secondary treatment, i.e., antihistamine preparations and respiratory analeptic (the neuroleptic is injected intramuscularly, phenothiazine derivative 25 mg). A syringe should be immediately prepared containing 1 mg aqueous adrenalin.

Desensitization: After the previous treatment, the patient is injected with 0.1 ml antivenin subcutaneously. In 15 minutes another injection of 0.5 ml antivenin is given subcutaneously. After another 10 minutes, according to the skin reaction, 2.0 to 5.0 ml is injected. The whole dose of antivenin, indicated for the case, is injected 5 minutes after the last injection, if no general reaction occurs.

The first symptom of anaphylactic shock is a general malaise and a feeling of fullness in the head; the aqueous adrenalin is then injected intramuscularly. It must be given by slow intravenous injection if facial cyanosis has already appeared.

After elimination of the shock, serum therapy is continued subcutaneously. This type of reaction is not dependent on the volume of antivenin injected.

Serum sickness may appear on the fourth to the seventh day after serum therapy, with its usual symptomology, such as fever, exanthema, micropolyadenia, and articular pain. This complication is easily controlled by antihistamines and intravenous injection of calcium gluconate.

Cases of serum sickness are now very rare since antivenins prepared today do not contain albumin only globulin.

F. Comments on the Therapy

I. Antitetanic Serum

It does not seem to us that the systematic application of antitetanic serum is of any value. The observation of thousands of patients showed only one case of tetanus which was easily cured; even so, the infectious focus did not seem to be the site of the bite.

2. Corticosteroids

In spite of the divergence of opinion on the effectiveness of corticosteroids, some authors using clinical data approve their application (Arora *et al.,* 1962; Morales *et al.,* 1961), others are against their use (Allam *et al.,* 1956; F. E. Russell and Emery, 1961); we do not use them. Experimental data (Rosenfeld and de Langlada, 1964a) showed that in addition to the fact that no advantage may be gained they may aggravate an acute envenomation. However, they may be useful in protecting against necrosis induced by necrotic and coagulant venoms (Rosenfeld and de Langlada, 1965); the same effect is obtained with Isoxsuprine (Rosenfeld and de Langlada, 1964b). However, there is no reason for not using corticosteroids for fighting shock after venom neutralization by antivenin.

3. Ligature-Cryotherapy

Or L-C as it is known, was suggested by Stahnke (1953), but McCollough and Gennaro (1963) present data and comments on the topic as an inadvisable method.

4. Revulsive Ointments

Pain around or at the site of the bite can be relieved by the activation of local peripheral circulation. This procedure also favors venom diffusion and the local effects are lessened. Unfortunately, these ointments are no longer available. It is necessary to prescribe ointments with menthol or camphor or vasodilators such as nicotinic acid and its derivatives.

5. Other Treatments

Prostigmine is ineffective in reversing the flaccid paralysis produced by neurotoxic venoms as observed by us and also experimentally as related by Barrio and Brazil (1951).

Exsanguineo-transfusions and decortication of the kidneys are ineffective in the treatment of the intermediate nephrotic syndrome due to hemolytic venom (Wajchenberg *et al.,* 1954).

6. Standardization of Therapeutics

There is no substitute for clinical experimentation, as stated by Paget. Even after exhaustive and expensive pharmacologic and pharmacodynamic experiments in laboratory animals, all therapeutic agents should be investigated clinically. But new attempts will be hampered if a standardization limits the number of the agents to be used by the physician in a hospital.

VII. SEQUELAE

Frequently, patients as well as physicians suspect certain disturbances to be a belated result of envenomation. Proteolytic venoms may cause destruction of tissue, vessels, tendons, which may have late consequences. But those, like the cerebral lesions due to hemorrhages (Teixeira, 1943), appear during the acute stage of envenomation and never as a late symptom. Cases with renal calcification are the exception, as mentioned by Oram *et al.* (1963). No sequelae follow envenomations by coagulant, hemolytic, and neurotoxic venoms.

VIII. PROPHYLAXIS

A. Habitats of Snakes

Circumstances of bites and the frequency of their localization may lead to the conclusion many accidents could be avoided by wearing proper shoes or boots. Very few bites occur on the hand and even these could be avoided with some caution and knowledge of snakes' habits.

Crotalus snakes prefer dry places and are seldom found in permanently wet ground. They are fond of empty termite holes, glades, cereal and coffee plantations, and farmhouse areas, especially where the cereal is stored and where rats can be found.

Bothrops snakes are more ecletic. In spite of preferring humid ground, they are also found in dry places. They frequently stay near houses, sheltered in the woodpile, holes, etc. They irritate more easily than the *Crotalus* snakes and consequently strike quickly.

Lachesis snakes live in tropical forests and frequent areas with many leaves and trunks.

Micrurus snakes live underground and usually come out only at night; there are some exceptions to this though.

In general, poisonous snakes are not arboricole, unlike the nonpoisonous snakes. There are some exceptions like the *Bothrops insularis* found only in the Queimada Grande Island at the São Paulo seashore, Brazil, and the *Bothrops bilineata,* found in north Brazil.

When there is heavy rainfall and wet ground, all the snakes seek dry areas and some of them climb bushes or trees. They are all good swimmers and are able to bite while swimming.

B. Protection and Precautions

Besides protective shoes, trousers are also some protection when not too tight; being away from the skin, they will hinder the fangs.

Snake bite is a problem closely related to the economic and cultural level of the population. It is a serious problem where people walk barefoot; it is not so serious where people wear shoes but even here it exists due to carelessness and ignorance.

C. Active Immunization

Active immunization of man was tried with success (Haast and Winer, 1955; Wiener, 1960, 1961; Flowers, 1963), but an *a priori* affirmation can be expressed that it is not satisfactory due to several factors. Instituto Butantan employees, working at the serpentarium have been bitten repeatedly and showed no immunity to subsequent bites. Horses immunized for antivenin production have a shorter life-span than normal. Necroscopy showed tissue degeneration of organs, such as the turbid swelling of the liver (Vaz and Araujo, 1948) probably as a result of the repeated venom injections needed to maintain high antibody content.

The injection of small and repeated doses of venom induces a good level antibody formation. Soon after stopping the immunizing injections, the antibody titer decreases and comes almost to zero. When a new immunization series is started in these animals, the high antibody titer is reached sooner than the first time, but this state will take days and weeks. In infection, the infectious agent will proliferate during the incubation period. There is a latent period. If the individual was previously immunized, the organism will feel the toxin's presence during this incubation period; after some days it will defend against the pathogenic agent by its immunity level. In snake envenomations, even if the animal or individual has been previously immunized, his antibody titer is low or nil after some time. But the venom is inoculated at once by the snake in one amount of the maximal dose it can inoculate. There is no time for latency, as there is even for the most virulent infections. There will not be sufficient antibodies unless it is immediately after a series of immunizing injections.

IX. ERRONEOUS CONCEPTIONS

For centuries snakes have impressed man. For this reason the folklore of many people propagated false and erroneous information. As most envenomations occur in rural zones where the cultural level is low, many fatal cases or lesions have occurred that could have been avoided if these false beliefs did not exist.

Unfortunately, as we stated frequently, sometimes even the physicians collaborate in maintaining an avoidable mortality, only by ignoring the fact that antivenin does not cure, but merely neutralizes venom, and does not cure lesions. They ignore that it is useless to inject an insufficient volume of antivenin, unable to neutralize the venom amount inoculated by the snake, and that one injection of the whole dose, as with the venom, is fundamental. And more, they are afraid of serum therapy without having had any experience of it.

X. SNAKE BITE IN DOMESTIC ANIMALS

Sensitivity to snake venoms has been known for some time (Fontana, 1781). It is variable not only in relation to the animal species that has been bitten but also to the snake species (Araujo and Belluomini, 1960–1962). Greater or lesser resistance was considered as a natural immunity (Wiener, 1960). Both conclusions are correct but their utilization has not been applied very logically. Many inferences have been drawn without experimental knowledge.

The great herbivores are known to be very sensitive to snake venom (Brazil and Pestana, 1909; Wiener, 1960; Araujo et al., 1963a) but no data showed the time of an efficient serum therapy or the premonitory symptoms of the animal's death. Araujo and Rosenfeld (1967) gave definitions and described experimental data giving an objective knowledge on the subject. One, of great practical utility, is that the antivenin in crotalic envenomation in bovides is only efficient if given intravenously within 3 hours after the bite. It should be mentioned that only proteolytic and coagulant venoms produce edema. No reaction at the site of the bite is caused by neurotoxic and hemolytic venoms. Some authors (Calmette, 1907) supposed that the pig was more resistant to venoms because of its adipose tissue. Others supposed that it was due to a natural immunity (Brazil and Pestana, 1909; Wiener, 1960). Araujo et al. (1963b) demonstrated that no protection is conferred by adipose tissue. They also proved that there was a resistance and not a natural immunity, since serum from pigs did not neutralize venoms in the proportions to the resistance observed.

Carnivores seem to be more resistant to snake venoms than other animals. However, the cat, for instance, is resistant to bothropic venom (proteolytic and coagulant) but very sensitive to South American crotalic venom (hemolytic and neurotoxic), while the hamster, herbivore, is more resistant than the cat to crotalic venom.

REFERENCES

Ahuja, M. L., Veeraraghavan, N., and Menon, L. G. K. (1947). Indian J. Med. Res. 35, 227.
Allam, M. W., Weiner, D., and Lukens, F. D. W. (1956). In "Venoms", Publ. No. 44, pp. 393–397. Am. Assoc. Advance. Sci., Washington, D.C.

Amorim, M. de F., and de Mello, R. F. (1952). *Mem. Inst. Butantan (São Paulo)* **24**, 281.

Amorim, M. de F., de Mello, R. F., and Saliba, F. (1951). *Mem. Inst. Butantan (São Paulo)* **23**, 63.

Araujo, P., and Belluomini, H. E. (1960–1962). *Mem. Inst. Butantan (São Paulo)* **30**, 143.

Araujo, P., and Rosenfeld, G. (1967). Unpublished data.

Araujo, P., Rosenfeld, G., and Belluomini, H. E. (1963a). *Arquiv. Inst. Biol. (São Paulo)* **30**, 43.

Araujo, P., Rosenfeld, G., Rosa, R. R., and Belluomini, H. E. (1963b). *Arquiv. Inst. Biol. (São Paulo)* **30**, 49.

Arora, R. B., Wig, K. L., and Somani, P. (1962). *Arch. Intern. Pharmacodyn.* **137**, 299.

Balozet, L. (1966). *Arch. Inst. Pasteur Tunis* **43**, 9.

Banks, H. H., Seligman, A. M., and Fine, J. (1948). *J. Clin. Invest.* **28**, 548.

Barrio, A., and Brazil, O. V. (1951). *Acta Physiol. Latinoam.* **1**, 291.

Boquet, P., Bussard, A., and Izard, Y. (1952). *Ann. Inst. Pasteur* **83**, 640.

Bouabci, A. S. (1964). *Hospital (Rio de Janeiro)* **66**, 213.

Braga, N. de P. (1964). Personal communication.

Brazil, V. (1901). *Rev. Med. São Paulo* **4**, 255; also in *Colet. Trab. Inst. Butantan* **1**, 1 (1918).

Brazil, V., and Pestana, B. R. (1909). *Rev. Med. São Paulo* **12**, 415; also in *Colet. Trab. Inst. Butantan* **1**, 160 (1918).

Calmette, A. (1896). "Le venin des serpents." Editions Scientifiques, Paris.

Calmette, A. (1907). "Les venins, les animaux venimeux et la serothérapie antivenimeuse." Masson, Paris.

da Fonseca, F. (1949). "Animais Peçonhentos." Instituto Butantan, Sao Paulo.

Fidler, H. K., Glasgow, R. D., and Carmichael, E. B. (1940). *Am. J. Pathol.* **16**, 355.

Flowers, H. H. (1963). *Nature* **200**, 1017; *U.S. Army Med. Res. Lab.*, Fort Knox, Ky., Rept. 594.

Fontana, F. (1781). "Traité sur le venin de la vipère, sur les poisons américains, sur le laurier-cerise et sur quelques autres poisons vegetaux." Firenze.

Gennaro, J. F., Jr. (1963). *In* "Venomous and Poisonous Animals and Noxious Plants of the Pacific Region" (H. L. Keegan and W. V. Macfarlane, eds.), p. 427. Pergamon Press, Oxford.

Gennaro, J. F., Jr., Leopold, R. S., and Merriam, T. W. (1961). *Anat. Record* **139**, 303.

Haast, W. E., and Winer, M. L. (1955). *Am. J. Trop. Med. Hyg.* **4**, 1135.

Hanut, C. J. (1936). *Compt. Rend. Soc. Biol.* **123**, 1232.

Henriques, O. B., Fichman, M., and Henriques, S. B. (1960). *Biochem. J.* **75**, 551.

Houssay, B. A., and Sordelli, A. (1919–1921). *Rev. Inst. Bacteriol. Malbran* **2**, 151.

Jackson, D., and Harrison, W. T. (1928). *J. Am. Med. Assoc.* **90**, 1928.

Keegan, H. L., Whittemore, F. W., Jr., and Flanigan, J. F. (1961). *Public Health Rept. (U.S.)* **76**, 540.

Keegan, H. L., Whittemore, F. W., and Maxwell, G. R. (1962). *Copeia* No. 2, 313.

Keegan, H. L., Weaver, R. E., and Matsui, T. (1964). *406th Med. Lab. Res. Rept.*

Kelen, E. M. A., Rosenfeld, G., and Nudel, F. (1960–1962). *Mem. Inst. Butantan (São Paulo)* **30**, 133.

Lima, J. P. R. (1966). Thesis, University of São Paulo.

McCollough, N. C., and Gennaro, J. F., Jr. (1963). *J. Florida Med. Assoc.* **49**, 959.

Machado, J. C. (1968). Personal communication.

Mendes, E., Ulhôa Cintra, A., and Corrêa, A. (1960). *J. Allergy* **31**, 68.

Minton, S. A., Jr. (1967a). *Toxicon* **5**, 47.

Minton, S. A., Jr. (1967b). *1st Intern. Symp. Animal Toxins, Atlantic City, 1966* p. 211.

Minton, S. A., Jr. (1967b). *In* "Animal Toxins," p. 211, Pergamon Press, Oxford and New York.

Morales, F., Root, D. H., and Perry, J. F., Jr. (1961). *Proc. Soc. Exptl. Biol. Med.* **108**, 522.

Moura Gonçalves, J. (1951). Thesis, Instituto de Biofísica, Universidade do Brasil, Rio de Janeiro.

Moura Gonçalves, J. (1956). *In* "Venoms," Publ. No. 44, p. 261. Am. Assoc. Advance. Sci., Washington, D.C.

Obiglio, A. E. (1930). *Compt. Rend. Soc. Biol.* **104**, 1022.

Ohsaka, A., Omori-Satoh, T., Kondo, H., Kondo, S., and Murata, E. (1966). *Mem. Inst. Butantan (São Paulo)* **33**, 193.

Oliveras, E. J. (1858). *Oglethorpe Med. Surg. J.* **1**, 224.

Oram, S., Ross, G., Pell, L. H., and Winteler, J. C. (1963). *Brit. Med. J.* **1**, 1647.

Paget, Sir J. (cited by Sir T. Lewis (1935)). *Brit. Med. J.* **1**, 631.

Parrish, H. M. (1956). *In* "Venoms," Publ. No. 44, p. 399. Am. Assoc. Advance. Sci., Washington, D.C.

Peck, S. M. (1932). *Proc. Soc. Exptl. Biol. Med.* **29**, 579.

Rocha e Silva, M., Beraldo, W. T., and Rosenfeld, G. (1949). *Am. J. Physiol.* **156**, 261.

Rosenfeld, G. (1950). *Ciencia Cult. (São Paulo)* **2**, 46.

Rosenfeld, G. (1951). *Proc. 3rd Intern. Soc. Hematol. Cambridge, 1950* p. 84. Grune & Stratton, New York.

Rosenfeld, G. (1960). *In* "Doenças infecciosas e parasitárias" (R. Veronesi, ed.), Vol. 2, pp. 1269–1289. Livraria Luso-Espanhola e Brasileira, São Paulo.

Rosenfeld, G. (1963a). *In* "Die Giftschlangen der Erde," p. 161. Behringwerk-Mitteilungen, Marburg.

Rosenfeld, G. (1963b). *Proc. 7th Intern. Congr. Trop. Med. Malaria, Rio de Janeiro, 1963* Vol. 4, p. 166.

Rosenfeld, G. (1964). *Sangre* **9**, 353.

Rosenfeld, G. (1965). *Pinheiros Terapeut. (São Paulo)* **17**, 3.

Rosenfeld, G. (1966). *In* "Trattato Italiano di Medicina Interna" (P. Introzzi, ed.), Part XI, p. 502. Sadea & Sansoni, Roma.

Rosenfeld, G. (1968). *Proc. 3rd Intern. Congr. Pharmacol. Sao Paulo, 1966* p. 119. Pergamon Press, Oxford.

Rosenfeld, G., and Belluomini, H. E. (1960). Unpublished data.

Rosenfeld, G., and de Cillo, D. M. (1956). *Rev. Clin. São Paulo* **32**, 27.

Rosenfeld, G., and de Cillo, D. M. (1958). *Rev. Clin. São Paulo* **34**, 51.

Rosenfeld, G., and de Langlada, F. G. (1964a). *Mem. Inst. Butantan (São Paulo)* **31**, 171.

Rosenfeld, G., and de Langlada, F. G. (1964b). *Ciencia Cult. (São Paulo)* **16**, 217.

Rosenfeld, G., and de Langlada, F. G. (1964c). *Mem. Inst. Butantan (São Paulo)* **31**, 185.

Rosenfeld, G., and de Langlada, F. G. (1965). *Ciencia Cult. (São Paulo)* **17**, 302.

Rosenfeld, G., and Kelen, E. M. A. (1963). *Ciencia Cult. (São Paulo)* **15**, 288.

Rosenfeld, G., and Kelen, E. M. A. (1966). *Toxicon* **4**, 7.

Rosenfeld, G., and Sawaya, P. Jr. (1957). Unpublished data.

Rosenfeld, G., and Takeuti, I. (1968). Unpublished data.

Rosenfeld, G., Nahas, L., de Cillo, D. M., and Fleury, C. T. (1957). *In* "Atualização Terapêutica" (F. Cintra do Prado, J. Ramos, and J. Ribeiro do Valle, eds.), p. 931. Livraria Luso-Espanhola e Brasileira, Rio de Janeiro.

Rosenfeld, G., Nahas, L., Schenberg, S., and Beraldo, W. T. (1958a). *Mem. Inst. Butantan (São Paulo)* **28**, 229.

Rosenfeld, G., Kelen, E. M. A., and Nahas, L. (1958b). *Rev. Clin. São Paulo* **34**, 36.

Rosenfeld, G., Hampe, O. G., and Kelen, E. M. A. (1959). *Mem. Inst. Butantan(São Paulo)* **29**, 143.

Rosenfeld, G., Martins, L. F., and Grecchi, R. (1961). *Ciencia Cult. (São Paulo)* **13**, 200.

Rosenfeld, G., Kelen, E. M. A., and Nudel, F. (1960–1962). *Mem. Inst. Butantan (São Paulo)* **30**, 117.

Rosenfeld, G., Kelen, E. M. A., and Nudel, F. (1962). *Ciencia Cult. (São Paulo)* **14**, 254.

Rosenfeld, G., Nahas, L., and Kelen, E. M. A. (1968). *In* "Venomous Animals and Their Venoms" (W. Bücherl *et al.,* eds.), Vol. I, p. 229. Academic Press, New York.

Russell, F. E., and Emery, J. A. (1961). *Am. J. Med. Sci.* **241**, 507.

Russell, P. (1796). "An Account of Indian Serpents." London.

Rzeppa, H. W., and Rosenfeld, G. (1966). *Acta Physiol. Latinoam.* **16**, Suppl. 1, 118.

Schöttler, W. H. A. (1951). *Am. J. Trop. Med. Hyg.* **31**, 836.

Slotta, K., and Fraenkel-Conrat, H. (1938). *Nature* **142**, 213.

Stahnke, H. L. (1953). *Am. J. Trop. Med.* **2**, 142.

Swaroop, S., and Grab, B. (1956). *In* "Venoms," Publ. No. 44, p. 439. Am. Assoc. Advance. Sci., Washington, D.C.

Tateno, I., Sawai, Y., Makino, M., Kawamura, Z., and Ogonuki, T. (1964). *Japan. J. Exptl. Med.* **34**, 125.

Teixeira, N. L. (1943). *Rev. Med. Militar, Rio de Janeiro* **32**, 296.

Trethewie, R. R. (1939). *Australian J. Exptl. Biol. Med. Sci.* **17**, 145.

Vaz, E., and Araujo, P. (1948). *Mem. Inst. Butantan (São Paulo)* **21**, 275.

von Klobusitzky, D. (1938). *Anais Inst. Pinheiros* **1**, 3.

von Klobusitzky, D., and König, P. (1936a). *Mem. Inst. Butantan (São Paulo)* **10**, 223.

von Klobusitzky, D., and König, P. (1936b). *Mem. Inst. Butantan (São Paulo)* **10**, 237.

Wajchenberg, B. L., Sesso, J., and Inague, T. (1954). *Rev. Assoc. Med. Brasileira* **1**, 179.

Wiener, S. (1960). *Am. J. Trop. Med. Hyg.* **9**, 284.

Wiener, S. (1961). *Med. J. Australia* **1**, 658.

VENOMOUS SAURIANS AND BATRACHIANS

Chapter 35

The Biology of the Gila Monster

ERNEST R. TINKHAM

INDIO, CALIFORNIA

I. INTRODUCTION

Surely one of the most interesting and bizarre of all reptiles among the saurians or lizards is the Gila monster. This strangely marked creature is unique in the reptile world, for it is the only known poisonous lizard. There is a possibility that the rare and poorly known lizard of the Bornean jungles, known to science as *Lanthanotus borneensis*, may be poisonous too, as claimed by Bornean natives. At one time it was considered a member of

FIG. 1. Sahuaro desert typical of the distribution area of the Arizona Gila monster (*H. s. suspectum*). The view showing the southern slopes of the Santa Catalina Mountains in the background depicts the Sahuaro (*Carnegeia gigantea*) with palo verde (*Carcidium microphyllum*) the dominant tree. The few scattered small shrubs in the foregound are burrobush (*Franseria deltoides*) and creosote (*Larrea divaricata*).

the Helodermatidae (Smith, 1946), the lizard family which contains the two species of *Heloderma* or Gila monsters. In more recent years it has been placed in its own family, the Lanthanotidae (Loveridge, 1945). Actually, the two are closely related, as Bogert and others have shown. This is not at all surprising as we find many faunal relationships between southeastern Asia and tropical America, especially in regions where an ancient jungle fauna persists. Proof of this is found in such creatures as the tapir, bamboo pit vipers, crocodiles, sun grebes, anhingas, trogons, and many others. Previous to the Pleistocene or Ice Age, a tropical flora and fauna existed in what is now the North Temperate Region of the world, so that in northeastern Colorado, in Logan County, the fossil remains of a true Gila monster have been found and named *Heloderma matthewi*. It was found in Oligocene beds, and roamed the earth some forty to fifty million years ago.

The two known Gila monsters are the Arizona Gila monster and the Mexican Gila monster, the latter of which is much larger and is also known as the Mexican beaded lizard or vernacularly in Mexico as the escorpión. Actually, both kinds are beaded lizards, so named from the "coat of mail" appearance of their scaly hides. Naturally, Americans have a partiality for their own Arizonan species, not because it is more imposing, for the Mexican Gila monster is truly an awesome creature considering its poorly understood potentialities, but because all fatalities and bite records, with one exception, relate to the Arizona Gila monster. As so much more is known about the American species, we will discuss it first.

II. ARIZONA GILA MONSTER [Heloderma suspectum suspectum (Cope)]

Until 1956, the Arizona Gila monster was known scientifically as *Heloderma suspectum* (Cope), but now that Bogert and Del Campo (1956) have described a new geographic or eremographic race from northwestern Arizona and adjacent regions known as *Heloderma suspectum cinctum*, we must use the trinomial or full name as indicated above.

The Arizona Gila monster is a member of the Sahuaro Desert (see Fig. 1) as typified by the sahuaro or giant cactus (*Carnegeia gigantea*), palo verde (*Cercidium microphyllum*), and creosote (*Larrea divaricata*) with associated numerous less dominant plants which find their optimum development in south central Arizona. It can be said that wherever one sees a good development or stand of sahuaros that that is also the domain of the Arizona beaded lizard. The Sahuaro Desert covers an elevation from about 1000 feet up to about 3500 feet and is primarily a summer rainfall desert with 9 to 15 inches of rain a year; it has living conditions which are optimum or satisfactory for the development of the Gila monster. Since adjacent areas in southeastern Arizona, such as the Great Chihuahuan Desert, possess an annual rainfall of about 15 inches and are thus somewhat similar to the Sahuaro Desert despite a different flora, we find that the distribution of the Gila monster extends into this desert. Along the southwestern border of New Mexico, the Gila monster extends its range up to the Gila River valley to the eastern limits of the creosote bush and into similar areas along the flanks of the Peloncillo Mountains where creosote abounds. Farther south, in the extreme southwestern corner of that state in Guadalupe Canyon, creosote is found and so is the Gila monster. In July 1959, while talking to a filling station attendant at Deming, I was informed that a Gila monster had been taken in the low mountains some miles south of town, which would put it into creosote desert areas that extend east from Rodeo to Las Animas and Hachita. This then would mark the eastern limits of the Helodermatid in Luna County, New Mexico. It is absent from the grassland areas of southern

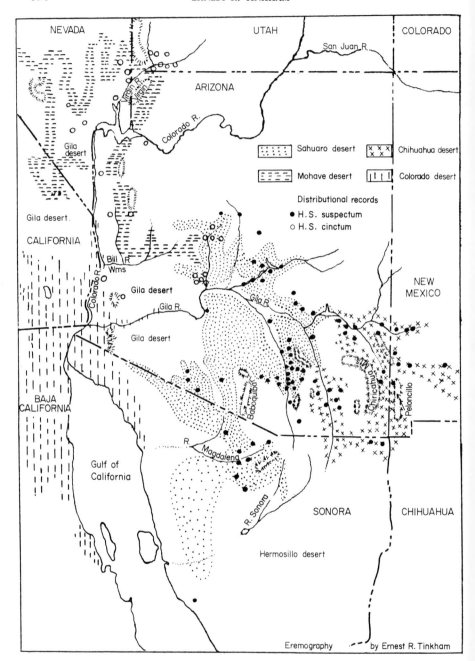

FIG. 2. Eremography of *Heloderma suspectum. H. s. suspectum* records. *H. s. cinctum* records.

New Mexico because the elevation is too high and temperatures too low (in winter) for its survival.

A perusal of the records submitted by Bogert and Del Campo, however, demonstrated that the greatest number of collecting records came from Pima County, Arizona (see Fig. 2), the heartland of the Sahuaro Desert. In Sonora, Mexico, records from Bavispe, Imuris, Magdalena, and Ilano and farther west are all Sahuaro Desert locations.

In west central Arizona in the region of Wickenburg and Congress Junction, the Arizona Gila monster meets the southeastern representatives of the newly described race *H. s. cinctum* which Bogert and Del Campo (1956) call the banded Gila monster (Fig. 2).

I find I must disagree, however, with the statement of those authors on page 51 where it is said concerning the range of *H. s. cinctum* that "the range of *cinctum* like that of other members of the genus does not conform to boundaries, mapped for desert or biotic provinces. It inhabits portions of the Sonoran, Mohave and Great Basin Deserts as mapped by Shreve (1942), and inhabits the central portion of the "Sonoran Biotic Province" as well as the eastern edge of the "Mohavian Province" as mapped by Dice (1943)."

I believe this incongruity is a man-made enigma. The problem is based on whether the Mohavian biotic province of Dice truly exists or exists only in the minds of certain scientists. To clarify the situation, we must briefly consider a few facts. The Mohave Desert (Fig. 3) as typified by the Joshua trees (*Yucca brevifolia*), creosote bush (*Larrea divaricata*), cotton thorn (*Tetradymia spinosa*), dwarf yucca (*Yucca baccata*), Mohave mound cactus (*Echinocereus mohavensis*), and many other lesser plants, is at most an ecotone between the creosote desert shrub of lower elevations (below 2800 feet) and the piñon-juniper zone at elevations above 3900 to 4100 feet. The question is whether this winding band of Joshua trees, seldom more than 10 miles wide, deserves to be called a desert. This band, which has its southern terminus in the Joshua Tree National Monument, winds north-westerly to Antelope Valley, where, looping around its western end, it ranges northeasterly to Death Valley. Here, the band swings around the north end to undulate up and down valleys across south central Nevada to the bajadas at the southwestern base of the Beaverdam Mountains in extreme southwestern Utah. From here, the zone passes south on the east slope of the Virgin Mountains to finally terminate about 13 miles southwest and west of Congress Junction, Arizona. The desert enclosed by this vast meandering band of Joshua trees and associated flora and fauna is almost entirely creosote mesas and bajadas constituting the Gila Desert. The Joshua tree belt which is the true Mohave Desert is invaded along its lower altitudinal margin (ca. 2800 feet) by creosote and other Gila Desert vegetation and

FIG. 3. Mohave Desert typical of the distribution of the banded Gila monster (*H. s. cinctum*) in southwestern Utah and northwestern Arizona. Photo shows forest of Joshua trees (*Yucca brevifolia*) in background with ground cover of typical shrubs such as spiny menodera (*Menodera spinescens*), squawberry (*Lycium spp.*), thamnosma (*Thamnosma montana*), cotton thorn (*Tetradymia spinosa*) and many other shrubs and plants. In the foreground at right is dwarf yucca (*Yucca baccata*) and in center is darning needle cactus (*Opuntia ramosissima*).

along its upper limits (3900–4100 feet) by vegetation of the piñon-juniper zone. To treat this bandlike Mohave Desert as a complete biotic province is a fantastic misconception. It is hardly a desert and much of its fauna and flora are derived from the Gila Desert. Thus the six eremological components of the Great Sonoran Desert are: Colorado, Gila, Mohave, Sahuaro, Hermosillo, and Gulf Coastal deserts. The inquiring student is referred to Tinkham (1957). Having disposed with the discrepancy caused by human error, we can definitely say that the Arizona and banded Gila monsters are members of the Great Sonoran Desert fauna.

The banded Gila monster is largely restricted to the northeastern sections of the Joshua tree belt or Mohave Desert because the Gila Desert, enclosed by the band, is too hot and dry (3–8 inches of rainfall a year; Fig. 2). As such its distribution extends northwesterly from the Congress Junction area to Hualpai Valley and Detrital Wash north to the Colorado River. From here it ranges up Grand Wash and the east side of the Virgin range to the vermilion rock escarpments some miles north of St. George, thence down the Virgin River Valley into Mesquite, Gypsum Cliffs, Overton, and Las Vegas and areas some miles north of that city. There is no evidence that anywhere in its range does it become a member of the Great Basin Desert fauna as typified by the northern sage (*Artemesia tridentata*) and its associated flora and fauna.

There are no authentic records to date for the Gila monster in California, nor is there ever likely to be now, since the drought years of the past two or more decades have been so severe that an examination of the shrubs in almost any location will reveal 50–75% of the desert shrub vegetation as dead or dying. Likewise, many rare species of desert animal life have perished since the 1930's. There is, however, some evidence that the Gila monster had been seen in California prior to 1945. The most likely places are those mountain ranges in the Searchlight, Nevada, region. Some years ago I made notes from statements made by my friend Lyle Howell. About April 25, 1943, while General Patton's tank corps was on maneuvers in the northeastern section of a branch of Chuckwalla Valley, some 25 miles northeast of Desert Center, a Gila monster was brought into headquarters by some of the men. As Mr. Howell was well acquainted with the chuckwallas, his statement that this specimen was a Gila monster in typical yellow and black markings, is probably correct.

A. Habitat

As a member of the Sahuaro Desert (Fig. 1), the Arizona Gila monster inhabits the bajadas or detrital slopes of desert mountains vegetated with creosote, palo verde, sahuaro, blood-of-the-dragon (*Jatropha cardiophylla*), ocotillo (*Fouquieria splendens*), chollas (*Opuntia* 4 spp.) prickly pears (*Opuntia* spp.), with the common shrub interspersed with creosote being burrobrush (*Franseria deltoides*). Naturally, there is a host of associated plants and flowers. Ironwoods (*Olneya tesota*), mesquite (*Prosopis juliflora*), palo verde, catclaw (*Acacia Greggii*), and many other shrubs and plants line the arroyos and their canyons. The desert floor is usually rocky and boulder strewn and called the "desert pavement." Two Gila monsters that I encountered in the early morning on the west side of the Tucson Mountains in early July were headed for, and a few feet from the top edge of an arroyo canyon which is about 20 feet deep and 50–60 feet across. There were many small caves or recesses in the walls of these arroyos. A third Gila monster, photographed on the morning of April 7, 1966, in the Growler Mountains at the northwestern corner of the Organ Pipe Cactus National Monument, was likewise within a few feet of a larger canyon, some 30 or more feet deep. Here, the vegetation representing the northern tip of the Hermosillo Desert showed many new plant elements that would not be present in the Sahuaro

Desert around Tucson. Organ pipes (*Lemaireocereus Thurberi*), ocotillos, palo verdes, jumping bean (*Sapium biloculare*), *Trixis californicus,* creosote, blood-of-the-dragon (*Jatropha cunata*), and many other plants grew on the granitic mountain slopes.

In Sonora, Mexico, Gila monster habitats would be somewhat similar to southern Arizona Sahuaro Desert habitats, since sahuaros, creosote, palo verde, ironwoods, and many other plants are typical to both areas in the range of the lizard.

In southeastern Arizona and adjacent parts of extreme southwestern New Mexico, no sahuaros are to be found since it is the Great Chihuahuan Desert, and such areas as are warm enough for the Gila monster to survive are mesas and bajadas covered with creosote, southern blackbrush (*Flourensia cernua*), ocotillos, and other plants.

The banded Gila monster in the southeastern sections of its range, as in the Congress-Wickenburg area, inhabits territory closely similar to that occupied by the Arizona Gila monster. The dominant plant life of this area is creosote, palo verde, mesquite, sahuaros on south-facing slopes of the low volcanic mountains, and other plants as well with soapweed (*Yucca elata*) scattered on the valley floors. On June 8, 1941, I picked up one of these banded fellows, about 10 miles west of Wickenburg, that had been run over by a car. Farther west and southwest of Congress, about 13 miles, the southeastern terminus of the Joshua tree (*Yucca bravifolia*) band will be observed, bringing in the so-called Mohave Desert element. This belt ranges northwest and north to the Virgin River in extreme southwestern Utah, as had already been mentioned.

Bogert and Del Campo (1956, p. 50) comment on the fact that the banded Gila monster, according to the late Dr. Angus Woodbury (1931), "lives mostly around the ledges and rocky places," and my observations also corroborate this.

During 1958 and 1959, while engaged in sand dune biota studies as a National Science Foundation Grantee, I stopped several times at Ernie's Auto Wrecking place on the northwest side of St. George. The proprietor told me that he sold about two dozen Gila monsters every summer to collectors back east and these were collected mostly by high school boys in the vermilion cliff areas visible quite some miles north of St. George, Utah. He was doing a very good job of exterminating the banded Gila monster in Utah. I reported the incident to Dr. Vasco M. Tanner and his herpetologist brother Dr. Wilmer Tanner both of Brigham Young University, hoping that they could initiate a movement to get protective legislation enacted but do not yet know whether the banded Gila monster receives protection as it does, commendably, in Arizona. On September 13, 1959, I met a piñon nut picker on top of the Toyabis, just east of Austin, Nevada, who informed me he could collect for me Gila monsters in the caves and cliffs north of

Mesquite, Nevada. From these observations, we note that a preponderance of evidence indicates that *H. s. cinctum* seems to occupy rocky cavernous areas as its preferred habitat. Evidence may also prove, eventually, that its Arizona relative also inhabits somewhat similar areas along the banks of arroyos where mesquites, ironwoods, and palo verdes lend their cooling shade to such habitats during the heat of summer.

The spatial habitat preference of a desert creature can be expressed by a chart plotting mean monthly temperature against mean monthly precipitation. Such a chart has been given various names but is here called an eremograph, perhaps for the first time, since it deals with climatic conditions relating to deserts (see Figs. 2 and 4).

Care must be taken in selecting these records from weather stations

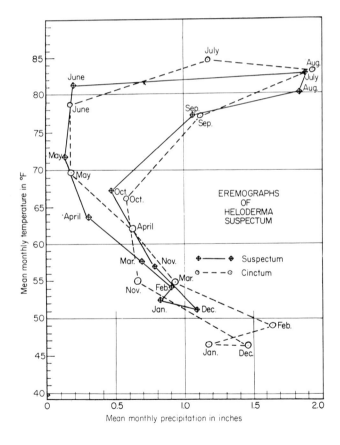

FIG. 4. Eremographs of *Heloderma suspectum*. +———+ *H. s. suspectum*. ⊙ - - - ⊙ *H. s. cinctum*.

representing the optimum range of the creature. For the Arizona Gila monster, ten stations representing some of the oldest were chosen. These were: Ajo, Baboquivari, Benson, Casa Grande, Chandler, Phoenix, Sacaton, Silver Bell, Tucson, and Vail. For the banded Gila monster only five were available: Aguila, Bagdad, Congress, Kingman, and Wickenburg.

Referring to the eremograph of the Arizona Gila monster (solid lines), we note that the Sahuaro Desert is chiefly a summer rainfall desert where July and August and, to a lesser extent, September, receive the maximum precipitation at a fairly high evaporative mean monthly temperature with lesser amounts in the winter months.

The eremograph expressing the climatology of the habitat of the banded Gila monster (dotted line) is closely similar for mean monthly temperatures and precipitation (in summer) but we note that the precipitation in the winter months of December, January, and February is considerably greater and at a lower mean monthly evaporative temperature.

B. Feeding Habits

Bogert and Del Campo (1956, pp. 79–83) quote statements from a great many writers concerning the food of the Gila monster in the wild and in captivity. While some of these are original observations, many more are simply copied from the writings of others. Statements that the Gila monster eats centipedes, insects, worms, frogs, and crickets seem without basis of fact. Since they live largely on eggs in captivity, many have assumed that the eggs of the Gambel quail (*Lophortyx gambelii*) and other desert birds are their chief diet in nature. However, the fact that Gila monsters prefer eggs in captivity does not substantiate the claim that they subsist on eggs in nature. Gila monsters love water and in the absence of eggs in captivity will drink copiously of water when it is offered, as it should be, frequently. It is natural, then, that eggs, the white of which is largely water, should prove attractive now and then to the thirsty Gila monsters.

Another often-quoted assumption is that because a Gila monster was once observed trying to climb up into a bush, all subsequent writers quote this as proof that it was hunting for bird eggs. No one seems to consider the fact that since it was daylight and the Gila monster is largely nocturnal, the heat-sensitive Helodermatid could have been trying to climb up into the bush to escape the heat of the ground. One Gila monster that I collected on July 4, 1947, at 8:00 AM on the west side of the Tucson Mountains was crawling rapidly across the hot desert pavement heading for the coolness of a nearby small canyon. I have observed sidewinders crawl into the tops of bushes to escape the hot sands, for most reptiles are killed by a steady exposure to 105°F.

While it is true that Gila monsters would not reject quail eggs in the wild, the purpose of the venomous teeth remains largely unexplained by such a statement. The egg-eating theory certainly does not explain the presence of venomous teeth in the Helodermatid. To elucidate the problem, consider the rattlesnakes which feed almost wholly on rodents such as gophers, ground squirrels, kangaroo rats, and mice as well as small rabbits which are killed in a few seconds by the injection of venom from the hypodermiclike fangs. Now the snakes that feed chiefly on bird eggs and young birds in their nests are the racers. In the Sahuaro Desert these are the red racer [*Coluber flagellum frenatum* (Stejneger)], sonoran racer [*C. semilineatus* (Cope)], and red-bellied black racer [*C. flagellum piceus* (Cope)]. These speedy reptiles the spend their active daylight hours exploring mesquites, palo verdes, ironwoods, and the oaks in the mountainous canyons, hunting for the nests of white-winged doves, mourning doves, towhees, thrashers, orioles, mockingbirds, and the like. I have caught and photographed them in the act of eating eggs and catching young birds. From this we can see that the avivorous or bird-eating snakes, such as the racers, are strictly nonpoisonous. Extending this analogy, it would indeed be strange, if in the lizard world, a reptile such as the Gila monster needed poisonous teeth (ca. 45) to feed on eggs and young birds. There is obviously much wrong with the statement that the Gila monster feeds largely on eggs in nature. Nature is far wiser than the thoughts of men, and since nature is conservative, the Gila monster has venomous teeth for a purpose—but not to eat eggs.

Within the past two decades considerable proof has been forthcoming to prove this statement. During the Easter vacation of 1948, my friend, the late Dr. Charles T. Vorhies, eminent zoologist of the University of Arizona, collected several Gila monsters on a cone-shaped knoll and placed them in a gunny sack. About half an hour later, he found on looking in, that several small jackrabbits had been disgorged into the bottom of the sack. Here was evidence, at last, that the Gila monsters used their poisonous teeth for a purpose. It is for this purpose that the openings of the three or four venom sacs are directed forward to envenom the anterior teeth of the lower jaw since these are the first to seize the animal. Apparently the chief foods of the Gila monster are the young of jackrabbits, cottontails, ground squirrels, and the like. Some years after this incident, Dr. Herbert L. Stahnke of Arizona State University at Tempe, captured a Gila monster that regurgitated an antelope ground squirrel. Here is proof that the Gila monster and the rattlesnake occupy the same relative niches in nature. The rattlesnake, a venomous snake, the Gila monster, a venomous lizard, are cogs in nature's balance wheel which maintain a control on the rodent population of the desert. Among the birds, hawks, falcons, owls, and eagles perform the same type of predatory control to keep the preyed-on populations virile. Man comes along, the greatest and most rabid of all the predators, and kills them all, and then wonders why he has to fight rodents and grasshoppers eating up his crops.

C. Food Storage

It is commonly known that the Arizona Gila monster, and to a lesser extent, the Mexican species, stores food in the form of fat in its tail. On this point all workers seem to be in agreement. Thus the condition of a Gila monster can be judged by the plumpness of its tail. If quite tumid it is very well fed, whereas Gila monsters taken from desert areas where extreme xeric conditions may have prevailed for some years usually have a relatively slender tail.

Food storage is a fairly common phenomenon in eremicolous animals. In our Nearctic deserts, the banded gecko (*Coleonyx variegatus* and *brevis*), rock gecko (*Phyllodactylus tuberculosus*), species of night lizards (*Xantusia*), the chuckwallas (*Sauromelas*), and the leaf-nosed snakes (*Phyllorhynchus*), and perhaps others, all store fatty foods in their tails. In the Gerontogeic deserts similar habits are observed in eremophilic reptiles, mammals, some scorpions, and perhaps other creatures. Thus the geckos of the genera *Pelamtogecko, Gymnodactylus, Pachydactylus, Eublepharis, Tarentola, Diplodactylus* and others, mammals such as the fat-tailed gerbil (*Pachyuromys*) and also deadly scorpions of the genus *Androctonus* of the northern Sahara region all store food in their anal appendages.

D. Ecdysis

During the past century it seems that hardly a single writer has agreed on the Gila monster's method of ecdysis (molting or shedding of skin). Some claim that the shed skin never comes off in great patches, others that it does; that shedding occurs at different times in the year, others that it is a continuous process; and others say that molting is over by October or November.

From years of observation extending over two decades, the writer wishes to clarify the record regarding ecdysis. Molting occurs only once a year, but it is a slow process, beginning on the flanks of the abdomen and tail and gradually progressing upwards, the last portions to loosen being along the middorsal line of back, tail, and the head. If the Gila monster is kept in a relatively cool, dry habitat, the skin will eventually come free in great semitranslucent sheets often half the size of the body. If the Gila monster has available a pan of water to lie in, naturally the shedding skin will get wet and adhere to the new "coat of mail" so that it will come off or rub off in small pieces. Shedding begins about midsummer (August) and is complete in midfall or in late October or early November.

In the Mexican Gila monster, personal observations indicate that ecdysis begins later in the year and is of shorter duration. The method is the same as in the American species but molting begins in late fall or early winter and is complete in late winter or early spring.

E. Hibernation

Nothing definite has been written about hibernation in the Arizona Gila monster. Most collectors are too prone to stick a hypodermic needle into their collected snakes, lizards, and amphibians to make observations on their habits in the "living" environment. Although the size of the Gila monster may help prevent this fate, nevertheless, few herpetologists seem to have studied the Gila monster in the wild or even in captivity.

Bogert and Del Campo (1956) state (p. 190) that "presumably it retires to a burrow, perhaps usurping that of some animal, but it is possible that the Gila monster sometimes digs its own or modifies one taken over." Such statements are certainly not scientific but as these authors state, nothing definite has been written on this subject up until the present.

Fig. 5. Arizona Gila monster sunning at the mouth of its hibernation burrow.

The information presented here is new and unpublished. In the spring of 1948, I had occasion to study a cone-shaped hill, about 200 feet in height that rested by itself and was surrounded on all sides by flat desert mesas. This conical knoll was the locale where Gila monsters congregated in the fall of the year for hibernation. For the preservation of the Gila monster, I will not identify the location other than as being in southern Arizona. It would appear that the lizards select such steep-sided hills where on the southern slopes as well as on the southeastern and southwestern sectors, they find protection from the cold northern winds of winter. On the warmer sunny days of winter when the sun's rays strike the steep declivity at approximately 90 degrees, the Gila monsters are able to bask safely in the sun at the mouths of their excavated burrows (see Fig. 5). One burrow about halfway up the incline was most advantageously selected for the protection of its owner. The mouth of the burrow entered directly into hard granitic soil of the hillside under a small boulder (see Fig. 6) and after penetrating down 2 feet angled to the right so that the chamber rested on granitic rock and under the protecting bight of a huge boulder many tons in weight (see Fig. 7).

FIG. 6. Entrance to the hibernation tunnel (at pickax) of an Arizona Gila monster on the steep southern slope of a hill before being disturbed by digging.

FIG. 7. The same after excavation. The rocks by the pickax in Fig. 6 have been removed and the tunnel excavated down to the hibernation chamber. The piece of shed *Heloderma* skin held by the man is just above the chamber area.

The chamber was large enough to accommodate a large Gila monster and the telltale evidence of patches of *Heloderma* skin (Fig. 7) testified that it had recently been occupied. No predator could have exhumed out the chamber and it took two men quite some time with their pickaxes to excavate down to the rocky chamber level. The evidence was that a Gila monster had excavated this chamber. The time of our visit was the first week of May. At this time all burrows found on the slope were empty but a burrow at the very base of the hill was occupied. This burrow went around to the right side of the big rock with the chamber behind the boulder. Several times during the morning I endeavored to sneak upwind on rubber-cushioned soles but all I obtained was a photo of the reptile disappearing into its hole (Fig. 8). Nor did I have the courage to seize its tail to try to pull out the creature. One other chamber found on the slope that morning was under a slab of rock that was elevated a few inches off the ground surface by

FIG. 8. Arizona Gila monster entering its hibernation burrow.

supporting rocks. Here sloughed-off skin was found. It is believed that most of the Gila monsters had departed for their summer peregrinations on the mesas. Some weeks earlier, the late Dr. Charles T. Vorhies had collected several Gila monsters in this location and it is possible that he may have captured some of the inhabitants of the chambers already mentioned.

Thus, in the first week of May, the only occupied burrow was at the very base of the mountain from which location the Gila monster could easily make its nocturnal prowlings for mammalian food out onto the desert.

F. Mating

The only definite mating observations were published 5 months after Bogert's and Del Campo's revisionary studies were published in the spring of 1956.

On June 4, 1955, Gerald O. Gates of Wickenburg, Arizona, observed his two captive males (*H. s. cinctum*) fighting over a female in the cage. He removed the smaller of the males. "Later in the evening" (no time given) "the male was mating with the female. The male had his head pressed down on top of the female's neck, while the trunks of the bodies were slightly separated from each other. The right (left in text) leg of the male was over the base of the female's tail, and his right (left in text) hand was crossed over the base of that of the female. The cloacas were in juxtaposition, with most of the torsion being supplied by that of the male's body (see Fig. 1). Copulatory movements were in slow, convulsive movements occurring every three or four seconds. The mating lasted nearly a half hour."

"The same Gila monsters were again observed mating in identical positions on June 9 and on September 30, 1955" (Gates, 1956).

Unfortunately the female died November 2, 1955, and dissection revealed it contained five eggs, four of which were dehydrated and the fifth measured 26.8 mm in length by 14 mm in diameter.

Apparently Mr. Gates had collected and killed three females (*H. s. cinctum*) that spring, two in April and one in May. The two specimens collected in April had 40 and 47 undeveloped ova, and the May female contained 53 undeveloped eggs. Hensley reports (1950) that a female collected in May and preserved in October contained five large and 25 small unfertilized eggs.

Oviposition

Considering the many years of rather indifferent observations, information on the oviposition methods and eggs laid by the Gila monster is rather sparse. Ditmars (1945) states the number of eggs to range from 6 to 13 while Arrington (1930) gives the number from 6 to 10 and Bogert and Del Campo (1956) list the count as 3 to 7. I have found from years of keeping chuckwallas that the complement of eggs varies from 11 to 13. It would thus appear from the above observations that the number of eggs for each ovary in the Arizona Gila monster may be 3 to 6 or 7. Whether these are deposited at the same time or the contents of one ovary are deposited in late July after the rains come and the other ovary cluster in August is not known at present.

When mature, the eggs measure approximately 3 inches long by 2½ inches in breadth and are covered with a white leathery epidermis typical of lizard eggs. Ditmars (1945, pp. 91–92) reports that a Mr. Ralston of Arizona who had done considerable study on the biology of *Heldoderma* claimed that freshly laid ova contained "minute but well formed embryos." Both Arrington and Bogert and Del Campo refer to Ralston's observations but note that his statements have not been verified by later workers. However, since rattlesnakes and certain northern and high-altitude species of horned lizards (*Phrynosoma*) give birth to living young, nothing exceptional must be credited to Ralston's observations. As a matter of fact, the initial development of an embryo within an amniotic sac, or within an egg within the body, or just being laid, has a great advantage over one in which no development takes place until after oviposition. Thus preovipositional development of an embryo helps assure the production of young, especially in desert regions where desiccation of the ova is a constant threat.

As for the time and location of oviposition, Ditmars again quotes Ralston as saying that the "eggs are laid in July and August. The female scoops out a hole in damp sand and deposits her eggs therein, when the sand is shovelled back again, entirely covering the eggs. Several nests were said to have been discovered. . . ." and "the majority of the eggs were buried to a depth of five inches."

As has been stated before, the Sahuaro Desert is a summer rainfall desert and the first rains usually arrive between July 15–18. As with other lizards, this rainy season is undoubtedly the oviposition period, as Arrington states they are deposited in the latter half of July and Ralston says July and early August. Ralston's statement could also be correct if the contents of one ovary were laid in late July and of the other in the early part of August. We also must remember that climatic conditions are never constant and that creatures in nature or on the desert do not follow too rigid a pattern.

There is some evidence, too, that oviposition does not occur every year but every second year. From my observations of the chuckwalla, this bi-annual cycle appears to hold and it may likewise be true of the Gila monster. Then too, there may be no oviposition during a series of drought years, for if the year is unfavorable for food it certainly would be so for development of eggs within the female and also for the eggs after they were laid in what would be desert earth.

Since the Gila monster is largely nocturnal, it is believed that courtship, mating, and oviposition of eggs may take place at night, thus accounting for our lack of information on these subjects. It is believed that mating may occur in early summer.

In summary we can say that *Heloderma suspectum* lays from 3 to 6 to 11 eggs deposited about 5 inches down in a hole dug by the female in damp sands of arroyos or other similar situations shortly after the beginning of the rainy season.

G. Growth

Here again nothing very definite seems to have been published on the growth rate of the Arizona Gila monster, Bogert and Del Campo (1956, p. 123) report that Charles A. Hewitt of Bueno Park, California, kept many saurians, including three Gila monsters and that these three "grew slowly with an average increase of .47 inches (ca. 11.0 mm) per year." The writer still has one female which measured 15 inches in length in June 1956, and was 19¼ inches in length in June 1967. This was a yearly growth increment of 10.0 mm. As my female was mature and at least 10 years old in 1956, a mean annual growth of 10.0 mm proves that growth is indeed slow after maturity is reached.

Lizards appear to fall into two general groups according to rate of growth to maturity. Horned lizards (*Phrynosoma*), collared lizards (*Crotophytus*), scalys (*Sceloporus*), umas, utas, and many others mature in 3 years, while chuckwallas (*Sauromelas*), desert tortoises (*Gopherus* or *Testudo*), Gila monsters, and even such arthropods as scorpions and tarantulas, take 10 years to mature. For such creatures, growth is slow the first 3 or 4 years, then the annual growth rate speeds up to the age of maturity (10 to 12 years) and then slows down so that from 15 to 20 years and beyond it is almost stationary. During the middle growth years in the Gila monster, the rate of growth is probably about 1 inch a year. Such growth curves, when graphed, usually exhibit the typical "sigmoid curve" of growth acceleration.

H. Longevity

The longevity of the Arizona Gila monster is not definitely known but surely reaches 25 to 30 years. One specimen reported in the literature was kept 19.4 years in captivity and if this one was adult when obtained, it was at least 30 years old at death. My female specimen is certainly beyond a score of years and should live many more with good care. Two other Gila monsters of mine lived $9\frac{1}{2}$ and $10\frac{1}{2}$ years, respectively. In May 1950, I obtained a 4-foot Yuma king snake that was full grown and which laid six eggs that month. This specimen was kept and fed in a large snake bag and died in April 1963, being at least 16 years of age. An adult rattlesnake (*Crotalus mitchelli pyrrhus*) I captured in April 1952, and which was at least 3 years old, died in April 1963, so that it was at least 14 years of age. Some of my chuckwallas are over 20 years of age and desert tortoises over 50 years of age show no growth rings after 20 or 25 years of age. Thus we see that 30 years of age is not especially old for a Gila monster.

I. Color Change with Growth and Age

Certain eremicolous saurians exhibit a color pattern change with attendant age and growth. Thus, Gila monsters, chuckwallas, and banded geckos all portray defined solid-banded patterns, especially on the tail, when young. The young of the chuckwallas, perhaps, exhibit the brightest tail coloration; from emergence from egg up to approximately 5 years, the tail is black, brightly and evenly banded with yellow. Gradually these yellow bands pale to yellowish white after 5 years of age so that at maturity the tail is almost white or a pale stone gray. When the body temperature is low, the area of the darker bands become slightly discernible; at higher temperatures around 100°F, the tail is almost uniformly white. In the young of both species of the banded gecko (*Coleonyx*), the body and tail are cross-banded with narrow, defined bands. As the creature grows in age and size, these bands break up and expand into indistinct, blotched, cross-bands.

In the Gila monster, the newly hatched young of both Arizona and Utah races possess a black tail crossbanded with three or four yellowish crossbands, thus delineating four or five black bands of slightly greater length. The banding of the body is less defined, the central portions of each band bearing a chainlike area of yellowish centers. After 4 or 5 years, when the growth increment is greatest, the black bands of the tail break up and the yellow bands develop streaks or dashes of black. At the same time the body bands enlarge, the yellowish areas within the bands expand, and the yellowish areas develop some linear blotching, the whole making a type of petroglyphic design that certain Indian tribes have used in the past in their basketry.

As we have noted in Figs. 2 and 4, the habitat of the banded Gila monster is slightly colder in winter than that of the Arizona subspecies and this phenomenon may be responsible or control somewhat the banding of the banded variety. As its name indicates, the coloration of the adult, in *H. s. cinctum* is more juvenile in character than that of the Arizona race and Bogert and Del Campo (1956, p. 47) comment on this feature. The coloration of the adults in the Arizona Gila monster is quite distinctive, so much so that one familiar with the markings can almost identify the locality where it was collected by the coloration. Thus the darkest specimens come from considerably west of Tucson, in the large area occupied by the Papago Indian reservation. Specimens from this region are predominantly black with a small amount of yellowish orange. This intensification of coloration for this region I have also observed in certain species of grasshoppers, especially the genus *Conalcea*; those coming from the Papago district are more brilliantly colored than those from the Tucson area and farther east.

It has also been noted by various authors that adults just brought from the desert are usually salmon pink, sometimes orange in general coloration, but that after a year or so in captivity this coloration has paled toward yellowish. This "wild" coloration may not only be due to the actinic rays of the sun but also to a more nutritious and varied diet on the mesas and bajadas of the desert. We find further corroboration of this in the most interesting bird of our deserts—the roadrunner. From years of close observation I know that the brilliant orange patch on the bare patch on the back of the neck in the male roadrunner fades to a pale pink in captivity. This change is undoubtedly due to a change in diet from grasshoppers, insects, lizards, snakes, and worms captured on the desert to a plain meat diet in captivity.

J. Sex Determination

It is strange that though most lizards are easily sexed by herpetologists in the living state, nothing seems to have been written about sex identification in the living Gila monster, or for that matter our largest iguana—the chuckwalla. Most sexing of the Gila monster appears to have been accomplished by dissection of the preserved specimen. Despite lack of published material, I have learned from years of study that the female chuckwalla can be separated from the male by the following features: narrower head, longer claws of forefeet (to dig eggs into the rocky desert ground), shorter and slenderer tail, more somber coloration with no brick red on the bellies, and also by the gentler and shyer disposition. Similarly, the Arizona and banded Gila monster can be sexed in the living condition by the following characteristics: males slightly smaller than females in size, claws of forefeet slightly shorter in the males than the females, tail more tumid in female and slightly shorter than in the male; and lateral postanal areas on the ventral side just back of the vent in the male showing evidence of the hidden peni. In addition, the preanal scales also differentiate the sexes. In the male there are four large quadrate, median preanal scales as shown in Fig. 9A, but as can be observed, there is no break in the continuity of the three rows of preanal scales crossing the preanal area, despite the large size of the four median quadrate preanal scales. In the female these four large median preanal scales are replaced by two (see Fig. 9B) so that the second anterior row of scales crossing the preanal area is broken in half by the two large median ones.

In summary we can say that the use of various characters, including the

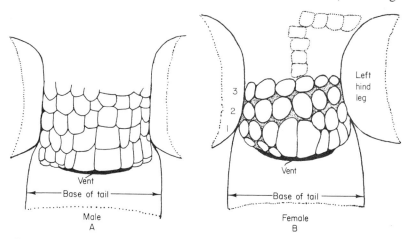

Fig. 9. Sex differentiation in *Heloderma suspectum suspectum*. Drawings by Ernest R. Tinkham.

preanal scales can be used to successfully determine sex in the northern species, *H. s. suspectum*.

K. Parasites

Bogert and Del Campo (1956, p. 140) state that the Gila monster is "remarkably free of nematodes, tapeworms, linguatulids, or other large internal parasites so commonly observed in the coelomic cavity of reptiles." I have never seen a parasite in a Helodermatid, although one of my females developed a sort of skin cancer, less than the size of a quarter, but this in no way seemed to affect the health of the individual. It was probably the result of lack of sunlight. Perhaps the nature of the diet accounts for the lack of parasites.

L. Enemies

Man and his invention, the automobile, is the greatest enemy of the Gila monster. Man is the greatest predator of all time, having exterminated countless species of wildlife within the past 100 years. Most of us are familiar with the extirpation of the passenger pigeon, great auk, labrador duck, ivory-billed woodpecker, Carolina parakeet, and other species near extinction, such as the condor and whooping crane in the United States, not to mention what is going on in the rest of the world. Most of our grizzlies, some of our big horn sheep and our wolves are forever gone from the American scene. The average American would be happy to know that all Gila monsters and rattlesnakes had been destroyed in America and would give no thought to the exploding rodent population that would inevitably engulf us as a result. Most Americans would destroy any rattler or Gila monster wherever found, even if in some isolated desert mountain range uninhabited by man. The killing instinct in man is far greater than tolerance for all creatures. The former is natural and inherited; the latter must be instilled by education and training into the unholy heart of man.

Probably more Gila monsters are killed by the automobile than by man himself. For years I have kept records on the number of coyotes, desert gray and kit foxes, skunks, ground squirrels, badgers, even raccoon on our deserts as well as quail, doves, roadrunners, owls of all kinds, nighthawks, poorwills and many other animals that have been killed on our highways, and the numbers are alarming. Considering the millions of miles of roads in our nation, and the wildlife killed in a single night on a 50-mile section of it, the figures for the wildlife killed in one night on the roads of our nation must reach a million or two. In one year countless millions are destroyed.

The slaughter of reptilian life is even more enormous since in the spring of the year, lizards by day and reptiles by night find the pavements warm to their chilled bodies and lying there they are soon run over by the countless cars speeding by. On one occasion when I was running back to retrieve a

Gila monster on the highway, the driver of an approaching car purposefully swerved to run over it before I could get to it.

The slaughter is so serious that all forms of wildlife should have full yearround protection from the hunter, for no species can long withstand the depredation that is constantly occurring on our highways by day and night.

Fortunately, in the early 1950's Arizona legislators brought some measure of protection to the Gila monster in Arizona. However, such protective laws will not stop many people from killing any Gila monster whenever or wherever they meet them, especially if they can do so beyond the pale of the law. Constant education of the ranks of the people and of the younger generations is one of the greatest means to achieve protection by destroying fear and substituting respect and interest in all wildlife. Those who find the Gila monster the most interesting of our reptiles will certainly not go out and destroy them.

M. Care of Gila Monsters

During the past three decades, I have had many reptiles and several Gila monsters and have learned a great deal from them. On one occasion, in early November of 1947, I lost a 19½-inch Gila monster by feeding her several raw eggs late one evening. Although she was in the house at Benson, Arizona, the night turned quite cold and I found her choked to death on regurgitated egg material the next morning. Consequently, I have made it a rule not to feed my Gila monsters during the hibernation period, that is, between late October and late March. Of course, if they are kept at a warm room temperature it may be safe to give them an egg every few weeks in addition to the water that should be given at least once a month. In the spring and fall I give them two eggs every two weeks and in the hot summer months the same every week. The larger Mexican Gila monster should have three eggs instead of two. Usually when the Gila monsters get restless, it is a sign they need food. After feeding, the Gila monsters remain quiet for a number of days, depending on the time of year and the room temperature.

III. THE MEXICAN GILA MONSTER

In 1829, A. F. A. Wiegmann described the Mexican Gila monster or beaded lizard and Cuernavaca, in Morelos, is considered the type locality. Up until 1956, the Mexican beaded lizard was a single species known scientifically as *Heloderma horridum* but after the revision by Bogert and Del Campo two additional races were recognized. The most northern race was named *H. h. exasperatum* and a totally black race in the southern Mexican state of Chiapas was called *H. h. alvarezi*.

The northern race was described from the Rancho de Guiracoba, some 30 miles east of Alamos, Sonora, and this variety has been called the Rio Fuerte Gila monster or Escorpión del Rio Fuerte. At present it is known only from the drainage system of the Rio Fuerte. On September 15, 1953, I saw a specimen of this subspecies at the Rio Cuchujaqui, 8 miles east of Alamos by a very rocky road through the thorn forest. It was in rather dense shrubbery just off the edge of the narrow road and my insect net was unavailable. After my 1948 experience with a baby Gila monster, I lacked the nerve to catch this slender long-tailed reptile by the tail and yank it out into the road, for I knew only too well the swiftness of their sidewise movements. Within several minutes we made a search for the creature but it was astonishing how such a large reptile could have vanished from the neighborhood.

The northern race *H. h. exasperatum* occupies the northern portions of the Sinaloan thorn forest which Shreve (1951) and Gentry (1942) and my studies (Tinkham, 1957) have shown to possess much the same floral element as the southern portions of the Hermosillo Desert immediately north of the Rio Mayo. The chief factor creating the change is that the Thorn Forest receives about 30 inches of rainfall a year, most of which falls in the summer months. Hence, in the Thorn Forest, the height of the shrubbery largely conceals the great forest of giant Organ Pipe Cacti characteristic of this region. The change from desert to Thorn Forest is most marked immediately south of the Rio Mayo.

The nominal race *H. h. horridum* Wiegmann has the greatest distribution of the three subspecies. It ranges from the Thorn Forest areas of central and southern Sinaloa, south through the palmetto palm savannahs of Nyarit south through varied floral areas to southern Oaxaca. Inland, its distribution extends almost one-third of the way across the narrowing peninsula.

No one seems to be able to explain, in the present status of our knowledge, why the nominal race apparently ignores many floral and faunal areas in the long stretch of its distribution along the coastal regions from central Sinaloa to southern Oaxaca. The writer believes that the answer will probably be found to be connected with the nocturnal temperature-habitat relationships of a nocturnal reptile. Despite the many floral areas represented, the nocturnal temperature of the habitat will probably show little variation in this long Pacific coastal strip of Mexico.

The southern race, called the Chiapan black Gila monster, has a rather confined distribution in north central Chiapas, although further studies may possibly extend its distribution southward into Guatemala.

A. Coloration

The southern race in Chiapas is entirely black. The northern race *H. h. exasperatum* of the Rio Fuerte drainage system exhibits the largest amount of yellow in its coloration and the nominal race, the Mexican Gila monster, apparently darkens in coloration as we progress southward in its range. All races possess a black tail in the juvenile with either five or six yellow crossbands, thus delineating six or seven black bands on the tail. *H. h. horridum* adults from Oaxaca have the yellow tail bands divided into halves by a narrow, somewhat irregular black central stripe and the black of the body is broken up with small irregular blotches and some larger areas of black-centered yellow rosettes. In Morelos, *H. h. horridum* adults usually exhibit, in addition to the above-mentioned markings, a narrow yellow stripe dividing the long black bands of the tail, and the black of the body is more conspicuous, with larger blotchings of yellow. The northern race of the Rio Fuerte has a much more generally blotched body and the tail has not only black streaking of the yellow bands but a narrow double-streaking of yellow in the black bands.

B. Some Differences

Although many habits and characteristics of the two species are closely similar, there are some striking differences. The tongue of my specimen in the Mexican species is a long, straplike rose-pink organ with a deep triangulate emargination at its tip, whereas the tongue of the Arizona species is blackish and broadly pear-shaped at base, narrowing abruptly to a slender tip which likewise is deeply incised. When eating or drinking, the Mexican form protrudes its long tongue into water or food material and then by quickly withdrawing it carries liquid or food into its mouth; often sort of gulping it into its mouth and down. The notched apex does not narrow during this act but retains its configuration at all times, the notching increasing the lapping areas of the apical regions of the tongue. The Arizona species cannot protrude its tongue to the same relative extent and when drinking or eating, the tips of the split tongue are usually touching the moment contraction begins.

Although it is beyond the scope of this work to make a detailed comparison of the two species, it does seem worthy of note that there is a difference in attitude of the two species. Anyone who has studied rattlesnakes knows that there is a difference in the temperament of various species. For instance, the diamondback is a highly irritable species whereas the black-tailed rattler is generally of a mild disposition. Likewise, it appears that the Arizona Gila monster, inured to xeric conditions and exposed to the hot bright desert sun, is quite responsive in its actions when met in the wild. Its movements when disturbed may be quite rapid forward or back and alarming, as it throws its mouth open with a loud hissing sound to warn would-be captors or enemies. On the other hand, the Mexican Gila monster seems to be much more docile in nature, often assuming a "play-oppossum" attitude to some extent. However, if held, its long, lithe body is very powerful as it tries to escape by twisting and turning. If one does not have a good hold it is best to drop the creature and then start over. Probably these very habits may account for the single case of Gila monster bite in Mexico as compared with numerous bite accounts in our country. Perhaps too, the fact that so many creatures are "muy muerte" or very deadly to the Mexican may breed a greater respect for this large venomous reptile.

There is probably little or no hibernation in the Mexican beaded lizard because of the generally milder temperatures of its habitat.

Both species walk in the same manner with a sort of sinuous ambulatory motion perhaps due to the fact that both legs are short and the feet palmate with all five toes of rather equal length. Gila monsters cannot run swiftly like iguanas and other eremicolous genera of the Iguanidae, but their movement is quite similar to that of the banded gecko (*Coleonyx*) and to the salamanders, both of which have similarly formed feet. It is worthy of note that Gila monsters can back up as well as they can walk forward.

As in the Arizona species, facts concerning mating, incubation, and oviposition, etc., are even more uncertain.

Facts concerning growth and growth rate are even more indefinite than with the American species, although present evidence tends to indicate that the life span in the more tropical species may be less than that of the Arizona form. In the Mexican Gila monster the tail equals the length from tip of head to the pelvic girdle, whereas in the Arizona Gila monster the length of the tail is about half of the length from snout to the hind legs.

Little definite information is available on the food of the Mexican species. Zeifeld and Norris found 13 eggs, resembling those of a quail, in one stomach and one young squirrel cuckoo (*Piaya cayana*) in another stomach of the Rio Fuerte race. A Señor Miguel Alvarez del Toro, who collected most of the black Chiapan Gila monsters, states that its food is "rats, lizards and eggs." The eggs could be of ground nesting birds or of lizards and snakes.

IV. SUMMARY

Much is known but much more remains to be discovered through further studies on all aspects of North America's two species and five races of the Gila monster or beaded lizard. These large and unique lizards, the only venomous saurians in the world, are naturally ranked as the most interesting lizards in the world.

Man and his automobile pose the most serious threat to their continued existence on this planet. The slaughter on the highways by day and night from the constant stream of thousands of speeding cars cannot be averted, but from Utah and Nevada south through Arizona and the Mexican states from Sonora to Chiapas, man can pass laws and educate the people to respect and protect this fascinating creature. Only by enacting the latter aspects of protection can the destruction on the highways be somewhat compensated. Perhaps with interest and effort man can prevent the name *Heloderma* from being added to the ever-increasing lists of his exterminated species.

REFERENCES

Arrington, O. N. (1930). *Bull. Antivenin Inst. Am.* **4**, 29–35, Figs. 1–6.
Bogert, C. M., and Del Campo, R. M. (1956). *Bull. Am. Museum Nat. Hist.* **109** [No. 1], 1–238.
Bogert, C. M., and Oliver, J. A. (1945). *Bull. Am. Museum Nat. Hist.* **83** [No. 6] 297–426.
Dice, L. R. (1943). "The Biotic Provinces of North America." Univ. of Michigan Press, Ann Arbor, Michigan.
Ditmars, R. L. (1945). "The Reptiles of North America." Doubleday, New York.
Gates, G. O. (1956). *Herpetologica* **12** No. 3, 184.
Gentry, H. S. (1942). *Carnegie Inst. Wash. Publ.* **527**.
Grant, C. (1952). *Herpetologica* **8**, 64.
Hensley, M. M. (1950). *Trans. Kansas Acad. Sci.* **53** [No. 2], 268–269.
Loveridge, A. (1945). "Reptiles of the Pacific World." Macmillan, New York.
Nichol, A. A. (1937). *Univ. Ariz. Tech. Bull.* **68**, 181–222.
Shreve, F., and Wiggins, I. L. (1951). *Carnegie Inst. Wash. Publ.* **591**.
Smith, H. M. (1946). "Handbook of Lizards," Comstock, Ithaca, New York.
Smith, H. V. (1945). *Univ. Ariz. Expt. Sta. Bull.* **197**, 1–112.
Tinkham, E. R. (1957). *Proc. 8th Pacific Sci. Congr. Pacific Sci. Assoc., Quezon City, 1953 Vol. 4*, pp. 139–147. Univ. of the Philippines, Quezon City, Philippines.
Turnage, W. V., and Mallery, T. D. (1941). *Carnegie Inst. Wash. Publ.* **529**, 1–29.
Woodbury, A. M. (1931). *Bull. Univ. Utah* **21**.
Zweifel, R. G., and Norris, K. S. (1955). *Am. Midland Naturalist* **54** [No. 1], 230–249.

Chapter 36

The Venom of the Gila Monster

ERNEST R. TINKHAM

INDIO, CALIFORNIA

I. INTRODUCTION

For many years there has been controversy over the reputed toxicity of Gila monster venom.

Certainly the case, briefly discussed in "Venoms" (Tinkham, 1956) and in the *Desert Magazine* (Tinkham, 1957) as well as in Bogert and Del Campo (case No. 27), fully demonstrates that a healthy person could require hospitalization from even a minute fraction of the venom imposed by the two front lower teeth of a juvenile Gila monster.

It is well known that the venom of poisonous reptiles kept in captivity becomes somewhat attenuated or weakened by confinement and improper diet. Add to this the fact that the baby Gila monster in question was in such poor health that there was a question of its survival, because it had not eaten in over a month in the summertime, and one must conclude that the venom of the Gila monster must be very potent indeed.

The mouth of this venomous lizard (Fig. 1) contains a formidable array of numerous small sharp teeth. The lower jaw has 9 dentary teeth on each side, of which the central 4 or 5 are the largest and the most anterior pair the smallest. The longest dentaries in *Heloderma horridum* are 6.0 mm and in *Heloderma suspectum* 5.0 mm and the anteriormost about half that size. In a one-third-grown juvenile, with which this account will deal, these anteriormost dentaries are in the neighborhood of 1.0 mm. The upper jaw

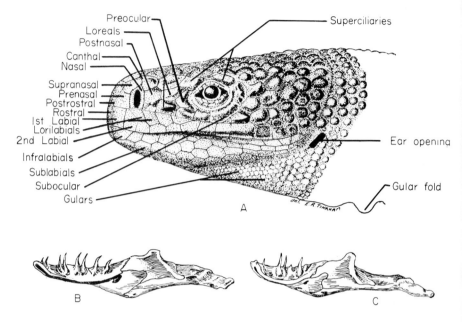

FIG. 1. A. Lateral view of the head of the Mexican Gila monster (*Heloderma h. horridum*) showing the scutellation and its nomenclature. (Redrawn and enlarged one-half after Dumeril, Bocourt and Mocquart in Bogert and Del Campo.) B. Lingual view of the right mandible of *Heloderma horridum*. (Redrawn by Tinkham after McDowell in Bogert and Del Campo, 1956.) C. Lingual view of the right mandible of Arizona Gila monster (*Heloderma suspectum*). (Redrawn by Tinkham after McDowell in Bogert and Del Campo, 1956, p. 130.)

likewise has with 8–9 maxillaries, slightly smaller than the dentaries, the largest of which measure 4.5 mm in *H. horridum* and 3.2 mm in *H. suspectum*. In addition, there are 7–9 premaxillary teeth across the front of the upper jaw, the largest of which measures approximately 2.3 mm in *H. horridum* and 2.0 mm in *H. suspectum*, so that in a one-half or one-third-grown individual the largest could be no more than 1.0 mm long. Thus, we have in an adult Arizona Gila monster a total of 18 dentaries, 16–18 maxillaries, and 7–9 premaxillaries, to make a grand total of 41–45 teeth. As the juvenile in this story had only 5 premaxillaries, the sum total of teeth in this small juvenile may have been about 30–36.

Unlike rattlesnakes, with their single pair of long, reinforced, hypo- dermiclike fangs situated near the front of the upper jaw, the teeth in the Gila monster are grooved along each anterior and posterior edge (Fig. 2A, B). This canaiculation or grooving is strongest on the anterior margin in the dentaries and maxillaries; the inside of the lateral premaxillaries are only slightly grooved; the central premaxillaries usually are without grooving.

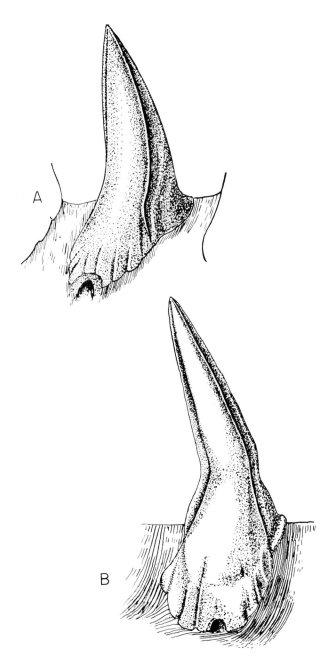

Fig. 2. A. Fourth mandibular tooth on left side of *Heloderma suspectum* showing stouter tooth than in *H. horridum* and with well-defined grooves for conduction of venom. (Redrawn by Tinkham and reduced one third after S. B. McDowell in Loeb *et al.,* 1913). B. Lingual view of sixth mandibular tooth on left side of *Heloderma horridum* showing grooves for venom conduction. (Reduced one third and redrawn by Tinkham after McDowell.)

Gray was the first to observe this grooving, in 1845, and commented in 1857 that the lizard was said "to be noxius."

Further examination reveals the swollen appearance of the lower jaws. This tumidity is due to the fact that there lies within, on each side, a large poison gland whose lower and inner edge curves in under the lower jaw. This gland is entirely encased by a fibrous capsule from which two or three septa which divide the gland into three or four poison sacs are extended inward. Contrast this total of eight venom sacs with the single pair in rattlesnakes and other venomous snakes and one cannot help but be impressed with the efficiency of this poisonous lizard.

The sacs rest in a strongly oblique position with their upper and anterior ends terminating in funnellike enlargements which open at the outer base of the five to six anterior most pairs of dentaries. This fact alone should indicate that the anterior teeth are used for envenoming their prey or enemies, a fact that earlier workers seem to have overlooked. As the largest maxillaries of the upper jaw have their apices resting in these funnels, the contraction of the muscle fasciæ covering the sacs expels venom into these receptacles, and since venom has great spreading qualities, it quickly fills the grooves of all or most of the teeth. On the first bite of an enraged Gila monster, approximately 35 envenomed teeth of both lower and upper jaws are able to introduce a lethal dose of poison into a victim; fortunately, this seldom happens.

II. ANATOMY OF THE VENOM SACS

The anatomy of the poison sacs, is explained in detail in Loeb *et al.* (1913). Briefly, we can say that each of the three or four sacs of the pair of venom glands are encased in a fibrous sheath and each sac is a distinct independent entity. Each sac bears a central cavity or lumen which narrows at its upper outer end to form a narrow canal or duct leading to the external funnellike openings (Fig. 3). The lumen is excretory in nature, since it collects the contents of the cells peripheral to it which liberate the poison by contraction of the muscle fasciæ at behest of the reptile. When thus forced to the external funnellike openings, both lower and upper rows of grooved teeth quickly pick it up and it is thus ready for almost instantaneous use. This central cavity is not a single collecting tube, but numerous fine lateral tubules extend peripherally into the surrounding protoplasm (Fig. 3) to break it up into numerous lobular masses. In this way the venom-secreting surface of these cells of the sacs is greatly amplified.

The histology of the intralobular masses is complicated and diverse. The cells of the lobes are irregularly columnar in form, with rather large granular nuclei, each containing a dark central nucleolus. The nuclei are surrounded by dense homogeneous masses of cytoplasm which are finely granular in

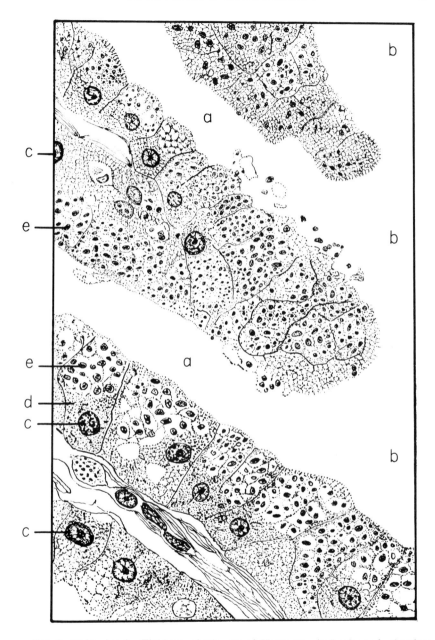

FIG. 3. Redrawing by Tinkham of Fig. 10 of Henry Fox in Loeb *et al.*, showing: a, intralobular tubule or lumen; b, central collecting lumen or duct; c. nuclei with dark central nucleoli; d, homogeneous mass of finely granular cytoplasm in the basal portion of the intralobular cells; and e, coarsely alveolar cells containing numerous "poison" granules which are liberated into the intralobular lumen.

appearance. The nuclei are located in the basal half of the cells, that is, that portion of the cell nearest the central core of each lobe, so that the outer half of each cell which is in contact with the intralobular lumen appears coarsely alveolar in nature. Each vacuole usually contains one to several or more granules surrounded by the variably sized alveolar structure of these cells. It is believed that the walls of these cells in contact with the intralobular lumen burst, thus liberating these granules into the lumen itself, where it is believed that the granules disintegrate, although this action is not definitely known. The granules contain the venom and once it reaches the lumen proper it is available for use.

In conclusion it should be obvious that the envenomation apparatus of the Gila monster is most efficient.

Fig. 4. The author holding a pair of Arizona Gila monsters from the western slopes of the Tucson Mountains. In the right hand is a female and in the left, Poncho, a male, whose bite hospitalized the author for a week. (Photograph taken March, 1957.)

FIG. 5. The author holding a powerful, 26-inch Mexican Gila monster (*Heloderma h. horridum*) from the barranca country about 20 miles east of Mazatlan, Sinaloa. (Photograph taken August 27, 1967.)

REFERENCES

Anonymous. (1879). *Sci. Am.* **41**, 399.

Anonymous. (1882). *Am. Naturalist* **16**, 842.

Anonymous. (1893). *Homeopathic Recorder* **8**, 318–322.

Arrington, O. N. (1930). *Bull. Antivenin Inst. Am.* **4**, 29–35, Figs. 1–6.

Bogert, C. M., and Del Campo, R. M. (1956). *Bull. Am. Museum Nat. Hist.* **109**, No. 1, 1–238.

Goodfellow, G. (1907). *Sci. Am.* **96**, 271.

Grant, M. L., and Henderson, L. J. (1957). *Proc. Iowa Acad. Sci.* **64**, 686–697.

Gray, J. E. (1845). "Catalogue of the Specimens of Lizards in the Collection of the British Museum." Edward Newman, London.

Gray, J. E. (1857). *Proc. Zool. Soc. London* p. 62.

Loeb, L. *et al.* (1913). *Carnegie Inst. Wash. Publ.* 177.

Phisalix, M. (1911). *Compt. Rend.* **152**, 1790–1792.

Phisalix, M. (1922). "Animaux vénimeux et venins," Vol. 2. Masson, Paris.

Shannon, F. A. (1953). *Herpetologica* **9**, 125–126.

Shufeldt, R. W. (1882). *Am. Naturalist* **16**, 907–908.

Shufeldt, R. W. (1901). *J. Homeopathics* **5**, 42–45.

Snow, F. H. (1906). *Trans. Kansas Acad. Sci.* **20** [No. 2], 218–221.

Storer, T. I. (1931). *Bull. Antivenin Inst. Am.* **5**, 12–15.

Terron, C. C. (1930). "Los Reptiles Ponzoñosos mexicanos. I. El Escorpion (*Heloderma horridum* Weigmann)." Imprenta del Inst. de Biologia, Chapultepec, D.F. Mexico.

Tinkham, E. R. (1956). *In* "Venoms," Publ. No. 44, pp. 59–63. Am. Assoc. Advancem. Sci., Washington, D.C.

Tinkham, E. R. (1957). *Desert Mag.* **10**, 11–12.

Van Denburgh, J. (1922). "The Reptiles of Western North America," Vol. 1, pp. 470–476, Plates 44–48.

Vorhies, C. T. (1917). *Univ. Ariz. Agr. Expt. Sta. Bull.* **38**, 357–392.

Woodson, W. D. (1943). *Desert Mag.* pp. 11–14.

Woodson, W. D. (1950). *Desert Mag.* pp. 19–22.

Chapter 37

Venomous Toads and Frogs[1]

BERTHA LUTZ

MUSEU NACIONAL, RIO DE JANEIRO, CONSELHO NACIONAL DE PESQUISAS,
(NATIONAL RESEARCH COUNCIL OF BRAZIL),
AND FEDERAL UNIVERSITY OF RIO DE JANEIRO

I. THE ANURANS OR SALIENTIANS

Toads and frogs are members of the class Amphibia and the order Anura, or Salientia, an equivalent term.

The Amphibia are characterized by a dual mode of life, expressed in their name. They are mostly aquatic in the early, larval, phase and nonaquatic as adults. There are some exceptions to this rule. The Amphibia are of great evolutionary significance since they were the first animals with a backbone that were able to leave the water and conquer the land. All other vertebrates, reptiles, mammals, and birds derive from certain types of fish, that had developed lungs and had transformed their lateral fins into limbs with articulated segments ending in digits. As there are usually four limbs, the terrestrial forms with four limbs are lumped together under the name Tetrapoda.

The Amphibia are a small class containing four orders, all of them small,

[1] This research was supported by a grant from the Conselho Nacional de Pesquisas of Brazil.

the largest being the Anura. According to Goin and Goin (1962), the order Apoda (coecilians) is composed of approximately 75 species; the order Trachystomata (sirens), separated off from the salamanders, does not contain more than 4 species. The real salamanders, or Caudata, number some 280 species, whereas the Anura supposedly contain 1,800 species; they are probably nearer 2,000.

The emancipation of the Amphibia from water is incomplete; most of them return to it to breed, and some live in it permanently, like the sirens. Even the terrestrial adults remain in damp environments because of the respiratory functions of their skin. Some forms spawn outside water, and some have achieved embryonic development, but the eggs do not have hard impermeable shells and consequently also need a damp environment. The Amphibia are ectothermal or heterothermal animals whose body temperature is bound to that of their environment. They do not have to maintain the high metabolism required for a high body temperature such as that of mammals and birds and consequently can fast more easily and rest longer, going into a lethargic state during unfavorable seasons. Contemporary amphibians are very small animals compared to their extinct ancestors, the Labyrithodontia, which dominated the animal world millions of years ago.

A. Anurans in General

This order contains toads, frogs, and tree frogs, which are subdivided into different families. They are characterized by the absence of a tail in the adult stage and by long hindlimbs that enable them to leap. From these basic characteristics, the names of the order were derived: Anura (tailless) from the Greek and Salientia (leapers) from the Latin.

The body is generally short and the forelimbs shorter and weaker than the hindlimbs. The latter have four segments, instead of the usual three, the tarsus having become elongated and constituting a separate segment from the rest of the foot. Most frogs are good jumpers.

1. The Adult Organism

In agreement with the great difference in size between the modern amphibians and their ancestors, the present forms have a much reduced skeleton. The head is generally flat and the skull has fewer bones and larger openings (fontanelles and fenestrations). The backbone is made up of few vertebrae, ten at most, often eight presacral and one sacral, which is more or less expanded into diapophyses at the sides. A long bone, the urostyle, beyond the sacral vertebra, is thought to have resulted from the fusion of several postsacral elements. Some forms of anurans have the two sides of the pectoral girdle firmly attached in front (Firmisternia), whereas in others one of the two cartilages that bring the two halves together in front is able to pass over the other cartilage, which confers greater elasticity (Arcifera). The nervous system is simple and the brain is produced into two olfactory lobes. There are only ten pairs of cranial nerves, as in fish. The heart is composed of three chambers. The lungs are very simple, alveolar structures, not subdivided into lobes. They are not able to carry out the entire respiratory function and are supplemented by cutaneous respiration. The digestive system follows the usual pattern; the liver seems very large. The excretory system ends in a large cloaca, of which the urinary bladder is a diverticulum. The cloaca eliminates the catabolic products, the sperm, and the eggs. The female has paired ovaries, generally oval in shape. The oviducts are long and convoluted and have a funnel-shaped opening to receive the eggs. They secrete the gelatinous envolucres. The testes of the males are also oval, though in some anurans they are rounder. Both sexes have fat bodies above the gonads, shaped like the fingers of a glove. These fat bodies store reserves and are larger at the beginning of the reproductive period than at the end.

2. Life History

Most tailless amphibians produce a great many very small eggs, which are laid in water and do not contain much yolk, but there are exceptions. The eggs are fertilized as they are laid, the male being present during spawning. He sits on the back of the female and clasps her with his arms, either below the shoulders (axillary amplexus) or above her thighs (inguinal amplexus). Fertilization is external. The males vocalize to attract the females; each kind has its specific call.

Very rudimentary minute embryos of only a few millimeters in length are generally hatched out of the eggs. They gradually develop into larvae, or tadpoles, which live in water for varying times, longer in the temperate zones than in the tropical belt and longer in some genera than in others. Eventually the larvae undergo an extremely complex metamorphosis, and they leave the water as miniature replicas of the full-grown adults.

3. The Skin and Its Secretions

The skin is dealt with separately in amphibians because it is a very important organ. It provides them with ample contacts with the outside environment. Initially it subserves the respiratory processes. It houses the glands which produce the main poisonous secretions of toads and frogs. It is a naked skin, having neither scales, feathers, or fur. The integument of the amphibian must be kept moist so as to be permeable to the gases that effect the respiratory exchanges in the numerous blood vessels that irrigate the skin. The glands and their secretions play an important part in the defense of the organism.

The skin has a narrow outer layer, the epidermis, and a deeper, inner one, the dermis. The outermost layer is composed of dead cells which protect the living cells beneath them. The skin contains pigments, disposed in chromatophores, either at the junction of epidermis and dermis or in the upper layers of the latter. The secretions are produced by glands generally having a rounded cavity and a tubular duct opening on the outer surface of the skin. Practically all tailless amphibians release secretions from their skin glands. They are, however, only venomous in proportion to the poisonous nature and the degree of virulence of their products. The skin of certain regions of the body may also produce warts, horny excrescences or spines, especially in breeding males.

B. Toads and Frogs with Poisonous Secretions

Venomous substances have occasionally been extracted from the blood or the inner organs of amphibians, and the eggs of some toads have also been found to contain them; but, generally speaking, the venom of toads and frogs is derived from the skin and its secretions.

Unicellular glands, which are common in fishes but rare in amphibians, occur in small patches on the head of hatching embryos, secreting a substance that enables them to open a way out of their gelatinous envolucres. However, in adults, multicellular glands are the rule. Since the work of the pioneers on venomous secretions, such as Charles and Marie Phisalix (1922), G. Bertrand and venomous secretions, such as Charles and Marie Phisalix and G. Bertrand and others (Phisalix, 1922), two main types of skin glands are known in tailless amphibians. Suffice it to say that these glands are organs of defense.

I. Mucous Glands

Generally the mucous glands are smaller and shallower than the granular glands. They are less localized although they may be more plentiful in regions of the body that are likely to come into contact with predators. Their secretions are relatively fluid and slimy, and under normal conditions they are sufficient to keep the body moist. When captured most frogs respond by immediately increasing the lubrication of the body making it difficult to maintain a grasp on them. This is usually accompanied by intense struggling. The secretions are especially abundant in some aquatic or semiaquatic forms like *Cyclorhamphus*. The increased secretion is often accompanied by a strong odor of musk *(Pithecopus rohdei)*, of crushed plants (many Brazilian species of *Hyla*), of garlic *(Pelobates fuscus)*, or of other substances. The secretions and accompanying odors are sometimes irritating to the eyes and nostrils *(Hyla musica)*.

2. Granular Glands

The products of the granular glands are generally thicker and creamier in appearance than the secretion of the mucous glands. The glands also show greater localization in certain regions of the body, especially at the sides of the head and shoulders or at the dorsolateral edges. This may have arisen as a defense against predators likely to seize frogs by the head; the parotoid glands of toads are the classic example of this type of gland. A similar localization of active parotoid glands also occurs in some other Salientia.

Toads seldom release their poisonous secretions voluntarily, and in laboratories it may have to be artificially provoked. Occasionally secretions may arise from gross mishandling, which would explain cases of dogs killed or made very ill by the poison of toads.

II. THE BUFONIDAE OR TRUE TOADS

From the point of view of venomous secretions, the most important tailless amphibians are the true toads of the genus *Bufo,* family Bufonidae.

The fossil record of the amphibians is very incomplete. Nevertheless, the family Bufonidae is known since the Eocene epoch of the Tertiary era. Besides the genus *Bufo,* the family contains about a dozen other living genera of toads. Some of the neotropical genera of small, toadlike anurans that have been misclassified probably also belong to the Bufonidae.

The descriptive characteristics of the Bufonidae are: the absence of teeth on both jaws and also in most, if not in all genera, the absence of odontoids on the vomers. (Most frogs have teeth on the upper jaw and many also on the vomerine bones in the roof of the mouth.) The vertebral column is procoelous, which means that the vertebrae present a concavity in front and a convex surface behind. The sacral vertebra is somewhat dilated on the sides. The pectoral girdle is arciferal. The main anatomical character of the family is the presence of Bidder's organs in the males (Fig. 1). These organs seem to be rudimentary ovaries than can develop into functional ones if the male glands are removed or incapacitated (Witschi, 1933; Davis, 1936). This permits an inversion of sex. When Bidder's organs develop, the rudimentary oviducts are also stimulated and the male toad becomes a female and lays eggs. Ponse (1926, 1927) raised little toads from eggs laid by former males; the only difference from the normal found was the unequal ratio of the sexes because the "mother" was originally a male.

A. The Genus *Bufo*

Bufo is the most important member of the family from every point of view.

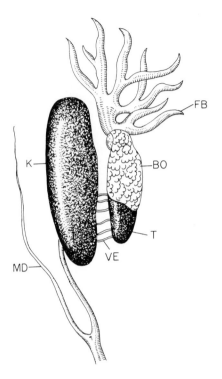

Fig. 1. Diagram of urogenital system of male toad. BO, Bidder's organ; FB, fat bodies; K, kidney; MD, Müllerian duct; T, testis; VE, vasa efferentia. (After Malcolm Smith.)

It constitutes the type genus, which is the most typical one. It includes more species than all the other genera put together, perhaps nine-tenths of all the Bufonidae.

I. Description

a. Differential Morphological Characters. Besides the general characteristics of the family, *Bufo* presents the following morphological characteristics of its own: it has a stout build, a generally squat body, relatively short hindlimbs that do not permit long leaps, consequently its gait consists of small hops, or walking, very rarely of running along the ground. Both fingers and toes are free or else there is only a moderate web between the digits. The tips of the digits are simple and sometimes end in small disks; the terminal phalanx is never claw shaped. The skin is very seldom smooth and usually is rough or glandular, often with corneous warts and spines. Poison glands are generally numerous and form masses on the sides of the head and shoulders, known as parotoid glands. The pupils are horizontal. The chromosome count in the species known is $2 n = 22$ metacentrals. Its life history is normal.

b. Specialized Anatomical Features. The main anatomical features of true toads are the bony ridges or crests on the head that are present in many species, although absent in some. These crests are moderate in most species, forming veritable wings or a complete skullcap that is concrescent with the skin in a few other species. The parotoid glands, mentioned previously, and the organs of Bidder are family characteristics, and, in this genus, they have been carefully studied by Swiss investigators.

2. Classification

Classification consists of the systematic arrangement of objects or beings in categories of different rank in order to gain a clear view of the entire group. There are always two kinds of characteristics: those that unite similar or identical beings and those that separate them. In the higher brackets of taxonomy, the emphasis is on the characteristics common to large groups; in the lower brackets, the factors that separate a unit from others of the same rank are the ones stressed. For example, with the toads, we reach them by a gradually narrowing subdivision of the ranks: animal kingdom, vertebrate phylum, class Amphibia, order Anura, family Bufonidae, genus *Bufo* and then go on to the different kinds of toads, such as *Bufo calamita, Bufo crucifer, Bufo superciliaris,* and *Bufo melanostictus.* Thus, we arrive at the species, which is the basic unit in evolutionary theory and in the classification of living organisms. Below this there are only subspecies or geographical races, populations and individuals.

At any degree of the scale and particularly when describing species minutely, it is always better to use characteristics that are easy to observe. Often one has to deal with museum specimens or with living animals of which only one or a few are available. The labels of the former and the information that accompanies the latter when sent in are often very inadequate. It is not always feasible, or desirable, to dissect the specimen, as this is time consuming, destroys the living animal under observation, and spoils the museum specimen. For this reason the main characteristics on which species are based are morphological and only occasionally anatomical. This is especially so in tailless amphibians of which the rarer forms and those from inaccessible places are often known from only one or very few specimens. As knowledge increases, physiological, biological, ecological, and geographical factors are gradually drawn into definitions of species. These factors are very important and will be dealt with below. At the moment, however, we shall examine only the structures used in determining toads.

a. Bony Ridges or Crests. The ancient amphibians had completely roofed skulls, as denoted by the name Stegocephalia, formerly in vogue. Most recent, i.e., living, amphibians have a greatly reduced, flattened skull perforated by fontanelles and windows, or fenestrations. A certain trend in the opposite direction, however, also occurs in several anuran families; in the true toads, it manifests itself by the development of thick ridges or of high crests on the crown of the head and the concrescence of the skin and bone of the skull into a dermatocranium, or cap. The outer appearance of the head thus provides some of the main characters used for the classification of toads. There are four main groups: toads without bony ridges on the crown, toads with thick ridges, toads with high crests, and toads with a skull cap.

i. Ridges absent. The heads of the toads of this group are like the heads of other anurans except for the more granular or warty appearance of the skin of their dorsal surface. This is due to the presence of warts and glands, especially to the grouping of the glands into parotoids which are a feature of the genus. A certain number of species of *Bufo*, probably less than half, are devoid of distinct crests. They are more numerous in the Old World than in the New World. The following species may be indicated as examples of this group: the three palaearctic toads of Europe, *Bufo bufo, Bufo calamita,* and *Bufo viridis* and their subspecies in Africa and Asia, the well-known African toads *Bufo regularis* and *Bufo mauritanicus; Bufo raddei* from Asia, *Bufo boreas* of North America, the Andean *Bufo spinulosus,* the Chilean *Bufo variegatus, Bufo guttatus, Bufo glaberrimus* and *Bufo blombergi* from northern South America. Some of the toads without bony ridges are warty, like *Bufo raddei* and the circumediterranean *Bufo bufo spinosus,* whereas others are quite smooth, like the *Bufo guttatus–Bufo glaberrimus* group.

ii. Bony ridges or crests. In the New World, especially in the Neotropical region, toads with bony ridges or crests on the head are very numerous and include some of the best-known forms, such as *Bufo marinus.*

iii. Forms of transition. Some South American species of *Bufo* serve as a transition from species with average ridges and species with a complete dermatocranium and exhibit various degrees, or modes, of concrescence of the skin of the head with the bones of the skull. They include *Bufo granulosus,* a small species with a number of extremely widespread races, from Trinidad, British West Indies, Venezuela, and the Guianas through Eastern, Central, and Southern Brazil, into Argentina. The northern form, *B.g. granulosus,* has a very distinct, thorny, crest around the eyes, which is more marked in the southern *B.g. d'orbignyi.* The skull is partly roofed.

iv. Skullcaps. Bufo guentheri of Puerto Rico has all the ridges present except the parietal one but, besides this, the derm of the head is concrescent with the bones of the skull. The same condition obtains to an even more marked degree in the two Cuban species *Bufo peltocephalus* and *Bufo empusus. Bufo peltocephalus* has the canthal ridge separated from the supraorbital and thickened into a knob posteriorly. The supraorbital forms an angle with the postorbital and both are considerably reinforced. In *Bufo empusus,* there is a bony process from the mastoid region to beyond the ear opening, which is thus reduced to a small perforation in a bony shield. These forms are phragmotic; they rest in burrows in the ground, probably dug by themselves, and use the crown of their heads to plug the openings of the burrows. This fact was discovered by Barbour in *Bufo empusus.* Subsequently, I also observed phragmosis in bromeliad-dwelling hylids. A more rudimentary stage of this habit was observed by Professor Gliesch (verbal communication) in *B. granulosus d'orbignyi,* which hides in hollows in the dunes of Rio Grande do Sul. At this stage, the protection afforded is inadequate since Glesch saw snakes yank the small toads out of their unstable and horizontal sand burrows by their heads. At a more advanced stage and on firmer ground, the reinforcement of the skull should have survival value.

v. Winged crests. Two other neotropical toads have a slightly different reinforcement of the skull: *Bufo lemur* of Puerto Rico and *Bufo typhonius* of the Amazonian rain forest. In *Bufo lemur* all the orbital ridges are prominent, the supraorbital ridge is some 7 mm high and with the orbitotympanic it buttresses the eye. There is also a reinforcement on the upper jaw between nostril and eye. In *Bufo typhonius,* the head is bluntly triangular and the jaws are thickened at their outer corners; the ridges on the crown are high, especially the orbitotympanic, which continues growing until it forms a veritable wing in old and large individuals. The name *alatus* was suggested for them but the races of *B. typhonius,* if any, are not clearly defined. Bony reinforcements of the skull, and sometimes of the back, of tailless anurans are often looked upon as a "lusus naturae" devoid of significance. I consider them adaptive devices of value to small animals, whose skulls are probably their best and almost their only protection against being seized by the head and swallowed. The roofed skull was probably also of use to their extinct stegocephalian ancestors.

b. Parotoid Glands. This is the name given to a conglomeration of tumefied, granular glands with a creamy, poisonous, secretion, forming a well-defined structure behind the eyes, on the sides of the head, and/or on the shoulders. The name of these glands might well be disputed since they have no similarity to the parotid glands of mammals. The parotoid glands are not generally used in the major keys for grouping species of toads, probably because their shapes are recurrent so that they have to be taken into account together with other characters. Nevertheless, they form an important element in descriptions of species and are very useful for distinguishing either different sympatric toads, i.e., those living in the same places, or localized geographic races. In the common European toad, *Bufo bufo bufo,* for instance, the parotoids are more regular in shape and more divergent at the neck than in *Bufo calamita,* whose range is partially similar. Lutz (1934) points out that in

Bufo m. paracnemis the longitudinal axis of the parotoids is more oblique and forms a greater angle to the longitudinal axis of the body than in *B.m. ictericus* (Fig. 2). The nominal form, *Bufo marinus marinus,* has enormous heart-shaped glands occupying a large surface on the shoulders. In *Bufo peltocephalus* of Cuba, the longitudinal axis of the parotoids is practically perpendicular to the axis of the body, so that the glands are transverse (Boulenger, 1882).

The position of the parotoids also varies. In most toads with well-defined ridges, they are placed just after the postorbital crests. Some species without ridges, like *B. guttatus,* have the parotoids deflected onto the sides. In *Bufo ceratophrys* the glands are far back on the shoulders, above the arms. The parotoids also vary greatly in size, although they are generally larger in large or medium-sized species than in small ones. The glands may be elongate, rounded, triangular, or obtriangular, etc. The elongate ones are sometimes quite elliptic *(Bufo funereus),* oval *(B. alvarius),* or kidney-shaped. In some species they form an elongate triangle; it is quite regular in *Bufo typhonius* (Fig. 3), very irregular in *arenarum*; and in both they continue on in a series of smaller glands down the dorsolateral edges of the body. In *B. typhonius* their pearl- or beadlike form led Laurenti to give the species the name *margaritifer.* In small species, like *Bufo granulosus,* the glands are often round or slightly distorted into bluntly subtriangular shape. In *Bufo lemur* of Puerto

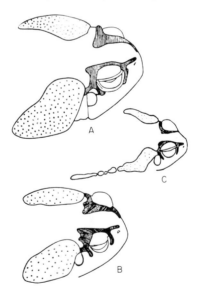

FIG. 2. Parotoid Crests. A. *Bufo m. paracnemis* female (Pindamonhangaba, state of São Paulo, Brazil). B. *Bufo m. ictericus,* female (Nova Friburgo, state of Rio de Janeiro, Brazil). C. *Bufo arenarum,* female (Buenos Aires, Argentina). (After Lutz, 1934.)

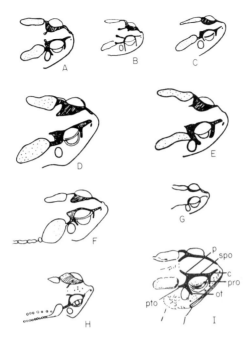

FIG. 3. Parotoid glands and crests. A. *Bufo g. d'orbignyi,* female (São Francisco de Paula, state of Rio Grande do Sul, Brazil). B. *Bufo g. d'orbignyi,* male (Buenos Aires, Argentina). C. *Bufo crucifer,* female (Gramado, Rio Grande do Sul). D. *Bufo crucifer* (Rio de Janeiro, Brazil). E. *Bufo crucifer* (Pernambuco, Brazil). F. *Bufo rufus* (Minas Gerais, Brazil). G. *Bufo g. granulosus* (Natal, Rio Grande do Norte, Brazil). H. *Bufo typhonius* (Kartabo, British Guiana). I. Diagram of ridges or crests. c, canthal; ot, orbitotympanic; p, parietal; pro, preorbital; pto, postorbital; spo, supraorbital. (After Lutz, 1934.)

Rico, the components of the parotoid glands are large spiny warts that do not coalesce into a single unit. Good examples of the shape, position, and appearance of the parotoid glands and crests are found in the beautifully illustrated books by Mertens (1959): *La Vie des Amphibiens et Reptiles* and by Cochran (1961) on *Living Amphibians of the World,* or in the works on regional amphibian faunas mentioned in the Bibliography.

Additional Glands. A few species have additional glands, similar to the parotoids in structure, but located on the limbs. These species include the North American *Bufo alvarius,* the South American *B.m. paracnemis,* the Asiatic, desert toad *Bufo raddei,* the palaearctic *Bufo calamita* and some Asiatic subspecies of *Bufo bufo,* namely *B. b. wrighti, B. b. minshanicus,* and *B. b. gargarizans.* In the last three, the extra glands are relatively small and more or less rounded and located on the dorsal aspect of the leg, or tibia. The location is the same in the others, but in *B. calamita* the gland is larger and placed on the upper side of the calf and similar glands are found on the outer side of the forearm. *Bufo alvarius* and *Bufo paracnemis* have their paracnemid gland on the upper edge of the leg that is nearer to the thigh when folded. *Bufo alvarius* has a similar gland on the thigh, a smaller one on the tarsus, and a very large one on the forearm. This toad seems to be richest in poisonous glands. As a rule *B. m. paracnemis* has only the tibial gland; however, of late, Zelnik, of the Instituto Butantan in São Paulo has discovered some specimens from the arid northeastern area of Brazil that have a similar, though less marked gland on the forearm. *Bufo raddei* is a desert toad and to a certain extent so is *Bufo alvarius*. *Bufo m. paracnemis* inhabits the Continental area of South America, the northeastern subarea of which is semiarid and subject to drought. A dry environment may be a stimulant to glandular response. Many toads have larger individual glands on the dorsolateral edges. The dorsal aspect of some toads is also more granular than that of others. Certain montane races of the andine *Bufo spinulosus* and *Bufo horribilis* of Mexico, which are probably near to *B. marinus*, have spiny warts surmounting the glands. A few toads are quite smooth like *Bufo guttatus* and *Bufo glaberrimus*, which is probably an altitudinal race of *B. guttatus*, a forest toad from the Amazonian Hylaea. Male toads are generally more spinescent than females. *Pseudobufo* and the bufonoid *Holoaden* are entirely glandular on their upper portions.

c. *Other Structural Characters.* A number of other characteristics are used in describing toads for the first time and in determining subsequent specimens of the same species. The best ones are those that are easy to evaluate, for instance, the extent of webbing on the feet. This permits separating *Bufo boreas boreas* from *Bufo boreas halophilus*; in the nominal subspecies the feet are fully webbed, whereas in *B. b. halophilus* the web is considerably shorter. Another good characteristic is the relative size of the first and the second finger. It is very easy to observe and offers at most three alternatives: first and second finger equal, first longer, or first shorter than second. This, and most other characters used refer to proportions between parts, such as the interorbital space and the width of the upper eyelid, the diameter of the tympanum compared to the diameter of the eye, the relative length of the hindlimb to the body, etc. Unfortunately, the last is not easy to use in living specimens since it is measured by bringing the hindlimb forward along the body and marking the point reached by the tibiotarsal articulation.

d. *Size.* Living frogs and toads range from slightly below 20 mm to somewhat above 200 mm total length, measured from the tip of the snout to the vent.

The largest toads are probably those from the northern part of South America, roughly included in the concept of Amazonian Hylaea, or equatorial rain forest. Spix indicates 9 inches for *Bufo marinus*; the American Museum of Natural History and the Museu Nacional in Rio have specimens with 230-mm snout-vent length. *Bufo marinus paracnemis,* from the Continental area, is somewhat smaller, but still very large, females often attaining 190-mm snout-vent length. The Atlantic form, *B. m. ictericus,* averages 130–140-mm snout-vent length. The recently described *Bufo blombergi,* from Colombia, is almost as large as *B.m. marinus,* some 200 mm in length. The description states that the natives of the region insisted that still larger examples exist. The smallest toad, *Bufo rosei,* from the Table Mountain at Capetown, is only 1 inch long (24 mm). Mertens calls his *Bufo fenoulheti damaranus* and *Bufo f. fenoulheti* dwarftoads, though they attain, respectively, 38 and 37 mm in length. The males of *Bufo pygmaeus* from Brazil, which is probably a member of the *Bufo granulosus* group, are 41–45 mm long. The other members of the group are larger; *Bufo g. granulosus* attain 70 mm and its southern representative, *B. g. d'orbignyi,* some 60 mm. *Bufo variegatus,* from the most southern part of South America inhabited by anurans and which reaches Puerto Bueno, Magallanes, Chile, is also small; females reach 45 mm in length, males less. This is in agreement with the rule that female toads and frogs are generally larger than males, though not invariably so. Size is frequently a racial rather than a specific characteristic. This is not surprising since growth depends in part on the food available and other conditions of the environment. Racial differences of size are very much in evidence in the Chilean subspecies of *Bufo spinulosus.* The males of the largest, *Bufo spinulosus arunco,* from Concepción and the Cordillera, attain 101–102 mm, whereas *B. s. rubropunctatus,* the smallest and most southern form, is only 45–52 mm. Most toads fit into a range of 75–100 mm with the smaller sizes probably more frequent than the larger. A number of toads reach 125 or 130 mm, for instance *Bufo melanostictus* of Asia, *Bufo superciliaris* of Africa and the North American *Bufo boreas boreas* and *B. b. halophilus.* In the Brazilian *Bufo crucifer,* this size is only attained by the largest individuals.

  e. Coloration. Toads are not brightly colored like some of the tree frogs of the Old and the New World and some of the small neotropical frogs of the families Dendrobatidae and Atelopodidae. Their dorsal color is generally rather dark, brownish, or grayish, varying in tone and especially in detail from one kind of toad to another. A few are more greenish, like the Colorado River toad, *Bufo alvarius,* and a few really green, like the palaearctic *Bufo viridis.* The brightest spot in the dorsal aspect is the golden, coppery, or greenish iris, generally with a metallic glint and often with dark venation. The ventral surface is mostly pale, whitish or grayish; sometimes it shows darker vermiculations, which may be pale or fairly vivid. Some of the toads have the color scheme inverted, especially forest dwellers. The dorsal aspect is then lighter than the ventral one and very clearly delimited by the dorsolateral margins. These may even simulate the outline of a leaf, as shown in a Cott photograph (1940) of a *B. typhonius*. This individual even had two dark spots on the sacrum, like holes in the leaf. This is mimetic or procryptic coloration. In such forms, the sides of the body are very dark and so are the limbs. *Bufo glaberrimus, Bufo guttatus,* and their beautiful Colombian relative, *Bufo blombergi* have this type of coloring.

Bufo blombergi is one of the largest and most beautiful toads; Dr. Cochran (1961) published a magnificent color photograph of it in which the back is a golden bronze and the sides of the body and limbs almost olivaceous black. The coloring of *Bufo guttatus,* a smaller form of the Amazonian Hylaea, is similar but inclined to gray. *Bufo glaberrimus,* from the eastern Andes, is figured by Guenther with a beautiful rose band of flash-color on the groin and smaller rose spots on the posterior aspect of the thigh. Bright or flash-colors on the surfaces concealed in repose are unusual in toads, though common in tree frogs. A few toads have modest ones; thus the color phase *B. stellatus* of *Bufo crucifer* has yellow spots on the thighs and groins. *Bufo viridis* of Europe, a very handsome species with large green marks on a neutral gray or cream ground, may have the warts tipped with rose. *Bufo punctatus* of North America and *Bufo spinulosus rubropunctatus* of Chile have red spots on the warts, and *Bufo caeruleostictus* from the Andes of Ecuador has blue dots on the warts on the sides of the body.

Toads occasionally show a reddish suffusion. In *Bufo rufus,* from the mountains of Minas Gerais, in Brazil, this suffusion is variable but generally present and very marked on the webs of the toes, which may be bright red. Dr. Cochran (1961) kept a *Bufo fowleri* that had "a brick red complexion" that made it very conspicuous on moss. In her vivarium, he was protected; but outside of it natural selection would probably have worked against him, making him too conspicuous to predators to live long enough to transmit his coloring to many descendants. This phenomenon is called erythrism from the Greek word for red. Miranda Ribeiro (1926) named colormorphs of *Bufo crucifer,* calling one of them *roseanus* because of the reddish suffusion on its back. The opposite phenomenon is also seen, namely melanism, when the toad is unusually dark, almost black. This coloring is common among females of *Bufo m. ictericus* from the southern part of Brazil. Sometimes juvenile specimens are brighter than adults, as noticed by Dr. Cochran in North American forms. I have seen very small specimens of *Bufo m. ictericus* with bright yellow tubercles on the palm and the sole and a relatively gay dorsal livery.

In reptiles the male is often more ornate than the female. In most species of anurans, including toads, this is not the case. In some toads the female has the same pattern or livery as the male but in brighter and more contrasting tones. In other species, the coloring differs according to sex. *Bufo marinus ictericus,* just mentioned, is a beautiful example of this sexual dichroism.

FIG. 4. *Bufo m. ictericus,* male. In a garden at Teresópolis, state of Rio de Janeiro, Brazil. Natural size, approximately 130 mm. (Photo by G. A. Lutz.)

The male is olivaceous brown or yellow and the name *ictericus* is very well chosen for him, since he really looks jaundiced (Fig. 4). The female, on the other hand, is beautifully marbled or spotted in dark brown on a creamy white ground, either with a broad light vertebral stripe and similar bands toward the edges and dark areas between them, or with insular spots disposed more irregularly (Fig. 5). *Bufo spinulosus atacamensis* exhibits very similar liveries (Cei, 1962). In both cases the character is peculiar to one geographic race and not to the whole species. However, in *Bufo m. paracnemis* from the Continental area, the males are sometimes very slightly marbled on a dark olivaceous ground, apparently exhibiting an incipient degree of female livery. It is not known whether this phenomenon ties to the changes of sex, inter-sexes, and hermaphroditism which occur in true toads. The specimens of *B. marinus marinus* from the Amazonian Hylaea seen by me were all dark, not marbled, nor have I been able to obtain information as to sexual dichroism in them from others. In certain species some races are merely more handsome than others. This seems to be the case in some of the Asiatic subspecies of *Bufo bufo* such as *B. b. formosus, B. b. japonicus,* and *B. b. asiaticus.* However, even at their brightest, true toads cannot rival the brilliant coloring exhibited by other salientians, especially tree frogs.

Fig. 5. *Bufo m. ictericus,* female. From Teresópolis, Rio de Janeiro, Brazil. Natural size, approximately 140 mm. (Photo by G. A. Lutz.)

3. Behavior

a. Rest. Toads, like most other recent anurans, are crepuscular and nocturnal animals, probably because of their small size and relatively weak defenses. Despite the poisonous glands of most and the prohibitive aspect of some, they prefer to go out of their way to avoid their enemies. They rest during the day and go about their normal business at dusk.

b. Food. In the evening, toads leave their homes and start to forage for food. Their squat bodies and generally short legs condition their gait to hopping, or even to walking, more rarely to running on the ground. They can swim but generally seek water only at the nuptial period. Prey is only recognized as such when it moves. Consequently, only living animals are taken, preferably beetles and juicy caterpillars, but also other insects, including many kinds of ants. Budgett (1899) tells of a half-grown toad from the Paraguayan Chaco that sat by a man's naked foot and flicked off 52 mosquitoes in a minute, as they alighted. Very small fare for a large toad. Snails are sometimes eaten but not easy to manage; worms are much better, especially earthworms. Some large kinds of toads can be reared in laboratories on a diet of newborn mice. Several juvenile *B. marinus,* sent from Port-au-Prince by me to Rio, grew very large in a few months on this diet. Dr. Smith (1951) mentions a big *Bufo marinus* that thrived in the zoo in London on full-grown mice. Gadow (1901) also mentions that a smaller *Bufo b. spinulosus* kept by him also accepted mice. Some kinds of insects are rejected; caterpillars with a nauseous smell or with hairs and bristles may be caught but are speedily ejected. Cott made some rather unkind experiments on toads, letting them starve for a while and then putting them near hives with stinging bees. Some learned at once to avoid them, others required some six trials, but after a week every one of them refused to catch the bees. The memory of the stings lasted about a fortnight.

Toads are difficult to keep because they require a great deal of food. Boulenger states that one of his captives would eat so many earthworms, one after the other, that they were finally passed alive.

Toads are often attracted to the vicinity of strong artificial lights, including those set up by entomologists to attract nocturnal insects. *Bufo m. ictericus* has been foregathering under street lamps in small eastern Brazilian towns since public lighting was introduced. Other toads do likewise in other continents and the *Bufo marinus* introduced into distant lands continue this habit.

The actual catching of food is interesting; the toad approaches, sits still for a time, sometimes with the toes twitching. Suddenly, the tongue, which is fixed in front and free behind, darts out and returns with the prey adhering to it. Action is almost too fast to be followed by the eye. If the prey is small it is immediately gulped down; hard-shelled snails may be cracked by the toothless jaws, or swallowed whole, making a lump in the stomach. Earthworms generally dangle out of the mouth; the fingers may then be used to wipe and push them in. When two toads seize the same worm, a tug-of-war may ensue, the wretched worm either breaks in two or is gradually released by one and swallowed by the other.

c. Voice. Toads and frogs have the first real voices in the vertebrate world; it is only the males who sing. The voice is produced by the larynx, or voice-box, at the anterior end of the trachea, the air passing to and fro through a slitlike glottis, supported by a pair of lateral cartilages. The mouth is kept closed. The sound may be increased by the presence of a vocal sac, which is an outgrowth from the mouth. Some species have a vocal sac, others do not. The sac may be rounded and not very large, as in *Bufo granulosus,* where it forms a small balloon covered by glaucous membrane; it may be relatively large and sausagelike, as in the very small North American oak toad *Bufo quercicus; Bufo cognatus,* of the Great Plains, also has a sausage-shaped but enormous vocal sac. Each species has a distinctive call. In South America the different forms of *Bufo marinus* are known far and wide by the name of *"sapo cururu."* The word "cururu" really sounds something like the liquid tremolo call of the males of this species. The small *Bufo granulosus,* on the other hand, produces shrill notes like the whistle of a night watchman. *Bufo cognatus* is stated by Goin to emit an "ear-splitting blast." Several authors compare the call of other toads to the peeps of chicks; personally, I have never heard such notes. Naturalists with a good ear for a music have always been able to distinguish the different species by their calls even when they are similar, like that of *Bufo crucifer* and of *B. m. ictericus.* Of late, good methods of recording voices on tapes and transcribing them as sonographs have introduced less subjective methods of distinguishing species and evaluating degrees of relationship between different kinds of toads. The method is largely due to the initiative of Dr. Bogert (1960) of the American Museum of Natural History and his associates. Other American and foreign authors are making good use of the method, especially A.P. and F. W. Blair of Texas.

A series of pioneer publications by W. F. Blair (1954, 1956a, 1956b, 1957a, 1957b, 1958) established the study of the mating call as a major method of investigating the status of toads. It permitted the separation of cryptic species, *Bufo speciosus,* from *Bufo compactilis* with which it was previously confounded (Bogert 1960). "Difference of mating call seems to be the most important of a complex of isolating mechanisms in most species of toads" according to W. F. Blair (1962).

The specific "mating" song or call of the male is a means of calling the female toads to mate. Other sounds are, however, occasionally produced. One of them is given out to stave off other males, who, in their sexual ardor, may attempt to clasp a toad of their own sex. This is called the "release call" by Martof and Thomson (1948) (Bogert, 1960) and the "preventive vibration" by Rengel (1948, 1949). Female toads give a "release call" when they have no ripe eggs to lay. Rengel (1949) and Cei (1962) publish graphs of release calls of female *Bufo spinulosus.* The work of Aronson (1944) indicates that the "warning vibrations" of the body flanks of toads are as important as the call in obtaining release. As early as 30 years ago, Lutz (1934) differentiated the sex of his *B. paracnemis* by titillating the belly, which causes the males to emit a clucking sound. Both sexes of tailless amphibians are apt to scream out loud, with the mouth open, when seized by a predator, generally a snake.

    d. Male Reproductive Behavior. Sexual activity brings a few new factors into play in the usually stereotyped behavior of toads. Cei (1949) found that some species of anurans have a continuous sexual cycle, denoted by the continuous production of spermatozoa, whereas in many others sexual activity is periodic. When the nuptial period approaches, males begin to differ from females even in species without sexual dichroism. Their glandular warts fill out and their spines become more evident. Horny excrescences develop on the outside of the two inner fingers and sometimes on the breast, which increase friction and permit a better hold under the arms and on the breast of females during amplexus. The forearms do not become so robust as in males of the neotropical frogs of the genus *Leptodactylus*.

In climates with very marked seasons, reproductive activities are triggered by spring or by the onset of the rains. It is quite "explosive", as I had the opportunity of witnessing at Miranda in southern Mato Grosso some years ago. The behavior of male *B. m. paracnemis* was quite typical of the behavior of male toads in general. For each female present there were a number of males, every one of them eager to embrace her and fertilize her eggs. They vied with each other, trying to butt, push, and kick their rivals out of the way and to mount the much-oppressed female from any point, even climbing over her head or limbs. The most strategically placed generally won the day. Goin tells of "balls of toads," milling around a centrally located female, struggling to reach her and not letting go until long after they have killed her by their collective ardor.

Sexual instinct is so strong and so undiscerning in male toads that they will clasp anything, from a male of their own species, who then issues the release call, to females of other species, frogs, and even inanimate objects of similar shape and surface. In the Maritime Mountains I have repeatedly had male *B. m. ictericus* clasp my feet, which were inside gum-boots, and exert enough pressure to permit me to lift them out of the water attached to the foot. In locales having permanently humid climates, reproduction does not begin so explosively. Males go to the artificial ponds in the National Park at the Organ Mountains, sit in the water, and serenade the opposite sex for many nights before obtaining a mate. The females only approach when their eggs are ripe. Hearing must play an important part in sexual selection by them since they only seek males of their own species. There is not much evidence of hybridism in Brazilian toads. When it does occur, it is probably due to the frantic behavior of sexually excited males.

In anurans fertilization is external. Nevertheless, the male is present during spawning. Amplexus is axillary in toads. The male mounts the female and passes his arms under her, reaching either the axillae or right across the breast.

e. Spawning. The process of spawning was carefully observed in *Bufo bufo bufo* by Savage (1934). After the male has mounted the female, the couple swims around for a time. The female stops when she finds a suitable place, in this species one with water weeds. She then stretches her body slowly, straightens her back and extends her legs behind her. The male stretches his also, with the heels together and the knees dug into the flanks of the female, so as to touch her cloaca. If he is small his toes are under it, if larger they are hooked round her belly. The eggs emerge slowly, passing over the toes or the tarsi of the male. He responds with pumping movements, interpreted as emissions of sperm. In *B. b. bufo* this goes on for some 10 seconds, during which a portion of egg-strings emerge. After that the couple rests a while. The female then starts moving around again, although somewhat anchored by the egg-strings. When she comes to rest, once more, spawning and fertilization are resumed. Savage (1934) indicates half-hourly intervals and several hours' time for the entire process. After the female has finished laying the eggs she continues to move around a bit and the pair again goes through the motions. The fact that no more eggs pass over his feet probably warns the male that spawning is finished and causes him to release the female and swim off.

4. Life History

a. Eggs. Unlike most other anurans, toads lay separate gelatinous strings of eggs from each oviduct. The strings are quite long, some 3 to 7 feet long in the European toads. The eggs are dark, very small, 1–2 mm in diameter, and disposed in one or two rows along the egg-string. Each has its own gelatinous envolucre but this is not perceptible as the whole string swells in water and becomes hyaline. After it has been immersed for a while, it collects debris and no longer calls attention to itself. Phisalix states that the eggs of toads contain the venom of the adults though the larvae do not. They are certainly unpalatable, or they would often not last long enough for the embryos to hatch. I once saw a horse thrust his head into a bucketful of *Bufo* egg-strings collected by me and hung on a hedge, which he must have taken for fodder. He seized a mouthful but immediately let them go and started to rub his mouth on the grass vigorously.

b. Embryos. The developing embryos are still very dark; in temperate zones they may take a few days to hatch, in tropical belts less. They are elongate, with a globose head and a narrower tail end. On the ventral aspect of the head, a triangular opening soon develops, the stomodaeum, or primitive mouth, and under it a horseshoe-shaped fold of tissue that soon divides into a pair of rounded cement organs. When the embryo leaves the capsule it fastens itself to the rest of the egg-string or to plants by the cement glands. Later it develops eyes on the sides of the head and external gills below them. The tail becomes more shapely. Very young embryos do not move spontaneously but soon they can be stimulated by touch into taking very short swims and promptly refastening themselves. Later they begin to swim on their own initiative. They do not have extensive yolk reserves and as soon as their organs function they have to feed themselves.

c. Larvae. In the second period of aquatic life the embryo becomes a larva or tadpole. The tissue around the mouth expands into a disk with an upper and a lower labial half. The stomodaeum develops into a mouth, which acquires a beak, composed of two horny mandibles, also upper and lower, generally with serrated edges. On each half of the labial disk rows of horny larval teeth, develop above and below the beak. In the majority of anurans there are two rows on the upper and three on the lower labial; these are also found in true toads. With the exception of *Bufo b. wrighti,* known toad tadpoles only have the inner upper labial row divided. Clusters of papillae develop on the edges of the mouth; in toads these are found only on the sides. Meanwhile the eyes have enlarged and become prominent. The external gills have long disappeared under a fold of skin that produces a branchial chamber; a small opening, the spiraculum, remains on the left side. Its function is to allow the water taken in by the mouth or nostrils to flow out through the gills. Toward the end of larval life, lungs develop. When the gills are no longer present, the larva must have access to air or it will drown; therefore, it comes to the surface to breathe.

In small translucent larvae, the heart and sometimes the liver can be seen; the rest of the abdomen is filled by intestinal coils that open on the outside by a median anus placed under the lower fin of the tail. This is now neatly divided into a central muscular part, an upper and a lower crest. The tip of the tail is generally blunt. The tadpole grows for a while and then metamorphoses. In toads, larval growth is very limited and metamorphosis supervenes at a very small size. During the larval period, the tadpole is very exposed to predators. Insect larvae and adults take a large toll; aquatic bugs and dragonfly larvae are probably their worst enemies. The tadpoles are often quite black and somewhat gregarious and, consequently, very conspicuous. There may be safety in numbers but a great many are eaten before they leave the water.

d. Metamorphosis. The change from an aquatic larva to a miniature terrestrial organism is conditioned mainly by secretions of the thyroid, triggered by the hypophysis. An interesting study could be made on the biochemical reasons why giant species of *Bufo* transform with little more than a tenth of their adult size, whereas some large frogs, like *Pseudis paradoxa* and *Megalelosia goeldii,* have tadpoles as large or even larger than the adults. The first sign of metamorphosis is the appearance of minute hindlimb buds that are held straight out. As they grow, the tips first become lobed and then develop into shapely toes; meanwhile, the joints also appear and the legs are not held straight back any longer but are movable. The arms develop at the same time but remain hidden in the branchial chamber; shortly before the little toad leaves the water, they emerge. The left one emerges through the spiraculum and the right through a perforation in the skin. Meanwhile, other changes have supervened. The tail dwindles, being absorbed, until only a little stump is left; this finally disappears also, often after the little toad has left the water. The mouth undergoes much greater changes; the larval beaks and teeth are shed; the bones of the adult jaw develop and soon the mouth opening reaches from side to side. Inner changes are even more pronounced but discussion of these would take us too far afield. The larval period is estimated at from 1 to 3 months for the European toads.

    e. Juveniles. Juvenile toads come ashore in large numbers. They are very minute and very smooth, have weak limbs, and have no protection at all against their enemies. They generally hide in daytime under stones or vegetation and only move inland at night or during rains. The mass migrations of little toads in wet weather have given rise to the quaint notion of showers of toads.

    Metamorphosed juveniles of *Bufo b. bufo* may be only 11 mm long for adults of 47–70 mm, if males, and females up to 87 mm. Those of *B. viridis* may be from 7 to 10 mm in length for adults of from 70 (males) to 80 mm (females). It is possible that the larger ones are future females.

    f. Growth. There is not much information about the growth of little toads. Those that are not destroyed during their early terrestrial life may double their size by their first autumn and double it again by the second. Dr. M. Smith (1951) gives the following figures: *Bufo b. bufo* 20 mm at the first hibernation, 40 mm at the second. *Bufo calamita* 26–40 mm in the second year averages 30 mm; males of this species are supposed to attain sexual maturity at the end of the third year and females in the fourth or fifth year. Anurans continue to grow after they begin to reproduce. An unusually large size denotes an individual that has had a good span of life.

    Growth was also observed in a series of baby Mexican toads (*Bufo valliceps*) by W. F. Blair (1953). They were marked at sizes from 11 to 55 mm. Eleven were recovered as participants of the first breeding aggregation of the next year, 1952. At 10 months of age they ranged from 61 to 78 mm. Growth appears to be rapid in the summer of hatching, to slow down in winter, show some acceleration the following spring and to slow once more as mature size is approached. Blair estimates a survival of 11 % of the metamorphosed toads.

    In toads the most interesting aspect of growth is the periodic shedding of the skin. This is necessary since the outer layer of the epidermis does not accompany a general increase in size. Toads about to moult hunch up their backs so that the skin splits down the middle. At the same time it splits behind at the junction of the legs, and down the midventral line. The skin is then jerked forward by movements, the toad opening and shutting its mouth. The legs are withdrawn from the old skin by bringing them sharply forward and then straightening them out. This is made easier because the old skin splits down the underside. The parts of skin that are moved forward are swallowed. The last bits to be removed are the skin of the arms and the underside of the throat, attached to the edge of the mouth.

g. *Span of Life.* There is very little information on this point also. The best-known individual life span is that of a common European toad, *B. b. bufo,* mentioned by Pennant, in 1776, and quoted in the literature ever since. This toad was found in an English garden when the family moved into the house and was then already adult. The toad, a female, remained with them for 36 years. It became very tame and allowed them to put it on a table and feed it by hand. At the end of that time, the record was unfortunately interrupted by a fatal accident. M. Smith (1951) estimates its life span at 40 years since female *B. b. bufo* take 4 years to become adult. C. J. Goin gives a span of 31 years for *Bufo terrestris,* a North American species, but does not quote his source. These examples show that toads which manage to escape the intense predation of their embryonic, larval, and juvenile periods and their other enemies as adults may live longer than many mammals.

5. Ecology

a. *Habitat Preferences.* Toads are terrestrial animals; except for the nuptial period, their life is generally spent on the ground. Since they are also nocturnal and are not well equipped for climbing, they generally seek damp, quiet, and protected resting places near the ground. The phragmotic Antillean forms dig their own burrows, others use shallow burrows made by other animals and natural or artificial hollows. Any suitable place is promptly put to use. Many toads live in the pipes sunk into the ground that drain rainwater from gardens, others under porches; greenhouses are favorite homes. A number of *B. m. paracnemis* were collected by me in the greenhouses of the National Park at the Paulo Affonso Cataracts. Over a year later one female was found in my garden where it had probably been living in the rainpipes. Budgett (1899) mentions that every outhouse and shed in the Paraguayan Chaco was inhabited by toads. Forest toads of southeastern Brazil sometimes climb laboriously up embankments to find shelter, especially if they can find a support for the body like those used in crevices and chimneys by mountaineers.

b. *Homing.* Toads become attached to their resting places and convert them into home sites. In Teresópolis we sometimes find the same toad night after night on the terrace of the house. The longest-known record is that of a *Bufo b. bufo* mentioned above, which lived on the same premises for over 30 years. Toads are not always solitary. In Rio Grande do Sul we found groups of *Bufo arenarum* and *Bufo granulosus* in the same hollows, several times. Bogert (1947) made some interesting experiments on homing, with 444 *Bufo terrestris* or Carolinean toads. The toads were marked and carried to various distances from their usual home site, in open, or not open terrain. A good percentage of them returned to their home site, the percentage naturally being larger among those that were released nearer than in those that had been carried further away; it was 60% for those taken 300 yards away and 18% for those liberated a mile from home in terrain not known to them. The chorus of the other males at the pond in the homesite seems to have guided them, as none of those released where sounds from the pond were no longer audible ever got back.

c. *Spawning Sites.* Toads breed in lentic, i.e., standing water, as a rule. Some are more particular than others. *Bufo b. bufo* is stated to prefer ponds and pools that are not very shallow whereas *Bufo calamita* will spawn in much smaller and shallower water—even in ditches. In the mountains of southeastern Brazil, *Bufo m. ictericus* is also contented with ditches and even with quite shallow trenches dug at the sides of roads and streets to drain off heavy rains. Under these circumstances larval life has perforce to be short. Even so, many egg-strings degenerate and the embryos perish. No species of *Bufo,* whose life history is known, spawns regularly in swiftly running, i.e., lotic waters. Nevertheless, I have seen *Bufo m. ictericus* in amplexus at the shallow edges of mountain streams in southern Brazil and found the tadpoles in the same places. The natterjack, *Bufo calamita,* will spawn in slightly brackish water. This may also be the case of *B. paracnemis* in southern Mato Grosso where some lagoons are slightly salty. Toads seem to make a choice of one or another pond or lagoon when several are available. Boulenger (1912) moved toads, singly and in pairs, from one pond to another nearby, but they returned to the original one.

d. *Hibernation. Aestivation.* At certain seasons toads may go into a lethargic state with metabolic changes reduced to a minimum and without intake of food. In the colder zones this happens during the winter. Hibernation consists in finding an appropriate place at the approach of cold weather and holing down in it until spring. Toads may bury themselves some 18 inches below the surface. This habit seems rather general and many examples are given in the literature. In warm and moist belts, this habit becomes unnecessary; but toads may become torpid when exposed to very low temperatures. On a short air trip, from southern Bahia to Rio de Janeiro, I transported several large toads in the pockets of my coat. During the passage over the Maritime Range the plane rose very high and the cabin became very cold. On arrival, soon afterward, the toads were quite cold and stiff but not dead, as I had feared. After a while they warmed up and began to move around.

In warm belts, with a very marked dry season or subject to drought, toads also hide away and go into aestivation. They are sometimes dug out of the muddy or barely damp remnants of periodic streams in northeastern Brazil from similar depths as those used for hibernation. These habits have given rise to many curious legends to the effect that toads can survive for centuries without air or food and have been recovered intact from foundations of old buildings.

e. Predator-Prey Relationships. Toads do not have aggressive weapons; they have neither size nor speed, claws or teeth. The poisonous secretions of their glands constitute their best, and almost their only, means of warding off their enemies. Even so, they rarely take the initiative. In over 40 years, only once have I seen a toad release the contents of its parotoids spontaneously, on being dropped to the ground from above a yard. Nor can toads asperge their foes from a distance. The liquid they release from the cloaca, on being seized, is not urine but water stored there. Toads are only lethal or really dangerous to small animals that bite them, including the young of poisonous snakes which have not yet learned to respect them (Brazil and Vellard, 1925).

When they cannot get away, toads try to save themselves by striking defensive attitudes. One of them, which is very much used by juveniles and other small anurans, consists in turning over on the back and lying quite stiff with the limbs close to the body and the palms and soles held up, feigning death. Whether this really is a hypnotic state, as some think, or a conscious attitude is not quite clear. I have seen small bromeliad tree frogs lie on their backs for half an hour, then open an eyelid, turn over, and leap away. The best-known defensive attitude consists in filling the lungs and swelling the body to its utmost capacity, sometimes increasing girth by almost 50% (M. Smith). At the same time the body is raised from the ground on the hind-limbs, the head is lowered and a side presented to the enemy. The increase of surface may be sufficient to prevent being swallowed. This attitude was discovered in European toads by Hinsche (1923) but others, for instance, *Bufo alvarius* (Hanson and Vial, 1956), use it too. I have seen *Holoaden bradei,* a small bufonoid toad get up on its little legs, bend over, and hiss while guarding the eggs in its terrestrial nest.

These attitudes exhaust the toads' very inadequate arsenal. Loveridge once saw a rat rip the skin off the back of a toad and proceed to devour the flesh. Other small predatory mammals have been seen tumbling toads over and attacking them from the belly.

f. Parasites. Like most other animals, toads are subject to parasites. Large neotropical toads often have ticks on them. The really destructive parasites, however, are internal. The larvae of the greenbottle fly, *Lucilia bufonivora.* and perhaps of *L. sylvarum,* prey on toads (Spence quoted by M. Smith, 1951). The eggs are deposited on the thighs of the toads or the hind part of the back of the toad, in a cluster of 60–70, or more. When the larvae hatch, they wriggle up the body of the toad to the head and enter the nostrils or, more often, the eye, behind the lower lid. Then they find their way into the nasal cavity, reached in a few minutes by the vanguard. The nostrils get blocked and the toad wanders around in great distress, opening its mouth continually. The larvae infiltrate the soft tissues, leaving only the cranial cavity and the skin of the head intact. The toad perishes after 2 or 3 days. It is unusual for parasites to kill their hosts but these are carrion eaters and they continue to feed on the carcass of their victim.

    g. *Human Contacts.* Contact with human beings is more often harmful, or even fatal to animals than beneficial. My colleague, Mr. A. Leitão de Carvalho, tells me that some years ago, a foreign company operating in Brazil distributed leaflets along the Amazonian river, offering to buy the skin or the bodies of toads. Many poor people, who had left the toads at their doors unharmed for years, started out to catch them and to skin them alive, leaving only the hands, feet, and head covered. The flayed toads were then released and were devoured alive by fire ants and other insects, undergoing great agony. This is only one of many instances of immigrants who abuse the hospitality of new countries to enrich themselves by torture and extinction of the fauna for the fabrication of the vulgar "native industry," exhibited all the way from Buenos Aires to Miami. A law was passed to protect the toads but law enforcement is not easy along the net of waterways of the Amazons.

    Not all human contacts are harmful. Some are charming, like those of the English family whose toad shared their garden for 36 years and has been an example of longevity for nearly 2 centuries. Mertens points out that the great toads, especially *Bufo marinus*, have almost become domestic animals, being transported to distant lands to keep down insect pests and allowed to live more or less unmolested; they seem able to adapt themselves to their new homes and to proliferate there. One of the unpleasant results of their introduction and of their wandering about is the great increase in the number of toads killed by automobiles.

    Generally speaking, it is a sad day for an animal when man discovers that it can be useful to him. Toads have been used in all sorts of shocking ways, from witches' cauldrons to application of the entrails of living animals to sore spots in human anatomy. A far milder use is that of the Somali porters, who after exhausting days of safari, rub a toad for a few minutes on their foreheads to remove fatigue, headache, and incipient inthermation, and then let it go. Since the Galli-Mainini (1948) test for pregnancy was discovered, millions of toads are kept prisoners and starved in analytical laboratories, This test consists in injecting urine of the patient in the dorsal or ventral lymph sac of a male toad. If she is pregnant the injection provokes a spermatorrhaea. In Brazil this test is so well-known that wayward girls from slums have learned to seek good-natured servants in medical laboratories to learn from them whether their indiscretions have had deeper consequences than they bargained for.

Scientists who use toads for medical or other experimental purposes should treat them in a humane way. Unfortunately this is not always observed; many persons seem oblivious of the fact that all animals, especially vertebrates, are sentient beings, subject to fear and pain.

It is not a good plan to keep toads indefinitely. They are nocturnal, they only take live food and they are heavy feeders. There are only two ways of providing for them. The first is to install them in a large place and to use lights at dusk to attract insects and devise some means of making them fall to the ground; this is not easy. The other way is to try to induce captive toads to eat newborn or even adult, mice. It is a diet only suitable for large toads and it is hard on the adult mice too. The best plan is to retain captive toads only for short periods and then set them free in suitable surroundings where they can survive and live out their lives peacefully.

6. Distribution

The genus *Bufo* has a very wide distribution. It is cosmopolitan, except for Australia, New Zealand, most of the Pacific Islands, and Madagascar. Some species of *Bufo* have been recorded from islands near their mainland distribution and a few forms have been artificially introduced by human agency into distant lands.

a. Range. Some species are successful and have acquired a very ample range. This is the case of the palaearctic species *Bufo bufo*; besides occupying a large part of Europe it has formed a number of subspecies in Asia; Okada (1931) has even described a series of geographic races from different islands of Japan. *Bufo marinus*, the large neotropical toad, also occupies a vast territory, beginning with most of South America, and extending far north into Central America and Mexico, even reaching the United States in Texas. It is also found in the Antilles, not only on the larger islands like Hispaniola, Jamaica, and Puerto Rico, but also in some of the lesser Antilles —St. Kitts and the Bermudas, for instance. During the Second World War *B. marinus* was carefully observed in the Mariannas by A. G. Smith (1949), a herpetologist on Guam. Much of its island and overseas territory is due to artificial introduction, often in the interests of sugar plantations, as in Hawaii. It is not always clear whence the first toads came, and this is regrettable as in cisandine South America there are three very well-defined forms, *Bufo marinus marinicus* in the north, *B. m. ictericus* in the Atlantic area of Brazil, and *B. m. paracnemis* in the continental depression, in Argentina, Brazil, Paraguay, and eastern Bolivia. These are considered subspecies by Lorenz Mueller and the author. full species by Lutz and Cei. Other examples of true toads with very wide ranges are *Bufo regularis* in Africa and *Bufo melanostictus* in Asia.

   b. Climatic Factors. Some toads are very tolerant of environmental differences, due not only to latitude and longitude but also to altitude. Thus, *Bufo m. ictericus* is found from the seaboard as far up the Brazilian coastal Sierras as small ponds for breeding occur; at Itatiaia this means over 2200 meters altitude. It seems to prefer the montane environment to that of the plains. In Argentina, *Bufo arenarum*, another large toad, goes all the way from the coastal plains to relatively high places like Mendoza. In Brazil it is found only at sealevel, in the most southern state. In the mountains of Minas Gerais it is supplanted by another, very similar form, *B. rufus.* According to Gallardo, the latter also occurs in the Argentine Misiones.

   Other species are strictly tied to greater altitudes. The andine toad *Bufo spinulosus* serves to illustrate this point. Guenther and Boulenger described a number of other andine toads from Peru and Ecuador. Perhaps the most noteworthy are the Asiatic forms, *Bufo raddei* and *B. himalayanus.*

   Some toads will tolerate semi arid or even quite arid conditions. One of these is the Asiatic species *Bufo raddei*, which is considered synonymous with *B. himalayanus* or *B. tibetanus* by some authors. *Bufo alvarius* of the southwestern United States and Mexico is also considered a desert toad. *Bufo m. paracnemis* thrives in climates which are not exactly desert like but have a very long dry season, including the northeastern subarea which is occasionally subjected to drought. Other forms are denizens of the forest. This is the case of the neotropical *B. guttatus,* of *Bufo typhonius,* and apparently also of *Bufo superciliaris* from Africa.

   The name of *B. boreus halophilus* suggests a preference for salt environments although this may be a simplified view. *Bufo variegatus* of Chile occurs at Magallanes, almost at the tip of Continental South America and as far south as anurans occur. *B. granulosus* and *B. arenarum* live on the dunes of Rio Grande do Sul.

   c. Adaptability. Noble (1931) points out the relatively well-developed lungs of *Bufo* compared to those of other salientians or tailless amphibians. The glandular skin must also be of some use. These factors may contribute to the success of many species of *Bufo,* and to their ability to establish themselves and form large populations once they are introduced. However, the early stages of *Bufo* are not well protected. Enormous toads will spawn in quite small sheets of water, or even ditches, releasing thousands of eggs. The larvae hatch at a very rudimentary stage and for some time they form dense clusters which make them an easy prey for their enemies; even the largest forms metamorphose at an incredibly small size. The toll taken at every stage must be very heavy. It seems likely that the toads that manage to conquer and to occupy large areas, at different altitudes, are those whose larvae are the most tolerant of differences in their environment. Larval exigencies are probably an important limiting factor in the expansion of forms with a restricted range.

d. Checklist. This is presented below because persons interested in toads as animals, or as scientific subjects, may wish to know which forms occur where they live and work.

A study of the checklist will also show that besides the widespread and well-known forms, a great many kinds of *Bufo* have been described on the basis of very few specimens, often of only one, the type, and are also only known from the place where they were originally found, i.e., the type locality. Some of these forms are very characteristic and one or two specimens suffice to show that they are true species. Others are much less well defined and leave one in doubt as to whether they are really new or only somewhat aberrant individuals, or populations, of widespread and well-known forms.

For correct naming of toads it is preferable to resort to a competent herpetologist. This may not be necessary in countries where zoological study is advanced and the fauna is known, as in Europe or North America. Any competent amateur or professional zoologist may then be able to name the specimens correctly.

The checklist was compiled on the basis of the regional and general lists of species contained in the following publications: Boulenger (1882), Bourret (1937), Cei (1962), Cochran (1955), Dickerson (1920), Gorham (1957, 1962), Liu (1950), Lutz, A. (1934), Nieden (1923), Okada (1931), Pasteur and Bons (1959), K. Schmidt (1941), M. Smith (1951), Smith and Taylor (1948), Stewart (1967), Stuart (1963), Taylor (1952), Vellard (1959), and A. H. Wright and A. A. Wright (1949).

Checklist of Bufo Laurenti, 1768

Canada, U.S.A.

Bufo alvarius Girard, 1859
   Colorado, Arizona, Southern California, valley of the Colorado River, Imperial Valley, into Mexico
Bufo americanus americanus (Holbrook), 1836
   Eastern North America, Hudson Bay to Alberta & Labrador; Maine, through all the Atlantic states
Bufo americanus charlesmithi Bragg, 1954
   Oklahoma dwarf toad
Bufo americanus copei Yarrow and Henshaw, 1878
   James Bay, northern Ontario
Bufo boreas boreas Baird and Girard, 1852
   Columbia River and Puget Sound; Northeast California, Colorado and Western Montana, British Columbia; Southwest Alaska

Bufo boreas exsul Myers, 1942
  Deep Springs Valley, Inyo County, California
Bufo boreas halophilus (Baird and Girard), 1853
  Most of California (not northeast), into high Sierra Nevada about Lake
  Tahoe (not in desert), western Nevada; Baja California
Bufo boreas nelsoni Stejn., 1893
  Nevada, restricted range
Bufo canorus Camp, 1916
  High central Nevada to Fresco County, California, Yosemite
Bufo cognatus (Say), 1889
  West of Mississippi River, Texas to Arkansas, Western states, Imperial
  Valley California; Mexico
Bufo debilis debilis Girard, 1854
  Kansas east of 100th meridian; Mexico
Bufo debilis insidior Girard, 1854
  New Mexico, southeast Arizona
Bufo hemiophrys Cope, 1886
  Alberta, Manitoba, Northwest Territory; North Dakota
Bufo houstonensis Sanders, 1953
  Central Prairie, near Houston, Texas
Bufo marinus (Linnaeus),
  Enters U.S.A. from the south
Bufo microscaphus microscaphus Cope, 1866
  Utah, Arizona, eastward to New Mexico, into Mexico (Durango)
Bufo microscaphus californicus Camp, 1915
  Southern California and northern Baja California
Bufo punctatus Baird and Girard, 1852
  Western states, Kansas to Utah, Nevada, Colorado, deserts of California
  to Mexico (Hidalgo, Jalisco)
Bufo quercicus Holbrook, 1840
  North Carolina to Alabama and Florida, west to Louisiana
Bufo retiformis Sanders and Smith, 1951
  Southwest Arizona to Mexico (northwest Sonora)
Bufo speciosus Girard, 1854
  Western Texas to northeast Mexico
Bufo terrestris (Bonnaterre), 1789
  Lowlands of Northern Carolina to Florida, along the Gulf to Mississippi,
  Louisiana
Bufo valliceps Wiegmann, 1933
  Louisiana, eastern Texas to Mexico (Chiapas)
Bufo woodhousi woodhousi Girard, 1854
  Texas, Kansas, Nebraska, Missouri, western Iowa, Montana, Idaho,
  western Washington, eastern Oregon, south to Nevada, northern Arizona

Bufo woodhousi australis Shannon and Lowe, 1955
  South Tucson, Arizona
Bufo woodhousi fowleri Hinkley, 1882
  New England, New York, southern Georgia to Missouri, Michigan,
  and southeast Iowa; Atlantic and Gulf coastal plains to central Texas
Bufo woodhousi velatus Bragg and Sanders, 1951
  Eastern Texas, Oklahoma, McCurtain Co.

Mexico

Bufo alvarius Girard, 1859
  Northwest Sonora etc. into U.S.A.
Bufo bocourti Brocchi, 1859
  Mexico; Guatemala
Bufo canaliferus Cope, 1877
  Mexico, Pacific slopes of Oaxaca and Chiapas, Tehuantepec to Guate-
  mala, possibly to El Salvador
Bufo cavifrons Firschein, 1950
  Volcán San Martin, San Andres, Tuxtla
Bufo coccifer Cope,
  Oaxaca, Guerrero (f. Taylor and Smith)
Bufo cognatus Say, 1823
  Chihuahua, Coahuila, Durango
Bufo cristatus Wiegmann, 1833
  Distrito Federal Veracruz, Oaxaca, Pacific slopes of Chiapas
  Synonymous with *valliceps*?
Bufo debilis debilis Girard, 1854
  Tamaulipas, Matamoros; into U.S.A.
Bufo debilis insidor Girard, 1854
  North and central Mexico, apparently widespread
Bufo gemmifer Taylor, 1939
  Guerrero, El Limoncito, near La Venta
Bufo horribilis Wiegmann, 1833
  Very widespread, including coastal Sierras, Michoacán (Considered
  synonymous with *B. marinus* by some authors; perhaps a race of it.)
Bufo kelloggi Taylor, 1936
  Nayarit, Acaponeta, 2 miles east of Mazatlán
Bufo marmoreus Wiegmann, 1833
  North Atlantic drainage, Pacific slope to Tehuantepec
Bufo mazatlanensis Taylor, 1939
  Nayarit, Tepec, 2 miles east of Mazatlán (*B. nayaritensis* is synonymous
  with this, fide Porter)
Bufo microscaphus microscaphus Cope, 1866
  Mexico into U.S.A.

Bufo microscaphus californicus (Camp), 1915
Baja California, Salinas Valley to Northern Sierra San Pedro Mártir
Bufo monksiae Cope, 1879
Guanajuato, Michoacán
Bufo occidentalis Camerano,
Michaoacán, Sierra de Coalcamán, 3,000–6,400 feet (*B. simus* is considered synonymous with this by Firschein)
Bufo perplexus Taylor, 1943
Guerrero, Balsas River near Mexcala, Pacific slopes to Chiapas
Bufo punctatus Baird and Girard, 1852
Baja California, Tamaulipas, Coahuila, Nuevo Leon, Chihuahua, S. Luis Potosi, Sinaloa, Sonora, Guanajuato
Bufo retiformis Sanders and Smith, 1951
Northwest Sonora
Bufo speciosus
Northeast and central Mexico, D.F. Veracruz, recorded from elsewhere
Bufo tecanensis P. W. Smith, 1952
Bufo valliceps Wiegmann, 1833
Mexico, U.S.A.

Central America

Bufo bocourti Brocchi, 1877
Guatemala; Mexico
Bufo canaliferus Cope, 1877
Guatemala, possibly to El Salvador
Bufo coccifer Cope, 1866
Costa Rica; Mexico, Guatemala, Tehuantepec, Oaxaca
Bufo coniferus Cope, 1862
Costa Rica, Atlantic and Pacific drainages; Nicaragua; ?Colombia, ?Ecuador
Bufo haematiticus Cope, 1862
Costa Rica, drainage to Pacific Ocean and to Caribbean Sea
Bufo ibarrai Stuart, 1954
Guatemala, Dept, Jalapa, alt. 1,000 meters Volcan de Jimay
Bufo luetkenii Blgr., 1891
Costa Rica; southern Guatemala; rare
Bufo marinus
Bufo melanochlorus Cope, 1875
Costa Rica
Bufo politus Cope, 1862
Nicaragua

Bufo simus O.Schm., 1858
  Panama, Chiriqui; Mexico, Colima, Sierra Madre
Bufo tecanensis Smith, 1952
  Western Guatemala; Mexico
Bufo valliceps valliceps Wiegm., 1833
  Northern Guatemala; British Honduras; Mexico; U.S.A.
Bufo valliceps wilsoni Stuart and Taylor, 1961
  Jacaltenango, Huehuetenango

Antilles

Bufo dunni Barbour, 1926
  Cuba, Cumanayagua, Mina Carlota, Guatemala
Bufo empusus (Cope), 1862
  Cuba
Bufo longinasus Stejn., 1905
  Cuba
Bufo peltocephalus Bibr. (Mss in Tschudi, 1838)
  Cuba
Bufo ramsdeni Barbour, 1914
  Cuba
Bufo guentheri Cochran, (*B. gutturosus* nec Daudin)
  Hispaniola
Bufo lemur (Cope), 1868
  Puerto Rico

Cisandine (E.) South America

Bufo andersoni Melin, 1941
  Taracuá, Uapés, Brazil
Bufo arenarum Hensel, 1867
  Continental and Platine areas, Southern Brazil; Uruguay; Argentina
Bufo blombergi Myers and Funkhouser, 1951
  Colombia. Closely allied to *B. guttatus, B. glaberrimus*
Bufo ceratophrys Blgr., 1882
  Ecuador, eastern face of Andes; Venezuela? Brazil
Bufo crucifer Wied, 1821
  Brazil, very wide range probably to adjacent republics
Bufo dapsilis Myers and Carvalho, 1945
  Bom Jardim, near Benjamin Constant, Brazil near Peru
Bufo diptychus Cope, 1858
  Paraguay

Bufo granulosus granulosus Spix, 1824
Brazil; Guianas, Venezuela, Trinidad, possibly still wider range
Bufo granulosus d'orbignyi D. & B., 1841
Argentina; Uruguay; southern Brazil
Bufo granulosus fernandezae Gallardo
Argentina, Uruguay
Bufo glaberrimus Gthr., 1868
Colombia, Ecuador, Venezuela (possibly an altudinal form of *guttatus*, f. Rivero)
Bufo guttatus Schneider, 1799
Brazil, Guianas, Venezuela, possibly southeast Colombia
Bufo manicorensis Gallardo, 1961
Brazil, Manicoré, rio Madeira
Bufo marinus L. 1734
Neotropical, widespread in Central and South America, reaching Mexico and U.S.A. Introduced into many places, including the Philippine Islands and the Mariannas
Bufo m. marinus (L.) 1734
Amazonian Hylaea: Brazil, Guianas, Venezuela, etc.
Bufo m. ictericus Spix, 1824
Atlantic area of Brazil, less common on coastal lowlands than in the Sierras
Bufo m. paracnemis Lutz, 1925
Argentina, Bolivia, Continental and Northeastern areas of Brazil. Very widespread
Bufo m. mertensi Cochran, 1950
Brazil, Sta. Catarina, Nova Teutonia. (Possibly an aberrant population of *m. ictericus*)
Bufo melini Andersson, 1945
Hylaea: Brazil, Amazonas, Taracuá
Bufo missionum Berg, 1896
Argentina, Misiones
Bufo ocellatus Gthr., 1858
Central Brazil
Bufo pygmaeus Carvalho and Myers,
Brazil, state of Rio de Janeiro, S. João da Barra, Marambaia (possibly a small race of *B. granulosus*)
Bufo rufus Garman, 1876
Brazil, Continental area: mountains of Minas Gerais. Argentinian Misiones, f. Gallardo (possibly a subspecies of *arenarum*)
Bufo schneider Werner, 1894
Paraguay

Bufo sternosignatus Gthr., 1858
   Venezuela, Colombia
Bufo typhonius (L.), 1734
   Northern South America

Andes and Pacific Coast

Bufo chanensis Fowler, 1913
   Western Ecuador
?*Bufo*?*coeruleocellatus* Fowler, 1913
  Western Ecuador
Bufo coeruleostictus Gthr., 1858
   Ecuador
Bufo cophotis Blgr., 1900
   Peru 2,100–2,700 meters alt.
Bufo fissipes Blgr., 1903
   Peru, 1,800 or 1,950 meters alt.
Bufo intermedius Gthr., 1858
   Andes of Ecuador
Bufo leptoscelis Blgr., 1912
   Southwest Peru, 1,950 meters alt.
Bufo molitor Tschudi, 1845
   Peru? (Synonymous with *poeppigei*?)
Bufo ockendeni ockendeni Blgr., 1902
   Peru; Bolivia
Bufo ockendeni inca Stejn., 1913
   Peru, Huadquinia, 1,500 meters alt. Subspecies of *ockendi* f. Gallardo
Bufo poeppigei Tschudi, 1845
   Peru, Columbia; Ecuador (Small form of *B. marinus* group, f. Vellard)
Bufo quechua Gallardo, 1961
   Bolivia, Depto. Cochabamba, Incachaca
Bufo spinulosus spinulosus Wiegmann, 1835
   Chile, mountains of Antofagasta; Bolivia, altiplano; Peru, southern Andes, Cuzco to 5,000 meters alt.
Bufo spinulosus altiperuvianus Gallardo, 1961
   Bolivia, Oruro (Alto Peru), Challapata 3,700 meters alt.
Bufo spinulosus arequipensis Vellard, 1959
   Peru, Arequipa region and some eastern valleys
Bufo spinulosus arunco Garnot and Lesson, 1826
   Chile, Cordillera Central, mountain valleys
Bufo spinulosus atacamensis Cei, 1938
   Chile, rio Huasco, Vallemar, desert of Atacama, Antofagasta

Bufo spinulosus flavolineatus Vellard, 1959
    Peru, central Andes
Bufu spinulosus limensis (Werner), 1892
    Peru, western valleys, Panama to Trujillo
Bufo spinulosus orientalis
    Peru (upper Marañon)
Bufo spinulosus rubropunctatus Guichénot, 1848
    Southern Chile, Valdivia, Chile
Bufo spinulosus trifolium (Tschudi), 1845
    Peru, Eastern Andes, high valleys
Bufo variegatus Gthr., 1870
    Southern Chile, Cordillera de Valdivia; Argentina; Patagonia, Neuquen, Magallanes

Europe

Bufo bufo bufo (L.), 1754
    Most of central and northern Europe. Subspecies in Asia.
Bufo bufo spinosus Bocage, 1867
    Southern Europe, Spain, Portugal
Bufo calamita Laur., 1768
    Western species: Iberian Peninsula; Britain, Ireland; Switzerland, Belgium; Holland, France, Germany, Denmark, southern Sweden, western Russia, Poland; Czechoslovakia. In Alps and Jura respectively up to 3,500 and 4,000 feet alt.
Bufo viridis viridis Laur., 1768
    Central and southern Europe, except Iberian Peninsula, most of France, Belgium, Britain to Denmark, southern Sweden, central Russia. Into Asia and North Africa

Asia

Bufo andersonii Blgr., 1871
    India; ?Arabia
Bufo andrewsii Schmidt, 1927
    China, Yunnan
    (Not very characteristic, fide Pope)
Bufo asper Grav., 1829
    Eastern India; Malaya, sealevel to 4,500 feet; southern Burma; Peninsular Thailand; Borneo; Java; Sumatra
Bufo bankorensis Barbour, 1908
    Central Formosa, montane
Bufo bedommii Gthr., 1875
    India, Malabar, Travancore

Bufo biporcatus Grav., 1829
Malayan Peninsula into Java; Borneo ?Lumbok
Bufo brevirostris Rao, 1937
Hassan, Mysore, India
Bufo bufo asiaticus Stdnr., 1867
China (f. Pope), Mongolia, Manchuria, Manchukuo, Amur, Korea (f. Okada)
Bufo bufo formosus Blgr., 1883
Formosa
Bufo bufo gargarizans Cantor, 1842
China; Japan Riu-Kiu Islands, Miyako; Korea
Bufo bufo hokkaidoensis Okada, 1927
Japan: Hokkaidô
Bufo bufo japonicus Schlegel, 1838
Honshô, southern part; more widely distributed than *B. b. formosus* (f. Okada)
Bufo bufo minshanicus Stejn., 1926
China: Choni, Kansu, Sungpan, Szechwan, 8,000–10,000 feet
Bufo bufo miyakonis Okada, 1931
Japan, Miyakojima, small island of Riu-Kiu group
Bufo bufo yakushimensis Okada, 1927
Japan, Yakushimi Island south of Kyûshû
Bufo bufo wrighti Schmidt and Liu, 1940
Western China, Szechwan mountains, 2,500–9,000 feet below Sungpan, Chêtotang, Kangting
Bufo burmannus Andersson, 1939
Northeast Burma, Kambarti, near China
Bufo dhufarensis Parker, 1931
Southeast Arabia
Bufo fergusonii Blgr., 1892
Ceylon, Travancore
Bufo galeatus Gthr., 1864
Cambodia
Bufo himalayanus Gthr., 1864
Himalaya, Nepal, Sikkim, Darjeeling, etc.
Bufo hololius Gthr., 1875
India, Malabar
Bufo kelaartii Gthr., 1858
Ceylon
Bufo latastii Blgr., 1882
India, Malabar
Bufo luristanicus Schmidt, 1952
Iran, Shah Bazab, Luristan

Bufo macrotis Blgr., 1887
    Burma, Thailand
Bufo melanostictus Schneider, 1799
    Cambodia; Siam; Java; S. China; Indochina; Formosa; India north to
    Himalayas; Burma; Malaya; Thailand; Borneo; Philippine Islands;
    Mauritius (imported)
Bufo microtympanum Blgr., 1882
    India, Malabar
Bufo nouettei Mocq., 1910
    Central Asia; western China
Bufo olivaceus Blanford, 1874
    Baluchistan, Iran
Bufo pageoti Bourret, 1937
    Fan-si Pan (2,500 meters), ex-French Indochina
Bufo parietalis Blgr., 1882
    India, Malabar
Bufo parvus Blgr., 1887
    Malaya; Malacca; western Burma; southern Thailand
Bufo persicus Nikolsky, 1900
    Southeast Persia (considered synonymous with *B. viridis viridis* by For-
    cart)
Bufo pulcher Blgr., 1882
    Brahmagiri Hills, India
Bufo quadriporcatus Blgr., 1887
    Malaya; Sumatra; Borneo
Bufo raddei Strauch, 1876
    Northwest China, Amur to Tibet, Manchoukuo, Jehol, desert toad
Bufo sachalinensis Nikolski, 1095
    Japan, Sakhalin Island only (fide Okada)
Bufo sikkimensis (Blyth), 1855
    Himalayas (=*himalayanus*?)
Bufo stomaticus Luetk., 1863
    India
Bufo stuartei Smith, 1929
Bufo surdus Blgr., 1891
    Baluchistan; southeast Iran
Bufo tibetanus Zarevsky, 1925
    Tibet, Kham Plateau; western China, high plateau (synonymous with
    raddei, fide Pope)
Bufo tuberculatus Zarevsky, 1925
    Tibet, Kham Plateau (synonymous with *B. b. gargarizans* fide Pope and
    Boring)

Bufo viridis viridis Laur., 1768
Southeast Iran; northeast Irak; Syria; northern Mesopotamia; Arabia
Petrea; perhaps synonymous with next form
Bufo viridis arabicus Heyden, 1827
Arabia; Iran
Bufo viridis orientalis Werner, 1896
Southern Hedjaz, fide Parker

Africa

Bufo amoenus Rochebr., 1884
Senegambia
Bufo angusticeps A. Smith, 1841
South Africa, Cape Peninsula
Bufo anotis Blgr., 1907
Mashoanaland; Tanganyika
Bufo berghei Laurent, 1950
Rwanda-Burundi-Kivu (Congo)
Bufo blanfordii Blgr., 1882
Ethiopia; Abyssinia; British Somaliland
Bufo brauni Nieden, 1910
Usambara, Tanganyika; Kenya
Bufo brevipalmatus Ahl., 1924
Somaliland
Bufo buccinator Rochebr., 1884
Senegambia
Bufo buchneri Peters, 1882
Congo region
Bufo camerunensis Parker, 1926
Gold Coast to Ituri (f. Loveridge) Nigeria, Congo forest form, common
in the Cameroons
Bufo carens A. Smith, 1849
East and South Africa; Bechuanaland; Rhodesia; Tanganyika
Bufo chappuisi Roux, 1936
East Africa
Bufo chevalieri Mocq., 1908
West Africa, Ivory Coast
Bufo cristiglans Inger, 1961
Sierra Leone, southern Kambui Forest Reserve
Bufo chudeaui Chabanaud, 1921
Bufo dodsoni Blgr., 1895
Western Somaliland

Bufo dombensis Bocage, 1895
Dombé, Benguella, Angola; southwestern Africa, Twyfelfontein west of Kakaoveld
Bufo esgoodi Loveridge, 1932
Ethiopia, Usandawi
Bufo fenouilheti fenouilheti Hewitt and Meth., 1913
Southwest Transvaal
Bufo fenouilheti damaranus Mertens, 1954
Southwest Africa, Damaraland, Ombigomatemba, Erongo; dwarf form
Bufo funereus funereus Bocage, 1866
Duque de Bragança, Angola; Uganda
Bufo funereus fuliginatus De Witte, 1932
Katanga (Congo)
Bufo funereus gracilipes Blgr., 1899
Spanish Guinea
Bufo garmani Meek, 1897
Somaliland (probably synonymous with *B. regularis*)
Bufo gardoensis Scortecci, 1933
Somalia
Bufo gariepensis A.Am., 1849
Southwest Africa southern part, central district, Cape Province
Bufo granti Blgr., 1896
South Africa
Bufo hoeschi Ahl,
Southwest Africa Damaraland, Okahandja Kaiser Wilhelm Berg
Bufo jordani Parker, 1936
Great Namaqualand, Satansplace 1,300 meters south of Vogligrund
Bufo katanganus Loveridge, 1932
Katanga
Bufo kisoloensis Loveridge
Mountain ranges in Central Africa
Bufo latifrons Blgr., 1900
Cameroon, Mt. Cameroon, Idenau. Perhaps synonymous with *polycerus*
Bufo lindneri Mertens, 1955
Dar es Salaam
Bufo lemairii Blgr., 1901
Congo, Pweto, Meru Lake
Bufo lönnbergi lönnbergi L. G. Anderss., 1911
East Africa, Kenya, Mt. Kenya
Bufo lönnbergi nairobensis Loveridge, 1932
Kenya, Nairobi

Bufo lughensis Loveridge, 1932
Italian Somaliland
Bufo maculatus Hallowell, 1854
Liberia to eastern Congo (rain forest)
Bufo mauritanicus Schl., 1836
Northwest Africa, Algiers; Tunis; Morocco
Bufo micranotis micranotis Loveridge, 1925
Tanganyika
Bufo micranotis rondoensis Loveridge, 1942
Tanganyika Rondo Plateau
Bufo mocquardi Angel, 1924
Kenya, Mt, Kenya
Bufo parkeri Loveridge, 1932
Tanganyika
Bufo pentoni J. Anders, 1893
Cameroon to South Arabia
Bufo polycerus Werner, 1897
West Africa, Congo; Gaboon; Cameroon; Uganda
Bufo preussi Matschie, 1893
West Africa, Cameroon, Buea
Bufo pusillus Mertens Sahara south to SW Africa, NE Transvaal, N Zululand
Bufo regularis regularis Reuss, 1834
Egypt; Abyssinia; eastern Cameroon; southern Spanish Guinea; Sierra Leone; Ivory Coast; Kilimanjaro etc., Tanganyika

All Africa except in regions where represented by the following races:

Bufo regularis gutturalis Power, 1927
Bechuanaland
Bufo regularis ngamiensis Fitz Simmons, 1932
Bechuanaland
Bechuanaland, Lake Ngami
Bufo rosei Hewitt, 1926
Cape Peninsula, Muizenberg Mt.
Bufo sauvagii Rochebr., 1884
Senegambia
Bufo sibilai Scortecci, 1929
Erythrea
Bufo somalicus Calabresi, 1927
Somalia
Bufo steindachneri Pfeff., 1893
East Africa, Kenya; Tanganyika, etc.

Bufo superciliaris Blgr., 1887
  Cameroon, Belgium Congo, Rio del Rey
Bufo taitanus taitanus Peters, 1878
  Kenya, Taita; Uganda; Tanganyika; Mafia Island
Bufo taitanus beiranus Loveridge, 1932
  Moçambique, Beira
Bufo taitanus nyikal Loveridge only Nyika Plateau, Nyassaland
Bufo taitanus uzungwensis Loveridge, 1932
  Tanganyika? Uzungwe Mts. Kigogo
Bufo togoensis Ahl. 1924
  Togo
Bufo tradouwi Hewitt, 1926
  South Africa
Bufo tuberculosus Bocage, 1896
  Bechuanaland
Bufo tuberosus Gthr., 1858
  West Africa, Cameroon, Gaboon to Ituri
Bufo uruguensis Loveridge, 1932
  Tanganyika, Urungu
Bufo ushoranus Loveridge, 1932
  Tanganyika, Ushora
Bufo vertebralis A. Smith, 1848
  South Africa
Bufo viridis Laur., 1768
  Tunisia (f. Mertens)
Bufo vittatus Blgr., 1906
  Egypt; Uganda, Entebba

e. Hybridization. Work on hybridization between toads of different species has been carried out for some years especially in North America. A. P. Blair (1941) performed the first hybridization experiments with members of the group of *Bufo americanus*. He obtained male hybrids from *Bufo woodhousei* females and *Bufo americanus* males (also 1955). Cory and Manion (1955) studied hybridization in the genus *Bufo* in the Michigan Indiana area. Thornton (1955) reported hybridization between *Bufo woodhousei* and *Bufo valliceps*. Volpe (1952) found physiological evidence for natural hybridization of *Bufo americanus* and *Bufo fowleri*. He studied the intensity of reproductive isolation between sympatric and allopatric populations of these two species (1955), hybridized *Bufo terrestris* and *Bufo fowleri* experimentally (1959) and wrote on the evolutionary consequences of hybrid sterility and vigor (1960). W. F. Blair (1956a) first reported on the relative survival of hybrid toads.

f. Genetic Compatibility and Incompatibility. Evolution in Bufo. From there

W. F. Blair went on to continued work on genetic compatibility or incompatibility and evolution in the genus *Bufo* at the University of Texas. This work is setting up standards for similar work elsewhere and has shed much light on natural relationships and the evolution of toads. W. F. Blair (1956c) first made a tentative arrangement of the U.S. species of *Bufo* into natural groups, which he reorganized next (1958). He next studied Genetic Compatibility and the Species Groups (1959) with a view to getting new, genetic, evidence of evolutionary divergence and so as to assess the possible role of genetic incompatibility in the complex of isolating mechanisms between different toads. He found that some groups (*B. americanus* and *Bufo boreas* groups) are composed of closely related species whereas others (f.i. *Bufo valliceps*) represent distinct evolutionary lines. Genetic incompatibility may be a postmating mechanism of greater significance when the premating isolation mechanisms break down. In 1961, Blair published continued evidence of high interspecific compatibility among the 6 members of the *Bufo americanus* group and the 3 members of the *Bufo cognatus* group. Various intergroup crosses gave evidence against compatibility between representatives of different groups. Intragroup genetic compatibility in the *B. americanus* group was published in 1963. Qualitative data permitted the conclusion that the potential for genetic exchange exists in the 6 species of the group. Quantative data suggest that one species, *Bufo woodhousei*, shows greater incompatibility than the others. This is interesting because *B. woodhousei* can exist together with 4 of the others in nature. In 1964 work was done on the *Bufo boreas* group. It was carried out on different and more ample lines. Females of *B. boreas* in a nuptial condition were crossed with males of 3 species of the *B. americanus* group and 12 other North American, South American and European species. High incompatibility of the progeny of the crosses with *B. americanus* suggested less close relationship than had been surmised. High degree of incompatibility was demonstrated with the Palaearctic species. Crosses with males of 3 species of the *B. valliceps, punctatus* and *cognatus* groups and the Andean *Bufo spinulosus* showed a high percentage of metamorphosis in the progeny. Morphological evidence suggests closer affinity between the *boreas, americanus* groups and *spinulosus* and the Palearctic toads *Bufo bufo* and *Bufo calamita* than with other North American groups. The similarities in morphology are attributable to parallel selection. Genetic similarity permitting compatibility has not been maintained as shown by the high incompatibility with the European toads. The crossing attempts with the more tropical and lowland group of *Bufo valliceps* (1966) were even more ample. They were made both within the group and with 12 North and Central American species groups, seven South American species, three Eurasian and seven African species of *Bufo*. High compatibility with *Bufo arenarum* connects the group with the South American radiation of *Bufo*; high compatibility with the *B. marmoreus* and *cognatus* groups connects it with the North American radiation. The intragroup work lead W. F. Blair to postulate two time levels in the differentiation of the *Bufo valliceps* group: separation into eastern and western segments, isolated by the barrier of the Mexican highlands; formation of the present allopatric species after isolation.

The more general papers by W. F. Blair are of special interest. They deal with nonmorphological data in anuran classification (1962), evolutionary relationships in North American toads (1963b) and on evolution at populational and intrapopulational levels (1964). These papers stress biological facts that have to be tested for their applicability elsewhere. For instance, in 1963 it was shown that most of the species groups reflect the geologically more recent speciations. Evidence gathered points out two major evolutionary lines of *Bufo* in North America. The existence of Asiatic counterparts of these two lines suggests that the separation between them occurred prior to the entrance of the genus *Bufo* into North America, possibly in the Oligocene. Essential restriction of one of these lines, both in North America and in Eurasia, to the temperate climates and similar restriction of the other line to tropical and subtropical climates suggests that the fixation of climatic adaptation of these lines dates far back in the evolutionary history of the genus. A majority of the North American species can be definitively associated with one or the other of the two lines. All evidence, derived from morphology, call-structure, genetical compatibility and geographical distribution is in agreement as to relationships although there are various instances of evidence of convergence or of parallel evolution between lines. The evolutionary history of the genus *Bufo* can be reconstructed with a considerable degree of confidence as regards the North American better-known groups of species. Work on similar lines will have to be undertaken elsewhere so as to bring about similar knowledge of the genus in the other regions where *Bufo* occurs.

B. Allied Genera

The true toads, belonging to the type-genus *Bufo*, are mostly similar enough to be correctly allocated taxonomically. However, some forms occur in the southern tips of Asia, in the former Dutch East Indies, and in groups of islands in the Pacific Ocean, which are now placed in allied bufonid genera, such as *Ansonia, Ophirophryne, Pedostibes, Pelophryne,* and *Pseudobufo*. A number of bufonid genera have also been created or separated off from *Bufo* in Africa, for instance *Didynamipus, Nectophrynoides, Werneria,* and *Wolterstorffina*.

Among the neotropical terrestrial anurans with a glandular skin and a toadlike habit, rather than a froglike outer appearance, there are a few genera that have been repeatedly moved around and mostly lumped together with others that present very different phenotypic characters. Some authors call them Atelopodidae, others Brachycephalidae. Unfortunately, anatomical study of these little frogs lags very much behind. They do not show the trend toward rudimentation of the digits which is characteristic of *Atelopus, Dendrophryniscus,* and *Brachycephalus.* They present a certain specialization of the vertebral column, with a tendency to fusion of elements and also, in some cases at least, a condition intermediate between the arciferal and the firmisternal pectoral girdle. They need to be carefully examined as to certain anatomical features, including the presence of Bidder's organs. Despite the deficiencies in knowledge respecting them, they probably have bufonid affinities rather than atelopodid ones. This applies to *Melanophryniscus* Gallardo, which should include the rough-skinned montane forms with ordinary tadpoles of the group *Phryniscus stelzneri, moreirae, fulvoguttatus, tumifrons (pachyrhinus)* etc. and those of the *Atelopus ignescens* group. *Oreophrynella* Boulenger and *Holoaden* Miranda Ribeiro also probably belong in here.

Some bufonid genera present interesting adaptations. *Pseudobufo,* for instance, is an aquatic toad with very fully webbed feet and the nostrils directed upward. *Ansonia,* though small, lays very large eggs and the tadpoles develop in mountain streams. *Nectophrynoides* is the only viviparous genus of tailless anurans known. The eggs develop for a time in the uterus but there is no placenta as in mammals. They are nourished from outside, probably through the vascular fins of the tail. *Holoaden* spawns on the ground, in rough little nests, and one of the parents guards the eggs, hissing when approached. All these genera live in special environments and are of great interest from an evolutionary point of view. Some have very glandular skins and probably poisonous secretions, but the probability of human contacts are so few that, at present, one must consider them as outside the scope of practical work on venomous anurans.

III. OTHER ANURANS WITH POISONOUS SECRETIONS

Most amphibians produce skin secretions that if not poisonous are at least irritating and sometimes sternutatory. Many collectors of frogs have found that when certain species are introduced into a collecting bag containing frogs of other kinds the secretion of the newcomer kills them. On frogs of their own kind the poison acts only if directly injected into the system, which obviously does not occur in nature.

A. Poisonous Frogs of Diverse Families

Rana palustris, the American pickerel frog, is one of these culprits. The Old World *Rana viridis* has an unpleasant odor and produces a burning sensation when handled. The curious Chilean frog *Batrachyla leptopus* also kills the frogs that come in contact with it (Cei, 1962). Cochran (1962) mentions the Australian "catholic frog" *Notoaden bennetti*, which has a cross of warts on its back, and exudes a milky secretion when handled.

   a. Discoglossidae. The toxicity of frogs was first observed in the tailless amphibians of Europe, the earliest to be thoroughly investigated. Gadow (1901) tells that he could keep bell toads, *Bombina*, in the same vivarium with turtles and even crocodiles. Turtles are very voracious but these veered off from the little toads as soon as they had touched them with the tip of the nose. There are a few species and varieties of this genus, all of them dull above but with beautiful coloring on the ventral surface. They belong to the family Discoglossidae, whose tongue is rounded and not free. The vertebrae are opisthocoel, i.e., hollowed behind, the second to the fourth vertebra generally bear very short ribs, an unusual feature in recent anurans. The secretion of the bell toad is accompanied by an acrid odor. Lataste (1876) quoted by Angel, (1947), in a very good chapter on Glands and Venoms, mentions that having dissected a few of these little toads, he was taken with violent fits of sneezing. They recurred for over a month whenever he entered the laboratory and ceased as soon as he left. The most interesting peculiarity of *Bombina* is the habit of turning up its head, hands, limbs, and feet and arching the body when in danger. The brilliant, apesematic, or warning, coloration comes into view and startles the enemy. This attitude is called the "Unkenreflex" from the German name "Unke" for *Bombina*.

   b. Pelobatidae. Other European frogs, of the family Pelobatidae, also have very irritating secretions, so much so that in Germany *Pelobates fuscus* is called the "Knoblauchkroete," meaning the garlic toad. The Pelobatidae have no ribs; their vertebrae are hollow in front or behind. The pupil is vertical. The inner metatarsal tubercle of the feet forms a shovel. According to Cochran (1962) *Pelobates fuscus* only gives out the secretion when roughly handled; she also mentions that the American spadefoot toads, *Scaphiopus*, belonging to this family, secrete a musty and peppery substance.

   The tropical tailless amphibians are not so well known as those from the Palaearctic and Neartic regions and probably contain many more forms with irritating secretions than allowed for.

B. Dendrobatidae

   The best-known neotropical poisonous frogs are the Dendrobatidae. They are small and very agile animals, generally with a dark, black, or indigo, dorsal surface, ornamented by a handsome pattern. They occur in the northern part of South America and in Central America. One species, however, has subspecies in eastern Bolivia and in the mountains of Goyaz and Minas Gerais. This one, *Dendrobates pictus,* has two longitudinal golden stripes on the back, black vermiculations on a pale blue background on the belly, in some races, and an orange-red spot on the concealed upper part of the thigh. The *Dendrobates* are largely diurnal forest dwellers found near little streams. The family, Dendrobatidae, is firmisternal, the pectoral girdle not being movable, an unusual condition in neotropical anurans. The disks of the fingers have two platelike scutes above that give them the look of a cloven hoof. They share this peculiarity and their diurnal habits with the Elosiinae of southern Brazil. They move their larvae about on their backs.

The skin of one of these beautiful little toads, *Dendrobates tinctorius,* is used by Amazonian Indians for dyeing the feathers of parrots. The green feathers are pulled out of the parrots, the toad is warmed to induce it to exude its secretion, and the secretion is rubbed into the feather sockets. The new feathers grow out blue or yellow and the parrots are known as "contrafeitos." The Colombian Indians of the Chaco area also use the secretion of the little frog to poison arrows for killing monkeys and other smallish game. The wretched little victims are impaled on sticks, from mouth to thigh, and the spit is turned over the fire, producing an agonic sweat of poison drops. The poison is collected in little vessels and stored so that the tips of the arrows can be immersed in it. The Indians are said to avoid contact with the poison and to remove a cone of flesh around the point of entrance of the arrow, from the animals killed with it. Posadas Arango, who studied the secretion, extracted an alkaloid from it which he called "batracine" and which paralyzes the victims. He found it innocuous, even in large doses, when taken by mouth, either by himself or by fowls, though the latter succumbed to a very much smaller injected dose.

C. Hylidae

Cutaneous secretions lethal to other frogs and irritating to mammals also occur in neotropical tree frogs of the family Hylidae. They differ from toads by the presence of large disks on the digits and of an intercalary cartilage between the penultimate and the last phalanx, which aids them in climbing by adhesion, one of their chief means of locomotion. The legs are long and permit great leaps. The skin is generally smooth and well lubricated. Many Brazilian species of *Hyla,* the type genus, with the characters above, and the usual life history of frogs, give out a strong odor when handled, often that of crushed plants. They do not generally affect the human skin though they cause irritation and watering of the eyes if inadvertently conveyed to the conjunctiva. Two extremely large forms, which are on the borderline between *Hyla* and other genera, have more virulent secretions which cause a rash like that produced by nettles. They have paired vocal sacs and the skull partly or entirely roofed; both are well over 100 mm long and correspondingly stout.

The first, *Hyla vasta*, from the Antillean island of Hispaniola, is almost a *Trachycephalus*, besides being one of the largest neotropical tree frogs. The Southeastern Brazilian *Trachycephalus nigromaculatus*, which is much smaller, has a very slimy skin and causes a slight burning of the palms and fingers when very much handled, for instance while being photographed.

The other, *Hyla venulosa*, is sometimes put in the genus *Phrynohyas* or *Acrodytes*. It is very widespread, from Mexico to Argentina and eastern Bolivia. There are many forms, most of which are probably geographic races, while a few may have become separate species. The rash caused by handling them has been mentioned in the literature a number of times. The following examples are derived from personal experiences. In the Upper Amazons, my former assistant, Miss G. R. Kloss, caught a specimen of the form of *H. venulosa* called *H. zonata* by Spix. Being much plagued by gnats, she passed a hand over one eye soon after handling the frog. The eye immediately became very much inflamed and she had a violent headache the whole of the next day, I handled several specimens of the form called *H. zonata* by Spix, from Belem do Pará, repeatedly, while they were being photographed; they gave out a rubbery secretion that had to be washed away but caused no irritation. The frogs we got out of the water reservoirs above the water closets in Matto Grosso, *Hyla hebes* Cope, did not cause the slightest irritation, even when I rubbed the secretion on the inside of the lower lip. And yet the bromeliad collectors Mr. and Mrs. Foster, who got some for me in the same region, called them the "india-rubber frog" because the secretion was so abundant that it could be worked into pellets. My first specimen of the form from the Maritime Range, *Hyla imitatrix* Miranda Ribeiro, was dropped from the sky by a bird, after a scuffle with it, and caught by me when it landed, unharmed, after a fall of 10 meters or more. It was put with a few other frogs collected previously. On arriving at the house, I found them all dead and stuck together except the newcomer. Meanwhile, a rash broke out on my hand and forearm. It smarted very much but subsided after about an hour without leaving any trace. Next day, J. Venancio, our assistant, caught another, asleep in a tree fern. On my warning him, he broke out into loud laughter, saying that his hefty Negro hand was quite different from that of a white lady with a delicate skin. Nevertheless, the same rash appeared. Many years later another specimen caught by me in the same place, laid a hindlimb along the back of my hand while struggling. It left a weal that lasted many days but did not burn. I am still uncertain as to whether these differences are regional and subspecific or not. I am inclined to believe that the secretion is more virulent when the frogs are in the pink of condition and at the nuptial period.

Poisonous secretions also occur in large hylids with a vertical pupil, the habit of spawning outside of water in folded leaves, a trend toward a grasping foot and a monkeylike manner of climbing by grasping the support. These frogs belong to the genera *Phyllomedusa* and *Pithecopus*. They are green above, with bright flash-colors on the hidden surfaces. *Phyllomedusa bicolor,* the largest of the species, is well over 100 mm long and lives in the Amazonian Hylaea. On her return from there Miss Kloss informed me that the Indians used this frog to induce vomiting as a relief from gastric discomfort, for instance, after a bout of drinking. The frog, which has very long and thin dorsolateral parotoids, is rubbed on the inside of the wrists. Later, another indication was offered me by Mr. A. L. Carvalho, from a verbal account given by a German anthropologist, Kurt Nimuendaju, who lived long among the Amazonian Indians, adopted a tribal name, and took an Indian wife; Nimuendaju told him, and later published a short note, about the use of *Phyllomedusa* poison by the Tukuna Indians. The Phyllomedusas are caught and kept in little cages like birds. When their services are needed, shallow excoriations are made on the inside of the wrists and on the temples of the patient and the parotoid region is rubbed on them. This induces not only vomiting but a general catharsis, which is interrupted when convenient by a plunge into the river. Phisalix (1922) mentions that the secretion is soapy and soluble in water.

This observation brings us to the end of the chapter. Its subject matter shows that recent amphibians are small and nonaggressive animals. Their secretions have a purely defensive role and are not brought into action unless provoked. Under the circumstances, it hardly seems fair to call them active venomous animals. To do so is but another instance of the old French proverb that says: "Cet animal est très méchant, si on l'attaque, il se défend."[1]

For the author's interpretation of the biological significance of cutaneous secretions in toads and frogs see Lutz, B. (1966).

ACKNOWLEDGMENTS

Many thanks are to Drs. Charles M. Bogert and Richard Zweifel for help with the lists of North American and of Mexican toads and to Dr. R. Laurent with that of the African ones; also to Mr. Antenor Leitão de Carvalho for other bibliographical help and interesting information.

REFERENCES

Angel, F. (1947). "Vie et moeurs des amphibiens." Payet, Paris.
Aronson, L. H. (1944). *Amer. Mus. Novitates* **1250**, 1–15.
Blair, A. P. (1941). *Genetics* **26**, 398–417.
Blair, A. P. (1955). *Amer. Mus. Novitates* **1722**, 38.

[1] This is a wicked beast; if attacked, it defends itself.

Blair, W. F. (1953). *Copeia* **4**, 208–212.
Blair, W. F. (1954). *Texas J. Sc.* **6**(1), 72–77.
Blair, W. F. (1956a). *Copeia* **4**, 259–260.
Blair, W. F. (1956b). *Texas J. Sc.* **8**, 87–106.
Blair, W. F. (1956c). *Texas J. Sc.* **8**, 350–355.
Blair, W. F. (1957a). *Texas J. Sc.* **9**, 99–108.
Blair, W. F. (1957a). *Copeia* **3**, 208–212.
Blair, W. F. (1958a). *Amer. Nat.* **92**, 27–51.
Blair, W. F. (1958b). *Bull. Ecol. Soc. Amer.* **39**, 75.
Blair, W. F. (1959). *Texas J. Sc.* **11**, 427–453.
Blair, W. F. (1960). *Texas J. Sc.* **12**, 216–227.
Blair, W. F. (1961). *Texas J. Sc.* **13**, 163–175.
Blair, W. F. (1962). *Systematic Zool.* **11**, 72–84.
Blair, W. F. (1963a). *Texas J. Sc.* **15**, 15–34.
Blair, W. F. (1963b). *Evolution* **17**, 1–16.
Blair, W. F. (1964a). *Texas J. Sc.* **16**, 181–192.
Blair, W. F. (1964b). *Quart. Rev. Biol.* **39**, 334–344.
Blair, W. F. (1966). *Texas J. Sc.* **18**, 333–351.
Blair, W. F. and Pettus, D. (1954). *Texas J. Sc.* **6**, 7277.
Bogert, C. M. (1947). *Am Museum Nevitiates* **1355**.
Bogert, C. M. (1960). *Am. Inst. Biol. Sci.,* Publ. **7**.
Boulenger, G. A. (1882). "Catalogue of the Batrachia Salientia s. Ecaudata in the Collection of the British Museum," London.
Boulenger, G. A. (1912). *Proc. Zool. Soc. London* 19–22.
Bourret, R. (1937). Notes herpétologiques sur l'Indochine Française. Annexe au Bull. Général de l'Instruction Publique No. 12.
Brazil, V., and Vellard, J. (1925). Contribuição ao estudo do veneno de batrachios do genero *Bufo. Publ. Brazil-Med.* **20**, No. 9.
Budgett, J. S. (1899). "Budgett Memorial Volume." Cambridge Univ. Press, London and New York.
Cei, J. M. (1949). *Acta Zool. Lilloana* **7**, 527–544.
Cei, J. M. (1962). "Batracios de Chile." Ediciones de la Universidad de Chile, Santiago de Chile.
Cochran, D. M. (1955). *U.S. Natl. Museum Bull.* 206.
Cochran, D. M. (1962). "Living Amphibians of the World." Doubleday, New York.
Cory, L., and Manion, J. J. (1955). *Evolution* **9**, 42–51.
Cott, H. B. (1950, 1957). "Adaptive Coloration in Animals." Methuen, London.
Davis, D. D. (1936). *Field Museum Nat. Hist., Zool. Ser.* 20, No. 15.
Dickerson, M. C. (1920). "The Frog Book. North American Toads and Frogs." Doubleday, New York.
Gadow, H. (1958, repr. of 1901). Amphibia and Reptiles, Cambridge Univ. Press **8**, 668 pp.
Galli-Mainini, C. (1948). *In* "El diagnostico del embarazo con batracios machos" (Impaglione, ed.), Artecnica, Buenos Aires.
Goin, C. J., and Goin, O. B. (1962). "Introduction to Herpetology." Freeman, San Francisco, California.
Gorham, S. W. (1957). *Can. Field Naturalist* **71**, No. 4, 182–192.
Gorham, S. W. (1962). *Can. Field Naturalist* **77**(1), 13–48.
Hanson, J. A., and Vial, J. L. (1956). *Herpetologica* **21**, 141–149.
Hinsche, G. (1923). *Biol. Zentr.* **43**, 16–26.
Liu, C.-C. (1950). *Fieldiana, Zool.* **2**.

Lutz, A. (1934). *Mem. Inst. Oswaldo Cruz.* **28**, No. 1, 111–159 and Plates XIII-XXVII (Portuguese and German texts).

Lutz, B. (1966). *Mem. Inst. Butantan Simp. Intern.* **33**(1), 55–59.

Martof, B. S., and Thomson, E. E. (1948). *Bull. Ecol. Soc. Amer.* **39**, 92 (reprint. *Animal Behavior* **6**, 224 and in *Behaviour* **13**, 243–248.).

Mertens, R. (1959). "La vie des amphibiens et reptiles." Horizons de France, Paris.

Miranda Ribeiro, A. de (1926). *Arch. Museu Nac., Rio de Janeiro* **27**, 22 pls.

Nieden, F. (1923). "Tierreich" **46**.

Noble, G. K. (1931). "The Biology of the Amphibia." McGraw-Hill, New York.

Okada, Y. (1931). Imper. Agri. Expe. Stat. (Japan).

Pasteur, G., and Bons, J. (1959). *Trav. Inst. Sci. Cheriffien, Ser. Zool.* **17**

Philsalix, M. (1922). "Animaux venimeux et venins." Vol. 2, pp. 1–174, Pl. III. Masson, Paris.

Ponse, K. (1926). *Compt. Rend. Soc. Phys. Hist. Nat. Geneve* **43**, 19–22.

Ponse, K. (1927). *Rev. Suisse Zool.* **34**, 217–220.

Rengel, D. (1948). *Acta Zool. Lilloana* **6**.

Rengel, D. (1949). *Acta Zool. Lilloana* **7**.

Savage, R. M. (1934). *Proc. Zool. Soc. London,* 49–98.

Schmidt, K. (1941). *Field Museum Nat. Hist., Zool. Ser.* **22**, No. 8, Publ. 512.

Smith, A. G. (1949). *Nat. Hist. Misc. Chicago Acad. Sci.,* **37**.

Smith, H. M., and Taylor, E. H. (1948). *U.S. Natl. Museum Bull.* **194**.

Smith, M. (1951). "The British Amphibians and Reptiles." Collins, London.

Stewart, M. M. (1967). Amphibians of Malawi.

Stuart, L. C. (1963). *Misc. Publ. Museum Zool., Univ. Mich.* **122**.

Taylor, E. H. (1952). *Univ. Kansas Sci. Bull.* **35**, Part 1, No. 5, 577–942.

Thornton, W. A. (1955). *Evolution* **9**, 455–468.

Vellard, J. (1959). *Mem. Museu Hist. Nat. "Javier Prado,"* Lima.

Volpe, E. P. (1952). *Evolution* **6**, 393–406.

Volpe, E. P. (1955). *Tulane Studies on Zool.* **4**(2), 61–75.

Volpe, E. P. (1959). *Texas J. Soc.* **11**, 335–342.

Volpe, E. P. (1960). *Evolution* **14**, 181–193.

Witschi, E. (1933). *Am. J. Anat.* **52**, 461–515.

Wright, A. H., and Wright, A. A. (1949). Handbook of Frogs and Toads of the U.S. and Canada. Comstock Publ. Co., Ithaca, N.Y.

Chapter 38

The Basic Constituents of Toad Venoms

VENANCIO DEULOFEU AND EDMUNDO A. RÚVEDA

DEPARTAMENTO DE QUÍMICA ORGÁNICA, FACULTAD DE CIENCIAS EXACTAS Y NATURALES, AND
CÁTEDRA DE FITOQUÍMICA, FACULTAD DE FARMACIA Y BIOQUÍMICA, UNIVERSIDAD DE BUENOS
AIRES, ARGENTINA

I. INTRODUCTION

The secretion of the parotoid glands of the toads, usually known as toad venom, contains two principal classes of pharmacologically active constituents. One class is formed by substances belonging to the steroids, the bufadienolides (bufogenins) and their derivatives, the bufotoxins (Chapter 40). To the other class belong different basic compounds. In some cases these compounds have been isolated from excised parotoid glands, in other cases the whole skin, including adhering parotoid and other small glands, has been extracted. Finally, the secretion of the glands has been collected and used as prime material for the extraction.

Although slight differences have been described, for some species, in the bases present in the glands, the skin, or the secretion, we will consider in this chapter the basic substances derived from all three sources and use for them the general denomination of toad venoms.

We also include information on the secretion of the glands used in China, in popular medicine, with the name of Ch'an Su, known in Japan as Senso, and which seems to be prepared from *B. gargarizans*.

II. HISTORY

The presence in the secretion of the parotoid glands of basic substances giving a positive alkaloid reaction was detected by Phisalix and Bertrand in 1893, working with the European toad *Bufo bufo bufo* (*B. vulgaris*). They obtained an amorphous base which was named bufotenine (Phisalix and Bertrand, 1902).

In 1912, Abel and Macht isolated crystalline adrenaline from the secretion of *B. marinus*, a widely dispersed American tropical toad. In 1920, Handovsky crystallized the oxalate and the picrate of a base present in the secretion of *Bufo bufo bufo* and kept for it the name of bufotenine, given earlier by Phisalix and Bertrand to the amorphous preparation.

From that time studies on the distribution, characterization, isolation, and determination of the chemical structure of the basic constituents present in the toad venoms were carried out in several laboratories.

The venoms of the genus *Bufo* contain bases of two different chemical types: (a) derived from phenylethylamine, (b) derived from tryptamine.

It was made clear after several species were studied that although there were differences in the bases that could be isolated from each, one could hardly speak of species specificity.

The systematic application of different methods of chromatography has widened our knowledge of the distribution of the bases in many species of toads but has not clarified our understanding of the biological significance of them, and the role they play in the animals.

Adrenaline was the first crystalline base isolated and for many years the only representative of the bases of phenylethylamine type. That other bases were indolic was suspected in their early chemical studies by H. Wieland *et al.* (1932) and by Jensen and Chen (1932). For the older literature see Tschesche (1945), Deulofeu (1948), and Kaiser and Michl (1958).

III. THE PHENYLETHYLAMINE BASES

A. Introduction

The following bases derived from phenylethylamine have been identified in toad venoms: dopamine (I), *N*-methyldopamine (epinine) (II), noradrenaline (norepinephrine) (III), adrenaline (epinephrine) (IV). All belong to the class of the catecholamines.

(I)

(II)

(III)

(IV)

B. Distribution

The distribution of the catecholamines in venoms from different species of toads is given in Table I.

Only adrenaline (IV) has actually been isolated. It was found to be the

TABLE I

CATECHOLAMINE BASES FOUND IN TOAD VENOMS[a]

Dopamine (I) HCl, mp, 240°–241°C; picrate, mp 189°C (Barger and Ewins, 1910; Wasser and Sommer, 1928).
Identification: *B. marinus* (Märki *et al.*, 1962).

N-Methyldopamine; Epinine (II) mp 186°–187°C, HCl, mp 177°–178°C (Bretschneider, 1947).
Identification: *B. marinus* (Märki *et al.*, 1962).

(—)-Noradrenaline; (—)-Norepinephrine (III) mp 216.5°–218°C $[\alpha]_D^{25°}$–37.3°C (H_2O,1 equiv. HCl); HCl, mp 145.2°–146.4°, $[\alpha]_D^{25°}$–40.0°C (H_2O) (Tullar, 1948).
Identification: Ch'an Su, (Lee and Chen, 1951), *B. marinus*. (Lasagna, 1951; Märki *et al.*, 1962). 0.20 μg per gram of glands (Östlund, 1954).

(—)-Adrenaline; (—)-epinephrine (IV) mp 216°–218°C; $[\alpha]_D^{20°}$–50.7°C (HCl,H_2O).
Identification: *B. marinus*, 5–7% in secretion (Abel and Macht, 1911–1912), 6–11.6% (Fischer and Lecomte, 1950) (Märki *et al.*, 1962). *B. regularis*, 4.6% in secretion (Chen and Chen, 1933). *B. arenarum*, 5.1% in glands, *B. mauretanicus*, 1% (Fischer and Lecomte, 1950). *B. vulgaris*, 3.7 μg per gram of glands (Östlund, 1954). *B. formosus* (S. Ohno and Komatsu, 1957). *B. crucifer* (Pereira and de Oliveira, 1961).
Isolation: *B. marinus*, 4.5% dried secretion (Abel and Macht, 1911–1912); 1.35% *idem* (Slotta *et al.*, 1937), Ch'an Su (Jensen and Chen, 1929; Chen *et al.*, 1931); *B. arenarum* (Deulofeu, 1935), 4 mg/gm dried secretion (Jensen, 1935); *B. regularis*, 3 mg/gm (Jensen, 1935); *B. paracnemis* (Deulofeu and Mendive, 1938).

---

[a]Tables I and II give information on the distribution of catecholamines (Table I) and indolethylamines (Table II) in the genus *Bufo*. Information on species belonging to other genera is sometimes added for comparison. The data include results from skins, excised parotoid glands, and secretions.

Isolation means that the base or at least one derivative, has been isolated in crystalline condition. Under identification, the data quoted have been obtained either by pharmacological methods or, mainly in recent years, by paper chromatography, and in those papers, information of interest, especially about systems of solvents and developing reagents, is found.

same stereoisomer obtained from the adrenal of higher animals: R-(—)-adrenaline (Pratesi *et al.,* 1958).

Noradrenaline (III) has been detected pharmacologically, chemically, and by paper chromatography. Lasagna (1951) determined by chemical reactivity and by paper chromatography that it was present in the old preparation of adrenaline from Abel and Macht (1911–1912) which derived from *B. marinus.* It has been detected in the extract from excised parotoid glands and from the secretion of the same species by Märki *et al.* (1962). Also, in an old preparation of adrenaline from Ch'an Su, Lee and Chen (1951) detected noradrenaline pharmacologically. The method employed verifies also that it is the same stereoisomer found in higher animals: R-(—)-noradrenaline (Pratesi *et al.,* 1959).

N-Methyldopamine (II) and dopamine (I) have been detected by paper chromatography in *B. marinus* (Märki *et al.,* 1962), the former for the first time in animals. It is evident that we can expect the finding of these catecholamines in other species of toads by the use of methods of greater sensibility, like those employed in this particular case.

Adrenaline (IV) and noradrenaline (III) (von Euler, 1956) are widely distributed in animals where they play the role of hormones. Dopamine (I), which is a biochemical precursor of noradrenaline, has also a large distribution.

On the other hand *N*-methyldopamine (II) has only been found in the venom of *B. marinus* and not in higher animals, although in them it can be enzymically transformed into adrenaline (Bridgers and Kaufman, 1962).

With the exception of adrenaline, the remaining catecholamines have also been found in products of vegetable origin. Noradrenaline and dopamine are the most common (Udenfriend *et al.,* 1959). The banana is a fruit containing usually rather large amounts of them (Waalkes *et al.,* 1958). *N*-Methyldopamine has been found in extracts of *Spartium scoparium* (Jaminet, 1959) and seeds of *Vicia faba* (Piccinelli, 1955).

C. Biosynthesis

A considerable amount of work has been done on the biosynthesis and metabolism of the catecholamines in higher animals (for reviews see Axelrod, 1959; Daly and Witkop, 1963; Weiner, 1964).

On the other hand, no data is available on their formation in toads. Pending further experimental work it can be assumed that in toads, the production of catecholamines follows, if not exactly the same, similar paths as those found in other animals, although species differences could be expected.

In higher animals the catecholamines are produced following the paths indicated in Fig. 1.

Phenylalanine (V) and tyrosine (VI) are the starting points of their biosynthesis, an intermediate being 3,4-dihydroxyphenylalanine (dopa, VIII) which on enzymic decarboxylation produces dopamine (I). This in turn can

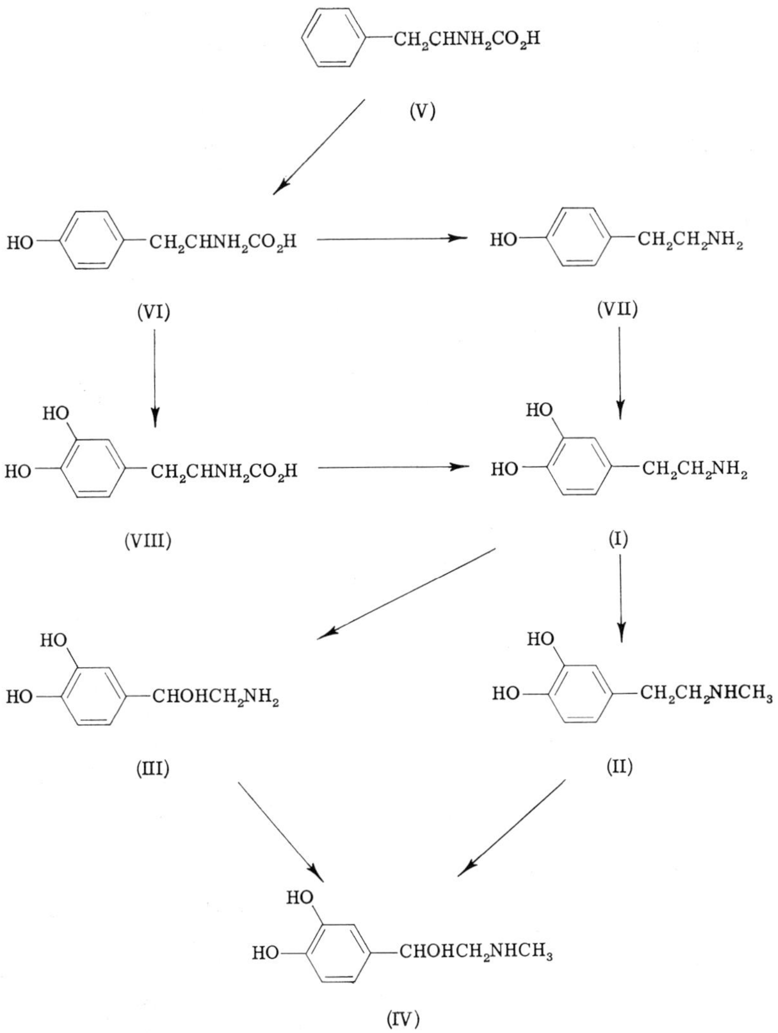

FIG. 1. Biosynthesis of the catecholamines.

be *N*-methylated to *N*-methyldopamine (II) or hydroxylated to noradrenaline (III). Oxidation of *N*-methyldopamine (II) or *N*-methylation of noradrenaline, gives in both cases adrenaline (IV). A minor pathway in the formation of dopamine (I) is tyramine (VII), formed by decarboxylation of tyrosine (VI).

It is interesting that all the four catecholamines (I–IV) involved in the metabolic pathways indicated in Figure I have been found in toad venoms. Besides, the work of Märki *et al.,* (1962) on the venom of *B. marinus,* has shown that it contains two enzymes which play a role in the *N*-methylation of catecholamines. One is a phenylethanolamine *N*-methyl transferase, which catalyses the *N*-methylation of phenylethanolamines by employing *S*-adenosylmethionine as methyl donor. It is similar, although not identical, to an enzyme found in adrenal glands (Axelrod, 1962). The other is an unspecific *N*-methyl transferase, which uses the same methyl donor and methylates a series of phenylethylamines and tryptamine derivatives. It is similar in action to an enzyme described by Axelrod (1961, 1963) present in rabbit lungs.

D. Metabolism

In man and higher animals the metabolism of noradrenaline (III) and adrenaline (IV) follows very similar pathways, the main steps being indicated in Fig. 2. The importance of some of the metabolic steps varies with the species and leads to differences in the main products excreted.

There are two interesting aspects to this metabolism. One is the oxidation of the side chain, initiated by a monoamine oxidase, which in the case of noradrenaline (III) produces an aldehyde; the latter in man, by further oxidation forms mandelic acid (XII) which with its 3-*O*-methyl ether (X) are the main products excreted. In rats, the aldehyde is reduced to the glycol (XIII), which on *O*-methylation produces the ether (XIV), an important metabolite in that species. Vanillic acid (XI), a further product of oxidation, has been found to be also a metabolic product of noradrenaline in man (Rosen and Goodall, 1962).

The other aspect is the enzymic *O*-methylation of the phenolic group at carbon 3. This methylation is catalyzed by a catechol *O*-methyl transferase which is widely distributed in a variety of species and tissues. It can 3-*O*-methylate the catecholamines and many of their metabolites. The methyl group is transferred from *S*-adenosylmethionine and Mg^{++} is needed for its activity (Axelrod and Tomchick, 1958).

It is responsible for the direct or indirect production of several of the metabolites indicated in Fig. 2: 3-*O*-methylnoradrenaline (IX); 4-hydroxy-3-methoxymandelic acid (X); 4-hydroxy-3-methoxyphenylethyleneglycol (XIV) and vanillic acid (XI). (Axelrod *et al.,* 1958a,b). It is also responsible for the *O*-methylation of adrenaline (IV) to 3-*O*-methyladrenaline (XV), the metabolism of both bases being parallel to that of noradrenaline and 3-*O*-methylnoradrenaline, as can be seen in Fig. 2.

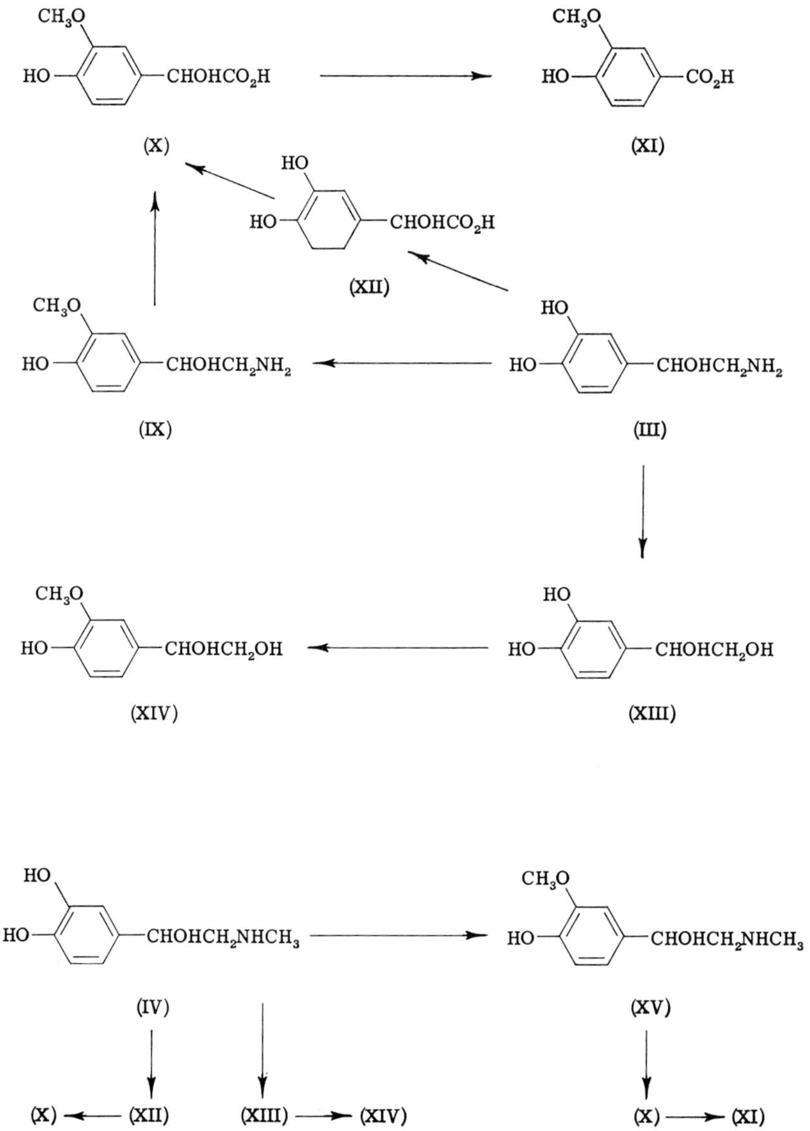

FIG. 2. Metabolic pathways of noradrenaline (III) and adrenaline (IV).

The metabolites of the catecholamines are usually excreted by the urine, although a minor excretion can also take place through the bile (rats). Several of the metabolites excreted are conjugated with glucuronic or with sulfuric acid (Hertting and La Brosse, 1962).

IV. THE TRYPTAMINE BASES AND DERIVATIVES

Introduction

The following bases derived from tryptamine have been isolated or identified with certainty from toad venoms, skins, or excised parotoid glands (Table II): (a) 5-hydroxytryptamine (serotonin) (XVI); (b) N-methyl-5-hydroxytryptamine (XVII); (c) N-methyl-5-methoxytryptamine (XVIII); (d) bufotenine (N,N-dimethylserotonine) (XIX) and its sulfuric acid ester, known as bufoviridine (XX); (e) O-methylbufotenine (XXI); (f) bufotenidine (N,N,N-trimethylserotonin) (XXII); (g) dehydrobufotenine (XXIII) and its sulfuric acid ester bufothionine (XXIV). The two last compounds are not indolethylamine bases but because of their chemical and biological relation to the former it is convenient to consider them together.

TABLE II

TRYPTAMINE BASES FOUND IN TOAD VENOMS

5-Hydroxytryptamine, serotonin, enteramine, thrombocytin, thrombotonin (XVI).
  mp 212°–214°C (Rapport et al., 1948); HCl, mp 167°–168°C (Hamlin and Fischer, 1951). Oxalate, mp 194°–196°C. Picrate, mp 196°–197°C (Harley-Mason and Jackson, 1954). Creatinine sulfate complex, mp 214°–216°C (Speeter et al., 1951). UV spectrum: λ 274 mμ (logϵ 3.77), shoulder 293 (3.63) (H_2O, pH 5.3). At pH 11.9 a small shift is observed in the 274 maximun, the shoulder disappears and a new maximum is observed at 322 mμ (Rapport, 1949; Asero et al., 1952).
  Identification: B. marinus (Udenfriend et al., 1952). B. bufo bufo (Spandrio, 1961), B. americanus, B. arenarum, B. bergei, B. calamita, B. fowleri, B. gargarizans, B. kisoloensis, B. mauretanicus, B. regularis, B. viridis (Erspamer, 1954, 1961). B. alvarius, 4–6 μg per gram of dried skin (Erspamer et al., 1965).
  Isolation: B. arenarum (Frydman and Deulofeu, 1961). In other genera: Xenopus laevis (van de Veerdonk et al., 1961). Bombinator pachypus, B. igneus, Discoglosus pictus, Hyla arborea, H. aurea, Rana esculenta, R. pipiens, R. palustris, R. madagascariensis, R. labrosa, Salamandra maculosa (Erspamer, 1954, 1961).

N-Methyl-5-hydroxytryptamine (XVII). Oxalate, mp 153°–159°C (Stoll et al., 1955).
  Identification: B. bufo bufo, B. americanus, B. calamita, B. fowleri, B. gargarizans, B. marinus, (Erspamer, 1954); B. viridis, (Erspamer, 1959). B. alvarius, 30–40 μg per gram dried skin (Erspamer et al., 1965).

N-Methyl-5-methoxytryptamine (XVIII). HCl, mp 166°–167°C; picrate, mp 220°–221°C (Wilkinson, 1958).
  Identification: B. alvarius, 20–23 μg per gram dried skin (Erspamer et al., 1965).

Bufotenine; N,N-dimethyl-5-hydroxytryptamine (XIX), mp 125°–126°C (Barlow and Khan, 1959; Iacobucci and Rúveda, 1964); 146°–147°C (H. Wieland et al., 1934). Oxalate, mp 178°C (H. Wieland et al., 1932). Monopicrate, mp 177.5°C. Dipicrate, mp 174°C (H. Wieland and Wieland, 1937; Hoshino and Shimodaira, 1935). Picrolonate, mp 120°–121°C (Deulofeu and Berinzaghi, 1946); 183°–184°C (Iacobucci and Rúveda, 1964). Flavianate, mp 130°–131°C (Jensen and Chen, 1932). UV spectrum; λ_{max} 225 mμ

Table II (*Continued*)

(logε 1.35), 280 (3.83) shoulder 303 (3.71) (ethanol). λ_{max} 277 (3.74), 296 (3.67) (0.1 N HCl). λ_{max} 218 (4.37), 376 (3.74), 323 (3.65) (0.1 N NaOH) (Stoll *et al.*, 1955).
Identification: B. americanus, B. calamita, B. fowleri (Erspamer, 1954). *B. viridis,* 630 μg per gram of fresh skin (Erspamer, 1959). *B. crucifer* (Pereira and de Oliveira, 1961).
Isolation: B. bufo bufo (Handovsky, 1920), 510 μg per animal (H. Wieland *et al.*, 1934); 47 μg per animal and 0.3% in dried secretion (males); 90 μg per animal and 0.33% in dried secretion (females) (H. Wieland and Behringer, 1941). Ch'an Su (Jensen and Chen, 1930). *B. arenarum* (Jensen and Chen, 1932); 5.1 mg per dried skin (H. Wieland *et al.*, 1934). *B. viridis* (Jensen and Chen, 1932) 0.06% in fresh skin (Erspamer, 1959). *B. paracnemis* (Deulofeu and Mendive, 1938). *B. chilensis, B. crucifer* (Deulofeu and Duprat, 1944). *B. formosus* (S. Ohno *et al.*, 1961). *B. alvarius,* 0.8–5 mg per gram dried glands, 0.33–2.15 mg per gram dried skin (Erspamer *et al.*, 1965).

Bufoviridine, bufotenine *O*-sulfate (XX), mp 210°–212°C (Erspamer, 1959).
Identification: B. calamita (Erspamer, 1959).
Isolation: B. viridis (Erspamer, 1959).

O-Methylbufotenine; *N,N*-dimethyl-5-methoxytryptamine (XXI), mp 66°–67°C; picrate, mp 176°–177°C; methiodide, mp 183° (Hoshino and Shimodaira, 1936).
Identification: B. alvarius, 60–160 mg per gram of dried glands, 1.0–3.5 mg per gram dried skin (Erspamer *et al.*, 1965).

Bufotenidine; cinobufagin; *N,N,N*-trimethyl-5-hydroxytryptamine (XXII). Iodide, mp 216°–217°C (H. Wieland and Wieland, 1937). Oxalate, mp 96.5°C; picrate, mp 198°C, flavianate, mp 195°–200°C (H. Wieland *et al.*, 1934). Picronolate, mp 255°C (Deulofeu and Berinzaghi, 1946).
Identification: B. americanus, B. calamita, B. gargarizans, B. paracnemis, B. viridis (Erspamer, 1954).
Isolation: Ch'an Su (Chen *et al.*, 1931, H. Wieland *et al.*, 1932) *B. gargarizans, B. fowleri* (Jensen and Chen, 1932). *B. bufo bufo,* 170 μg per animal (H. Wieland *et al.*, 1934), 84 μg per animal and 0.53% in dried secretion (males), 170 μg per animal and 0.62% in dried secretion (females) (H. Wieland and Behringer, 1941). *B. formosus,* 2% in dried secretion (A. Ohno and Komatsu, 1957).
Other genera: isolation from *Xenopus laevis* (Jensen, 1935).

Bufothionine, dehydrobufotenine *O*-sulfate (XXIV). mp 250°C (H. Wieland and Vocke, 1930).
Isolation: B. formosus, 2 mg per dried skin (H. Wieland and Vocke, 1930). *B. arenarum,* 7.1 mg per dried skin (H. Wieland *et al.*, 1934). *B. chilensis, B. crucifer, B. paracnemis, B. spinulosus* (Deulofeu and Duprat, 1944).

Dehydrobufotenine (XXIII), mp 218°C (H. Wieland and Wieland, 1937). HCl, mp 242°C (H. Wieland and Vocke, 1930). Picrate, mp 186°C (H. Wieland and Wieland, 1937). Picronolate, mp > 300°C (Deulofeu and Berinzaghi, 1946). Flavianate, mp 260°–265°C (H. Wieland *et al.*, 1934).
Identification: B. bufo bufo, B. americanus, B. calamita, B. fowleri, B. paracnemis, B. viridis (Erspamer, 1954).
Isolation: B. marinus (Jensen and Chen, 1932), 6 mg per animal (Märki *et al.*, 1961). *B. valliceps* (Jensen and Chen, 1932). *B. arenarum* (H. Wieland *et al.*, 1934). *B. regularis* (Jensen, 1935). Ch'an Su (Chen *et al.*, 1931). *B. chilensis, B. crucifer, B. paracnemis, B. spinulosus* (Deulofeu and Duprat, 1944).

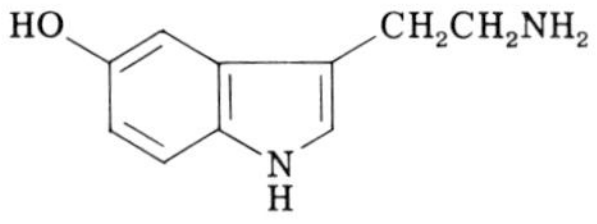

(XVI)

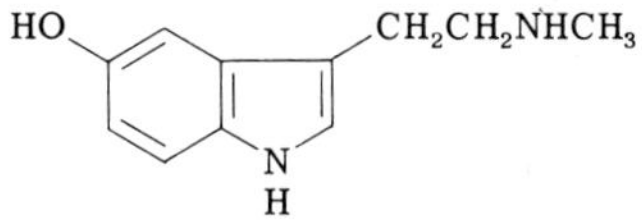

(XVII)

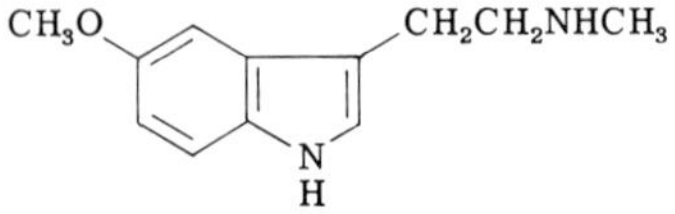

(XVIII)

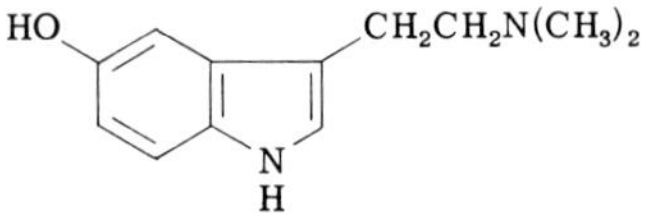

(XIX)

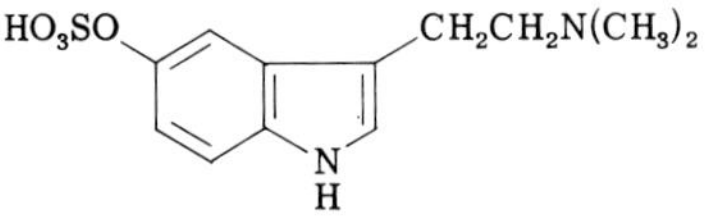

(XX)

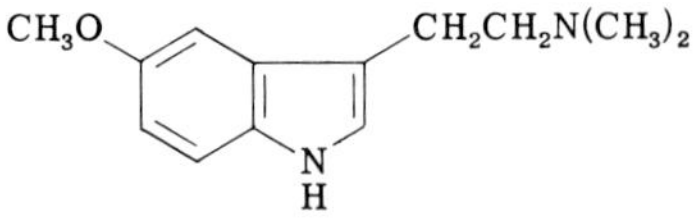

(XXI)

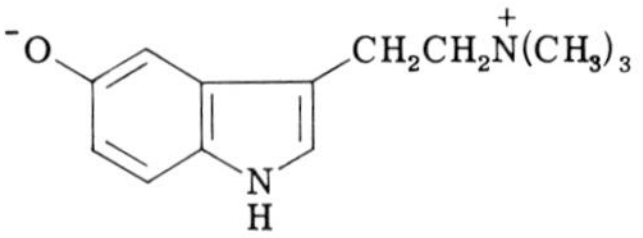

(XXII)

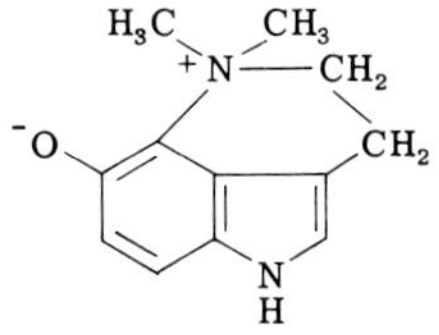

(XXIII)

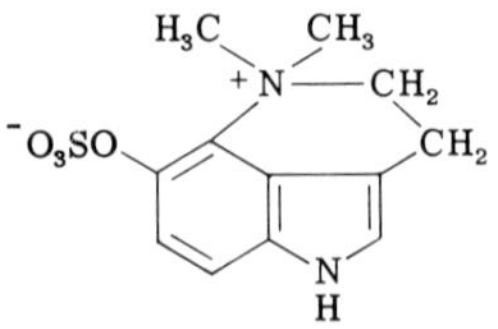

(XXIV)

I. 5-Hydroxytryptamine (Serotonin, Enteramine, Thrombocytin) (XVI)

The identification of 5-hydroxytryptamine in different animal tissues and its responsibility for several physiological activities was particularly the result of researches carried out by two groups of workers. In Italy, Erspamer and co-workers studied the substance responsible for the typical histochemical properties of the enterochromaffin cells, initially of the gastrointestinal mucosa and later of many other organs. In America, Rapport and Page were investigating the principle responsible for the vasoconstrictor properties of serum (for reviews on 5-hydroxytryptamine and related bases, see Erspamer 1954, 1961; Erspamer 1966; Page, 1954, 1958; Daly and Witkop, 1963; Lewis, 1964).

When both substances were isolated in pure form as salts or complexes (Rapport *et al.,* 1948; Erspamer and Asero, 1952) and their structure determined, they were found to be identical to 5-hydroxytryptamine (Rapport, 1949).

Afterwards, 5-hydroxytryptamine was found to have a widespread distribution in organs of many species of animals and to be of important significance, especially because of its relation to brain function and metabolism.

a. Chemistry. That the substance which determines the vasoconstrictor properties of serum was identical to 5-hydroxytryptamine was rightly proposed by Rapport (1949) and confirmed early by synthesis (Hamlin and Fischer, 1951; Speeter *et al.,* 1951; Erspamer and Asero, 1952; Asero *et al.,* 1952). Several syntheses of serotonin were developed afterwards (Speeter and Anthony, 1954; Harley-Mason and Jackson, 1954; Ek and Witkop, 1954; Young, 1958; Noland and Hovden, 1959; Kondo *et al.,* 1959; Suvorov and Murasheva, 1960).

b. Biosynthesis. The main steps in the biosynthesis of 5-hydroxytryptamine are indicated in Fig. 3.

There is a large amount of direct and indirect evidence of the capacity of microorganisms and higher animals for hydroxylating tryptophan (XXV) at carbon 5, with formation of 5-hydroxytryptophan (XXVI). L-Tryptophan labeled with ^{14}C has been transformed in patients suffering from carcinoidosis into labeled 5-hydroxy-L-tryptophan (XXVI) (Sjoerdsma *et al.,* 1957). The same hydroxylation has been observed *in vitro* by the action of particulate fractions of intestinal mucosa cells (rat, guinea pig) (Cooper and Melcer, 1961). Phenylalanine hydroxylase from the liver of rats is also capable of 5-hydroxylating tryptophan (Reuson *et al.,* 1962).

The 5-hydroxytryptophan is then decarboxylated enzymically with production of 5-hydroxytryptamine (XXVII) (Buzard and Nytch, 1957). Administration of 5-hydroxy-DL-tryptophan (XXVI) to rats and other higher

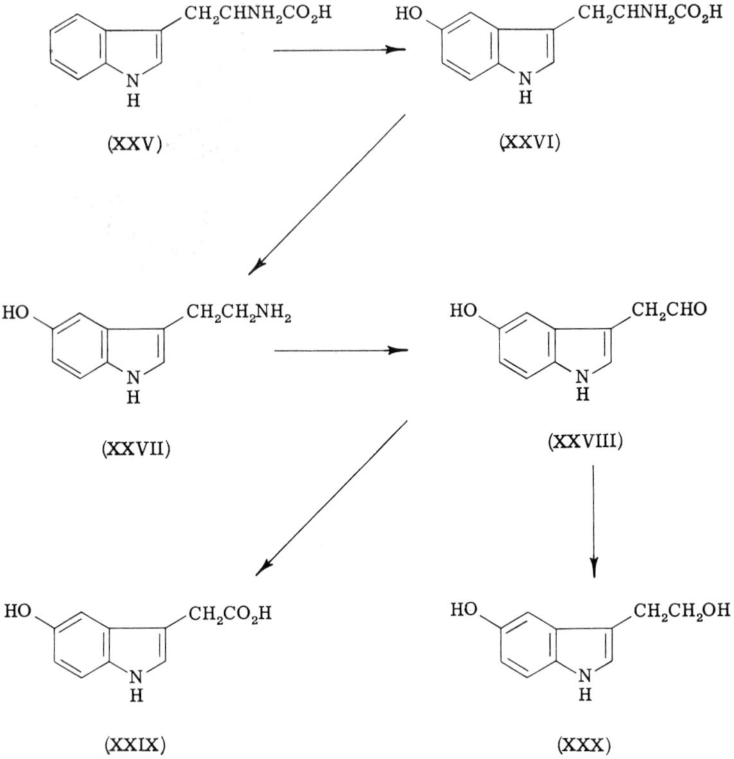

Fig. 3. Main pathways in the biogenesis and metabolism of 5-hydroxytryptamine.

animals produces an increase in the amount of 5-hydroxytryptamine present in tissues and in the excretion of 5-hydroxyindoleacetic acid (XXIX) in the urine, both deriving from the L-isomer (Davidson *et al.,* 1957; Udenfriend *et al.,* 1957). If large amounts are administered, 5-hydroxytryptamine appears (rats) in the urine.

In the toad *B. marinus,* 5-hydroxylation of tryptophan also takes place, because after administration of labeled L-tryptophan (XXV) to the whole animals, active 5-hydroxy-L-tryptophan (XXVI) was isolated from the venom glands. Labeled dehydrobufotenine (XXIII) was also isolated (Udenfriend *et al.,* 1956).

c. Metabolism. The products of 5-hydroxytryptamine metabolism vary with the species. Oxidation produces 5-hydroxyindoleacetic acid (XXIX), which is excreted in the urine, and was found to be an important metabolite in rats and rabbits.

It is partially conjugated with D-glucuronic acid, a small amount of conjugation with sulfuric acid being also found (Erspamer and Bertaccini, 1962). An intermediate in the oxidation of 5-hydroxytryptamine (XVI) to 5-hydroxyindoleacetic acid (XXIX) is 5-hydroxyindoleacetaldehyde (XXVIII). Its reduction product, 5-hydroxytryptophol (XXX), appears to be one of the major metabolites found in the urine of rats, receiving 5-hydroxytryptophan (Kveder *et al.,* 1962). Some *N*-acetylation (XXXI) of 5-hydroxytryptamine has also been observed in rats, which is interesting, because of the structural relation of *N*-acetyl-5-hydroxytryptamine (XXXI) to melatonin (XXXII) the substance isolated from the pineal gland of cattle by Lerner *et al.* (1960), which is active on the melanocytes and produces paling of the frog, toad, and fish skins.

$$HO—[indole] CH_2CH_2NHCOCH_3 \qquad CH_3O—[indole] CH_2CH_2NHCOCH_3$$

(XXXI)   (XXXII)

$$CH_3O—[indole] CH_2CO_2H$$

(XXXIII)

The formation of melatonin (XXXII) implies also the existence of a reaction of *O*-methylation of 5-hydroxyindoles and an enzyme catalyzing it has been found in the pineal body by Axelrod and Weissbach (1961). In this context the characterization in the skin of *B. alvarius* of 5-methoxyindoleacetic acid (XXXIII) is of interest (Erspamer *et al.,* 1965).

In the skin and venom of many species of toads and in the skin of other species of amphibians and in plants, compounds deriving from the *O*-methylation or/and *N*-methylation of 5-hydroxytryptamine have been isolated or characterized. They are considered as ulterior metabolic products of that amine and will be described individually in the following pages (see bufotenine, bufotenidine, dehydrobufotenine, and derivatives).

d. Distribution. 5-Hydroxytryptamine has a wide distribution in the animal kingdom. It has been found in the blood, gastrointestinal tract, nervous system, spleen, etc., of vertebrates. It is also present in the venom of certain reptiles, scorpions, and in the venomous apparatus of mollusks and coelenterates, the ganglia and nerve cords of several invertebrates.

Erspamer and Vialli (1951, 1952) investigated the presence of

5-hydroxytryptamine in the extracts from many amphibian skins by paper chromatography (Table II). The presence of 5-hydroxytryptamine in the venom of certain species of toads has been studied in particular for *B. marinus* (Udenfriend *et al.,* 1952), *B. bufo bufo* (Spandrio, 1961), *B. arenarum* (Frydman and Deulofeu, 1961), and *B. alvarius* (Erspamer *et al.,* 1965).

2. N-Methyl-5-hydroxytryptamine (XVII)

This amine was characterized for the first time by Erspamer and Vialli (1951, 1952) in the extracts from skins and parotoids from several species of *Bufo* by paper chromatography. It has not been found in other genera of amphibians.

3. N-Methyl-5-methoxytryptamine (XVIII)

In animals this base has been found only in the extracts of the skin of *B. alvarius* (Erspamer *et al.,* 1965). It has been isolated from several species of plants: *Phalaris arundinacea* (Wilkinson, 1958); *Piptadenia peregrina* (Legler and Tschesche, 1963; Iacobucci and Rúveda, 1964) and *Desmodium pulchellum* (Ghosal and Mukkerjee, 1965).

4. N,N-Dimethyl-5-hydroxytryptamine (Bufotenine) (XIX)

Bufotenine was not only the first indolic base isolated from toad venoms but also the first 5-hydroxytryptamine derivative isolated from organisms and synthesized. Its indolic structure was suspected by H. Wieland *et al.,* (1932) and by Jensen and Chen (1932). Further work settled the structure (H. Wieland *et al.,* 1934) and it was synthesized a year later by Hoshino and Shimidaira (1935). Recently Erspamer (1959) isolated from the secretion of *B. viridis* its sulfuric acid conjugate, which has been named bufoviridine (XX).

a. Chemistry. The interest in 5-hydroxytryptamine derivatives has produced in recent years a series of syntheses of bufotenine (Speeter and Anthony 1954; Harley-Mason and Jackson, 1954; Stoll *et al.,* 1955; Kondo *et al.,* 1959, 1960).

b. Distribution. In animals, with a few exceptions, bufotenine has been found only in the skin and secretion of species of the genus *Bufo* (toads, see Table II). It was not found by Erspamer and Vialli (1951) in the extracts of skins of other genera of amphibians, where 5-hydroxytryptamine was present.

Bufotenine and some of its derivatives have been found in plants. Some of the species containing them have been employed for smoking during tribal rites and no doubt the hallucinogenic properties of bufotenine played a part in their use. Stromberg (1954) was the first to isolate it from *Piptadenia peregrina* (Leguminoseae) (see also Alvares Pereira *et al.,* 1963); it was found in other species: *P. macrocarpa* (Fish *et al.,* 1955); *P. colubrina* (Pachter *et al.,* 1959); *P. excelsa* (Iacobucci and Rúveda, 1954), and *P. falcata* (Mennucci

Giesbrecht, 1960), in *Desmodium pulchellum* (Leguminosae) (Ghosal and Mukkerjee, 1964), and in *Phalaris tuberosa* (Gramineae) (Culvenor *et al.,* 1964). It was also isolated or identified in several species of *Amanita* although it is absent in others. (T. Wieland and Motzel, 1953; Tyler, 1961). The *N*-oxide of bufotenine has been isolated or identified in *P. peregrina, P. macrocarpa* (Fish *et al.,* 1955), *P. excelsa* (Iacobucci and Rúveda, 1964), and *D. pulchellum* (Ghosal and Mukkerjee, 1964).

5. N,N-Dimethyl-5-hydroxytryptamine-O-sulfate (Bufoviridine) (XX)

From the many species of amphibians investigated by Erspamer and Vialli (1951, 1952) bufoviridine (XVIII) was detected only in *B. viridis* and *B. calamita*. Later it was isolated from *B. viridis* by Erspamer (1959). On acid hydrolysis, bufotenine (XVII) and sulfuric acid were produced, showing that bufoviridine was the sulfuric ester of the base.

The formation of bufoviridine is an indication of the capacity for the sulfoconjugation of 5-hydroxyindole derivatives which is found in toads. It parallels the sulfoconjugation of dehydrobufotenine to bufothionine, which will be considered later.

6. N,N-Dimethyl-5-methoxytryptamine (O-Methylbufotenine) (XXI)

O-methylbufotenine has been isolated for the first time from an animal source from the skin and glands of *B. alvarius* (Erspamer *et al.,* 1965). It was already known to be present in some plants: *Dictyloma incanescens* (*Rutaceae*) (Pachter *et al.,* 1959); *P. peregrina* (Legler and Tschesche, 1963) and *Phalaris tuberosa* (Culvenor *et al.,* 1964). It is also present in *D. pulchellum* which also contains its *N*-oxide (Ghosal and Mukkerjee, 1964, 1965).

7. N,N,N-Trimethyl-5-hydroxytryptamine (Bufotenidine, Cinobufagine) (XXII)

Cinobufagine was the name given to a base isolated as flavianate from Ch'an Su by Jensen and Chen (1930). H. Wieland *et al.* (1932), who isolated it from the same source and found that it was also present in the secretion of *B. vulgaris*, named the base bufotenidine. He showed that bufotenidine resulted from the quaternization of bufotenine, on treatment with methyl iodide. It has been found in the skin and secretion of many species of toads investigated (Table II), but not in all of them, in spite of the fact that they contain bufotenine, which can be considered a logical precursor. It is present in the skin of *Xenopus laevis*.

a. Biosynthesis of the N-Methyl-5-hydroxytryptamines. The bufotenine (XIX) and bufotenidine (XXII) found on toad venoms are considered to be derived from 5-hydroxytryptamine by *N*-methylation. *N*-Methylation of tryptamines does not seem to be important in mammals and even in many other lower species of animals where 5-hydroxytryptamine has been detected and where only in very exceptional cases its *N*-methylated derivatives have been found. The data collected by Erspamer (1954, 1961) on the occurrence of indolethylamines in the amphibian skin shows this clearly.

Enzymes which can catalyze the N-methylation of indolethylamines have been described. Axelrod (1961, 1962) found in rabbit lungs an unspecific N-methyl transferase which methylates 5-hydroxytryptamine with production of N-methyl-5-hydroxytryptamine and also of this last compound, with production of bufotenine (XIX). The unspecific N-methyl transferase found in phenylethylamines can also methylate tryptamine derivatives. It seems to be similar although not identical to the enzyme present in the lungs.

The isolation from animals and plants of 5-methoxytryptamines in different stages of N-methylation is indicative that O-methylation is an enzymic reaction common to certain species of the two kingdoms. In this connection the isolation from the skin and glands of *B. alvarius* of several 5-methoxyindole derivatives (Erspamer *et al.,* 1965) must be noted.

There is no information available to decide if N-methylation takes precedence over O-methylation or vice versa.

b. Metabolism of the N-Methyl-5-hydroxytryptamines. We do not have any experimental information on the ulterior metabolism of bufotenine (XIX) and bufotenidine (XXII) in the toad. The presence of bufoviridine (XX) in two species shows that in some cases O-sulfonation can take place.

Some studies have been made in man and higher animals on the fate of administered bufotenine. Compared to 5-hydroxytryptamine, it is slowly metabolized by monoamine oxidase with the result that in rats, only a small percent of the administered amount is transformed and excreted as 5-hydroxyindoleacetic acid (XXX). A much larger proportion is found in the urine unchanged or conjugated with glucuronic acid (Gessner *et al.,* 1960).

It is interesting that 5-methoxyindoleacetic acid (XXXIII) has been found in the skin and glands of *B. alvarius* which also contains O-methylbufotenine and N-methyl-5-methoxytryptamine (Erspamer *et al.,* 1965).

8. Dehydrobufotenine (XXIII) and Bufothionine (XXIV)

Bufothionine (XXIV) was first isolated by H. Wieland and Vocke (1930) from *B. gama.* By acid hydrolysis, sulfuric acid and a basic substance which was named dehydrobufotenine (XXIII) were produced. The last compound was later found identical to a base, isolated as flavianate from several species of toad (Jensen and Chen, 1932; H. Wieland *et al.,* 1934). Bufothionine was obviously the sulfuric ester of the base.

(XXXIV)

a. Chemistry. On the basis of its transformation to bufotenine by hydrogenation, dehydrobufotenine was assigned the structure (XXXIV) by H. Wieland and Wieland (1937). This structure was put in doubt by Witkop (1956) who noticed that its ultraviolet spectrum was closely similar to that of 5-hydroxyindole. In 1961, two groups of workers (Märki *et al.,* 1961; Robinson *et al.,* 1961), proposed, on the basis of the nuclear magnetic resonance spectrum, a structure (XXIII) for dehydrobufotenine, from which is derived the structure (XXIV) for bufothionine.

Structure (XXIII) explains all the chemical reactions described for dehydrobufotenine and also the physical properties of the compound. Dehydrobufotenine as the free base or as the sulfoconjugate, bufothionine, has been found in many species of toads (Table II). It has not been found outside the genus *Bufo,* with the exception of the amphibian *Acris crepitans* in whose skin extracts Erspamer and Vialli (1952) detected bufothionine as the only indolic component, by paper chromatography.

b. Metabolism. That 5-hydroxytryptophan is a precursor of dehydrobufotenine was shown by Udenfriend *et al.* (1956) who found that after the administration of radioactive DL-tryptophan to *B. marinus,* labeled dehydrobufotenine could be isolated from the venom glands of the toad.

Although we have no evidence, it is plausible to postulate that dehydrobufotenine is formed from bufotenine by nucleophilic attack of the amino nitrogen atom on the 4-carbon atom of the indole nucleus, under enzymic catalysis. The ultimate fate of the base, except that it can be sulfoconjugated to form bufothionine (XXIV), is not known.

V. ANALYTICAL METHODS

The investigation of the bases present in extracts of toad skins, glands, or secretions can be done by paper chromatography. A table of R_f for several amines, in four different solvent systems, has been published by Reio (1960). Thin layer chromatography has been applied to dopamine derivatives by Kuehl *et al.* (1964).

Erspamer and Vialli (1951, 1952) and Erspamer (1959) have described several solvent systems which can be employed in paper chromatography for the identification of indolic bases. Chromatographic methods combined with the use of radioactive derivatives have been employed by Udenfriend *et al.* (1952) and by Märki *et al.* (1962).

Column chromatography has been used for the isolation of some bases, for example bufoviridine (Erspamer, 1959). Ion exchange columns have been used with success by Märki *et al.* (1961) for the isolation of dehydrobufotenine.

A gas chromatographic method for the separation and identification of the indolic bases present in plant material has been reported by Holmstedt *et al.* (1964) and has been successfully applied to plant products (Holmstedt, 1965).

By applying mass spectrometry to the fractions separated by gas chromatography, Holmstedt (1967) has developed a powerful method of identification, which he has applied with success to South American snuffs and plants.

Colored reactions for bufotenine and bufotenidine have been described by Hamet and Lelogeais (1954). The application and study of the colors given by a modified Keller reaction to indolic bases, including bufotenine, has been done by Rieder and Böhner (1959).

REFERENCES

Abel, J. J., and Macht, D. I. (1911–1912). *J. Pharmacol. Exptl. Therap.* **3**, 319.

Alvarez Pereira, N., Marins, I. C., and Moussatché, H. (1963). *Rev. Brazil. Biol.* **23**, 211.

Asero, B., Coló, V., Erspamer, V., and Vercellone, A. (1952). *Ann. Chem.* **576**, 69.

Axelrod, J. (1959). *Physiol. Rev.* **39**, 751.

Axelrod, J. (1961). *Science* **134**, 343.

Axelrod, J. (1962). *J. Pharmacol. Exptl. Therap.* **138**, 28.

Axelrod, J. (1963). *Science* **140**, 499.

Axelrod, J., and Tomchick, R. (1958). *J. Biol. Chem.* **233**, 702.

Axelrod, J., and Weissbach, H. (1961). *J. Biol. Chem.* **236**, 211.

Axelrod, J., Inscoe, J. K., Senoh, S., and Witkop, B. (1958a). *Biochim. Biophys. Acta* **27**, 210.

Axelrod, J., Senoh, S., and Witkop, B. (1958b). *J. Biol. Chem.* **233**, 696.

Axelrod, J., Wurtman, R. J., and Winget, C. M. (1964). *Nature* **201**, 1134.

Barger, G., and Ewins, L. (1910). *J. Chem. Soc.* p. 2257.

Barlow, R. B., and Khan, I. (1959). *Brit. J. Pharmacol.* **14**, 265.

Bretschneider, H. (1947). *Monatsh. Chem.* **78**, 82.

Bridgers, W. F., and Kaufman, S. (1962). *J. Biol. Chem.* **237**, 526.

Buzard, J. A., and Nytch, P. D. (1957). *J. Biol. Chem.* **227**, 225.

Chen, K. K., and Chen, A. L. (1933). *J. Pharmacol. Exptl. Therap.* **49**, 503.

Chen, K. K., Jensen, H., and Chen, A. L. (1931). *J. Pharmacol. Exptl. Therap.* **43**, 13.

Cooper, S. R., and Melcer, I. (1961). *J. Pharmacol. Exptl. Therap.* **132**, 265.

Culvenor, C. C. J., Dal Bon, R., and Smith, L. W. (1964). *Australian J. Chem.* **17**, 1301.

Daly, J. W., and Witkop, B. (1963). *Angew. Chem.* **75**, 552; *Angew. Chem. Intern. Ed. Engl.* **2**, 421 (1963).

Davidson, J. D., Loom, S., and Udenfriend, S. (1957). *J. Clin. Invest.* **36**, 1954.

Deulofeu, V. (1935). *Z. Physiol. Chem.* **237**, 171.

Deulofeu, V. (1948). *Fortschr. Chem. Org. Naturstoffe* **5**, 241.

Deulofeu, V., and Berinzaghi, B. (1946). *J. Am. Chem. Soc.* **68**, 1665.

Deulofeu, V., and Duprat, E. (1944). *J. Biol. Chem.* **153**, 459.

Deulofeu, V., and Mendive, J. R. (1938). *Ann. Chem.* **534**, 288.

Ek, A., and Witkop, B. (1954). *J. Am. Chem. Soc.* **76**, 5579.

Erspamer, V. (1954). *Pharmacol. Rev.* **6**, 425.

Erspamer, V. (1959). *Biochem. Pharmacol.* **2**, 270.

Erspamer, V. (1961). *Progr. Drug Res.* **3**, 151.
Erspamer, V. (1966). "5-Hydroxytryptamine and Related Indolealkylamines," Vol. 19, Heffter-Heubner *Handbuch of Experimental Pharmacology*. Springer, Berlin.
Erspamer, V., and Asero, B. (1952). *Nature* **169**, 800; *J. Biol. Chem.* **200**, 311 (1900).
Erspamer, V., and Bertaccini, G. (1962). *Arch. Intern. Pharmacodyn.* **137**, 6.
Erspamer, V., and Vialli, M. (1951). *Nature* **167**, 1033.
Erspamer, V., and Vialli, M. (1952). *Ric. Sci.* Suppl. **22**, 1421.
Erspamer, V., Vitali, T., Roseghini, M., and Cei, J. M. (1965). *Experientia* **21**, 504.
Fischer, P., and Lecomte, J. (1950). *Arch. Intern. Physiol.* **57**, 277.
Fish, M. S., Johnson, N. M., and Horning, E. C. (1955). *J. Am. Chem. Soc.* **77**, 5892.
Frydman, B., and Deulofeu, V. (1961). *Experientia* **17**, 545.
Gessner, P. K., Khairallah, P. A., McIsaac, W. M., and Page, I. H. (1960). *J. Pharmacol. Exptl. Therap.* **130**, 126.
Ghosal, S., and Mukkerjee, B. (1964). *Chem. Ind.* (*London*) p. 1800.
Ghosal, S., and Mukkerjee, B. (1965). *Chem. Ind.* (*London*) p. 793.
Hamet, R., and Lelogeais, P. (1954). *Bull. Soc. Chim. Biol.* **36**, 933.
Hamlin, K. E., and Fischer, F. E. (1951). *J. Am. Chem. Soc.* **73**, 5007.
Handovsky, H. (1920). *Arch. Exptl. Pathol. Pharmakol.* **86**, 138.
Harley-Mason, J., and Jackson, A. H. (1954). *J. Chem. Soc.* p. 1165.
Hertting, G., and La Brosse, E. H. (1962). *J. Biol. Chem.* **237**. 2291.
Holmstedt, B. (1965). *Arch. Intern. Pharmacodyn.* **156**, 285.
Holmstedt, B. (1967). *In* "Ethnopharmacologic Search for Psychoactive Drugs" (D. H. Efron, B. Holmstedt, and N. S. Kline, eds.). Public Health Service, Washington.
Holmstedt, B., Vandenheuvel, W. J. A., Gardiner, W. L., and Horning, E. C. (1964). *Anal. Biochem.* **8**, 151
Hoshino, T., and Shimodaira, K. (1935). *Ann. Chem.* **520**, 19.
Hoshino, T., and Shimodaira, K. (1936). *Bull. Chem. Soc. Japan* **11**, 221; *Chem. Abstr.* **30**, 5982 (1936).
Iacobucci, G. A., and Rúveda, E. A. (1964). *Phytochemistry* **3**, 465.
Jaminet, F. (1959). *Farmaco.* (*Pavia*), *Ed. Sci.* **14**, 120.
Jensen, H. (1935). *J. Am. Chem. Soc.* **57**, 1765.
Jensen, H., and Chen, K. K. (1929). *J. Biol. Chem.* **82**, 397.
Jensen, H., and Chen, K. K. (1930). *J. Biol. Chem.* **87**, 741.
Jensen, H., and Chen, K. K. (1932). *Chem. Ber.* **65**, 1310.
Kaiser, E., and Michl, H. (1958). "Die Biochemie der tierischen Gifte." Franz Deuticke, Vienna.
Kondo, H., Kataoka, H., Hayashi, V., and Dodo, T. (1959). *Ann. Rep. Itsuu Lab.* **10**, 1; *Chem. Abstr.* **54**, 492 (1960).
Kondo, H., Kataoka, H., and Dodo, T. (1960). *Ann. Rep. Itsuu Lab.* **11**, 53; *Chem. Abstr.* **55**, 17619 (1961).
Kveder, S., Iskrič, V. S., and Keglević, D. (1962). *Biochem. J.* **85**, 447.
Kuehl, F. A., Hichens, M., Ormond, R. E., Meisinger, M. A. P., Gale, P. H., Cirillo, V. J., and Brink, N. G. (1964). *Nature* **203**, 154.
Lasagna, L. (1951). *Proc. Soc. Exptl. Biol. Med.* **78**, 876.
Lee, H. M., and Chen, K. K. (1951). *J. Pharmacol. Exptl. Therap.* **102**, 286.
Legler, G., and Tschesche, R. (1963). *Naturwissenschaften* **50**, 94.
Lerner, A. B., Case, J. D., and Takahashi, Y. (1960). *J. Biol. Chem.* **235**, 1992.
Lewis, G. P. (1964). *In* "The Hormones" (G. Pincus, K. V. Thimann, and E. B. Astwood, eds.), Vol. 4, pp. 387–402. Academic Press, New York.
Märki, F., and Witkop, B. (1961). Quoted by Märki *et al.* (1961).

Märki, F., Robertson, A. V., and Witkop, B. (1961). *J. Am. Chem. Soc.* **83**, 3341.
Märki, F., Axelrod, J., and Witkop, B. (1962). *Biochim. Biophys. Acta* **58**, 367.
Mennucci Giesbrecht, A. (1960). *Anais Assoc. Brasil. Quim.* **19**, 117.
Noland, W. E., and Hovden, R. A. (1959). *J. Org. Chem.* **24**, 894.
Ohno, S., and Komatsu, M. (1957). *Ann. Rep. Itsuu Lab.* **8**, 31; (*Chem. Abstr.* **51**, 14984 (1957).
Ohno, S., Ohmoto, T., and Komatsu, M. (1961). *Yakugaku Zasshi* **81**, 1339; (*Chem. Abstr.* **56**, 7391 (1962).
Östlund, E. (1954). *Acta Physiol. Scand.* **31**, Suppl. 112, 167.
Pachter, I. J., Zacharius, D. E., and Ribeiro, O. (1959). *J. Org. Chem.* **24**, 1285.
Page, I. H. (1954). *Physiol. Rev.* **34**, 563.
Page, I. H. (1958). *Physiol. Rev.* **38**, 277.
Pereira, N. A., and de Oliveira, D. S. (1961). *Rev. Brasil. Farm.* **42**, 13; (*Chem. Abstr.* **56**, 6486 (1962).
Phisalix, C., and Bertrand, G. (1893). *Compt. Rend.* **116**, 1080.
Phisalix, C., and Bertrand, G. (1902). *Compt. Rend.* **135**, 46.
Piccinelli, D. (1955). *Boll. Soc. Eustachiana Ist. Sci. Univ. Camerino* **44**, 105; (*Chem. Abstr.* **53**, 8327 (1959).
Pratesi, P., La Manna, A., Campiglio, A., and Ghislandi, V. (1958). *J. Chem. Soc.* p. 2069.
Pratesi, P., La Manna, A., Campiglio, A., and Ghislandi, V. (1959). *J. Chem. Soc.* p. 4062.
Rapport, M. M. (1949). *J. Biol. Chem.* **180**, 961.
Rapport, M. M., Green, A. A., and Page, I. H. (1948). *Science* **108**, 329; *J. Biol. Chem.* **176**, 1243 (1948).
Reio, L. (1960). *J. Chromatog.* **4**, 458.
Reuson, J., Weissbach, H., and Udenfriend, S. (1962). *J. Biol. Chem.* **237**, 2261.
Rieder, H. P., and Böhner, M. (1959). *Helv. Chim. Acta.* **42**, 1793.
Robinson, B., Smith, G. F., Jackson, A. H., Shaw, D., Frydman, B., and Deulofeu, V. (1961). *Proc. Chem. Soc.* p. 310.
Rosen, L., and Goodall, M. (1962). *Proc. Soc. Exptl. Biol. Med.* **110**, 767.
Sjoerdsma, A., Weissbach, H., and Udenfriend, S. (1957). *Am. J. Med.* **23**, 5.
Slotta, K. H., Valle, J. C., and Neisser, K. (1937). *Mem. Inst. Butantan (São Paulo)* **11**, 101.
Spandrio, L. (1961). *Arch. Sci. Biol. (Bologna)* **45**, 337.
Speeter, M. E., and Anthony, W. C. (1954). *J. Am. Chem. Soc.* **76**, 6208.
Speeter, M. E., Heinzelmann, R. V., and Weisblat, D. I. (1951). *J. Am. Chem. Soc.* **73**, 5514.
Stoll, A., Troxler, F., Peyer, J., and Hofmann, A. (1955). *Helv. Chim. Acta* **38**, 1452.
Stromberg, V. L. (1954). *J. Am. Chem. Soc.* **76**, 1707.
Suvorov, N. N., and Murasheva, V. S. (1960). *Zh. Obshch. Khim.* **30**, 3112; (*Chem. Abstr.* **55**, 17620 (1961).
Tschesche, R. (1945). *Fortschr. Chem. Org. Naturstoffe* **4**, 1.
Tullar, B. F. (1948). *J. Am. Chem. Soc.* **70**, 2067.
Tyler, V. E. (1961). *Lloydia* **24**, 71.
Udenfriend, S., Clark, C. T., and Titus, E. (1952). *Experientia* **8**, 379.
Udenfriend, S., Titus, E., Weissbach, H., and Peterson, R. F. (1956). *J. Biol. Chem.* **219**, 335.
Udenfriend, S., Weissbach, H., and Bogdanski, D. F. (1957). *J. Biol. Chem.* **224**, 803.
Udenfriend, S., Lovenberg, W., and Sjoerdsma, A. (1959). *Arch. Biochem. Biophys.* **85**, 487.
van de Veerdonk, F. C. G., Huismans, J. W., and Addink, A. D. F. (1961). *Z. Vergleich, Physiol.* p. 22, **44**, 323.
von Euler, U. S. (1956). "Noradrenaline: Chemistry, Physiology, Pharmacology and Clinical Aspects." Thomas, Springfield, Illinois.

Waalkes, T., Sjoerdsma, A., Creveling, C. R., Weissbach, H., and Udenfriend, S. (1958). *Science* **127**, 648.

Wasser, E., and Sommer, H. (1928). *Helv. Chim. Acta.* **6**, 54.

Weiner, N. (1964). *In* "The Hormones" (G. Pincus, K. V. Thimann, and E. B. Astwood, eds.), Vol. 4, pp. 403–479. Academic Press, New York.

Wieland, H., and Behringer, H. (1941). *Ann. Chem.* **549**, 209.

Wieland, H., and Vocke, F. (1930). *Ann. Chem.* **481**, 213.

Wieland, H., and Wieland, T. (1937). *Ann. Chem.* **528**, 234.

Wieland, H., Hesse, G., and Mittasch, H. (1932). *Chem. Ber.* **64**, 2099.

Wieland, H., Konz, W., and Mittasch, H. (1934). *Ann. Chem.* **513**, 1.

Wieland, T., and Motzel, W. (1953). *Ann. Chem.* **581**, 10.

Wilkinson, S. (1958). *J. Chem. Soc.* p. 2079.

Witkop, B. (1956). *J. Am. Chem. Soc.* **78**, 2873.

Young, E. H. P. (1958). *J. Chem. Soc.* p. 3493.

Chapter 39

Chemistry and Pharmacology of Frog Venoms

JOHN W. DALY AND BERNHARD WITKOP

NATIONAL INSTITUTE OF ARTHRITIS AND METABOLIC DISEASES,
NATIONAL INSTITUTES OF HEALTH, BETHESDA, MARYLAND

I. INTRODUCTION

The pharmacologically active substances contained in the cutaneous secretions of various amphibians (Michl and Kaiser, 1963) appear to have evolved in this class of animals as agents which are employed more or less passively in defense against predators. These substances belong to many chemical classes and possess a wide range of pharmacological activities. Bufodienolides from toads of the genus *Bufo* (Chapters 37, 40), samandarine alkaloids from the European fire salamander (*Salamandra maculosa*) (Chapter 42), tetrodotoxin from newts of the genus *Taricha* (Mosher *et al.*, 1964), and batrachotoxin from the Colombian poison arrow frog (*Phyllobates aurotaenia*) (Märki and Witkop, 1963) are all extremely toxic compounds. Serotonin from various amphibians (Erspamer, 1954), bradykinin from the European grass frog (*Rana temporaria*) (Anastasi *et al.*, 1965), and the various histamines and leptodactyline from the giant leptodactylid frog of Central and South America (*Leptodactylus pentadactylus*) (Erspamer *et al.,* 1964d) are better considered local irritants than poisons.

These various substances may serve merely as passive protective venoms or they may have other, more physiological functions in amphibians. It is noteworthy that the occurrence of such active substances in amphibians is often accompanied by gaudy warning coloration. In the diurnal frogs of the family *Dendrobatidae*, only the genera *Phyllobates* and *Dendrobates* are brightly colored, and only these frogs contain toxic principles; the closely allied dull-colored frogs of the genus *Colostethus* contain relatively inactive substances (see below).

Attempts to correlate the bright coloration and the level of toxicity in skin extracts were, however, unsuccessful with some 7 populations of a small Panamanian frog, *Dendrobates pumilio*. The range in general coloration in these frog populations was from dark blue, through various greens, reds, and reddish orange to bright red (Daly and Myers, 1967).

Correlation of serotonin levels in frog skin with habits of frogs of the genus *Rana* was more successful (Welsh and Zipf, 1966). High levels of serotonin were found only in the semiterrestrial species which might be more exposed to predators than their aquatic relatives.

Toads of the genus *Bufo*, although of dull coloration, are mainly terrestrial and have developed well-defined parotid glands capable of discharging their contents with some force. In these toads, the pharmacologically active compounds of the gland include cardioactive bufodienolides (Chapter 40), a variety of 5-hydroxyindolic amines (Cei *et al.,* 1968), and epinephrine (Abel and Macht, 1912); these compounds probably provide good protection against many predators. It has been observed that dogs promptly drop toads because of the effects of the gland secretions. The profuse skin secretions of the brightly colored European toads of the genus *Bombina* are reported to afford protection against natural predators such as snakes (Kiss and Michl, 1962). Preliminary experiments with Dendrobatid frogs indicate that frog-eating snakes, e.g., *Rhadinaea* promptly reject *Dendrobates pumilio* and *D. auratus* and then attempt to cleanse their buccal tissue (C. W. Myers, personal communication).

II. GENERAL CLASSES OF PHARMACOLOGICALLY ACTIVE SUBSTANCES FROM AMPHIBIANS

A. Biogenic Amines

This class of compounds occurs in many amphibians and has been extensively studied by Erspamer, Cei and co-workers. These amines occur in various anurans at levels of greater than 1 mg/gm skin and may serve as a method for chemical taxonomy of certain amphibians (Cei *et al.,* 1967, 1968).

One of the most commonly occurring amines is serotonin and its *N*-methyl derivatives, *N*-methylserotonin, bufotenine, and bufotenidine (I). This type of compound is common in many anurans from the genera *Hyla*

(I)

(Erspamer *et al.,* 1966; De Caro *et al.,* 1968), *Leptodactylus* (Cei *et al.,* 1967), *Rana* (Welsh and Zipf, 1966), and *Bufo* (Cei *et al.,* 1968). Certain anurans of these genera, such as *Hyla pearsoniana,* and *Leptodactylus pentadactylus,* have levels of 5-hydroxyindolethylamines ranging from 1–10 mg/gm skin, while, in other genera of these compounds may be detected in low concentration or not at all. 5-Hydroxyindolic amines also have been reported (Cei *et al.,* 1967, 1968; Erspamer 1954; Erspamer *et al.,* 1966b; Cei and Erspamer, 1966; Welsh and Zipf, 1966) in *Rhinoderma, Odontophrynus, Discoglossus, Bombina, Xenopus, Pleurodema, Eleutherodactylus, Melanophrynicus, Cyclorana,* and *Lechriodus.* In most of these anurans, serotonin or bufotenine is the major indolic material. In toads of the genus *Bufo,* a wider spectrum of amines occur (Cei *et al.,* 1968), these include dehydrobufotenine (II), its sulfate ester, bufothionine, and the sulfate ester, bufoviridine, of bufotenine. A hallucinogenic substance (Gessner *et al.,* 1968), 5-methoxy-*N,N*-dimethyl-tryptamine, is present in extremely high levels in *Bufo alvarius* (Erspamer *et al.,* 1967), but not in other toads of the same genus.

(II)

Serotonin has been reported (Erspamer, 1954) to be present in the European fire salamander (*Salamandra maculosa*) but not in certain other caudates [*Notophthalmus viridescens* (Welsh and Zipf, 1966), *Triturus pyrrhogaster* (Erspamer *et al.,* 1964c), *Ambystoma tigrinum, Triturus cristatus, Pleurodeles waltlii, Euproctes rusconi* (Erspamer, 1954), *Necturus maculosus* (Welsh, 1964)]. Tryptamine was reported (Erspamer 1954) in both *Salamandra maculosa* and *Triturus cristatus.*

The phenolic quarternary amine, leptodactyline (III) is also widespread

HO
N(CH₃)₃⁺Cl⁻

(III)

among anurans. It appears in many species (Cei *et al.*, 1967) of the genera *Pleurodema* and *Leptodactylus* and is also reported in *Telmatobus, Calyptocephalella, Elosia, Thoropa, Eupsophus, Physalaemus, Ceratophrys, Lepidobatrachus,* and *Odontophrynus*. It was first discovered in *Leptodactylus ocellatus* (Erspamer, 1959) in levels as high as 9 mg/gm skin (Erspamer *et al.*, 1964c). In addition to leptodactyline, the skin of *Leptodactylus pentadactylus pentadactylus* contains the *p*-hydroxy isomer, candicine, and small amounts of tyramine (Erspamer *et al.*, 1963, 1964c).

Extracts from the glands of *Bufo marinus* and certain other toads of this genus contain catecholamines consisting, principally, of epinephrine with lesser amounts of norepinephrine, dopamine, and epinine (*N*-methyldopamine) (Chen and Chen, 1934, 1951, Oestlund, 1954, Henderson *et al.*, 1960, Märki *et al.*, 1962). Epinephrine was not detected in all toads of this genus (Chen and Chen, 1934; Henderson *et al.*, 1962).

Histamine is found in certain anurans from the genera *Leptodactylus* (Cei *et al.*, 1967) and *Hyla* (Erspamer *et al.*, 1966b). Skin extracts from *Leptodactylus pentadactylus labyrinthicus* contain, in addition to histamine, *N*-methylhistamine, *N*-acetylhistamine, and *N,N*-dimethylhistamine, 2 bicyclic amines, spinaceamine (IV), and 6-methylspinaceamine (Erspamer *et al.*, 1964c,d). A summary of the biogenic amines and other active substances

HN⌐―――NH
N

(IV)

in various anurans is presented in Tables I–IV. Most of the amines, because of their effects on buccal and mucous tissue, could serve as passive defensive agents against predators.

B. Peptides

A variety of peptides, many with extremely high toxic activity, have been isolated from extracts of various frogs. These include bradykinin (Arg-Pro-Pro-Gly-Phe-Ser-Pro-Phe-Arg), physalaemin (Pyroglutamyl-Ala-Asp-Pro-Asp(NH₂)-Lys-Phe-Tyr-Gly-Leu-Met-NH₂), and caerulein (Pyroglutamyl-Glu-Asp-Tyr-(SO₃H)-Thr-Gly-Tryp-Met-Asp-Phe-NH₂).

Physalaemin was first discovered in *Physalaemus fuscumaculatus* (Erspamer *et al.*, 1964a) and has also been identified in *P. centralis*. It produces an extreme vasodilatation and prolonged hypotension. It is characterized by its immediate stimulant action on extravascular smooth musculature which contrasts with the slower action of bradykinin-like peptides. Physalaemin-like polypeptides have also been detected in *Physalaemus bresslaui* and *P. cuvieri* and in *Phyllomedusa rohdei*, *P. hypochondrialis*, and 4 other species of *Phyllomedusa*, in *Telmatobius jelskii* (Bertaccini *et al.*, 1965; Erspamer and Anastasi, 1965), and in *Uperoleia rugosa* and *U. marmorata* (Erspamer *et al.*, 1966a). In the latter species the name "Uperolein" has been suggested for the hypotensive polypeptide.

Bradykinin was first discovered in amphibians in extracts from the skin of *Rana temporaria* (Anastasi *et al.*, 1965a). It is also present in other frogs of this genus (*R. nigromaculata*) (Nakajima, 1968). Bradykinin-like peptides have been noted in *Rana esculenta* (Erspamer and Anastasi, 1965) and other *Ranae* (Erspamer *et al.*, 1964c), in *Phyllomedusa rohdei* (bradykinyl-Ile-Tyr-(SO₃H) = phyllokinin) (Anastasi *et al.*, 1965b), in *Rana nigromaculata* (Val-Pro-Pro-Gly-Phe-Thr-Pro-Phe-Arg = 1-Val-6-Thr bradykinin) (Nakajima, 1968), and in *Ascaphus truei* (Erspamer and Anastasi, 1965). *Acaphus truei* contains very large amounts of a bradykinin-like peptide which may be identical with phyllokinin. The activity of bradykinin as a hypotensive agent, a smooth muscle stimulant, and an irritant is well known.

Caerulein was first isolated from *Hyla caerulea* (Anastasi *et al.*, 1968). Caerulein or caerulein-like peptides occur in many species of *Hyla* (De Caro *et al.*, 1968), in certain species of *Leptodactylus* (*L. rubido, curtus, pentadactylus, laticeps*) (Cei *et al.*, 1967), and in certain *Phyllomedusae* (Erspamer and Anastasi, 1965). Caerulein is a hypotensive agent and a smooth muscle stimulant and, in addition, it stimulates gastric and pancreatic secretions.

Certain other peptides with either novel or unknown pharmacological activity have been isolated from frog skin extracts. These include: carnosine from various species of *Eleutherodactylus* (*E. portoricensis, E. karlschmidti, richmondi, wightmanae, locustus, eneidae, gryllus,* and *longirostris*), *Leptodactylus albilagrius* (Daly and Heatwole, 1966), and *Dendrobates pumilio*; tryptokinins from *Phyllomedusa rohdei* and *P. hypochondrialis* (Erspamer and Anastasi, 1965); and a hexapeptide [Ala-Glu-His-Phe-Ala-Asp (NH₂)₂] from *Bombina variegata* (Kiss and Michl, 1962).

Because of their stimulation of smooth muscle, many of these peptides could serve the frog, in defense, as chemical irritants, but the possibility exists that they may also have a physiological function. The occurrence of active peptides in various anurans is presented in Tables I–III.

C. Proteins

Many anurans contain extremely toxic hemolytic proteins in their skin secretions. Earlier investigations also had demonstrated such proteins in extracts of certain salamanders (*Salamandra maculosa, Triturus vulgaris, T.*

TABLE I

The Occurrence of Pharmacologically Active Substances in Extracts of Skin from Various Anurans[a]

	Serotonin	N-Methyl-serotonin	Bufotenine	Bufotenidine	Bufoviridine	Dehydrobufotenine	Bufothionine	Leptodactyline	Histamine	Active peptides	Epinephrine	Hemolytic proteins	Steroidal alkaloids
Ascaphidae													
Ascaphus truei	—	—	—	—	—	—	—	—	—	B	—	—	—
Pseudidae													
Pseudis paradoxus	○	—	—	—	—	—	—	—	—	—	—	—	—
Discoglossidae													
Discoglossus pictus	+ +	○	○	○	○	○	○	—	—	—	—	—	—
Bombina variegata	+ +	○	○	○	○	○	○	—	—	×	—	×	—
B. bombina	+	○	○	○	○	○	○	—	—	—	—	—	—
Pipidae													
Xenopus laevis	+	○	○	+ +	○	○	○	—	—	—	—	—	—
Pelobatidae													
Pelobates fuscus	○	○	○	○	○	○	○	—	—	—	—	×	—
Bufonidae[b]													
Bufo regularis	+ + +[c]	○	○	○	○	○	○	○	○	○	×	—	—
B. mauretanicus	+ + +	○	○	○	○	○	○	○	○	○	×	—	—
B. kisoloensis	+ + +	○	○	○	○	○	○	○	○	○	—	—	—
B. berghei	+ + +	○	○	○	○	○	○	○	○	○	—	—	—
B. funereus	+ + +	○	○	+	○	○	○	○	○	○	—	—	—
B. melanostictus	+ + +	○	+	+	○	+	○	○	○	○	○	—	—

B. bufo bufo
B. bufo formosus
B. bufo gargarizans
B. viridis
B. calamita
B. boreas
B. alvarius
B. punctatus
B. marmoreus
B. perplexus
B. debilis
B. spinulosus
B. trifolium
B. variegatus
B. terrestris
B. americanus
B. fowleri
B. quercicus
B. woodhousei
B. microscaphus
B. hemiophrys
B. marinus
B. ictericus
B. paracnemis
B. arenarum
B. granulosus
B. pygmaeus
B. major
B. fernandezae
B. typhonius
B. valliceps
B. cognatus

TABLE I—(continued)

	Serotonin	N-Methyl-serotonin	Bufotenine	Bufotenidine	Bufoviridine	Dehydrobufotenine	Bufothionine	Leptodactyline	Histamine	Active peptides	Epinephrine	Hemolytic proteins	Steroidal alkaloids
B. speciosus	++	++	○	○	○	+	○	○	○	○	—	—	—
B. canaliferus	++	++	○	+	○	○	+	○	○	○	—	—	—
B. coccifer	++	++	○	++	○	○	○	○	○	○	—	—	—
B. luetkeni	+++	++	+	○	○	○	○	○	○	○	—	—	—
B. haematiticus	+++	○	○	○	○	+	○	○	○	○	—	—	—
B. crucifer	○	○	+	○	○	○	○	○	○	—	×	—	—
B. blombergi	—	—	—	—	—	—	—	—	—	—	×	—	—
B. peltocephalus	—	—	—	—	—	—	—	—	—	—	×	—	—
B. asper	—	—	—	—	—	—	—	—	—	—	○	—	—
Atelopódidae													
Melanophryniscus moreirae	+	+	+++	○	○	○	○	○ᶠ	○	○	—	—	—
M. stelzneri	○	○	++	○	○	○	○	○	○	○	—	—	—

ᵃ The data reported are taken from references cited in the text: — = No data available; ○ = not present or less than 1 μg/gm; + = 1–100 μg/gm; ++ = 100–1000 μg/gm; +++ = 1–10 mg/gm; × = present in extracts; P = physaelamin-like peptides; B = bradykinin-like peptides; C = caerulin-like peptides.

ᵇ Toads of the genus *Bufo* contain cardiotoxic bufodienolides (Chapters 38, 40).

ᶜ Contains the *O*-sulfate of serotonin.

ᵈ Contains *N*-methyl-5-methoxytryptamine, the N_1-sulfate of bufotenine, the *N*-sulfate of 5-methoxy-*N,N*-dimethyltryptamine, and enormous amounts of 5-methoxy-*N,N*-dimethyltryptamine (50–160 mg/gm skin).

ᵉ Contains the N_1-sulfate of bufotenine in large amounts (>500 μg/gm skin).

ᶠ Contains a phenolic amine

TABLE II

THE OCCURRENCE OF PHARMACOLOGICALLY ACTIVE SUBSTANCES IN EXTRACTS OF SKIN FROM VARIOUS ANURANS[a]

	Serotonin	N-Methyl-serotonin	Bufotenine	Bufotenidine	Bufoviridine	Dehydrobufotenine	Bufothionine	Leptodactyline	Histamine	Active peptides	Epinephrine	Hemolytic proteins	Steroidal alkaloids
Leptodactylidae													
Batrachophrynus macrostomum	O	O	O	O	O	O	O	O	O	O	—	—	—
B. patagonicus	O	O	O	O	O	O	O	O	O	O	—	—	—
Telmatobius montanus	O	O	O	O	O	O	O	O	O	O	—	—	—
T. hauthali	O	O	O	O	O	O	O	O	O	O	—	—	—
T. halli	O	O	O	O	O	O	O	O	O	O	—	—	—
T. jelskii	O	O	O	O	O	O	O	O	O	P	—	—	—
Calyptocephalella gayi	O	O	O	O	O	O	O	+	O	O	—	—	—
Elosia aspera	O	O	O	O	O	O	O	O	O	O	—	—	—
E. lateristrigata	O	O	O	O	O	O	O	O	O	O	—	—	—
Cyclorhamphus fuliginosus	O +	O	O	O	O	O	O	O	O	O	—	—	—
Thoropa miliaris	O	O	O	O	O	O	O	+	O	O	—	—	—
T. petropolitana	O	O	O	O	O	O	O	O	O	O	—	—	—
Eupsophus rosens	O	O	O	O	O	O	O	O	O	O	—	—	—
E. nodosus	O	O	O	O	O	O	O	O	O	O	—	—	—
Eleutherodactylus ranoides	O	O	O	O	O	O	O	O	O	O	—	—	—
E. bufoniformis	O	O	O	O	O	O	O	O	O	O	—	—	—

TABLE II—(continued)

Species	Serotonin	N-Methyl-serotonin	Bufotenine	Bufotenidine	Bufoviridine	Dehydrobufotenine	Bufothionine	Leptodactyline	Histamine	Active peptides	Epinephrine	Hemolytic proteins	Steroidal alkaloids
E. portoricensis	O	O	O	O	O	O	O	O	O	O[b]	—	—	—
E. martinicensis	+	—	—	—	—	—	—	—	—	—	—	—	—
Rhinoderma darwini	++	O	O	O	O	O	O	O	O	O	—	—	—
Physalaemus biligonigerus	O	O	O	O	O	O	O	+	O	P	—	—	—
P. fuscumaculatus	—	—	—	—	—	—	—	—	—	P	—	—	—
P. centralis	O	O	O	O	O	O	O	O	O	P	—	—	—
P. bresslaui	O	O	O	O	O	O	O	O	O	P	—	—	—
P. cuvieri	O	O	O	O	O	O	O	O	O	O	—	—	—
Eupemphix nattereri	O	O	O	O	O	O	O	O	O	O	—	—	—
Pleurodema cinerea	O	O	O	O	O	O	O	+	O	O	—	—	—
P. tucuma	O	O	O	O	O	O	O	+	O	O	—	—	—
P. quayapae	O	O	O	O	O	O	O	O	O	O	—	—	—
P. neubulosa	O	O	O	O	O	O	O	O	O	O	—	—	—
P. bibroni	+	O	O	O	O	O	O	+	O	O	—	—	—
P. bufonina	O	O	O	O	O	O	O	O	O	O	—	—	—
Leptodactylus bufonicus	O	O	O	O	O	O	O	+	O	O	—	—	—
L. prognathus	O	O	O	O	O	O	O	+	O	O	—	—	—
L. sibilatrix	O	O	O	O	O	O	O	+	O	O	—	—	—
L. gracilis	O	O	O	O	O	O	O	+	O	O	—	—	—
L. mystacinus	++	O	O	O	O	O	O	+	O	O	—	—	—
L. caligonosus	+	O	O	+	O	O	O	++	O	O	—	—	—
L. melanonotus	+	+	O	+	O	O	O	++	O	O	—	—	—
L. rubido	+	O	O	+	O	O	O	++	O	C	—	—	—

L. curtus	+	+	O	O	O	O	+++	O	C	—	—
L. pentadactylus	++	++	O	O	O	O	+++[d]	++[e]	C	—	—
L. laticeps	++	O	O	O	+[c]	O	+	++[f]	C	—	—
L. bolvianus	O	O	O	O	++[c]	O	++	O	O	—	—
L. ocellatus	O	O	O	O	O	O	+++	O	O	—	—
L. chaquensis	O	O	O	O	O	O	++	O	O	—	—
Ceratophrys ornata	O	O	O	O	O	O	+	O	—	—	—
Lepidobatrachus asper	O	O	O	O	O	O	+	O	—	—	—
L. salinicola	O	O	O	O	O	O	O	O	—	—	—
L. llanensis	O	O	O	O	O	O	+	O	—	—	—
Odontophrynus americanus	++	O	O	O	O	O	+	O	—	—	—
O. occidentalis	+++	O	O	O	O	O	O	O	O	—	—
Adelotus brevis	O	O	O	O	O	O	O	O	O	—	—
Mixophyes fasciolatus	O	O	O	O	O	O	O	O	O	—	—
Limnodynastes fletcheri	O	O	O	O	O	O	O	O	O	—	—
L. ornatus	O	O	O	O	O	O	O	O	O	—	—
L. peroni	+	O	O	O	O	O	O	O	O	—	—
Cyclorana alboguttatus	++	++	O	O	O	O	O	O	O	—	—
Lechriodus fletcheri	+	O	O	O	O	O	O	O	P	—	—
Uperoleia rugosa	—	—	—	—	—	—	—	—	P	—	—
U. marmorata	—	—	—	—	—	—	—	—	—	—	—
Pseudophryne corroboree	—	—	—	—	—	—	—	—	—	—	+

a The data reported are taken from references cited in the text: — = No data available; O = not present or less than 1 μg/gm; + = 1–100 μg/gm; ++ = 100–1000 μg/gm; +++ = 1–10 mg/gm; × = present in extracts; P = physaelamin-like peptides; B = bradykinin-like peptides; C = caerulin-like peptides.

b Contains carnosine.

c L. pentadactylus dengleri (not in other subspecies).

d L. pentadactylus pentadactylus also contains candicine and tyramine.

e L. pentadactylus labyrinthicus also contains N-methylhistamine, N,N-dimethylhistamine, N-acetylhistamine, and (6-methyl)spinaceamine, and amounts of 5-hydroxytryptamine over 1 mg/kg.

f Contains spinaceamine.

TABLE III

The Occurrence of Pharmacologically Active Substances in Extracts of Skin from Various Anurans[a]

	Serotonin	N-Methyl-serotonin	Bufotenine	Bufotenidine	Bufoviridine	Dehydrobufotenine	Bufothionine	Leptodactyline	Histamine	Active peptides	Epinephrine	Hemolytic proteins	Steroidal alkaloids
Dendrobatidae													
Phyllobates aurotaenia	○	○	○	○	○	○	○	○	○				×
P. vittatus	○	○	○	○	○	○	○	○	○				×
Dendrobates pumilio		○	○	○	○	○	○	○	○	[b]			×
D. granuliferus													×
D. auratus													×
D. histrionicus													×
D. leucomelas													×
D. minutus													×
Hylidae													
Hyla arborea	+	○	○	○	○	○	○	○	○	×		×	
H. aurea	+	○	○	○	○	○	○						
H. caerulea	±	○	○	○	○	○	○	○	±	C			
H. pearsoniana	+	○	±	○	○	○	○	○	○	○			
H. peroni	±	+	±	○	○	○	○	○	○	○			

Species	1	2	3	4	5	6	7	8	9	10	11	12	13
H. lesueuri	—	—	—	C	○	○	○	○	○	○	+	○	+
H. latopalmata	—	—	—	C	○	○	○	○	○	○	○	○	+
H. rothi	—	—	—	C	○	○	○	○	○	○	○	○	+
H. infrafrenata	—	—	—	C	‡	○	○	○	○	○	○	○	‡
H. gracilenta	—	—	—	C	○	○	○	○	○	○	+	○	‡
H. nasuta	—	—	—	○	○	○	○	○	○	○	○	○	+
H. gilleni	—	—	—	C	+	○	○	○	○	○	○	○	+
H. chloris	—	—	—	C	—	—	—	—	—	—	—	—	—
H. dentata	—	—	—	C	○	○	○	○	○	○	‡	○	‡
H. verreauxi	—	—	—	C	—	—	—	—	—	—	—	—	—
H. phyllochroa	—	—	—	○	—	—	—	—	—	—	—	—	—
H. bicolor	—	—	—	○	○	○	○	○	○	○	○	○	○
H. rubella	—	—	—	○	○	○	+	○	○	○	+	○	+
Acris crepitans	—	—	—	—	—	—	—	—	—	—	○	—	○
Trachycephalus nigromaculatus	—	×	—	P,B	—	—	—	—	—	—	—	—	—
Phyllomedusa rohdei	—	—	—	P	—	—	—	—	—	—	—	—	—
P. callidryas	—	—	—	P	—	—	—	—	—	—	—	—	—
P. annae	—	—	—	P	—	—	—	—	—	—	—	—	—
P. helenae	—	—	—	P	—	—	—	—	—	—	—	—	—
P. dachnicolor	—	—	—	P,B	—	—	—	—	—	—	—	—	—
P. hypochondrialis	—	—	—	—	—	—	—	—	—	—	—	—	—
P. burmeisteri	—	×	—	—	—	—	—	—	—	—	—	—	—
Ranidae													
Rana temporaria	—	—	—	B	○	○	○	○	○	○	+	+	‡
R. esculenta	—	×	—	B	—	—	○	○	○	○	○	○	+
R. pipiens	—	—	—	B	—	—	○	○	○	○	○	○	‡

TABLE III—(continued)

	Serotonin	N-Methyl-serotonin	Bufotenine	Bufotenidine	Bufoviridine	Dehydrobufotenine	Bufothionine	Leptodactyline	Histamine	Active peptides	Epinephrine	Hemolytic proteins	Steroidal alkaloids
R. palustris	+++	O	O	O	O	O	O	—	—	—	—	—	—
R. labrosa	+++	O	O	O	O	O	O	—	—	—	—	—	—
R. madagascariensis	++	O	O	O	O	O	O	—	—	—	—	—	—
R. fuscigola	O	O	O	O	O	O	O	—	—	—	—	—	—
R. latastei	++	—	—	—	—	—	—	—	—	O	—	O	—
R. dalmatina	++	—	—	—	—	—	—	—	—	O	—	—	—
R. sylvatica	+++	—	—	—	—	—	—	—	—	—	—	—	—
R. aurora	+++	O	O	O	O	O	O	—	—	—	—	—	—
R. sphenocephala	+++	—	—	—	—	—	—	—	—	B	—	—	—
R. catesbeiana	O	O	O	O	O	O	O	O	O	—	—	—	—
R. grylio	O	O	O	O	O	O	O	—	—	B	—	—	—
R. clamitans	O	O	O	O	O	O	O	—	—	B	—	—	—
R. nigromaculata	++	O	O	O	O	O	O	O	O	B	—	—	—
R. japonica	++	O	O	O	O	O	O	O	O	O	—	—	—
R. rugosa	+++	O	O	O	O	O	O	O	O	—	—	—	—
R. limnocharis	O	O	O	O	O	O	O	O	O	—	—	—	—
Arthroleptis adolphifriederici	O							—	—	—	—	—	—
Ptychadena mascariniensis	O							—	—	—	—	—	—
Rhacophoridae													
Rhacophorus madagascariensis	O	O	O	O	O	O	O	—	—	—	—	—	—
Chiromantis rufescens	O	O	O	O	O	O	O	—	—	—	—	—	—
Polypedates buergeri	O	O	O	O	O	O	O	—	—	O	—	—	—

a The data reported are taken from references cited in the text: — = No data available; O = not present or less than 1 µg/gm; + = 1–100 µg/gm; ++ = 100–1000 µg/gm; +++ = 1–10 mg/gm; × = present in extracts; P = physaelamin-like peptides; B = bradykinin-

cristatus, and *T. marmoratus*) (Michl and Kaiser, 1963). Such hemolytic proteins occur in various anurans, e.g., *Bombina variegata, Rana esculenta,* and *Hylo arborea* (Kiss and Michl, 1962). Partial purification of a hemolytic protein from *Pelotes fuscus* has been carried out (Lábler *et al.,* 1968). The occurrence of such hemolytic substances is presented in Tables I–III.

D. Steroidal Alkaloids

Extremely toxic steroidal alkaloids have now been found to occur not only in salamanders (Chapter 42) but also in anurans. The principal alkaloid, samandarine (V), from the European fire salamander *Salamandra maculosa* has now been demonstrated in an Australian anuran *Pseudophryne corroboree* (Habermehl, 1965). The steroidal alkaloids contained in a small dendrobatid frog (*Dendrobates pumilio*) were first regarded as similar in structure

(V)

to samandarine (Daly and Myers, 1967), but subsequent studies have shown that they are a new type of alkaloid (see below).

It has long been known that dendrobatid frogs contain very toxic skin secretions. In 1871, A. Posada Arango (1871) first described a poison used by the Noanama, Cuna, and Choco Indians of the Choco and Antioquia departments of Colombia for their blow darts. This poison was obtained from a milky secretion of a small frog which was given the name *Phyllobates chocoensis* by this author. Later studies by Santesson (1935) on this poison attributed its source to *Dendrobates tinctorius.* Santesson observed paralysis of the muscles and central nervous system, especially the respiratory centers, and finally systolic heart arrest. Mezey (1947), using venom from darts 15 years old, attributed it to a species of *Dendrobates,* observed dyspnea, bradycardia, paralysis of the hind limbs, and loss of equilibrium. He attributed death to respiratory failure and cardiotoxic effects resulting in hypotension. In our own studies (Märki and Witkop, 1963) on poison isolated directly from the skin of frogs, dyspnea and loss of equilibrium followed by violent intermittent convulsions and death were observed. Two varieties of frogs were used for these studies, and these were incorrectly identified first as *Phyllobates bicolor* (Märki and Witkop) and then as *P. latinasus* (Latham, 1966). The correct name is *Phyllobates aurotaenia* (Daly and Myers, 1967), a frog originally described by Boulenger (1913) as *Dendrobates aurotaenia.* It is possible that a variety of dendrobatid frogs from this region have been used by Indians to poison their blow darts, but at the present time only the two varieties of *Phyllobates aurotaenia* seem to be employed. Parenthetically, Breder (1946) mentions the use of *Dendrobates auratus* as the former source of poison for blow darts by the Choco Indians of the Darien region of Panama.

The chemical structure of these toxic principles has now been elucidated. The major venom of the Colombian poison arrow frog (*Phyllobates aurotaenia*) is batrachotoxin (VIb) (Märki and Witkop, 1963; Daly *et al.*, 1965; Tokuyama *et al.*, 1968, 1969). The structure of homobatrachotoxin, formerly called isobatrachotoxin, is shown below (VIc). Crystalline batrachotoxinin A (VIa), the much less active steroidal alcohol component of the esters (VIb, VIc) was used, as the *p*-bromobenzoate derivative, for a complete X-ray

(VIa) R = H

(VIb) R =

(VIc) R =

analysis. Such novel nitrogenous steroids have now been demonstrated not only in *Phyllobates aurotaenia* (Choco, Colombia) but also, in much smaller amounts, in *P. vittatus* (Southwestern Costa Rica) and are perhaps typical of this genus.

Dendrobates pumilio of the closely allied genus, *Dendrobates*, contains a different type of compound (Daly and Myers, 1967) whose simplest member is pumiliotoxin C, a bicyclic amine with the empirical formula $C_{13}H_{25}N$. Similar compounds have now been isolated from *D. auratus* and may be typical of this genus.

All these poisons are irritating to mucous and buccal tissues and in all likelihood provide a great deal of protection for these small brightly colored terrestrial diurnal frogs. Toxic steroidal alkaloids have been reported for a limited number of genera (Tables I–III).

It is noteworthy that there are similarities (3,14β-oxygen functions, 18,19β-alkyl groups, and the β-6H) among the steroid moiety of batrachotoxin, the most active cardiotoxin known, and the cardioactive bufodienolides, e.g., (gamabufotalin VII), found in toads of the genus *Bufo* (Chapter 40).

(VII)

III. STEROIDAL ALKALOIDS FROM THE COLOMBIAN POISON ARROW FROG (*Phyllobates aurotaenia*)

The Colombian frog, *Phyllobates aurotaenia*, is employed by Choco Indians in Colombia to poison their blow darts. Two color varieties are employed: one a gleaming black frog with bright yellow dorsolateral stripes, the other, from higher areas in the mountains, a slightly larger black frog with either broad red orange dorsolateral stripes or, when the stripes are fused, a solid red-orange back. Each frog contains less than 50 μg of the toxic principles, a quantity sufficient to kill approximately 1000 mice when injected subcutaneously.

The active principles are isolated by extracting the skins with methanol, concentrating the extract, and partitioning with chloroform and water. The chloroform layer is then extracted with 0.1 N HCl. The aqueous acid is made basic with 1 N ammonium hydroxide, and the toxic alkaloids are then re-extracted into chloroform. To curtail losses of the rather labile active compounds, these operations should be carried out at 5°C. Thin-layer or column chromatography in silica gel yields the 4 major bases batrachotoxin, homobatrachotoxin (which supersedes the earlier incorrect terminology, isobatrachotoxin), pseudobatrachotoxin, and batrachotoxinin A (Tokuyama *et al.,* 1968, 1969). Results from high resolution mass spectrometry indicate that these toxic materials are closely related in structure and that they are steroidal alkaloids. Batrachotoxin, homobatrachotoxin, and pseudobatrachotoxin all appear to be isomers ($C_{24}N_{33}NO_4$), while batrachotoxinin A ($C_{24}H_{35}NO_5$) differed in its hydrogen and oxygen content. Batrachotoxin and homobatrachotoxin exhibited characteristic ultraviolet spectra (λ_{max} 234, $\varepsilon = 9200$, 264, $\varepsilon = 5100$), an infrared absorption band at 1690 cm^{-1}, which indicated a carbonyl or vinyl ether grouping, and a positive Ehrlich test which indicated a (potential) pyrrole ring. Pure pseudobatrachotoxin was not isolated in sufficient quantity for n.m.r. or infrared analysis because it converts easily to batrachotoxinin A. Neither of these latter compounds exhibited ultraviolet absorption bands or a positive Ehrlich test.

When extracts from 5000 frogs (Latham, 1966) were processed, a total of 11 mg of pure batrachotoxin, 16 mg of homobatrachotoxin, 1 mg of pseudobatrachotoxin, and 42 mg of batrachotoxinin A were isolated. Batrachotoxin, homobatrachotoxin, pseudobatrachotoxin, and batrachotoxinin A appear to occur in this frog (*P. aurotaenia*) in a ratio of approximately 3:1:3:3. Most of the pseudobatrachotoxin is converted to batrachotoxinin A during isolation.

Because of the limited quantity of alkaloid available, attention was directed toward obtaining a suitable crystalline derivative for X-ray analysis. In 1967, batrachotoxinin A was converted with *p*-bromobenzoic anhydride, under Schotten-Baumann conditions, to a crystalline *O*-*p*-bromobenzoate which, on treatment with base, regenerated batrachotoxinin A. X-ray analysis of a single crystal of this derivative led to the complete structure of batrachotoxin A (VIa) (Tokuyama *et al.*, 1968).

Reexamination of n.m.r., ultraviolet, infrared, and mass spectral data of batrachotoxinin A, batrachotoxin, and homobatrachotoxin demonstrated that the ion at m/e = 399 ($C_{24}H_{33}NO_4$), formerly thought to be the parent ion of batrachotoxin and homobatrachotoxin, was, in fact, a fragment. These alkaloids appeared to contain the steroidal nucleus of batrachotoxinin A and an additional moiety which accounted for the characteristic ultraviolet spectra (infrared band at 1690 cm^{-1}), Ehrlich reaction, and the presence (n.m.r. spectra) of 2 additional methyl groups in batrachotoxin, or 1 methyl and 1 ethyl group in homobatrachotoxin. Examination of the mass spectra of batrachotoxin and homobatrachotoxin, from this point of view, led to the identification of a fragment due to this additional moiety at m/e 139 ($C_7H_9NO_2$) in batrachotoxin and m/e 153 ($C_8H_{11}NO_2$) in homobatrachotoxin. Since this additional moiety, $C_7H_9NO_2$, had to account for the ultraviolet chromophore, the infrared absorption band at 1690 cm^{-1}, the positive Ehrlich reaction, and the presence in the n.m.r. spectrum of peaks due to 2 additional aryl methyl groups, we concluded that batrachotoxin is a dimethylpyrrolecarboxylate of batrachotoxinin A and that homobatrachotoxin is the corresponding methylethylpyrrolecarboxylate. An attempt to demonstrate a true molecular ion for batrachotoxin was now successful and it was found, as expected, at m/e 538 ($C_{31}H_{42}N_2O_6$).

The ultraviolet absorption peaks of (homo)batrachotoxin were then compared with spectra of dialkylated pyrrolecarboxylates and were found to be identical with that of ethyl 2,4-dimethylpyrrole-3-carboxylate. The n.m.r. studies confirmed that batrachotoxin and homobatrachotoxin were pyrrole-3- rather than 2-carboxylates. Investigation of changes in the chemical shift of the aryl methyl groups in different solvents provided n.m.r. evidence that homobatrachotoxin is a 2-ethyl-4-methylpyrrole-3-carboxylate rather than the other possible isomer 4-ethyl-2-methylpyrrole-3-carboxylate.

The point of attachment of the pyrrole carboxylate was also investigated. Hydrolysis with a strong base had been found to regenerate batrachotoxinin A. Evidence that the 20α-hydroxyl group was involved in the ester formation in batrachotoxin came from a comparison of the mass spectral fragmentation patterns and n.m.r. spectra of (homo)batrachotoxin and batrachotoxinin A. The n.m.r. resonance peak for the C-20 hydrogen was at $\delta = 5.8$–5.9 in (homo)batrachotoxin, an indication that the 20α-hydroxyl group was esterified.

The structures of batrachotoxin (VIb) and homobatrachotoxin (VIc) were confirmed by partial synthesis from batrachotoxinin A and the mixed anhydride prepared from 2,4-dimethylpyrrole-3-carboxylic acid and ethyl chloroformate. The synthetic batrachotoxin was identical in all respects (toxicity, n.m.r., infrared and mass spectra, color reactions, and chromotographic properties) with natural batrachotoxin (Tokuyama *et al.*, 1969).

Comparative toxicities of (homo)batrachotoxin, batrachotoxinin A, and their analogs are presented in Table IV along with samandarine and pumiliotoxin A and B.

TABLE IV

Relative Toxicity in Mice of Nitrogenous Bases Isolated from Anurans and Various Synthetic Analogs

	LD_{50} (μg/kg s.c.)
Batrachotoxin	2
Homobatrachotoxin	3
Batrachotoxinin A	1000
Batrachotoxinin A-20-(2,5-dimethylpyrrole-3-carboxylate)	2.5
Batrachotoxinin A-20-(2,4,5-trimethylpyrrole-3-carboxylate)	1
Batrachotoxinin A-20-(2,4-dimethyl-5-ethylpyrrole-3-carboxylate)	8
Batrachotoxinin A-20(N,2,4,5-tetramethylpyrrole-3-carboxylate)	280
Batrachotoxinin A-20-pyrrole-2-carboxylate	1000
Samandarine	300
Pumiliotoxin A	2500
Pumiliotoxin B	1500

Batrachotoxin is an extremely active cardiotoxin that interferes with conduction in the heart and causes extrasystoles, ventricular fibrillation, and death. Blockage of neuromuscular transmission was observed in a variety of preparations (Märki and Witkop, 1963). Batrochotoxin does not affect the nerve action potential at concentrations which do block neuromuscular transmission. The poison is temperature-dependent with an optimum activity at 37°C. It first blocks muscle contraction induced by neural stimulation and later blocks contraction induced by direct stimulation of the muscle. A microelectrode study on the extensor digitorum muscle of the rat indicated an increase of miniature end plate potentials (mepp) from 6–8 to 600/second for the first 20 minutes after which their amplitude and frequency decreased. These effects, elicited by 1×10^{-8} M batrachotoxin, are similar to those caused by 5×10^{-4} M ouabain (strophantidin). Batrachotoxin does not affect the action potential generating mechanism of either muscle or nerve. It does not inhibit aectylcholine esterase nor does it inhibit the sodium/potassium-dependent ATPase of muscle. The effects of batrachotoxin, which are blocked by tetrodotoxin, appear due to the selective, irreversible increase in membrane permeability to sodium ions which is evoked by low concentrations of batrachotoxin. (Albuquerque et $al.$, 1970).

IV. ALKALOIDS FROM OTHER DENDROBATID FROGS

Extracts from a variety of other dendrobatid frogs (*Phyllobates vittatus, P. lugubris, Dendobrates auratus, D. histrionicus, D. pumilio, D. granuliferus, D. minutus, D. leucomelas, Colestethus talamancae, C. inquinalis, C. pratti,* and *C. nubicola*) from Costa Rica, Panama, Colombia, and Venezuela have been examined for steroidal alkaloids. Only *Phyllobates vittatus* (southwestern Costa Rica) and possibly the much smaller *Phyllobates lugubris* (Bocas del Toro, Panama) contained (homo)batrachotoxin and batrachotoxinin A. Basic extracts from frogs of the genus *Dendrobates* contained a different type of alkaloid. The basic extracts from frogs of the genus *Colestethus* were of low toxicity and contained small amounts of alkaloids of undetermined types.

The toxic principles from *Dendrobates pumilio* were isolated by thin-layer and column chromatography and analyzed by mass spectrometry (Daly and Myers, 1967). Three major alkaloids, pumiliotoxin A ($C_{19}H_{33}NO_2$, MLD 2.5 mg/kg mouse), pumiliotoxin B ($C_{19}H_{33}NO_3$, MLD 1.5 mg/kg mouse), and pumiliotoxin C ($C_{13}H_{25}N$) were present. Each of these frogs (*D. pumilio*) contains 50–500 μg of pumiliotoxin A and B. The structure of pumiliotoxin C (VIII) was determined by X-ray analysis of the crystalline carbonate

(J. W. Daly *et al.,* 1970). Pumiliotoxin A and B are closely related in structure but contain a 9-carbon side chain instead of the isopropyl group of pumiliotoxin C and one or more hydroxyl groups or double bonds.

(VIII)

Pumiliotoxin A and B have also been isolated from extracts derived from the Panamanian poison arrow frog (*Dendrobates auratus*), which contain, in addition, a number of other structurally related alkaloids. This general type of alkaloid showed mass spectra characterized by an intense fragment ion at m/e 168 or 152 ($C_{10}H_{16}NO$ or $C_{10}H_{18}N$) due to loss of the side chain.

Pumiliotoxin A or B, when injected subcutaneously in mice, cause locomotor difficulty and extensor movements of the hind limbs, followed by clonic convulsions and death.

V. SUMMARY

The investigation of pharmacologically active compounds contained in cutaneous secretions from various anurans has already led to the discovery of a number of compounds of novel structure such as dehydrobufotenine (II) (Märki *et al.,* 1961), spinaceamine (IV), samandarine (V), batrachotoxin (VIb), and pumiliotoxin C (VIII). Table I presents the results from some 170 anurans out of a total of approximately 2600 species; these data show that much more research is required. The steroidal alkaloid fraction, for example, has been investigated in only a few of these 170 species. Further chemical and pharmacological studies would be expected to lead to the discovery of other novel compounds, while biochemical studies on their biosynthesis and biological and biochemical studies on their function should allow further insight into the evolution of these compounds in amphibians.

ACKNOWLEDGMENT

The authors wish to thank Dr. Charles W. Myers of the American Museum of Natural History, for his invaluable advice during the preparation of this manuscript.

REFERENCES

Abel, J. J., and Macht, D. I. (1912). *J. Pharmacol. Exptl. Therap.* **3**, 319.

Albuquerque, E. X., Warnick, J. E., Daly, J., and Witkop, B. (1970). *Science* (in press).

Anastasi, A., Erspamer, V., and Bertaccini, G. (1965a). *Comp. Biochem. Physiol.* **14**, 43.

Anastasi, A., Erspamer, V., Bertaccini, G., and Cei, J. M. (1965b). *In* "Hypotensive Peptides" (Erdos, E. G., Back, N., and Sicuteri, F., eds.), p. 76. Springer-Verlag, New York.

Anastasi, A., Erspamer, V., and Endean, R. (1968). *Arch. Biochem. Biophys.* **125**, 57.

Bertaccini, G., Cei, J. M., and Erspamer, V. (1965). *Brit. J. Pharmacol.* **25**, 363.

Boulenger, G. A. (1913). *Proc. Zool. Soc., London* **1913**, 1029.

Breder, C. M., Jr. (1946). *Bull. Am. Mus. Nat. Hist.* **86**, 375.

Cei, J. M., and Erspamer, V. (1966). *Copeia* **1**, 74.

Cei, J. M., Erspamer, V., and Roseghini, M. (1967). *Syst. Zool.* **16**, 328.

Cei, J. M., Erspamer, V., and Roseghini, M. (1968). *Syst. Zool.* **17**, 219.

Chen, K. K., and Chen, A. L. (1934). *Arch. Intern. Pharmacodyn.* **47**, 297.

Chen, H. M., and Chen, K. K. (1951). *J. Pharmacol. Exptl. Therap.* **102**, 286.

Daly, J. W., Witkop, B., Bommer, P., and Biemann, K. (1965). *J. Am. Chem. Soc.* **87**, 124.

Daly, J. W., and Heatwole, H. (1966). *Experientia* **22**, 764.

Daly, J. W., and Myers, C. W. (1967). *Science* **156**, 970.

Daly, J. W., Tokuyama, T., Habermehl, G., Karle, I. L., and Witkop, B. (1969). *Ann. Chem.* **729**, 198.

DeCaro, G., Endean, R., Erspamer, V., and Roseghini, M. (1968). *Brit. J. Pharmacol. Chemotherap.* **33**, 48.

Erspamer, V. (1954). *Pharmacol. Rev.* **6**, 425.

Erspamer, V. (1959). *Arch. Biochem. Biophys.* **82**, 431.

Erspamer, V., Cei, J. M., and Roseghini, M. (1963). *Life Sciences* 825.

Erspamer, V., Anastasi, A., Bertaccini, G., and Cei, J. M. (1964a). *Experientia* **20**, 489.

Erspamer, V., Bertaccini, G., and Urakawa, N. (1964b). *Japan. J. Pharmacol.* **14**, 468.

Erspamer, V., Roseghini, M., and Cei, J. M. (1964c). *Biochem. Pharmacol.* **13**, 1083.

Erspamer, V., Vitali, T., Roseghini, M., and Cei, J. M. (1964d). *Arch. Biochem. Biophys.* **105**, 620.

Erspamer, V., and Anastasi, A. (1965). *In* "Hypotensive Peptides," (Erdös, E. G., Back, H., and Sicuteri, F., eds.), p. 63. Springer-Verlag, New York.

Erspamer, V., DeCaro, G., and Endean, R. (1966a). *Experientia* **22**, 738.

Erspamer, V., Roseghini, M., Endean, R., and Anastasi, A. (1966b). *Nature* **212**, 204.

Gessner, P. K., Godse, D. D., Krull, A. H., and McMullan, J. M. (1968). *Life Sciences* **7**, 267.

Habermehl, G. (1965). *Z. Naturforsch.* **20**, 1129.

Henderson, F. G., Welles, J. S., and Chen, K. K. (1960). *Proc. Soc. Exptl. Biol. Med.* **104**, 176.

Henderson, F. G., Welles, J. S., and Chen, K. K. (1962). *Science* **136**, 775.

Kiss, G., and Michl, H. (1962). *Toxicon* **1**, 33.

Lábler, L., Keilová, H., Sorm, F., Kornalik, F., and Styblová, A. (1968). *Toxicon* **5**, 247.

Latham, M. (1966). *Natl. Geographic* **129**, 682.

Lee, H. M., and Chen, K. K. (1951). *J. Pharmacol. Exptl. Therap.* **102**, 286.

Märki, F., Axelrod, J., and Witkop, B. (1962). *Biochim. Biophys. Acta* **58**, 367.

Märki, F., Robertson, A. V., and Witkop, B. (1961). *J. Am. Chem. Soc.* **83**, 3341.

Märki, F., and Witkop, B. (1963). *Experientia* **19**, 329.

Mezey, K. (1947). *Rev. Acad. Colombia Bogotá* **7**, 319.

Michl, H., and Kaiser, E. (1963). *Toxicon* **1**, 175.

Mosher, H. S., Fuhrman, F. A., Buchwald, H. D., and Fisher, H. G. (1964). *Science* **144**, 1100.

Nakajima, T. (1968). *Chem. Pharm. Bull.* **16**, 769, 2088.

Oestlund, E. (1954). *Acta Physiol. Scandanavia* **31**, Suppl. 112, 55.

Posada Arango, A. (1871). *Arch. Med. Nav.* **16**, 203.

Santessen, C. G. (1935). *Ethnological Studies* **1**, 105.

Tokuyama, T., Daly, J. W., Witkop, B., Karle, I. L., and Karle, J. (1968). *J. Am. Chem. Soc.* **90**, 1917.

Tokuyama, T., Daly, J. W., and Witkop, B. (1969). *J. Am. Chem. Soc.* **91**, 3931.

Welsh, J. H., and Zipf, J. B. (1966). *J. Cell. Physiol.* **68**, 25.

Welsh, J. H. (1964). "Comparative Neurochemistry" (D. Richter, ed.), p. 355. Pergamon Press, London.

Chapter 40

Collection of Toad Venoms and Chemistry of the Toad Venom Steroids

KUNO MEYER AND HORST LINDE*

INSTITUTE OF PHARMACEUTICAL CHEMISTRY, UNIVERSITY OF BASEL, SWITZERLAND

I. INTRODUCTION

Accounts of the chemistry of toad venom steroids have been given within the compass of a larger frame of reference (Shoppee, 1942, 1958; Tschesche, 1945, 1954; Kaiser and Michl, 1958; Michl and Kaiser, 1963; Fieser and Fieser, 1959) as well as in review articles dealing solely with this subject (Behringer, 1943; Deulofeu, 1948; Sellhorn, 1952; Komeya and Takemoto, 1957; Zelnik, 1965; Suga, 1968; Okada, 1968; Daly and Witkop, 1968).

Knowledge of the toxicity of toad venom goes back to ancient times. Physicians of antiquity mentioned medicines prepared from toads, and

* The authors wish to express their thanks to Dr. Robert A. Micheli for most valuable help in translating this article.

described their effect on the heart and respiration. In the seventeenth and eighteenth centuries, dried toads were temporarily listed as "bufones exsiccati" in the official pharmacopoeias and were used as diuretics against dropsy and other diseases, even before digitalis was introduced by Withering. In China and Japan, the dried venomous secretion of the Chinese toad, formed into round, smooth, dark brown discs and known as Ch'an Su or Senso, is still used today for various ailments. For historical details see Behringer (1943) and especially Chen and Jensen (1929).

An accumulation of skin glands of elliptical form, the so-called parotids, which are located behind the ear, contain the bulk of the venom. A smaller amount of venom is also secreted by the small verrucose skin glands that cover the entire back of the animal. Two groups of toxic substances are found in this secretion as well as others.

1. The heart-active principles, representing steriod derivatives, commonly known as bufogenins (bufagins), or systematically as bufadienolides, and bufotoxins (conjugates of the bufogenins with suberylarginine), which are primarily responsible for the pharmacological effect of the poisonous secretion.

2. The basic components of the bufotenine type and the hormones adrenaline and noradrenaline.

Elucidation of the chemical structure of the bufogenins was begun by H. Wieland and his associates in Germany, by Jensen and Chen in America, and by several groups in Japan such as Kotake and Kuwada, and Kondo and Ohno. It is only in the last three decades that unequivocal proof of the steroid nature of the bufogenins has been presented and the details of their structure completely elucidated.

In Table I the venomous secretions of those toad species that have been subjected to close chemical investigation are listed.

II. PROCEDURE FOR OBTAINING THE TOAD POISONS

The toad poisons can be obtained from the following sources: from the dried skins of the animals (Wieland and Vocke, 1930), from secretions of the skin glands which are stimulated by electrical irritation of the living animals, or best, from dried parotid secretion.* The following method is recommended for obtaining this secretion (Bolliger and Meyer, 1957): the animals are firmly held with one hand and pressed down on a board that is

---

* In this connection it should be mentioned that the poisonous substances have also been detected and isolated in other parts of the body, e.g., in the blood (Phisalix and Bertrand, 1893; Gessner, 1926), and the ovaries (Ohno and Ohmoto, 1961).

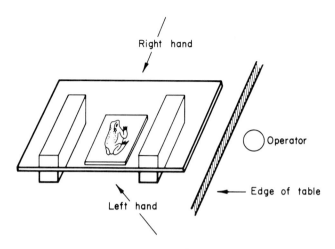

FIG. 1. Side view of the procedure for obtaining toad poison.

covered with a glass plate (Fig. 1). The raised, oblong gland accumulations located behind the ear are held near their base, between thumb and forefinger and squeezed firmly, so that the milky secretion squirts onto the underside of the glass plate. The glass plates charged with the venom secretion are kept in a horizontal position (secretion on top) at room temperature without exposure to direct sunlight until the secretion is dry and can easily be scraped off. This slow drying probably permits the enzyme contained in the raw secretion to split the bufotoxins into bufogenins and suberylarginine. The above procedure can be repeated after a rest period of 30–40 minutes and yields again a considerable quantity of secretion. After this the venom glands are completely empty and regeneration does not take place for between 4 to 6 weeks. An excellent source for obtaining a series of bufogenins (Ruckstuhl and Meyer, 1957; Hofer and Meyer, 1960) is Ch'an Su or Senso, mentioned above. This is still obtainable in large quantities from the pharmacies, for instance, of Hong Kong and Tokyo.

III. ISOLATION OF THE BUFOGENINS

1. From Dried Toad Skin

The dried toad skins are soaked in dilute alcohol for several months and the extract is evaporated in vacuum to dryness. Fats and cholesterol are removed by extraction with petroleum ether and then the bufogenins are isolated with chloroform, which is an excellent solvent for these substances.

TABLE I

Species	Source	Length of animals	Approx. amount of venom (dried) in mg per animal	References
Bufo alvarius Girard	Southern Arizona and Southern California to Mexico	80–165 mm	400	[a]
Bufo americanus Holbrook	Eastern part of North America from the Hudson Bay southward	54–110 mm	16	[b]
Bufo arenarum Hensel (*arenarius* Lutz)	Uruguay, Northern Argentina, Southern Brazil	75 mm	87	[c]
Bufo asper	Indonesia, Siam, Malayan Peninsula	260 mm	—	[d]
Bufo blombergi Myers and Funkhouser	Columbia, Venezuela, Amazonean Brazil	180 mm	1200	[a]
Bufo bufo bufo Linnaeus = *Bufo vulgaris* Laurenti	Europe, not including the Mediterranean, temperate zone of Asia	60–80 mm	13	[e]
Bufo bufo gargarizans = *Bufo gargarizans* Cantor = *Bufo asiaticus*	China	75–114 mm	19	[b]
Bufo crucifer Wied 1821	Coastline of Brazil and Argentina	up to 75 mm	18	[f]
Bufo formosus Boulenger	Japan	125 mm	75	[g]
Bufo woodhausi fowleri Hinckley	New England and New York south to Georgia and westward along the Great Lakes to Michigan. Along the Gulf Coast to central Texas.	51–82 mm	14	[a]
Bufo granulosus Spix subsp. *fernandezi* Gallardo	East-northeastern part of the South American continent from Panama to the southern part of Buenos Aires in Argentina	50–55 mm	—	[a]

Species	Distribution	Size		
Bufo ictericus Spix 1824 = Bufo marinus Boulenger 1882, part.	Brazil	up to 140 mm	190	h
Bufo marinus (Linnaeus 1758) = Bufo marinus (L.) Schneider	West Indies, Mexico, Central and South America	up to 200 mm	580	i
Bufo mauritanicus Schlegel	Morocco, Algeria, and Tunisia	122 mm	190	j
Bufo melanostictus	Southeast Asia, Indonesia	116 mm	90	d
Bufo paracnemis Lutz	Guiana, south and southwest of Brazil	130–220 mm	240	k
Bufo peltocephalus Tschudi	Cuba	130 mm	120	a
Bufo quercicus Holbrook	North Carolina to Florida west from Louisiana	19–32 mm	2	b
Bufo regularis Reuss	Africa, widespread	up to 136 mm	180	l
Bufo spinulosus Wiegmann (= B. chilensis Tschudi)	Chile, Peru	up to 100 mm	—	a
Bufo valliceps Wiegmann	Louisiana, east and south Texas to New Mexico and Costa Rica	53–125 mm	18	a
Bufo viridis viridis Laurenti	Europe, not including Iberian peninsula, North Africa, Near East, eastward to Mongolia, Tibet, and Himalaya area	80–140 mm	27	b

[a] Barbier et al. (1961).
[b] Chen and Chen (1933a).
[c] Rees et al. (1959).
[d] Iseli et al. (1964).
[e] Urscheler et al. (1955).
[f] Zelnik and Ziti (1963a).
[g] Iseli (1965).
[h] Zelnik and Ziti (1963b).
[i] Barbier et al. (1959).
[j] Linde and Meyer (1958).
[k] Zelnik et al. (1964).
[l] Bharucha et al. (1961a).

Investigators then concentrated the chloroform and added petroleum ether, thus precipitating most of the bufogenins. Repeated precipitation from various organic solvent pairs gave a series of residues from which crystals could be obtained under appropriate conditions. Homogeneous crystals are seldom obtained by this method. Separation is accomplished easier and more satisfactorily by chromatography of the chloroform-soluble steroid venom (see below).

2. From the Dried Parotid or Skin Secretion (e.g., from Ch'an Su)

The finely pulverized venomous secretion is mixed with an equal volume (or more) of sand and extracted in a Soxhlet apparatus (Tschesche and Offe, 1936; Ruckstuhl and Meyer, 1957; Hofer and Meyer, 1960) or in a percolator with chloroform. The yellowish-brown colored extracts, containing the sterols and the bufogenins, are submitted to an initial purification (Ruckstuhl and Meyer, 1957; Hofer and Meyer, 1960) in order to separate these two classes of compounds. The bufogenins so obtained are chromatographed on alumina or silica gel (Ruckstuhl and Meyer, 1957; Hofer and Meyer, 1960). Thus, relatively simply and in a short time, the major bufogenins can be obtained in a crystalline state. However, those substances present in small or minute quantities may require further application of chromatographic methods, for example, partition chromatography (Urscheler et al., 1955), which may be used directly for separation of the crude venom, instead of absorption chromatography on alumina or silica gel. Mixtures which are difficult to separate may be resolved by preparative paper chromatography (Linde and Meyer, 1958) or dispersion on long columns of silica gel (Duncan, 1962).

IV. PAPER CHROMATOGRAPHY OF THE BUFOGENINS

With the aid of this commonly used procedure it is possible in most cases to obtain an unequivocal identification of the bufogenins, thus enabling the chemist to analyze minute amounts of the venom. Furthermore, paper chromatography is the most reliable and easy method for determining the homogeneity of crystalline materials. The following systems have proved their effectiveness: benzene–cyclohexane (1:1)–formamide, benzene–formamide, benzene–chloroform (3:2), (7:5)–formamide and chloroform–formamide. The chromatographic systems propyleneglycol–water (4:1)–petroleum-ether–benzene (1:1) and propyleneglycol–water (4:1)–benzene–chloroform (1:1) are also very useful.* It is particularly worth using both types side by side, since in some cases two bufogenins give only one spot in one system, while in

---

* The stationary phases were diluted with acetone (1:4) before impregnation.

the other a distinctive separation is obtained (Ruckstuhl and Meyer, 1957). Another mixture for separating the bufogenins is the pyridine–water–acetone–(4:1:10)–heptane system (Ohno and Komatsu, 1956). The chromatograms are dried for half an hour at 90°C to get rid of the stationary phase. The migrated substances are detected by spraying with a solution of SbCl₃ in chloroform (20 gm/100 ml) and heating to about 80°C for several minutes. The different colors thus appearing in day- or ultraviolet light help further to characterize these substances. Some bufogenins (arenobufagin, bufarenogin) are only recognizable under ultraviolet light. Since the bufogenins have a strong ultraviolet absorption at about 290–300 mμ (see below) they can be located directly on the paper by a photo copy with filtered ultraviolet light (Bernasconi et al., 1955), or even better with a monochromator (Arnold et al., 1963); here 0.005 mg of bufogenin can be detected.

V. THIN LAYER AND GAS CHROMATOGRAPHY OF THE BUFOGENINS

These new methods have been used for the identification of bufogenins and for proving the homogeneity of crystals. It is said that as little as 0.001 mg of pure bufogenin is enough to achieve results by TLC (Zelnik and Ziti, 1962). The following solvent systems are used: ethyl acetate (water-free or saturated with water), ethyl acetate–cyclohexane (4:1) and ethyl acetate–acetone (9:1) (Zelnik and Ziti, 1962). Ethyl acetate–methanol (9:1) (Zelnik et al., 1964) and ethyl acetate–cyclohexane (1:2) (Zelnik et al., 1964) have given good results as well. With the bufogenins from Ch'an Su petroleum ether–acetone–benzene (1:1:1), chloroform–methanol (9:1), chloroform–acetic acid (4:1), and cyclohexane–isopropanol (7:3) proved its superiority (Meyer and Linde, unpublished data). Here again the migrated substances are detected by spraying the plates with a solution of SbCl₃ in chloroform (20 gm/100 ml) and subsequent heating to 100°–120°C. The different colors thus obtained are an additional aid to identification. In some cases (arenobufagin, bufarenogin) this reagent gives only extremely weak spots, which can be made more intense by spraying with a mixture of ethanol and H_2SO_4 (1:1, v/v). For further separations see Omoto (1964) and Komatsu et al. (1965). On gas chromatographic separations see Sakurai et al. (1964) and Sakurai et al. (1968).

VI. COLOR REACTIONS OF THE BUFOGENINS

The different colorations or the change of the colors which the bufogenins give with strong acids are very useful in identification. It is especially stressed here that reliable results are only obtained if pure crystals are used and if at the same time authentic substances are tested as well. With concentrated sulfuric acid (Ruckstuhl and Meyer, 1957; Meyer, 1949a) or 84% sulfuric acid (Urscheler *et al.,* 1955) it is best to use white spot plates, but color reactions with a solution of $SbCl_3$ in chloroform (20 gm/100 ml) are performed on filter paper. They need only minute amounts of substances (0.05 mg). The colored spots should also be examined under ultraviolet light. The Liebermann color reaction (Liebermann, 1885) and its modifications (Burchard, 1889, 1890; Stoll and Hofmann, 1935) have become obsolete.

VII. THE CHEMISTRY OF THE BUFOGENINS

The bufogenins are C_{24} steroids (compare the C_{23} steroids of the digitalis and strophanthus type; Fieser and Fieser, 1959). Their steroid nature was first deducted from dehydrogenation experiments: chrysene was obtained from bufotalin (Wieland and Hesse, 1935), and 3′-methyl-1.2-cyclopentenophenanthrene from a mixture of cinobufagin and cinobufotalin (Tschesche and Offe, 1935; Jensen, 1935b), pseudodesacetylbufotalin (Ikawa, 1935), and marinobufagin (Jensen, 1937). Direct transformation of a bufogenin into a steroid of known structure was not achieved until about 30 years ago.

I. Steroid Ring Skeleton of the Bufogenins (A/B cis and C/D cis)

Stereochemistry. The stereochemistry of the six asymmetric centers of the ring junctures at C-5, C-8, C-9, C-10, C-13, and C-14, making possible in principle 64 isomers, is the same for all naturally occurring bufogenins and is determined according to the projected formula given below. A similar situation exists in the bile acids series, with the exception of the C/D ring juncture.

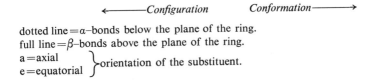

THE RING SKELETON OF THE BUFOGENINS

←———*Configuration*   *Conformation*———→

dotted line $=a$–bonds below the plane of the ring.
full line $=\beta$–bonds above the plane of the ring.
a$=$axial
e$=$equatorial $\Big\}$ orientation of the substituent.

2. The Lactone Group

Wieland and Weyland (1920) first formulated the lactone group as five-membered and later (Wieland and Hesse, 1935), on the basis of experiments with ozone (formation of formic acid), a structure with only one double bond in a six-membered lactone was proposed. In 1936 Wieland and associates correctly interpreted the ultraviolet spectrum [$\lambda_{max}^{alcohol}$ 290–300 mμ (log $\epsilon = 3.75$)] of bufogenins as derived from an α-pyrone structure (I). The same spectrum is found with cumalic acid methyl ester (II) (Tschesche and Offe, 1936) as well as with the scilla glycosides and their aglycones (Stoll

(I)   (II)

and Hofmann, 1935). Later Jensen (1937), Kubota (1938), and Kondo and Ohno (1938d) showed that on extensive ozonolysis H_2CO, $HCOOH$, $(COOH)_2$ and CHO-COOH was formed, while Schröter and Meyer (1959) established that on mild ozonolysis the first attack of ozone is at the C-20-C-21 double bond, giving rise to the formation of 20-keto-21-norcholanic acid and not an aldehyde, as could be assumed from Dean's speculations (1963). In detecting bufadienolides today, we are largely dependent on ultraviolet spectra and infrared spectra [1718–1736 cm^{-1} with a shoulder between 1751–1740 cm^{-1} (C=O), 1639 cm^{-1} and 1603 cm^{-1} (C=C)] (Jones and Herling, 1954; Jones and Callagher, 1959). Considering all the evidence, we believe the lactonic structure I to be established beyond doubt by analysis as well as synthesis (Bertin et al., 1961).

3. The Oxygen Functions

The bufogenins differ primarily in the number and position of the hydroxyl groups, which are scattered all over the skeleton: so far known, a primary hydroxyl group has been found in cinobufaginol and in hellebrigenol. Each bufogenin has at least at C-3 a secondary hydroxyl group [very recently (Iseli et al., 1965) 3-ketobufogenins have been isolated in minute quantities]. Additional hydroxyl groups are found at C-11 (gamabufotalin, arenobufagin), at C-12 (bufarenogin), and at C-16 (deacetylated products of bufotalin, cinobufagin, and cinobufotalin, probably artifacts and formed during chromatography on alumina). Acetylated secondary hydroxyl groups have up to now been found only at C-16 (bufotalin, cinobufagin, cinobufaginol, and cinobufotalin). Most bufogenins have as an integrant part a teritiary hydroxyl group located at C-14 (similar to the cardioactive aglycones of the digitalis-strophanthus group) and sometimes an additional one at C-5. A separate group is made up of those bufogenins carrying an oxido group at C-14/C-15 (resibufogenin, marinobufagin, cinobufagin, cinobufaginol, cinobufotalin, and bufotalinin). A keto group was detected at C-12 (arenobufagin), and C-11 (bufarenogin), an α-diketo group at C-11, C-12 (argentinogenin), and an aldehyde group at C-10 (hellebrigenin, bufotalinin).

4. Degradation of the Bufogenins

a. Hydrogenolysis with the Formation of Bile Acids. First attempts to determine the structure of the bufogenins were performed with bufotalin. It has the molecular formula $C_{26}H_{36}O_6$, or $C_{24}H_{34}O_5$ if one substracts the acetate group present. The bile acids have the same number of carbon atoms. From the very beginning a relationship between bufogenins and cholesterol, or the bile acids has been assumed, and attempts were made to interrelate these two groups by hydrogenating the unsaturated lactone ring. Circumstances made it inevitable that with bufotalin these experiments led to an acid which was isomeric but not identical with cholanic acid or any other of the known steroid acids (Wieland *et al.,* 1932). With anhydroscillaridin A, obtained as the aglycone part from the plant diglycoside scillaren A after acidic hydrolysis, it was found that an acid was formed during hydrogenolysis that proved to be identical with allocholanic acid (Stoll *et al.,* 1935). This showed a principal route from the bufogenins to the bile acids which could be used for structure determinations. Since relatively large amounts of substance are necessary, this method is no longer in use. Furthermore, the yields of the bile acids are, depending on the catalyst used, rather poor, and the formation of the saturated lactone ring is favored. These saturated compounds could not be used for further degradation.

b. Permanganate Oxidation with Formation of Etianic Acids. The peracetylated bufogenins are dissolved in acetone and are then oxydized with a slight excess of $KMnO_4$ (Steiger and Reichstein, 1938; Meyer, 1949b). Up to now this has been the only method of degradation which gives a relatively high yield (up to 50%) of crystalline etianic acid esters. (See Table III A and B, pp. 541–543.)

c. Ozonolysis with Formation of 20-Keto-21-norcholanic Acids. Initial attempts to degrade the unsaturated lactone ring by ozone were made with gamabufotalin diacetate (Kotake and Kubota, 1938). The acids thus obtainable were later (Schröter and Meyer, 1959) recognized as 20-keto-21-norcholanicacids. Since they cannot be degraded further, they are unsuited for a method intended to lead to steroids of known structure.

VIII. THE BUFOGENINS WITH KNOWN STRUCTURE

a. Arenobufagin (XX). (See Table II, pp. 532–533). This constituent of Ch'an Su (Chen and Chen, 1933e) and other toad venoms, was first isolated by

Chen and co-workers (1931–1932a) from the South American toad *B. arenarum* Hensel. The correct formula ($C_{24}H_{32}O_6$) was established by Deulofeu and associates (1940). Work on elucidating its constitution, undertaken by Hofer *et al.* (1960a), showed that arenobufagin has, except for the difference in the side chain, the same structure as the cardenolide aglycone sinogenin (Renkonen, *et al.*, 1957). Arenobufagin is therefore $3\beta,11\alpha,14\beta$-trihydroxy-12-keto-5β-bufo-20,22-dienolide.

 b. Argentinogenin (XXII). This bufogenin was first obtained by Rees and associates (1959) from the venom of *B. arenarum* Hensel, and is probably an artifact. It was observed (Rees *et al.*, 1959) that on prolonged contact with alumina, arenobufagin undergoes a change: an isomere bufarenogin and a dehydration product of arenobufagin, called argentinogenin, are formed. The analysis of the latter agreed approximately with $C_{24}H_{30}O_6$. The ultraviolet spectrum shows $\lambda_{max}^{alcohol}$ 289 mμ (log ϵ = 4.18), and the substraction curve from bufalin $\lambda_{max}^{alcohol}$ 298 mμ (log ϵ = 3.77) gives a maximum for the second chromophore of $\lambda_{max}^{alcohol}$ 287 mμ (log ϵ = 4.03). This corresponds to the absorption of the diosphenol group in ring C. Argentinogenin gives, furthermore, a gray-blue color with $FeCl_3$. The amorphous diacetate of argentinogenin shows, besides the normal bufadienolide ultraviolet maximum, a further maximum of $\lambda_{max}^{alcohol}$ 247 mμ (log ϵ = 3.65), which corresponds to the acetylated diosphenol group. These observations make it highly probable that argentinogenin has the structure XXII [see also Huber *et al.* (1967)].

 c. Bufalin (III). Kotake and Kuwada (1939) were the first to isolate this substance and to take up the task of elucidating its structure. They found the correct formula to be $C_{24}H_{34}O_4$ (Kotake and Kuwada, 1939), and by oxidative degradation were finally able to obtain a 3-hydroxyetianic acid (Kotake and Kuwada, 1942). Although this acid and the derivative prepared from it could not be compared with authentic samples, the physical data agreed so well with the values given in the literature that there could be no doubt that the Japanese authors had been the first to succeed in connecting a bufogenin with a known steroid. In the course of degradative studies in our laboratory, methyl 3β-acetoxy-14-hydroxy-$5\beta,14\beta,17\alpha H$-etianate was obtained (Meyer, 1949b), previously derived from digitoxigenin acetate (Hunziker and Reichstein, 1945), the structure of which was also proved by synthesis (Danieli *et al.*, 1962). This confirmed the findings of the Japanese, and in addition showed that bufalin is identical in nuclear structure with digitoxigenin.

 d. 3- Epibufalin. This bufogenin was obtained by Iseli *et al.*, in 1965 from an alcohol extract prepared from the fresh skins of *B. formosus* Boulenger. Its identification as the 3-epimere of bufalin was established by direct comparison with the product prepared by Tamm and Gubler (1959) from bufalon (by

TABLE II
Isolated Bufogenins

Bufogenin	No.	Formula	mp (°C)	Shape of crystals	Crystallized from[a]	$[\alpha]_D$[b]	Ref.	Acetate	mp (°C)	Shape of crystals	Crystallized from[a]	$[\alpha]_D$[b]	Ref.
Arenobufagin	XX	$C_{24}H_{32}O_6$	225–232°	Clusters of prisms	an	+55 me / +67.5	1	$C_{28}H_{36}O_8$	233–240°	Clusters of prisms	an/e	+43	1
Argentinogenin	XXII	$C_{24}H_{30\text{-}2}O_6$	227–230°	Colorless kernels	me/e	–20.6	2, 3		Amorphous				3
Artebufogenin 14α	XVIII	$C_{24}H_{32}O_4$	255–261°	Plates	an	+27.4	4	$C_{26}H_{32}O_5$	218–223°		an/e	+35.2	4
Artebufogenin 14β	XIX	$C_{24}H_{32}O_4$	124–130° impure	Prisms	me/w	–71.8	5	$C_{26}H_{32}O_5$	230–233°		an/e	–59.1	4/5
Bufagin A		$(C_{29}H_{42}O_7)$	173–175°				6						
Bufagin B		$(C_{22}H_{30}O_6)$	243–250°				6						
Bufagin C			238–242°				6						
Bufalin	III	$C_{24}H_{34}O_4$	244–248°	Thick prisms	me/e	–8.7	1	$C_{26}H_{36}O_5$	236–247°	Thin plates	an/e	–6	22
3-Epibufalin			268–272°	Needles	an / an/e	–12 me / –1.5	7	$C_{26}H_{36}O_5$	230–235°	Needles	an/e	+21	2/29
Bufalon	XXI	$C_{24}H_{32}O_4$	241–243°	Plates	an/e	+3	2, 8						
Bufarenogin		$C_{24}H_{32}O_6$	230–233°	Prisms	me/e	+11.3 me	7, 3	$C_{28}H_{36}O_8$	288–290°	Fine needles	an/e	+25.7	3
Bufo chilensis Genin D		$C_{24}H_{32}O_6$	223–224°				18						
Bufo mauritanicus Genin D		$C_{24}H_{32}O_6$	195–202°										
Bufotalidin = Hellebrigenin	IV	$C_{24}H_{34}O_6$	153–157° / 232–235°	Plates	an	+19.5	2	$C_{26}H_{34}O_7$	242–247°	Needles	an/e	+33.7	30
Bufotalin	V	$C_{26}H_{36}O_6$	156–158° / 227–231°	Prisms	an	+5.4	10/2	$C_{28}H_{38}O_7$	258–263°dec.	Plates	an/e	+4.4 / –2.1 d	17
Bufotalin F₁							11/12	$C_{26}H_{35}O_6 \times CO.C_6H_4.NO_2$	171–178°				31
Bufotalin F₂								$(C_{25}H_{32}O_5 \times CO.C_6H_4.NO_2)$	286–287°	Needles			31
Bufotalin F₃	X	$C_{24}H_{30}O_6$	243–245°	Rhombes	an/e	+14.08 at 80% / +19.2	13	$C_{26}H_{32}O_7$	235–243°dec.	Thin plates	an	+30.6	14, 3
Bufotalinin			213–233°				14/2/, 15/2		246–254°		me/e		
Bufotalon		$C_{26}H_{34}O_6$	275–280°dec.	Prisms	an/e	+6	1		Amorphous				1
Ch'an Su Genin H			248–262°	Prisms	an	+6 me							
Cinobufagin	XI	$C_{26}H_{34}O_6$	165–170° / 216–217°	Prisms	an/e	–3.6	7	$C_{28}H_{36}O_7$	204–205°	Fine needles	an/e	–1.6	7
Cinobufaginol	XII	$C_{26}H_{34}O_2$	239–242°	Clusters of prisms	an/e	–2 me	33	$C_{26}H_{38}O_9$	205–208°	Prisms	an/e	+9	33
Cinobufotalidin		$C_{24}H_5O_6$	217°dec.		et		16	$(C_{26}H_{34}O_8)$	209–210°		an/e		16

Name	No.	Formula	m.p.	Crystal form	Solvent[a]	$[\alpha]_D$[b]	Refs.	Formula	m.p.	Crystal form	Solvent[a]	$[\alpha]_D$[b]	Refs.
Cinobufotalin	XIII	$C_{26}H_{34}O_7$	257–259°/251–255°	Needles / Octaeder Tetraeder	an / an/e	+10.7	7/17, 2	$C_{28}H_{36}O_8$	215–218°	Prisms	an/e	+24.2 / +10.1 me	17/1
Desacetyl-bufotalin	VI	$C_{24}H_{34}O_5$	218–238°	Plates	me	+40 / +30 d / +19.5 / +33.6 me	2, 17	$C_{28}H_{38}O_7$	258–263°dec.	Plates	an/e	+4.4 / −2.1 d / −1.6 / +24.2 / +10.1 me	17, 7
Desacetylcinobufagin	XIV	$C_{24}H_{32}O_5$	179–181°	Pellets	an/e		17	$C_{28}H_{36}O_7$	204–205°	Fine needles	an/e	−11.4 / −13 me	17/1, 17
Desacetyl-cinobufotalin	XV	$C_{24}H_{32}O_6$	251–261°dec.	Long prisms	me/e		17/1	$C_{28}H_{36}O_8$	215–218°	Prisms	an/e		
Gamabufotalin	VII	$C_{24}H_{34}O_5$	260–269°	Prisms	me/e	−13.7 me	19/2	$C_{28}H_{38}O_7$	257–259°	Rosettes of prisms	an		
Gamabufotalininol / Hellebrigenol	VIII	$C_{28}H_{36}O_6$ / $C_{24}H_{34}O_6$	263–265°/158–165°/240–245°	Pellets	me/e	+1.8 et / +5.3 me / +14.6 chl/me	20, 3/21	$C_{28}H_{38}O_7$ / $C_{28}H_{38}O_8$	126–129°/178–181°	Clusters of needles / Long prisms	an/e / an/p	+36.3	32, 30, 22
Jamaicobufagin			167–172°/210–220°	Prisms	an/e	+10	22	$C_{26-28}H_{34-36}O_{7-8}$	181–186°/210–218°/162–178°	Thin needles	an/e	+52.7	22
								See also Viridobufagin and Gamabufotalin					
Marinobufagin / Quercicobufagin	XVI	$C_{24}H_{32}O_5$	220–224°/258–259°	Thin plates / Featherlike crystals	an/e / et/e/pe		3/19, 23, 24	$C_{26}H_{34}O_6$	200–220°	Needles	an/e	+24.9	22
Gamabufotalin and Arenobufagin													
Regularobufagin / Resibufogenin	XVII	$C_{24}H_{32}O_4$	113–140°/155–168°/120–130°/200–218°	Prisms / Cubes	an/w / an/e	−7.1	9, 22	$C_{26}H_{34}O_5$	223–234°	Thin Plates	an/e	−1.8	22
Telocinobufagin	IX	$C_{24}H_{34}O_5$	160–175°/210–211°	Flat prisms	an	+4.4	7	$C_{26}H_{36}O_6$	270–278°	Prisms	an/e	+23.6	3/7
Vallicepobufagin = Marinobufagin(?) Viridobufagin = Gamabufotalin(?)							25/23/26/27, 28/26						

[a] an = acetone; chl = chloroform; d = dioxane; e = ether; et = ethanol; me = methanol; p = pentane; pe = petroleum ether; w = water. rotations in chloroform, unless stated otherwise.

[b] $[\alpha]_D$ rotations in chloroform.

1. Hofer and Meyer (1960).
2. Iseli et al. (1965).
3. Rees et al. (1959).
4. Linde and Meyer (1959).
5. Ragab et al. (1962).
6. van Gils (1938).
7. Meyer (1949a).
8. Tamm and Gubler (1959).
9. Linde and Meyer (1958).
10. Wieland and Weil (1913).
11. Kondo and Ohno (1938c).
12. Kondo and Ohno (1938b).
13. Wieland et al. (1936).
14. Urscheler et al. (1955).
15. Wieland and Weyland (1920).
16. Kondo and Ohno (1938e).
17. Ruckstuhl and Meyer (1957).
18. Deulofeu (1940).
19. Wieland and Vocke (1930).
20. Ohno et al. (1961).
21. Katz (1957).
22. Barbier et al. (1959).
23. Chen and Chen (1933d).
24. Bharucha et al. (1961b).
25. Schröter et al. (1958a).
26. Chen et al. (1931–1932a).
27. Barbier et al. (1961).
28. Chen et al. (1933c).
29. Tamm (1960).
30. Schmutz (1949).
31. Kondo and Ohno (1938a).
32. Ohno and Ohmoto (1961).
33. Linde et al. (1966).

$NaBH_4$, Tamm and Gubler, 1959, or $LiAlH[OC(CH_3)_3]_3$ Tamm, 1960, reduction). The ratio of bufalin to 3-epibufalin was estimated at about 1:7.

   e. Bufalon. Although bufalon has long been known as a dehydration product of bufalin (Kotake and Kuwada, 1939; Kuwada, 1939), it was only recently isolated from natural material (Iseli *et. al.,* 1965), together with 3-epibufalin. Whether it is an artifact formed during the isolation process or a genuine substance cannot be said.

   f. Bufarenogin (XXI). The ketol bufarenogin ($C_{24}H_{32}O_6$), first isolated by Rees and associates in 1959 from the venom of *B. arenarum* Hensel, is isomeric with arenobufagin, and may be formed when the latter comes into contact with alumina (Rees *et al.,* 1959). Since arenobufagin has the same structure as the cardenolide aglycone sinogenin (see above), if one disregards the lactone ring, an analogous inversion of the ketol system may occur as in the case of sinosid, the glycoside of sinogenin. The latter affords caudosid, a 11-keto-12a-hydroxy compound (Renkonen *et al.,* 1959a,b). Rees and co-workers (1959) believe that bufarenogin is also a 11-keto-12-hydroxy steroid. The orientation of the hydroxyl group is still in question since molecular rotation measurements are not compatible with those of model substances [for new results see Huber *et al.* (1967)].

   g. Bufotalidin = Hellebrigenin (IV). Bufotalein, so called by Weil in his doctoral thesis (1913), was isolated from the common European toad. Wieland, Weil's teacher, later changed this name to bufotalidin (1920). The correct molecular formula ($C_{24}H_{32}O_6$) was only elucidated in 1936 by Wieland and associates. In a reinvestigation of the venom of *B. bufo bufo* in 1955 Urscheler and co-workers also obtained bufotalidin. Its identity could be established by direct comparison with Wieland's compound. At the same time the remarkable discovery was made that bufotalidin is identical with hellebrigenin, the aglycone of the glycoside hellebrin. The former corresponds in its structure to strophanthidin if one disregards the side chain (Schmutz, 1949).

   h. Bufotalin (V). In 1913 Weil and Wieland and Weil (1913) succeeded in isolating in crystalline form the main substance of *B. bufo bufo,* bufotalin, which had been described as an amorphous substance and found strongly heart-active by Faust as early as 1902. For three decades bufotalin has remained the most investigated bufogenin and the work undertaken on this steroid has been the subject of a series of publications by Wieland and his school. In 1920 (Wieland and Weyland, 1920) the correct molecular formula was found ($C_{26}H_{36}O_6$), and at the same time the presence of a lactone ring, a secondary acylable hydroxyl group and a tertiary hydroxyl group was proved. Under the influence of strong hydrochloric acid, water and acetic acid were split off and a yellow product (bufotalien) was formed. This was interpreted by Wieland as follows: ". . . das vierfach ungesättigte Bufotalien, dem sonst jedes Chromogen fehlt, kann seine intensiv gelbe Farbe *nur der koordinierten Lage seiner vier Doppelbindungen verdanken*" (Wieland and Weyland, 1920). In Wieland's publication (Wieland *et al.,* 1936), in which he gives correct interpretation of the lactone ring, he unfortunately overlooked his previous statement that the four double bonds must be in conjugation in order to afford a yellow product (bufotalien), and located the hydroxyl and the acetoxy groups at position C-14 and C-5, respectively. The correct structure formula was not proved until 1949. At this time permanganate degradation of bufotalin acetate afforded an acid, which proved to be identical with the etianic acid, obtained by the same procedure from oleandrigenin acetate (= gitoxigenin diacetate) (Meyer, 1949e).

i. Bufotalinin (X). Bufotalinin was isolated from *B. bufo bufo* by Wieland and Hesse in 1935. They stressed that this substance does not possess an acetoxyl group and readily decomposes. A later investigation (Wieland *et al.,* 1936) established the correct formula ($C_{24}H_{30}O_6$) and showed that bufotalinin forms only a monoacetate. With alkali in methanol the enol salt of an ester is formed, but no 14-isoproduct (14→21-enol ether). This led to the interpretation that a HO-group might be missing at C-14. Work in Reichstein's laboratory in 1955 has shown that on CrO_3 oxidation bufotalinin yields neutral and acidic material in a ratio of about 2:1 (Urscheler *et al.,* 1955). This was taken as evidence for the presence of an aldehyde group. It was later observed in the same laboratory (Schröter *et al.,* 1958b) that tetrahydrobufotalinin acetate shows $\lambda_{max}^{alcohol}$ 301 mμ (log ϵ = 1.68). This and the optical rotatory dispersion curves of tetrahydrohellebrigenin acetate, tetrahydrobufotalinin acetate and 3β, 20-di[dimethylamino]-18-oxopregnene-(5) were further convincing evidence for a 19-aldehyde group. Furthermore, the probable presence of an epoxide ring was indicated by a band in the infrared spectrum (of the tetrahydro compound), at about 3.32 μ. Several compounds were obtained by Wolff-Kishner reduction of tetrahydrobufotalinin acetate and since all of them (identified by paperchromatography) were produced by the same reaction from tetrahydromarinobufagin acetate it is evident that bufotalinin is the 19-oxoderivative of marinobufagin.

j. Bufotalon. Bufotalon was first obtained by Wieland and Weyland in 1920, by CrO_3 oxidation of bufotalin. In 1965 Iseli *et. al.,* described it as a constituent of *B. formosus* Boulenger. Whether bufotalon is real or an artifact is not known.

k. Cinobufagin (XI). The main bufogenin from Ch'an Su was named cinobufagin by Jensen and Chen in 1930 (1930a). It had been isolated as early as 1916 by Shimizu who referred to it as "substance B." The first thorough chemical investigations by Jensen (1932), and Chen and Chen

(1933f), showed the presence of a lactone ring (probably doubly unsaturated), an acetoxyl, and a secondary hydroxyl group. Furthermore, what they believed to be a tertiary hydroxyl was later shown to be an epoxide group (Hofer *et al.*, 1960b). The correct molecular formula ($C_{26}H_{34}O_6$) was established 1935–36 by Crowfoot (1935) and Crowfoot and Jensen (1936). Tschesche and Offe (1935) obtained Diels' hydrocarbon by selenium dehydration thus establishing the steroid nature of cinobufagin. Kotake and Kuwada (1937a) showed that the cinobufagin, mp 220°C gave by chromatography on alumina two different substances, mp 211°–213°C and 248°–249.5°C. The name cinobufagin was retained for the first, while the second substance was called cinobufotalin. If these two substances are recrystallized in the ratio 2:1 the "old cinobufagin," mp 220°C, is obtained again. It was only in 1960 that Hofer and co-workers (1960b) proved the structure of cinobufagin to be 3β-hydroxy-14,15β-epoxy-16β-acetoxy-5β-bufa-20,22-dienolide.

l. Cinobufaginol (XII). In 1960 Hofer and Meyer (1960) isolated from Ch'an Su a new, very polar bufogenin ($C_{26}H_{34}O_7$), which they named "Substanz G." It forms a diacetate ($C_{30}H_{38}O_9$). The infrared spectrum indicates the absence of a hydroxyl group and the presence of, most probably, an epoxide ring. The NMR spectrum of the diacetate, which has been taken recently, shows in comparison with the one taken of cinobufagin acetate that "Substanz G" must be 19-hydroxycinobufagin and was therefore named cinobufaginol (Linde *et al.*, 1966).

m. Cinobufotalin (XIII). In 1937 Kotake and Kuwada (1937a) were able to resolve the "old cinobufagin" (mp 220°C) into pure cinobufagin and the previously unknown cinobufotalin by chromatography on alumina (cf. cinobufagin). Although initial chemical investigation (Kotake and Kuwada, 1937a,b; Meyer, 1949a) suggested a molecular formula $C_{26}H_{34-36}O_7$, it was not until 1962 that Bernoulli and co-workers established the correct structure. That cinobufotalin has indeed the formula $C_{26}H_{34}O_7$ was proved by oxidation and dehydration to a known α,β-unsaturated ketone, also obtained from cinobufagon. The HO groups at C-3 and C5 were both shown to be β-orientated by conversion of etianic acid derivatives of cinobufotalin and telocinobufagin into the same steroid.

n. Desacetylbufotalin (VI). In an extensive reexamination of Ch'an Su, Ruckstuhl and Meyer (1957) isolated a nonhomogeneous product which they called desacetylbufotalin. Pure crystalline desacetylbufotalin is easily obtained from bufotalin by mild saponification (Ruckstuhl and Meyer, 1957). It has not been established whether desacetylbufotalin is present in Ch'an Su or whether it is formed from bufotalin during chromatography on alumina. Desacetylbufotalin might be a genuine substance, since it has been recently isolated without chromatography on alumina by Iseli *et. al.,* (1965) from a skin extract of *B. formosus* Boulenger.

o. Desacetylcinobufagin (XIV). This substance was isolated first by Ruckstuhl and Meyer (1957) and later by Ohno and co-workers (1961) from Ch'an Su, as an amorphous compound; but has since been obtained in crystalline form (Hofer and Meyer, 1960). It can be prepared artificially from cinobufagin by mild saponification (Hofer and Meyer, 1960). This, and the fact that XIV leads to cinobufagin acetate, are proof of its structure. It is not definite whether this bufogenin is present as such in Ch'an Su or whether it is formed during chromatography on alumina. Desacetylcinobufagin might be a genuine substance, at least of *B. formosus* Boulenger, since it has been recently isolated from this species by Iseli (1965) without resorting to chromatography on alumina.

p. Desacetylcinobufotalin (XV). Ruckstuhl and Meyer (1957) were the first to isolate this bufogenin and to describe it in detail. The structure was proved by transformation into cinobufotalin acetate, and from the fact that desacetylcinobufotalin is formed on mild saponification (Ruckstuhl and Meyer, 1957) or by chromatography of cinobufotalin on alumina (Meyer and Linde, unpublished). Since XV has been isolated from *B. formosus* Boulenger without coming into contact with alumina (Iseli, 1965), it might be regarded as a genuine substance, at least in this case.

q. Gamabufotalin (VII). This bufogenin from *B. formosus* Boulenger was first isolated and described in detail in 1928 by Kotake (1928b,c). An identical substance was later found in the same species of toad by Wieland and Vocke (1930) (= gamabufogenin) and Chen and associates (1933d) (= gamabufagin). Gamabufotalin has the molecular formula $C_{24}H_{34}O_5$ and forms a diacetate. Ohno (1940b) obtained from the latter, by permanganate degradation, an etianic acid unknown at the time. Meyer (1949d) repeated this degradation which afforded apparently the same acid. Its methyl ester was identical with methyl $3\beta,11\alpha$-diacetoxy-14-hydroxy-$5\beta,14\beta,17\alpha H$-etianate (previously obtained from sarmentogenin diacetate; Katz, 1948).

Somewhat later, Imamura and co-workers (1950) repeated the degradation of gamabufotalin acetate but mistakenly concluded that the etianic acid derivatives they obtained were 3,12-disubstituted.

r. Hellebrigenol (VIII). This bufadienolide was first obtained by $NaBH_4$ reduction of hellebrigenin (Katz, 1957) and has been later found by Rees and associates (1959) and by Barbier and co-workers (1959) to be present in the parotid secretion of *B. arenarum* Hensel and *B. marinus* Schneider, respectively.

s. Marinobufagin (XVI). The first member of the bufogenins obtained in crystalline form was previously named bufagin by Abel and Macht (1911–1912) and later renamed marinobufagin by Chen and associates (Jensen and Chen, 1933; Chen and Chen, 1933c). The work by Jensen led to the correct formula ($C_{24}H_{32}O_5$) in 1932. Pataki and Meyer (1955) established the structural formula in all respects, except for the location of one oxygen, which they suggested might be present in an oxide bridge. In a theoretical paper Thiessen (1958) pointed out that the oxide ring must be located at C-14/C-15; this was proved correct by the experimental work of Schröter and co-workers (1959). In connection with the structural elucidation of such oxides, Bharucha and associates (1959) claimed that by reductive opening of the oxide ring (with $NaBH_4$) marinobufagin is converted into telocinobufagin (IX).

t. Resibufogenin (XVII) and the Artebufogenins (XVIII and XIX). While investigating resinous fractions of Ch'an Su in 1952 Meyer isolated an amorphous substance which was characterized as a crystalline monoacetate. Previous workers had had the same substance in hand in varying degrees of purity. This substance described by each investigator under different names (Meyer, 1952) was designated as resibufogenin by Meyer and given the formula $C_{24}H_{32}O_4$. It could also be shown that the four oxygen atoms are distributed as follows: two in the α-pyrone ring, one in a secondary HO group, the fourth in an oxide ring. In a theoretical paper in 1958, Thiessen was able to state with conviction that resibufogenin differs from marinobufagin only in lacking the 5β-hydroxyl group. Shortly afterward Linde and Meyer (1959) independently furnished further results, establishing that this is indeed the case, and clarified the nature of a substance encountered by chance on evaporation of an acetone solution of crude resibufogenin. This artifact, artebufogenin, had already been obtained by Kotake (1934) by treating resibufogenin acetate ("Anhydroacetyl-bufalin") with concentrated hydrochloric acid, and can also be prepared by heating resibufogenin in acetone containing a trace of perchloric acid (Linde and Meyer, 1959). The acid-catalyzed isomerization of resibufogenin to artebufogenin was characterized as cis hydride shift forming the 15-ketone, which shows 14α-configuration (XVIII). This ketone (14α-artebufogenin) is labile and is rearranged by alumina to 14β-artebufogenin (XIX) (Ragab *et al.,* 1962).

u. Telocinobufagin (IX). Telocinobufagin was first isolated in 1949 (Meyer, 1949a) by careful chromatography on alumina of a chloroform extract of Ch'an Su. The name is based on this bufogenin being eluted at the end (Gr. telos = end) of the chromatographic procedure. In the same year the correct formula $C_{24}H_{34}O_5$ as well as the structure (IX) could be established (Meyer, 1949c).

STRUCTURAL FORMULAS OF THE BUFOGENINS

(III)   Bufalin R_1, R_4 = HO; R_2, R_3, R_5 = H; R_6 = CH_3

(IV)   Bufotalidin R_1, R_2, R_4 = HO; R_3, R_5 = H; R_6 = CHO
        (Hellebrigenin)

(V)   Bufotalin R_1, R_4 = HO; R_2, R_3 = H; R_5 = $OCOCH_3$; R_6 = CH_3

(VI)   Desacetylbufotalin R_1, R_4, R_5 = HO; R_2, R_3 = H; R_6 = CH_3

(VII)   Gamabufotalin R_1, R_3, R_4 = HO; R_2, R_5 = H; R_6 = CH_3

(VIII) Hellebrigenol R_1, R_2, R_4 = HO; R_3, R_5 = H; R_6 = CH_2OH

(IX)   Telocinobufagin R_1, R_2, R_4 = HO; R_3, R_5 = H; R_6 = CH_3

(X)   Bufotalinin R_1, R_2 = HO; R_3 = H; R_4 = CHO

(XI)   Cinobufagin R_1 = HO; R_2 = H; R_3 = $OCOCH_3$; R_4 = CH_3

(XII)   Cinobufaginol R_1 = HO; R_2 = H; R_3 = $OCOCH_3$; R_4 = CH_2OH

(XIII) Cinobufotalin R_1, R_2 = HO; R_3 = $OCOCH_3$; R_4 = CH_3

(XIV)   Desacetylcinobufagin R_1, R_3 = HO; R_2 = H; R_4 = CH_3

(XV)   Desacetylcinobufotalin R_1, R_2, R_3 = HO; R_4 = CH_3

(XVI)   Marinobufagin R_1, R_2 = HO; R_3 = H; R_4 = CH_3

(XVII) Resibufogenin R_1 = HO; R_2, R_3 = H; R_4 = CH_3

(XVIII) 14α - Artebufogenin
(XIX)   14β - Artebufogenin

Arenobufagin

(XX)

Bufarenogin

(XXI)

Argentinogenin

(XXII)

TABLE IIIA

R_1	R_2	R_3	R_4	mp (°C)	$[\alpha]_D^{chf}$	Reference
HO	H	H	$\beta COOCH_3$	150-150.5°	+1°	a
AcO	H	H	$\beta COOH$	192-194°		a
AcO	H	H	$\beta COOCH_3$	160-166°	+4°	a
HO	HO	H	$\beta COOCH_3$	189-192.5°	+21°	b
AcO	HO	H	$\beta COOH$	228-229° (dec.)		b
AcO	HO	H	$\beta COOCH_3$	161-162°	+32°	b, c
HO	H	HO	$\beta COOH$	193-196° (dec.)		d
HO	H	HO	$\beta COOCH_3$	158-161°	−18°	d
AcO	H	HO	$\beta COOCH_3$	208-211°	−11°	d
AcO	H	AcO	$\beta COOH$	197-201°		e
AcO	H	AcO	$\beta COOCH_3$	169-171°	+11°	d
HO	HO	HO	$\beta COOCH_3$	186-188°	+2°	f
HO	HO	AcO	$\beta COOCH_3$	215-222°	+27°	f
AcO	HO	HO	$\beta COOCH_3$	195-198°	+18°	f
AcO	HO	AcO	$\beta COOH$	215-217° (dec.)		f
AcO	HO	AcO	$\beta COOCH_3$	175-178°/185-186°	+36°	f
HO	H	H	$\alpha COOCH_3$	172-174.5°	+8°	a
AcO	H	H	$\alpha COOCH_3$	155°	+14°	a
HO	HO	H	$\alpha COOCH_3$	204-225°	+24°	b
AcO	HO	H	$\alpha COOCH_3$	78-108°	+42°	b
HO	H	HO	$\alpha COOH$	225-230° (dec.)		d
HO	H	HO	$\alpha COOCH_3$	225-229°	+29°	d
AcO	H	AcO	$\alpha COOCH_3$	186-187°	+72°	d
HO	HO	HO	$\alpha COOH$	230-235° (dec.)		f
HO	HO	HO	$\alpha COOCH_3$	219-228°/232-234°	+53°	f
AcO	HO	AcO	$\alpha COOCH_3$	199-201°	+87°	f

[a] Meyer (1952).
[b] Schröter et al. (1959).
[c] Pataki and Meyer (1955).
[d] Ruckstuhl and Meyer (1958).
[e] Hofer et al. (1960b).
[f] Bernoulli et al. (1962).

TABLE IIIB

R₁	R₂	R₃	R₄	R₅	R₆	R₇	R₈	R₉	mp (°C)	$[\alpha]_D^{ehf}$	Ref.
HO	H	H	H₂	βHO	H₂	H	CH₃	COOH	234°(dec.)		g
HO	H	H	H₂	βHO	H₂	H	CH₃	COOCH₃	70°		g
AcO	H	H	H₂	βHO	H₂	H	CH₃	COOH	205–245°		g,h
AcO	H	H	H₂	βHO	H₂	H	CH₃	COOCH₃	156–158°	+32°	h
HO	HO	H	H₂	βHO	H₂	H	CH₃	COOH	212–242°		f
HO	HO	H	H₂	βHO	H₂	H	CH₃	COOCH₃	235–237°	+45°	f, i
AcO	HO	H	H₂	βHO	H₂	H	CH₃	COOH	205–208°		f
AcO	HO	H	H₂	βHO	H₂	H	CH₃	COOCH₃	162–165° / 182–184°	+59°	f, i
HO	H	HO	H₂	βHO	H₂	H	CH₃	COOCH₃	140–142°	+18°	j
AcO	H	AcO	H₂	βHO	H₂	H	CH₃	COOH	256–259°		k
AcO	H	AcO	H₂	βHO	H₂	H	CH₃	COOCH₃	167–168°	+17°	k
HO	H	H	H₂	βHO	H₂	HO	CH₃	COOCH₃	166–174° / 174–176°	0°	l
AcO	H	H	H₂	βHO	H₂	AcO	CH₃	COOH	256–267°(dec.)		m
AcO	H	H	H₂	βHO	H₂	AcO	CH₃	COOCH₃	188–190°	−10°	m
HO	HO	H	H₂	βHO	H₂	H	CH₂OH	COOH	225–227°	+38° an	n
HO	HO	H	H₂	βHO	H₂	H	CH₂OH	COOCH₃	167–169°	+54°	n

							m.p.	$[\alpha]_D$	Ref.
HO	H	H_2	βHO	H_2	H	$COOCH_3$	$COOCH_3$	127–129°		[n]
AcO	H	H_2	βHO	H_2	H	$COOCH_3$	$COOCH_3$	157–159°	+75°	[n]
HO	HO	O	βHO	H_2	H	CH_3	COOH	245–249°(dec.)	+49°	[o]
HO	HO	O	βHO	H_2	H	CH_3	$COOCH_3$	156–159°		[p]
AcO	AcO	O	βHO	H_2	H	CH_3	COOH	242–247°	+67°	[p]
AcO	AcO	O	βHO	H_2	H	CH_3	$COOCH_3$	131–134°	+59°	[p]
=O	=O	O	βHO	H_2	H	CH_3	$COOCH_3$	160–165°	+67°	[p]
AcO	H	H_2	αH	O	H	CH_3	$COOCH_3$	180–186°		[q]
HO	H	H_2	βH	O	H	CH_3	COOH	245–249°		[r]
HO	H	H_2	βH	O	H	CH_3	$COOCH_3$	174–177°	−30°	[q]
AcO	H	H_2	βH	O	H	CH_3	$COOCH_3$	176–179°	−20°	[q]
AcO	H	H_2	βHO	O	H	CH_3	$COOCH_3$	201–206°	+60°	[q]
AcO	H	H_2	αHO	O	H	CH_3	$COOCH_3$	238–248°	+54°	[q]
HO	H	H_2	αH	O	H	CH_3	$COOCH_3$	215–232°	+81°	[b]
AcO	H	H_2	αH	O	H	CH_3	$COOCH_3$	195–198°	+101°	[b]
HO	H	H_2	βH	O	H	CH_3	$COOCH_3$	205–211°		[b]
AcO	H	H_2	βH	O	H	CH_3	$COOCH_3$	182–184°	+14°	[b]

[g] Meyer and Reichstein (1947).
[h] Meyer (1949b).
[i] Meyer (1949c).
[j] Meyer and Linde (unpl.).
[k] Meyer (1949d).
[l] Zingg and Meyer (1960).
[m] Meyer (1949e).
[n] Schmutz (1949).
[o] Renkonen et al. (1957).
[p] Hofer et al. (1960a).
[q] Linde and Meyer (1959).
[r] Lardon et al. (1959).

TABLE IV

Bufogenin obtained in crystalline form / as derivative / identified by paperchromatography	No. / (No.) / [No.]	Arenobufagin	Argentinogenin	Bufalin	3-Epibufalin	Bufalon	Bufarenogin	Bufotalidin = Hellebrigenin	Bufotalin
B. alvarius Girard								[4]	
B. arenarum Hensel		1, 5, 6 (2), [3]	6	6			6	6 [3]	
B. asper									[7]
B. bufo bufo L.								8, 9, 10, [3]	11, 12, 8, 9, 13, 10 [3]
B. bufo gargarizans		14		15, 16, 17, 18, 19, 14			14	19, 14	18, 19, 14
B. crucifer Wied 1821		[34]	[34]					[34]	
B. formosus Boulenger		35 [4]	[35]	36, 35, [4]	35	35		35, (36)	35 (37, 38) [4]
B. woodhausi fowleri Hinckley		[4]	[4]						
B. granulosus Spix subsp. fernandezi Gallardo									
B. ictericus Spix 1824		[44]						44	
B. marinus (L.) Schneider			[45]	45				45 [3]	
B. mauritanicus Schlegel		48	49, 48					48 (49), [3]	
B. melanostictus				?[7]				[7]	[7]
B. paracnemis Lutz			[50]	[50]				[50]	
B. peltocephalus Tschudi		[4]	[4]						
B. regularis Reuss		[3] [52] [53]	[52]					52, [3]	
B. spinulosus Wiegmann		[3] [4]							
B. valliceps Wiegmann		[4]							[4]

1. Chen et al. (1933b).
2. Jensen (1935a).
3. Schröter et al. (1958a).
4. Barbier et al. (1961).
5. Wieland and Behringer (1941).
6. Rees et al. (1959).
7. Iseli et al. (1964).
8. Wieland and Hesse (1935).
9. Wieland et al. (1936).
10. Urscheler et al. (1955).
11. Faust (1902).
12. Wieland and Weyland (1920).
13. Raymond-Hamet and Lelogeais (1954).
14. Hofer and Meyer (1960).
15. Kotake and Kuwada (1937b).
16. Kondo and Ohno (1938a).
17. Kotake and Kuwada (1939).
18. Meyer (1949a).
19. Ruckstuhl and Meyer (1957).
20. Jensen and Chen (1930a).
21. Jensen and Chen (1930c).
22. Chen et al. (1931).
23. Chen and Chen (1933e).
24. Jensen and Evans (1934).
25. Tschesche and Offe (1935).
26. Kotake and Kuwada (1937a).
27. Kondo and Ohno (1938e).
28. Kobayashi (1939).

DISTRIBUTION OF THE BUFOGENINS ON THE DIFFERENT TOAD SPECIES INVESTIGATED

Bufotalinin	Bufotalon	Cinobufagin	Cinobufotalin	Desacetyl-bufotalin	Desacetyl-cinobufagin	Desacetyl-cinobufotalin	Gamabufotalin	Hellebrigenol	Marinobufagin	Resibufogenin	Telocinobufagin
											[3]
6 [3]							[3]	6	6 [3]	6	6 [3]
			?[7]						[7]	?[7]	
9, 10 [3]									10 [3]		10 [3]
		20, 21, 22, 23, 24, 25, 26, 15, 27, 28, 29, 18, 14, (30)	26, 15, 27, 28, 31, 29, 18, 19, 14	19	19	19, 14	15, 32, 19, 14 (18)			14, (20), (21) (33)	18 19 14
[34]							[34]	[34]	34		34
35	35	35 (36)	36, 35	36	35 (36)	36, 35	37, 39, 40, 16, 30, 41, 42, 18, 36, 35 [4]		35 [4]	(43), (36) (35) [4]	35, (36), [3], [4]
							[4]		[4]		[4]
							[4]				
[44]							[44]	[44]	44	[44]	44
							(45), [3]	45, [3]	46, 47, 45, [3]	45	47, 45, [3]
48 (49) [3]									48, (49), [3]	48 (49) [3]	48
									[7]	?[7]	
[50]							[50]	[50]	51 [50]	[50]	[50]
							[4]		[4]		[4]
52 [3]							[3], [53]	[3], [52]	[3], [52]	52 [3]	52, [3]
[3]							[3], [4]		[3], [4]	[3], [4]	[3]
							[3], [4]	[3], [4]			[3], [4]

29. Kobayashi (1943).
30. Kondo and Ohno (1938b).
31. Kondo and Ohno (1940).
32. Kondo and Ohno (1939).
33. Meyer (1952).
34. Zelnik and Ziti (1963a).
35. Iseli et al. (1965).
36. Ohno et al. (1961).
37. Kotake (1928b).
38. Kotake (1928c).
39. Wieland and Vocke (1930).
40. Chen et al. (1933d).
41. Kotake and Kubota (1938).
42. Kubota (1938).

43. Ohno and Ohmoto (1961).
44. Zelnik and Ziti (1963b).
45. Barbier et al. (1959).
46. Chen and Chen (1933c).
47. Meyer (1951).
48. Linde and Meyer (1958).
49. Bolliger and Meyer (1957).
50. Zelnik et al. (1964).
51. Deulofeu and Mendive (1938).
52. Bharucha et al. (1961a).
53. Bharucha et al. (1961b).

IX. THE BUFOTOXINS

Toad venoms contain besides the bufogenins, a further type of cardio-active substances, the so-called bufotoxins. These represent conjugates of bufogenins with suberylarginine and were isolated and investigated in a number of laboratories in the 1930's. With the exception of Wieland's bufotoxin, all are rather poorly characterized due to the fact that these substances are extremely difficult to obtain in homogeneous crystals. This is even the case when modern chromatographic methods are applied. No doubt the bufotoxins so far described are mixtures, and it seems very probable that Wieland's bufotoxin is also impure. According to the definition of bufo-toxins given above, the "toxins" of Shimizu (1916) and Kodama (1919, 1921) do not belong to this type of cardioactive compounds but to the bufogenins.

   a. Bufotoxin (Wieland and Alles, 1922) = *Vulgarobufotoxin* (Chen *et al.*, 1931–1932b, 1933a). In 1922, Wieland and Alles were the first to isolate a nitrogen-containing substance from the venom of *B. bufo bufo*, which they named bufotoxin. On acidic hydrolysis this afforded bufotalien (Weil, 1913; Wieland and Weil, 1913) (= an anhydro compound of bufotalin) as the steroid residue and a compound with the formula $C_{13}H_{25}O_5N_4$. Further hydrolysis split the latter up into arginine and suberic acid. Some ten years later Chen and co-workers (1933a) obtained a similar product from the same species of toad and assumed that it was identical with Wieland's bufotoxin. They called it vulgarobufotoxin and retained the term bufotoxin as a general name for the suberylarginine esters of the bufogenins (Chen *et al.*, 1931–1932b). In 1936 Wieland and associates reported on further findings made with their bufotoxin [ultraviolet spectrum, and a new, correct empirical formula $(C_{40}H_{60}O_{10}N_4)$], postulating at the same time, that in bufotoxin the subery-larginine part is esterified with the hydroxyl group at C-14 of the steroid moiety. The view later advanced by Szemenzow (1939), that the suberyl-arginine part should be placed at C-5, is now untenable, following the elucidation of the constitution of bufotalin (see cf.), which at C-5 carries only hydrogen. In 1941, Wieland and Behringer were able to report on the results of further experiments (the indication of an acetoxyl group, and the formation of a dehydration product which was considered to be a 3-ketone); these seemed to be an unequivocal evidence that bufotoxin is a 14-suberylar-ginine ester. Regarding this question, Fieser and Fieser (1959, p.792) write: "In the absence of further characterization of the ketonic product, the evidence seems to us to lack conviction." We should like to associate our-selves with this view, for good reasons can be adduced that make it even more probable that the suberylarginine residue is linked to the 3-hydroxyl group of bufotalin (see marinobufotoxin).

   b. Cinobufotoxin. This bufotoxin, isolated from Ch'an Su, was described by Chen and co-workers as early as 1929 (Chen and Jensen, 1929, 1930; Jensen and Chen, 1930 a,c; Chen *et al.,* 1931). Japanese authors (Kondo and Ikawa, 1933a; Ohno and Komatsu, 1953) have also obtained it from the same source. In addition to suberic acid and arginine, acidic hydrolysis of cinobufotoxin yielded only an amorphous steroid residue. The empirical formula set up by the American authors (Jensen and Chen, 1930a; Chen *et al.,* 1931) was later confirmed by Japanese workers (Kondo and Ikawa, 1933a; Ohno and Komatsu, 1953) who likewise established it as $C_{40}H_{62}O_{11}N_4$. If the molecule of difficultly eliminated water of crystallization and suberylarginine are subtracted from this formula, it remains $C_{26}H_{36}O_6$, the formula for bufotalin. Though the close relationship with Wieland's bufotoxin was especially stressed (Chen *et al.,* 1931), these two bufotoxins do not seem to be identical (a direct comparison of them has never been carried out). As cinobufagin ($C_{26}H_{34}O_6$) is the main bufogenin of Ch'an Su, this suggests that this bufogenin is the steroid part of cinobufotoxin, which in turn would lead to the formula $C_{40}H_{58}O_{10}N_4$ for cinobufotoxin. Should this really prove to be the case, then the suberylarginine residue could be linked only at C-3, as only at this point is an HO group free for esterification. From the Chinese toad *B. gargarizans* Cantor Chen and co-workers (1931–1932b; Chen and Chen, 1933e) isolated a bufotoxin, the properties of which agreed favorably in melting point and analytical data with the values found for cinobufotoxin.

c. Gamabufotoxin. This bufotoxin was first isolated by Wieland and Vocke (1930) from the venom of the Japanese toad *B. formosus* Boulenger. On hydrolysis with 1 *N* HCl these investigators obtained a product ($C_{24}H_{32}O_4$ mp 261°C) which they named anhydrogamabufotalin II. Hydrolysis with 0.5 *N* HCl, however, yielded the isomeric anhydrogamabufotalin I (mp 204°C) which by the action of concentrated HCl could be rearranged into the higher melting isomer. Gamabufotalin itself, after heating with 0.5 *N* HCl, yielded anhydrogamabufotalin I as the only crystallizable product. The structures of anhydrogamabufotalin I and II were recently reported (Komatsu, 1964). This showed in fact that gamabufotalin ($C_{24}H_{36}O_5$) represents the steroid residue of gamabufotoxin. The formula $C_{38}H_{58}O_9N_4$ for this bufo-toxin, i.e., the same formula as that orginally given (Wieland and Vocke, 1930) minus 1 mole of water of crystallization, is thus established. Gamabufotoxin has also been isolated from the same species of toad, and described by Chen and associates (1933d), as well as by Kondo and Ohno (1938a,b).

d. Marinobufotoxin. This bufotoxin was first isolated by Jensen and Chen (1930b,c) from the venom of *B. marinus* Schneider in 1930. Jensen and Evans (1934) later showed that on acidic hydrolysis marinobufotoxin gives a dianhydro compound ($C_{24}H_{28}O_3$, mp 245°–246°C), as the steroid part. This has proved to be identical with the anhydro product obtained by the same procedure from marinobufagin (formerly bufagin). We have repeated this experiment with marinobufagin (Meyer and Linde, unpublished) and also obtained a product of the same melting point as that reported by Jensen and Evans. On the basis of the ultraviolet spectrum [$\lambda_{max}^{alcohol}$ 298, $\sim$241 sh, 232, 225, and 220 mμ sh (log ϵ = 3.77, $\sim$4.13, 4.33, 4.335 and 4.28)] and the infrared spectrum [strongly increased shoulder (as with the artebufogenins), at 1735–1740 cm^{-1} compared with marinobufagin and no hydroxyl band] this anhydro product is $\Delta^{3,5}$-15-keto-bufa-20,22-dienolide. This proves conclusively that the suberylarginine residue in marinobufotoxin must have been either at C-5 or C-3, since the oxide bridge in marinobufagin is located at C-14/C-15. As indicated above, only C-14 and C-3 are relevant as linkages in bufotalin. Thus, we believe it most probable that the suberylarginine part in marinobufotoxin, and in all other bufotoxins (also cinobufotoxin), should be localized at C-3. This would represent a certain similarity to the cardioactive glycosides which bear their sugar residue at C-3. After subtracting 1 mole of water of crystallization which is extremely difficult to remove, the formula given for marinobufotoxin by Chen and Chen (1933c) ($C_{38}H_{58}O_{10}N_4$) turns out to be correct. Marinobufotoxin has also been isolated from *B. paracnemis* Lutz, a variety of *B. marinus* (Deulofeu and Mendive, 1938).

X. MISCELLANEOUS

The name *americanobufagin* was given by Chen and Chen (1933d) for a bufogenin, which they assumed to be present in *B. americanus*.

From *B. arenarum* Hensel nitrogen-containing compounds were isolated by two different groups (Jensen and Chen, 1930c; Chen *et al.*, 1933d; Wieland and Behringer, 1941) and named by both *arenobufotoxin*. The empirical formula was determined as $C_{39}H_{60}O_{11}N_4$ and $C_{38}H_{54}O_9N_4$, respectively. The strong divergence in the analytical data from the different laboratories makes it obvious that these two bufotoxins cannot be identical (mixtures?).

The name *bufagin* was first given by Abel and Macht (1911, 1911-1912) to marinobufagin and also by Kodama (1919, 1921), Kotake (1928a, 1934), and Chen and Jensen (1929) to cinobufagin. The expression "bufagin" should be used only in the general sense, i.e., as a synonymous word for bufogenin. (See later under *B. melanostictus*.)

F_1–F_3 *Bufotalin* are bufogenins from *B. formosus* Boulenger, which have been described by Kondo and Ohno. F_1 and F_2 could only be isolated in crystalline form as *p*-nitrobenzoates (mp 171°–178°C and 286°–287°C, respectively (1938b). F_3, the source of which is not given clearly, was obtained in crystalline form (mp 243°–245°C) (Kondo and Ohno, 1938c).

In the earliest literature of toad venoms the name "*bufotoxin*" has been used for two different products isolated from Ch'an Su, which were relatively soluble compared with normal bufogenins. The first one was obtained by Shimizu (1916) and named substance C or "bufotoxin." It was found to be free of nitrogen and therefore cannot belong to the series of the bufotoxins in the sense of the name being used today. It was probably a mixture containing mainly resibufogenin. The second substance described as "bufotoxin" was isolated by Kodama (1919, 1921) by treating a mother liquor, which contained quite soluble material, with HCl gas. The compound that precipitated melted at 203°–204°C (corr.). Kodama assumed that this compound contained only C, H, and O. We believe that it was in fact the chlorohydrin of resibufogenin, since the values for C and H found for Kodama's "bufotoxin" agree very well with the ones calculated for resibufogenin chlorohydrin.

In the last few years (Hofer and Meyer, 1960) two new substances have been isolated from Ch'an Su and designed as G and H. From NMR, IR and UV data it was recently concluded (Linde et al., 1966) that substance G is a C-19 oxygenated cinobufagin and was therefore named cinobufaginol. From infrared data was deduced that substance H contains also an epoxide and an acetoxyl group. The little amount of H did not allow further investigation.

Cinobufotalidin ($C_{24}H_{34}O_6$) was described by Kondo and Ohno in 1938 (1938e) as a by-product of cinobufagin (from Ch'an Su). It was stated to contain no acetyl group and shows the expected bufadienolide ultraviolet spectrum. An "acetyl-anhydro-bufotalidin" ($C_{26}H_{34}O_6$, mp 209°–210°C) and a mono p-nitrobenzoate ($C_{31}H_{35}O_8N$, mp 236°–238°C) were obtained. On sublimation cinobufotalidin lost 2 molecules of H_2O, giving a "dianhydrocompound" ($C_{24}H_{30}O_4$, mp 125°–128°C). Two reports from the same school by Kobayashi (1939, 1943) deal only with the isolation of cinobufotalidin and do not give further chemical results. Although Ch'an Su has since been investigated very thoroughly (Ruckstuhl and Meyer, 1957; Hofer and Meyer, 1960) it was not possible to isolate the bufogenin in question. But one should bear in mind that so far it has not been possible to determine with certitude the exact source of Ch'an Su. According to Chen and Chen (1933e) the source of Ch'an Su might be B. gargarizans Cantor, while Kobayashi (1943) states that the Chinese drug is made from the dried skin secretion of various Chinese toads.

Fowlerobufagin was isolated by Chen and Chen (1933d) from B. fowleri and is probably not homogeneous. Lately Barbier and co-workers (1961) identified the bufogenins from this toad species by paper chromatography.

Gamabufotalininol. From a skin extract of B. vulgaris formosus Boulenger Ohno and co-workers (1961) isolated this new bufogenin with the mp 263°–265°C and the empirical formula $C_{26}H_{36}O_6$. In a very careful investigation of a skin extract from the same toad, Iseli et al. (1965) could not detect this substance. The ultraviolet spectral data $\lambda_{max}^{alcohol}$ 305 mμ; log ϵ = 4.81 given by Ohno and associates (1961) for gamabufotalininol are erroneous. A sample kindly supplied by Prof. Ohno enabled us to reproduce the ultraviolet spectrum. Our findings on the basis of the formula $C_{26}H_{36}O_6$ $\lambda_{max}^{alcohol}$ 298 mμ; log ϵ = 3.80, conform with the data normaly found for a bufogenin. According to the infrared spectrum, gamabufotalininol does not contain an acetoxyl group, therefore a C_{26}-formula is most unlikely for this bufogenin.

Isobufalin (Kuwada, 1938) from Ch'an Su was found to be identical with resibufogenin (Meyer, 1952; Linde and Meyer, 1958).

Jamaicobufagin was first obtained by Barbier and associates (1959) from *B. marinus* Schneider in two types of crystals, mp 164°–167°C and 217°–220°C (dec.), respectively. It was stated that (according to paper chromatography) these two crystal modifications were identical. The acetate crystallized from acetone ether as long needles, which showed first a double melting point of 181°–186°/210°–218°C. After recrystallization (from the same solvent) the mp changed to 162°–178°C. These may be due to a rearrangement. From the infrared spectroscopy of the acetate it was deduced that this bufogenin contains, besides the α-pyron ring, a nonacylable hydroxyl group and perhaps an epoxide ring. Lack of material prevented a closer chemical investigation.

From the skin extract of *B. mauritanicus* Schlegel six unknown bufogenins (A-F) were isolated (Linde and Meyer, 1958). Only one of them could be obtained crystalline (mp 195°–202°C), but the amounts obtained were too small for a chemical investigation.

From *B. melanostictus* have been isolated three cardioactive compounds named bufagin A, B, and C (van Gils, 1938). Reexamination of the crude venom secretion by paper chromatography showed that hellebrigenin and marinobufagin are the main constituents (Iseli *et al.,* 1964).

Pseudobufotalin and *pseudodesacetylbufotalin.* In 1933 Kondo and Ikawa (1933a) isolated from Ch'an Su a substance (mp 145°–146°C), which they assumed to be the same as Wieland's bufotalin. Later these authors proved that it was not identical and therefore they changed the name of bufotalin into pseudobufotalin (Kondo and Ikawa, 1933b). Because pseudobufotalin is lacking an acetoxyl group it was then renamed pseudodesacetylbufotalin (Kondo and Ohno, 1938a). In this work it was isolated as an amorphous product. The monoacetate was also amorphous, but a crystalline p-nitrobenzoate (mp 226°–227°C) and 3,5-dinitrobenzoate (mp 238°–240°C) could be obtained. By chromatography, Kobayashi (1939) isolated from the crude pseudodesacetylbufotalin besides cinobufagin, cinobufotalin, cinobufotalidin "pure" amorphous pseudodesacetylbufotalin ($C_{24}H_{34}O_5$). Finally pseudodesacetylbufotalin acetate was degraded by Ohno (1940a) with permanganate in acetone and afforded an etianic acid (mp 180°–183°C) unknown so far in its structure.

Regularobufogenin, which was isolated by Chen and Chen (1933b) and

Jensen (1935a) from *B. regularis* Reuss, was proved in Reichstein's laboratory (Bharucha *et al.*, 1961b) to be a mixture of mainly gamabufotalin and appreciable amounts of arenobufagin.

Vallicepobufagin. This bufogenin was isolated by Chen and co-workers (1931–1932a; Chen and Chen, 1933d) from *B. valliceps* Wiegmann, mp 212°–213°C (from ethanol). The analysis agreed very well with $C_{26}H_{38}O_6$, which can be resolved in $C_2H_5OH + C_{24}H_{32}O_5$. Since marinobufagin has not only the same formula but also shows the same melting point, it seems to us most probable that vallicepobufagin is identical with marinobufagin. This assumption is further supported by the fact that Barbier and associates (1961) and Schröter and co-workers (1958a) demonstrated by paper chromatography that marinobufagin is one if not the main constituent of the venomous secretion of *B. valliceps* Wiegmann.

Viridobufagin (Quercicobufagin). This bufogenin was obtained by Chen and associates (1931–1932a, 1933c) from *B. viridis viridis* Laurenti. The melting points of this substance and its diacetate (255°–255.5° and 253°–254° C respectively) agree with those observed for gamabufotalin and its acetate. We assume that these two compounds are identical. Chen and Chen (1933d) isolated from *B. quercicus* Holbrook a bufogenin, which they named quercicobufagin. These authors believe that this substance is identical with viridobufagin, which shows the same melting point. If their assumption is really correct, we can conclude that not only viridobufagin but also quercicobufagin are identical with gamabufotalin.

The following bufotoxins could be obtained in very small quantities and in an amorphous form only: *alvarobufotoxin* (Chen and Chen, 1933d), *americanobufotoxin* (Chen and Chen, 1933d), *fowlerobufotoxin* (Chen and Chen, 1933d), *quercicobufotoxin* (Chen and Chen, 1933d,f), *vallicepobufotoxin* (Chen and Chen, 1933d,f).

Chen and Chen (1933b) obtained from *B. regularis* Reuss a compound designated *regularobufotoxin*, in an apparently impure state. Jensen (1935a) later reported an mp of 205°C and assigned the empirical formula $C_{39}H_{60}O_{11}N_4$ to this substance.

For *viridobufotoxin*, which was isolated from *B. viridis viridis* Laurenti by Chen and co-workers (1933c), the following data have been given: mp 198°–199°C and molecular formula $C_{37}H_{60}O_{10}N_4$.

XI. THE BIOGENESIS OF THE TOAD POISONS

Very little is known about the biosynthetic pathway of the formation of the toad poisons. The close relationship in the structure of the bufadienolides and the bile acids indicate that these two types of steroids are formed from the same basic substance. Tschesche and Koîte (1953) assumed that the lactone ring of the bufadienolides is built up from oxalacetoacetic acid. The investigations so far undertaken with labeled substances, which could serve as building stones for the biosynthesis of the bufadienolides, have only shown, that radioactive carbonate (Doull *et al.*, 1951) or sodium acetate-1-^{14}C (Siperstein *et al.*, 1957)), and sodium acetate-2-^{14}C (Doull *et al.*, 1951) are not incorporated. On the other hand, by feeding *B. marinus* with ^{14}C—containing algae (Doull *et al.*, 1951) or by parenteral administration of cholesterol-4-^{14}C (Siperstein *et al.*, 1957) it was possible to obtain radioactive marinobufagin and marinobufotoxin, respectively. These results show that cholesterol is a precursor in the synthesis of the bufogenins and bufotoxins. This observation would suggest that cholesterol or a closely related compound is the major source of the cardiotonic sterols in the toad. The conversion of cholesterol to marinobufagin implies a number of biochemical transformations of which the most interesting appears to be the alteration of the cholesterol side chain to give the doubly unsaturated lactone ring. Studies with cholesterol labeled in the side chain should clarify this problem.

XII. THE SYNTHESIS OF BUFADIENOLIDES

While the synthesis of C_{23} steroids possessing a 17β-butenolide was realized in 1941 (Ruzicka *et al.*, 1941; Turner, 1948) a steroid with a 14β-hydroxyl substituent in addition to the 17β-butenolide, both of which are necessary requirements for a true cardenolide, was prepared only recently (Danieli *et al.*, 1962). Earlier attempts to synthesize C_{24} steroids of the bufogenin type failed (Fried and Elderfield, 1941) or contradicted each other (Fried and Elderfield, 1941, Ruzicka, 1944). Finally in 1961 Bertin and co-workers (1961) were able to obtain a C_{24} steroid lactone but with 14α-configuration. The method used by these investigators is not suitable for steroids containing the desired 14β-hydroxyl function, since this group would be lost during synthesis or would give rise to a rearrangement of the β-orientated side chain.

REFERENCES

Abel, J. J., and Macht, D. I. (1911). *J. Am. Med. Assoc.* **56**, 1531.
Abel, J. J., and Macht, D. I. (1911–1912). *J. Pharmacol. Exptl. Therap.* **3**, 319.
Arnold, W., Bührer, R., von Euw, J., Lüscher, E., Schindler, O., Stich, K., Zoller, P., and Reichstein, T. (1963). *Helv. Chim. Acta*, **46**, 178.
Barbier, M., Schröter, H., Meyer, K., Schindler, O., and Reichstein, T. (1959). *Helv. Chim. Acta* **42**, 2486.

Barbier, M., Bharucha, M., Chen, K. K., Deulofeu, V., Iseli, E., Jäger, H., Kotake, M., Rees, R., Reichstein, T., Schindler, O., and Weiss, E. (1961). *Helv. Chim. Acta* **44**, 362.
Behringer, H. (1943). *Angew. Chem.* **56**, 83.
Bernasconi, R., Sigg, H. P., and Reichstein, T., (1955). *Helv. Chim. Acta* **38**, 1767.
Bernoulli, F., Linde, H., and Meyer, K. (1962). *Helv. Chim. Acta* **45**, 240.
Bertin, D., Nédélec, L., and Mathieu, J. (1961). *Compt. Rend.* **253**, 1219.
Bharucha, M., Jäger, H., Meyer, K., Reichstein, T., and Schindler, O. (1959). *Helv. Chim. Acta* **42**, 1395.
Bharucha, M., Jäger, H., Weiss, E., and Reichstein, T. (1961a). *Helv. Chim. Acta* **44**, 651.
Bharucha, M., Chen, K. K., Weiss, E., and Reichstein, T. (1961b). *Helv. Chim. Acta* **44**, 844.
Bolliger, R., and Meyer, K. (1957). *Helv. Chim. Acta* **40**, 1659.
Burchard, H. (1890). *Chem. Zentr.* **61**, I, 25.
Burchard, H. (1889). Inaugural Dissertation, University of Rostock.
Chen, K. K., and Chen, A. L. (1933a). *J. Pharmacol. Exptl. Therap.* **47**, 281.
Chen, K. K., and Chen, A. L. (1933b). *J. Pharmacol. Exptl. Therap.* **49**, 503.
Chen, K. K., and Chen, A. L. (1933c). *J. Pharmacol. Exptl. Therap.* **49**, 514.
Chen, K. K., and Chen, A. L. (1933d). *J. Pharmacol. Exptl. Therap.* **49**, 526.
Chen, K. K., and Chen, A. L. (1933e). *J. Pharmacol. Exptl. Therap.* **49**, 543.
Chen, K. K., and Chen, A. L. (1933f). *J. Pharmacol. Exptl. Therap.* **49**, 561.
Chen, K. K., and Jensen, H. (1929). *J. Amer. Pharm. Assoc.* **18**, 244.
Chen, K. K., and Jensen, H. (1930). *Proc. Soc. Exptl. Biol. Med.* **26**, 378.
Chen, K. K., Jensen, H., and Chen, A. L. (1931). *J. Pharmacol. Exptl. Therap.* **43**, 13.
Chen, K. K., Jensen, H., and Chen, A. L. (1931–1932a). *Proc. Soc. Exptl. Biol. Med.* **29**, 905.
Chen, K. K., Jensen, H., and Chen, A. L. (1931–1932b). *Proc. Soc. Exptl. Biol. Med.* **29**, 907.
Chen, K. K., Jensen, H., and Chen, A. L. (1933a). *J. Pharmacol. Exptl. Therap.* **47**, 307.
Chen, K. K., Jensen, H., and Chen, A. L. (1933b). *J. Pharmacol. Exptl. Therap.* **49**, 1.
Chen, K. K., Jensen, H., and Chen, A. L. (1933c). *J. Pharmacol. Exptl. Therap.* **49**, 14.
Chen, K. K., Jensen, H., and Chen, A. L. (1933d). *J. Pharmacol. Exptl. Therap.* **49**, 26.
Crowfoot, D. (1935). *Chem. & Ind. (London)* **54**, 568.
Crowfoot, D., and Jensen, H. (1936). *J. Am. Chem. Soc.* **58**, 2018.
Daly, J., and Witkop, B. (1968) *Mem. Inst. Butantan, Simp. Internac.* **2**, 425–432.
Danieli, N., Mazur, Y., and Sondheimer, F. (1962). *J. Am. Chem. Soc.* **84**, 875.
Dean, F. M. (1963). "Naturally Occuring Oxygen Ring Compounds," pp. 98 and 99. Butterworth, London and Washington, D.C.
Deulofeu, V. (1940). *Bol. Soc. Quim. Peru* **6**, 27; *Chem. Abstr.* **35**, 1427 (1941).
Deulofeu, V. (1948). *Fortschr. Chem. Org. Naturstoffe,* **5**, 241.
Deulofeu, V. and Mendive, J. R. (1938). *Ann. Chem.* **534**, 288.
Deulofeu, V., Duprat, E., and Labriola, R. (1940). *Nature* **145**, 671.
Doull, J., Dubois, K. P., and Geiling, E. M. K. (1951). *Arch. Intern. Pharmacodyn.* **86**, 454.
Duncan, G. R. (1962). *J. Chromatog.* **8**, 37.
Faust, E. S. (1902). *Arch. Exptl. Pathol. Pharmakol.* **47**, 279; **49**, 1 (1902).
Fieser, L. F., and Fieser, M. (1959). "Steroids," p. 787. Reinhold, New York.
Fried, J., and Elderfield R. C. (1941). *J. Org. Chem.* **6**, 566.
Gessner, O. (1926). *Arch. Exptl. Pathol. Pharmakol.* **118**, 325.
Hofer, P., and Meyer, K. (1960). *Helv. Chim. Acta* **43**, 1495.
Hofer, P., Linde, H., and Meyer, K. (1960a). *Helv. Chim. Acta* **43**, 1950.
Hofer, P., Linde, H., and Meyer, K. (1960b). *Helv. Chim. Acta* **43**, 1955.

Huber, K., Linde, H., and Meyer, K. (1967). *Helv. Chim. Acta* **50**, 1994.

Hunziker, F., and Reichstein, T. (1945). *Helv. Chim. Acta* **28**, 1472.

Ikawa, S. (1935). *J. Pharm. Soc. Japan* **55**, 144.

Imamura, J., Nakagawa, M., and Kotake, M. (1950). *J. Inst. Polytech., Osaka City Univ.* C1, 15; *Chem. Abstr.* **45**, 4731 (1951).

Iseli, E., Weiss, E., Reichstein, T., and Chen, K. K. (1964). *Helv. Chim. Acta* **47**, 116.

Iseli, E., Kotake, M., Weiss, E., and Reichstein, T. (1965). *Helv. Chim. Acta* **48**, 1093.

Jensen, H. (1932). *Science* **75**, 53.

Jensen, H. (1935a). *J. Am. Chem. Soc.* **57**, 1765.

Jensen, H. (1935b). *J. Am. Chem. Soc.* **57**, 2733.

Jensen, H. (1937). *J. Am. Chem. Soc.* **59**, 767.

Jensen, H., and Chen, K. K. (1930a). *J. Biol. Chem.* **87**, 741.

Jensen, H., and Chen, K. K. (1930b). *J. Biol. Chem.* **87**, 755.

Jensen, H., and Chen, K. K. (1930c). *J. Biol. Chem.* **87**, Sc. Proceedings XXXI.

Jensen, H., and Chen, K. K. (1933). *J. Biol. Chem.* **100**, Proceedings LVII.

Jensen, H., and Evans, E. A., Jr. (1934). *J. Biol. Chem.* **104**, 307.

Jones, R. N., and Gallagher, B. S. (1959). *J. Am. Chem. Soc.* **81**, 5242.

Jones, R. N., and Herling, F. (1954). *J. Org. Chem.* **19**, 1252.

Kaiser, E., and Michl, H. (1958). "Die Biochemie der tierischen Gifte," p. 102. Deuticke, Vienna.

Katz, A. (1948). *Helv. Chim. Acta* **31**, 993.

Katz, A. (1957). *Helv. Chim. Acta* **40**, 831.

Kobayashi, Y. (1939). *Proc. Imp. Acad. (Tokyo)* **15**, 326.

Kobayashi, Y. (1943). *Folia Pharmacol. Japon.* **39**, 134; *Chem. Abstr.* **41**, 5217 (1947).

Kodama, K. (1919). *Acta Schol. Med. Univ. Imp. Kioto* **3**, 299.

Kodama, K. (1921). *Acta Schol. Med. Univ. Imp. Kioto* **4**, 201.

Komatsu, M. (1964). *Yakugaku Zasshi* **84**, 77; *Chem. Abstr.* **61**, 3167 (1964).

Komatsu, M., Kamano, Y., Suzuki, M. (1965). *Bunseki Kagaku* **14**, 1049; *Chem. Abstr.* **64**, 11028 (1966).

Komeya, K., and Takemoto, T. (1957). *Kagaku (Kyoto)* **12**, 453.

Kondo, H., and Ikawa, S. (1933a). *J. Pharm. Soc. Japan* **53**, 2.

Kondo, H., and Ikawa, S. (1933b). *J. Pharm. Soc. Japan* **53**, 62.

Kondo, H., and Ohno, S. (1938a). *J. Pharm. Soc. Japan* **58**, 15.

Kondo, H., and Ohno, S. (1938b). *J. Pharm. Soc. Japan* **58**, 37.

Kondo, H., and Ohno, S. (1938c). *J. Pharm. Soc. Japan* **58**, 102.

Kondo, H., and Ohno, S. (1938d). *J. Pharm. Soc. Japan* **58**, 232.

Kondo, H., and Ohno, S. (1938e). *J. Pharm. Soc. Japan* **58**, 235.

Kondo, H., and Ohno, S. (1939). *J. Pharm. Soc. Japan* **59**, 186.

Kondo, H., and Ohno, S. (1940). *J. Pharm. Soc. Japan* **60**, 230.

Kotake, M. (1928a). *Ann. Chem.* **465**, 1.

Kotake, M. (1928b). *Ann. Chem.* **465**, 11.

Kotake, M. (1928c). *Sci. Papers Inst. Phys. Chem. Res. (Tokyo)* **9**, 233.

Kotake, M. (1934). *Sci. Papers Inst. Phys. Chem. Res. (Tokyo)* **39**.

Kotake, M., and Kubota, T. (1938). *Sci. Papers Inst. Phys. Chem. Res. (Tokyo)* **34**, 824.

Kotake, M., and Kuwada, K. (1937a). *Sci. Papers Inst. Phys. Chem. Res. (Tokyo)* **32**, 1.

Kotake, M., and Kuwada, K. (1937b). *Sci. Papers Inst. Phys. Chem. Res. (Tokyo)* **32**, 79.

Kotake, M., and Kuwada, K. (1939). *Sci. Papers Inst. Phys. Chem. Res. (Tokyo)* **36**, 106; *Chem. Abstr.* **33**, 7304 (1939).

Kotake, M., and Kuwada, K. (1942). *Sci. Papers Inst. Phys. Chem. Res. (Tokyo)* **39**, 361.

Kubota, T. (1938). *J. Chem. Soc. Japan* **59**, 255; *Chem. Abstr.* **32**, 9092 (1938).

Kuwada, K. (1938). *J. Chem. Soc. Japan* **59**, 650.
Kuwada, K. (1939). *J. Chem. Soc. Japan* **60**, 335; *Chem. Abstr.* **35**, 5123 (1941).
Lardon, A., Sigg, H. P., and Reichstein, T. (1959). *Helv. Chim. Acta* **42**, 1457.
Liebermann, C. (1885). *Ber. Deut. Chem. Ges.* **18**, 1803.
Linde, H., and Meyer, K. (1958). *Pharm. Acta Helv.* **33**, 327.
Linde, H., and Meyer, K. (1959). *Helv. Chim. Acta* **42**, 807.
Linde, H., Hofer, P., and Meyer, K. (1966). *Helv. Chim. Acta* **49**, 1243.
Meyer, K. (1946). *Helv. Chim. Acta* **29**, 718.
Meyer, K. (1949a). *Pharm. Acta Helv.* **24**, 222.
Meyer, K. (1949b). *Helv. Chim. Acta* **32**, 1238.
Meyer, K. (1949c). *Helv. Chim. Acta* **32**, 1593.
Meyer, K. (1949d). *Helv. Chim. Acta* **32**, 1599.
Meyer, K. (1949e). *Helv. Chim. Acta* **32**, 1993.
Meyer, K. (1951). *Helv. Chim. Acta* **34**, 2147.
Meyer, K. (1952). *Helv. Chim. Acta* **35**, 2444.
Meyer, K. (1968). *Mem. Inst. Butantan, Simp. Internac.* **2**, 433–440.
Meyer, K., and Linde, H. Unpublished results. The thin layer chromatography method used was that described by Gamp, A., Studer, P., Linde, H., and Meyer, K. (1962). *Experientia* **18**, 292.
Meyer, K., and Reichstein, T. (1947). *Helv. Chim. Acta* **30**, 1508.
Michl, H., and Kaiser, E. (1963). *Toxicon* **1**, 175.
Ohno, S. (1940a). *J. Pharm. Soc. Japan* **60**, 226.
Ohno, S. (1940b). *J. Pharm. Soc. Japan* **60**, 236.
Ohno, S., and Komatsu, M. (1953). *J. Pharm. Soc. Japan* **73**, 796.
Ohno, S., and Komatsu, M. (1956). *J. Pharm. Soc. Japan* **76**, 770.
Ohno, S., and Ohmoto, T. (1961). *J. Pharm. Soc. Japan* **81**, 1341.
Ohno, S., Komatsu, M., and Ohmoto, T. (1961). *Yakugaku Zasshi* **81**, 1345.
Okada, M. (1968). *Mem. Inst. Butantan, Simp. Internac.* **2**, 589–602.
Omoto, T. (1964). *Kagaku No Ryoiki, Zokan* No. **64**, 115; *Chem. Abstr.* **62**, 10800 (1965).
Pataki, S., and Meyer, K. (1955). *Helv. Chim. Acta* **38**, 1631.
Phisalix, C., and Bertrand, G. (1893). *Compt. Rend. Soc. Biol.* **45**, 474.
Ragab, M. S., Linde, H., and Meyer, K. (1962). *Helv. Chim. Acta* **45**, 1795.
Raymond-Hamet, and Lelogeais, P. (1954). *Bull. Soc. Chim. Biol.* **36** 933.
Rees, R., Schindler, O., Deulofeu, V., and Reichstein, T. (1959). *Helv. Chim. Acta* **42**, 2400.
Renkonen, O., Schindler, O., and Reichstein, T. (1957). *Croat. Chem. Acta* **29**, 239.
Renkonen, O., Schindler, O., and Reichstein, T. (1959a). *Helv. Chim. Acta* **42**, 160.
Renkonen, O., Schindler, O., and Reichstein, T. (1959b). *Helv. Chim. Acta* **42**, 182.
Ruckstuhl, J. -P., and Meyer, K. (1957). *Helv. Chim. Acta* **40**, 1270.
Ruckstuhl, J. -P., and Meyer, K. (1958). *Helv. Chim. Acta* **41**, 2121.
Ruzicka, L., Reichstein, T., and Fürst, A. (1941). *Helv. Chim. Acta* **24**, 76.
Ruzicka, L. (1944). U.S. Patent 2362408; *Chem. Abstr.* **39**, 2848 (1945).
Sakurai, K., Yoshii, E., Kubo, K. (1964). *Yakugaku Zasshi* **84**, 1166; *Chem. Abstr.* **62**, 10293 (1965).
Sakurai, K., Yoshii, E., Hashimoto, H., Kubo, K. (1968). *Chem. Pharm. Bull.* **16**, 1140.
Schmutz, J. (1949). *Helv. Chim. Acta* **32**, 1442.
Schröter, H., and Meyer, K. (1959). *Helv. Chim. Acta* **42**, 664.
Schröter, H., Tamm, C., Reichstein, T., and Deulofeu, V. (1958a). *Helv. Chim. Acta* **41**, 140.
Schröter, H., Tamm, C., and Reichstein, T. (1958b). *Helv. Chim. Acta* **41**, 720.
Schröter, H., Rees, R., and Meyer, K. (1959). *Helv. Chim. Acta* **42**, 1385.
Sellhorn, K. (1952). *Arch. Pharmazie* **285**, 382.

Shimizu, S. (1916). *J. Pharmacol. Exptl. Therap.* **8**, 347.

Shoppee, C. W. (1942). *Ann. Rev. Biochem.* **11**, 103.

Shoppee, C. W. (1958). "Chemistry of the Steroids," p. 257. Butterworth, London and Washington, D.C.

Siperstein, M. D., Murray, A. W., and Titus, E. (1957). *Arch Biochem. Biophys.* **67**, 154.

Steiger, M., and Reichstein, T. (1938). *Helv. Chim. Acta* **21**, 828.

Stoll, A., and Hofmann, A. (1935). *Helv. Chim. Acta* **18**, 401.

Stoll, A., Hofmann, A., and Helfenstein, A. (1935). *Helv. Chim. Acta* **18**, 644.

Suga, T. (1968/69). *Mem. Inst. Butantan, Simp. Internac.* **3**, 965–972.

Szemenzow, A. (1939). *Pharm. J.* **12**, 19. (In Russian.) *Chem. Zentr.* **II**, 773. (1940).

Tamm, C. (1960). *Helv. Chim. Acta* **43**, 338.

Tamm, C., and Gubler, A. (1959). *Helv. Chim. Acta* **42**, 473.

Thiessen, W. E. (1958). *Chem. & Ind. (London)* p. 440.

Tschesche, R., and Korte, F. (1953). *Angew. Chem.* **65**, 81.

Tschesche, R., and Offe, H. A. (1935). *Ber. Deut. Chem. Ges.* **68**, 1998.

Tschesche, R., and Offe, H. A. (1936). *Ber. Deut. Chem. Ges.* **69**, 2361.

Tschesche, R. (1945). *Fortschr. Chem. Org. Naturstoffe* **4**, 12.

Tschesche, R. (1954). *In* "Über Sterine, Gallensäuren und verwandte Naturstoffe" (H. Lettré, H. H. Inhoffen, and R. Tschesche, eds.), Vol. 1, p. 287. Enke, Stuttgart.

Turner, R. B. (1948). *Chem. Rev.* **43**, 1.

Urscheler, H. R., Tamm, C., and Reichstein, T. (1955). *Helv. Chim. Acta* **38**, 883.

van Gils, G. E. (1938). *Acta Brevia Neer. Physiol., Pharmacol., Microbiol.* **8**, 84; *Chem. Abstr.* **32**, 8010 (1938).

Weil, F. J. (1913). Inaugural Dissertation, University of München.

Wieland, H., and Alles, R. (1922). *Ber. Deut. Chem. Ges.* **55**, 1789.

Wieland, H., and Behringer, H. (1941). *Ann. Chem.* **549**, 209.

Wieland, H., and Hesse, G. (1935). *Ann. Chem.* **517**, 22.

Wieland, H., and Vocke, F. (1930). *Ann. Chem.* **481**, 215.

Wieland, H., and Weil, F. J. (1913). *Ber. Deut. Chem. Ges.* **46**, 3315.

Wieland, H., and Weyland, P. (1920). *Sitzber. Bayer. Akad. Wiss. Math. Physi. Kl.* p. 329.

Wieland, H., Hesse, G., and Meyer, H. (1932). *Ann. Chem.* **493**, 272.

Wieland, H., Hesse, G., and Hüttel, R. (1936). *Ann. Chem.* **524**, 203.

Zelnik, R., and Ziti, L. M. (1962). *J. Chromatog.* **9**, 371.

Zelnik, R., and Ziti, L. M. (1963a). *Anais Acad. Brasil. Cienc.* **35**, 51.

Zelnik, R., and Ziti, L. M. (1963b). *Anais Acad. Brasil. Cienc.* **35**, 45.

Zelnik, R. (1965). *Ciencia Cult.* **17**, 10; *Chem. Abstr.* **63**, 3361 (1965).

Zelnik, R., Ziti, L. M., and Guimaraes, C. V. (1964). *J. Chromatog.* **15**, 9.

Zingg, M., and Meyer, K. (1960). *Helv. Chim. Acta* **43**, 145.

Chapter 41

Distribution, Biology, and Classification of Salamanders

WOLFGANG LUTHER*

DEPARTMENT OF ZOOLOGY, TECHNOLOGICAL UNIVERSITY, DARMSTADT, GERMANY

Externally the tailed amphibians or salamanders (Amphibia-Caudata-Urodela) resemble lizards and it is not surprising that these two groups of animals are sometimes confused by the layman. Actually the tailed amphibians, as far as anatomy and physiology are concerned, have very little in common with the reptiles. The long, slender body, the clearly delimited, articulated head, and the four articulated limbs are common features of all terrestrial vertebrates. On the other hand, the skin of a salamander is quite different from the dry, horny epidermis of a lizard or crocodile. The amphibian skin is moist and rich in glands, much like the slimy epidermis of fishes. As in certain fishes (weevers—Trachinidae, scorpion fish—Scorpaenidae) the poison of the so-called venomous salamanders and toads is produced by skin glands, whereas the poisonous snakes and the only poisonous lizard, *Heloderma,* form their venom in transformed salivary glands of the mouth cavity.

Compared with the immense variety of different species of frogs (Amphibia caudata, Anura or Salientia) and their world-wide distribution, the Urodela represent a relatively small group of some hundred species, almost entirely confined to the northern hemisphere. Only certain forms of Plethodontidae pass the equator in Central America. The fauna of southeast Asia is equally poor in salamanders. No salamanders are found in South Africa and in Australia, in spite of the fact that a considerable variety of frogs inhabit these regions.

As in frogs, the development of the salamander begins with an aquatic larval stage or tadpole. The typical urodele larva is born or hatched with three or four internal gill arches. For respiration, water enters the oral cavity

* Deceased.

557

through the nasal openings or choanae and leaves it, passing the gills, through a pair of gill slits at the caudal part of the head. As an additional respiratory organ, salamander larvae have several (usually three) external gills (Fig. 1A). Such external gills are also found in larvae of certain primitive fishes *(Polypterus, Ceratodus,* etc.). It seems, therefore, justified to conclude that these gills represent a primitive character which the amphibians have taken over from their ancestors the crossopterygians. The tadpoles of frogs and toads also develop external gills during embryonic life, but these are resorbed soon after the hatching of the embryo.

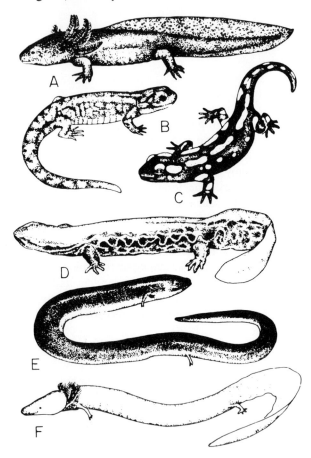

Fig. 1. Some characteristic forms of salamanders of Europe and North America. A. Axolotl-neotenic larva of *Ambystoma mexicanum.* B. *Ambystoma tigrinum,* North America. C. *Salamandra salamandra,* Europe. D. *Cryptobranchus alleganiensis,*"Hellbender," North America. E. *Amphiuma means,* "Congo eel," North America. F. *Proteus anguineus,* cave salamander "Olm," southeastern Europe.

A characteristic feature of salamanders and their larvae is the fact that with the different families and species the external gills disappear at very different times, according to life habits and environment. In many aquatic species the external gills persist until sexual maturity is reached. Such a persistence of larval characteristic is called neoteny by physiologists. Neoteny may be found exceptionally in species which normally undergo complete metamorphosis (e.g., *Ambystoma* and *Triturus*). The external gills of the European cave salamander *(Proteus)* and the American mudpuppies *(Necturus)* (Fig. 1F) must therefore be considered as an adaptation to aquatic life and not as a sign of an especially antique, primitive organization.

Another adaptive characteristic without systematic value is the reduction of the lungs in many terrestrial salamanders. It is typical for the family of Plethodontidae, but it does occur independently in certain Salamandrides and Hynobiides. The lungless salamanders are in general relatively small animals which live in very moist places, such as caves or the moss of small brooks. In this environment skin respiration is sufficient to ensure an adequate supply of oxygen, especially since these animals usually move slowly and lazily. There are some exceptions, however, as some Plethodontidae of the genus *Euricea* run and jump almost like lizards and have often been found at a considerable distance from water. The lungless salamanders take up oxygen not only through their skins, but also through the highly vascularized buccal epithelium. Rhythmic expansion and contraction of throat muscles ensures regular ventilation of the oral cavity. This mouth-breathing can be seen in almost all amphibians, even in species which have well-developed lungs.

The physiological and histological aspect of the larval skin differs considerably from the skin of the metamorphosized land salamander. A section through the larval tail of an axolotl (neotenic form of the *Ambystoma* of Central America, *Ambystoma mexicanum)* shows little or no cornification of the skin surface (Fig. 2). The epidermis is formed by several layers of living cells which are all growing and dividing, even in the last outer layers. Embedded between the actual cells of the epidermis there are glandular cells, called "Leydig's glands" after their discoverer. These are large, round or oval cells with a central nucleus. The plasma is filled with a secretion which appears granular after fixation and does not stain very well with the classic methods. Leydig's cells are ductless. In successfully stained preparations a fine net of capillaries is seen surrounding the cells externally. It probably serves to take up the secretion produced in the plasma (Fig. 3). Similar cells are found in the skins of a great number of bony fishes and in lampreys, where they are called "mucous cells" or "granular cells."

The secretion produced by the Leydig cells does not seem to have poisonous or burning effects. All salamander larvae are eaten by fishes and aquatic birds and even by other amphibians, including those which would

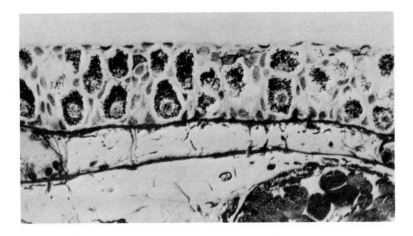

FIG. 2. Larval skin of *Ambystoma mexicanum*. Few horny changes, many mononuclear glands.

refuse the larvae of toads *(Bufo bufo)*. The Mexican axolotl, which reaches a length of about 30 cm, is widely eaten, boiled or fried, by Spaniards and Mexicans and is considered an excellent, nourishing dish.

The skin of the terrestrial forms undergoes important changes in all its layers when the animal adopts land life. The superior layers of the epidermis turn horny and become hard and leatherlike, especially in species which live in a dry environment. This "stratum corneum" protects the underlying living cells of the epidermis from drying out. The unicellular Leydig glands disapear, while large multicellular mucous or poison glands develop in the subepidermal layer of the corium. A long time before the very first external signs of metamorphosis become visible, plungerlike plugs grow from the epidermis down into the subepidermal tissue. These plugs develop into large multicellular glands, spherically shaped, with a narrow duct leading to the surface of the skin (Figs. 4–5). Usually there are two types of such glands. The so-called "mucous glands" are smaller and discharge a clear, thin fluid. The cavity of the gland is wide and the walls are formed by flat or cubiform epithelium. The secretion is apocrine, that is to say, the fluid is secreted by the surface of the epithelium of the walls. These so-called mucous glands are generally distributed very regularly over the whole body surface but may be lacking or be scarce at the fingertips or in the dorsal fin of male *Triturus*. Most authors agree that the mucous glands serve to keep the skin moist and elastic. In the aquatic forms they have probably also some function in osmoregulation.

The larger, so-called poison glands are usually concentrated in certain

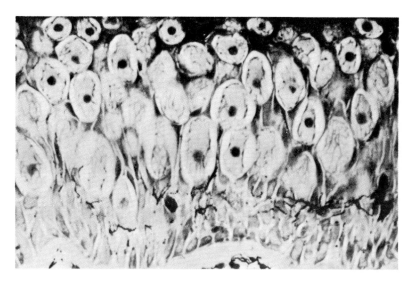

FIG. 3. Larval skin of *Ambystoma mexicanum*. The fine capillary meshwork on the surface of the glandular Leydig cells is clearly seen.

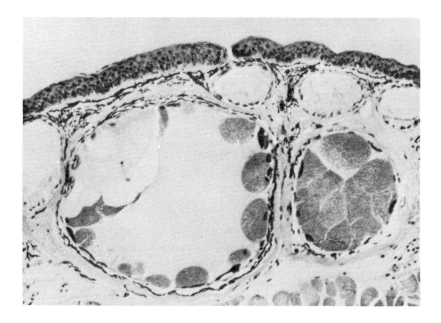

FIG. 4. Section from parotoid gland of a *Salamandra salamandra* after metamorphosis. Poison glands and mucous glands.

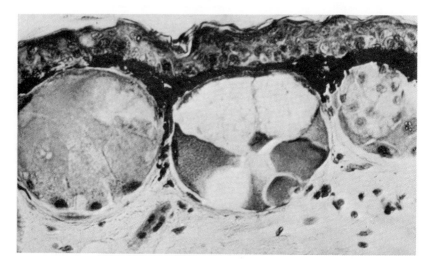

FIG. 5. Section from the body of a *Salamandra salamandra*. Poison glands at different stages of development.

areas of the body. They are especially well developed along the medial line of the back where they form one or several rows. The European earth salamanders *(Salamandra)* have two big clusters of poison glands at both sides of the head, the so-called parotoids. The outlet of the glandular ducts is often indicated by a row of warts on the surface of the skin. In the species *Triturus cristatus* a great number of poison glands are distributed very regularly over the whole surface of the body.

The secretion of the poison glands is whitish, slimy, and sticky. On stimulation the slime is squirted with great power from the glandular ducts. The European earth salamander (genus *Salamandra* L.) can spill the contents of its parotoid glands over a distance of several inches. The wall of a poison gland is formed by large, highly inflated cells, which fill up nearly the whole lumen of the glandular cavity. Secretion is holocrine, that is to say, the whole content of the cells is transformed into secretion and the cells, when ripe, are expelled as a whole into the outleading duct.

The secretion of the so-called mucous glands as well as the secretion of the actual poison glands produces a burning sensation on the mucous membranes of eyes and mouth. It is highly poisonous to most animals, especially to vertebrates. Its composition varies from one species to the other. There are even differences among geographic varieties of the same species. On the other hand, the sensitivity of animals living in the same environment with the salamanders can be rather different. Fishes, for instance, usually do not eat aquatic salamanders, whereas predaceous insects, such as the diving beetle

(Dystiscus marginalis) and its larva, the "water tiger" will attack any amphibian immediately after being put together in an aquarium. The European ringsnake *(Tropidonotus natrix)* feeds in springtime almost exclusively on water newts but does not eat the green frog *(Rana esculenta)* common in the same regions, though these frogs are eaten with pleasure by mammals (polecat, *Putorius putorius)* and birds (heron, stork). To our knowledge no comparative investigations of the skin secretions of different varieties of amphibians have been carried out yet, which would explain these differences.

Systematically the order of Caudata (Urodela) is classified into four groups, which show certain differences of anatomy and reproductive habits. The external aspect: short and stumpy or slender and limbless, like eels, must be considered as an adaptation to certain environmental conditions, just as the persistence of the external gills. It may therefore vary independently with different families and species.

The Hynobiidae are considered the most primitive group. They are found throughout the northern part of east Asia from Siberia to China and Japan. Hynobiidae are generally small or medium-sized animals and look "typically like salamanders." The adults are terrestrial but, as the *Triturus* of Europe and the *Ambystoma* of North America, have aquatic interludes for breeding in spring, going back to land afterward, where they may be found under roots of trees and in moist moss. The frost resistance of certain Hynobiidae is surprising. An Asiatic species, *Hynobius keyserlingi* (DYB), goes up to 66° of northern latitude in east Siberia and has been found at the "cold pole of the world" at Werchojansk, where temperatures below −40°C are registered for months during the winter.

Another curious representative of the Hynobiidae is the clawed salamander, *Onychodactylus japonicus* Houtt, which has sharp horny claws on the toes of its front and hind legs. Similar claws have only been observed in a South African species of frog *(Xenopus)*—they are unique in the group of Urodela.

The Cryptobranchidae (Fig. 1D) belong to the same suborder as the Hynobiidae. They include the giant salamander of Japan *(Megalobatrachus)* and the hellbender of North America. The cryptobranchids are the biggest living salamanders. *Megalobatrachus* is said to reach a length of 160 cm and a corresponding weight. Specimens of a length of 1 m are not at all unusual. Even the hellbender reaches a length of up to 56 cm. The Cryptobranchidae live in clear, fast-running waters. During the day they stay hidden under stones or roots of trees, while at night they are found crawling around at the bottom of the stream. It is most unlikely that they ever leave the water, except when forced to do so. Even in winter they can be observed underneath the ice where they remain active and lively in spite of the low temperature.

A typical external feature of the Cryptobranchidea is the folded, wrinkled skin at the flanks and the legs. The larvae have, as all salamander larvae, external and internal gills. After metamorphosis the external gills disappear, whereas the inner gills persist. Upon stimulation the skin of Cryptobranchus discharges a considerably amount of slime with a peculiar peppery smell. Apparently it has not yet been determined whether the secreted slime is poisonous or not.

The members of the family Sirenidae are still more specialized for aquatic life. They are found exclusively in the south of the United States of America. Sirens are eellike creatures with very small front limbs, too short to be used for walking. The hindlimbs are absent and the animals move like a snake or an eel. The external gills persist throughout its lifetime.

A common feature of all these species is their primitive form of reproduction. Eggs and sperm are spawned into the water, the same way as fishes do, and fertilization takes place outside the body. In higher specialized salamanders the eggs are always fertilized inside the body of the female. The sperm, enveloped in jelly and released in form of a so-called spermatophore, is retained by the female in a special pocket, the spermatheca, and fertilizes the eggs as they are laid. The European earth salamanders *(Salamandra maculosa* and *Salamandra atra)* and the cave salamander *(Proteus)* are viviparous. Fertilization takes place in summer and the developing larvae are retained in the uterus of their mother for a whole year, being born in the following spring at a body length of 3–4 cm.

Salamanders with internal fertilization are the numerous species of Ambystomoidae, living exclusively on the American continent (the best-known form is the Mexican axolotl, Fig. 1A) and the world-wide distributed families of Salamandroidea (including the genera *Salamandra, Triturus, Plethodon,* and many others). The salamanders of the genus *Ambystoma* (= Ambystoma) are predominately terrestrial and are rarely found in the water except at the time of breeding. A typical feature of the great majority of Ambystomae is the stumpy, plump body and the broad, round head with protruding eyes and a relatively short broad snout. The tail is not longer than the body and only slightly flattened at the sides, or even frankly circular in section. The eggs are laid in big clusters into still water or in moss. The larvae of certain species *(A. tigrinum, Dicamptodon)* may remain in the water for several years before they undergo metamorphosis. *Ambystoma mexicanum* (axolotl) reaches sexual maturity at a larval stage of organization. Metamorphosis can be induced at any time by the administration of supplementary thyroid hormone.

The family of Salamandroidea includes a great variety of very different species. Both forms of European earth salamanders (Fig. 1C) *(Salamandra maculosa* and *S. atra)* are strictly terrestrial and can be drowned in water. Both of them are viviparous. In spring *Salamandra maculosa* gives birth to about 25 to 40 gilled larvae of a length of 3 cm, which pass through an aquatic stage before metamorphosing. The larvae are well developed already in October but the female guards them inside her body during hibernation until April of the following year. Two or 3 months after birth the larvae leave the water. *Salamandra atra,* the alpine salamander, gives birth to two live, completely metamorphosed, young salamanders of a length of about 5 cm. Although about 20 to 30 eggs are released from the ovaries, all but one or two disintegrate in the oviducts and their yolk serves to feed the two developing larvae.

Another, not quite so startling, but in its final result equally effective adaptation to terrestrial life is found with certain North American sala-manders of the family Plethodontidae. The genera *Plethodon, Ensantina,* and *Aneides* lay their eggs, separately or in clusters, into hollow trees, mouse holes, or underneath the bark of dead trees, but not in water. The eggs are watched and defended by the female. The larvae develop external gills but do not leave the egg membrane until metamorphosis has occurred. The gills are closely applied to the inner wall of the egg and oxygen is taken up through the egg membrane, as occurs with the allantois of the chick embryo.

Morphologically as well as physiologically the family of Plethodontidae is of great variety. Besides the plump, clumsy, terrestrial salamanders (e.g., *Ensantina*) one finds slender, agile species, as for instance the Euriceae which run and jump like lizards. There are others with tails more than twice as long as their bodies and very short limbs, which at first sight resemble earthworms (worm-salamander, *Batrachoceps*). A common feature of all Plethodontidae is the absence of lungs. Respiration takes place exclusively through the skin and the mucous membranes of the oral cavity, which is well supplied with blood vessels.

The so-called water newts of Europe and their relatives in North America and Asia (genus *Triturus*) every year undergo a curious transformation related to the mating and breeding season. The animals hibernate on land under stones or in holes in the ground. In April during the first warm rainy days they leave their winter refuges and walk, often for miles, to their breeding quarters. They always find their way back, even to hidden ponds high up in the hills.

Ditches and brooks which might be considered as well fit for breeding are crossed without stopping there. Apparently, just as salmons migrate from the sea to the home rivers where they have been bred, the newts return to the water where they have lived as larvae.

Once the animals have reached the breeding place, the body of the male of many species undergoes important changes under the influence of the sexual hormone. The skin becomes more flaccid and relaxed and takes a bright brilliant hue of black, red, or blue pigments. *Triturus cristatus* and *Triturus taeniatus* develop a fringed crest which makes the body look almost twice as big as before. At the same time changes take place in the mucous membranes of the nose which enable the male to smell the female or to find food under water or hidden in the ground. The breeding season lasts for 3 or even 4 months. After this period the brilliant "wedding dress" of the male regresses and the animals go back to land. If this return to land is impeded artificially, certain species, such as *Triturus vulgaris,* may die from malnutrition since they refuse food during their constant efforts to climb up the walls of the aquarium and some may even become literally drowned. Other species, such as the *Triturus cristatus* and the American *Triturus (Diemyctilus) viridescens* remain in water for some time after the breeding season until the cold weather drives them back to their winter quarters.

The Spanish newt, *Pleurodeles,* that lives in wells and cisterns, is completely aquatic and never leaves the water.

The congo eels, *Amphiuma* (Fig. 1E), of North America are a small family with only two species, *A. means* and *A. tridactylum.* Their relationship with the Salamandridae and the Ambystomoidae is confirmed by their breeding habits (internal fertilization). *Amphiuma means* and *A. tridactylum* may reach a length of more than 1 m. The body is eellike and the limbs are reduced to tiny little stumps. Two open gill slits which persist throughout the lifetime are an adaptation to aquatic life and remind one of the cryptobranchids. The eggs, however, are laid on land in moist places where they are guarded by the female. This may indicate a relationship with the Plethodontidae (see p. 559). The congo eel, however, has well-developed lungs. Another interesting anatomic feature of *Amphiuma* is the fact that they have the biggest blood cells of all vertebrates. Their erythrocytes are so large that they may be seen as a fine dust when floating in water. Their volume is about 1000 times the volume of a human red blood cell, but their number per milliliter of blood is proportionately smaller.

The last, highly specialized group of salamanders with internal fertilization which should be mentioned here are the Proteidae, the mudpuppies of America and the *Proteus anguineus* of southeastern Europe. These salamanders are completely adapted to aquatic life and retain their external and internal gills throughout their lifetime. Their neoteny or persisting of larval characters has become, as in sirens, permanent. While in the neotenic axolotl *(Ambystoma mexicanum)* metamorphosis may be induced at any time by an injection of thyroid hormone, the gills and the skin of *Proteus* show no reaction to thyroxine. Evidently the cells have lost the property to react. The European cave salamander shows, moreover, many other adaptations to life in subterranean waters: lack of pigmentation, regression of the eyes, slim eellike body with reduced limbs. *Proteus* is viviparous and gives birth, as the alpine salamander, to only two, relatively big young, which resemble the adult animals in all details. It has been observed that under abnormal conditions (keeping the animals in daylight at an increased temperature) the European cave salamander may lay some 40–50 eggs which hatch into larvae in due course.

Two forms of blind cave salamanders which are found in the subterranean waters of North America show a certain resemblance to the European cave salamander but are probably neotenic Plethodontidae.

Salamanders have no commercial value to man. They are of interest to biologists and chemists because of the poisonous substances produced in the poison glands of their skin and because of their ability to regenerate not only limbs but quite a number of other organs, gills, eyes, and even parts of the skull. Furthermore, the eggs and larvae of aquatic salamanders are interesting test objects for experimental embryology. For the scientist who intends to use these animals for experimental investigations it may be of interest to add at the end of this chapter some information concerning the collection and keeping of common species of terrestrial and aquatic salamanders.

Salamanders, newts, and some species of toads are protected by the law in most civilized countries. They may be caught only for scientific purposes with a special license. In Europe as well as in America, there are licensed collectors who can provide interested scientists with a certain number of almost any fairly frequent species. The addresses of these specialists may be obtained from aquarium magazines or at the zoological departments of most universities.

The live animals are shipped in moist moss or water. If they are not to be killed and preserved immediately, the terrestrial species may be kept in a refrigerator at a temperature of $+4°$ to $+8°C$. They fall into a kind of hibernation and need no food for several months. Axolotls and aquatic salamanders are kept in low aquaria; the water must be changed fairly frequently and they should be fed with earthworms, mosquito larvae and, if these are not available, with very small pieces of meat or liver. It is very important that not too many animals be kept in the same receptacle. One dead specimen which is not discovered and removed immediately poisons the whole container in a very short time.

An excellent method for fresh conservation of salamanders is refrigeration at $-18°C$. The animals are anesthetized with a 2% solution of ethylurethane, and kept in a plastic bag with some water in a deep freeze. In the frozen block the tissues stay unaltered and fresh for years without drying out.

If living salamanders are to be observed in an active state at room temperature, care has to be taken that the container is kept in a cool place, protecting it from direct sunlight. A shelter made of a hollow stone or a fragment of a small flower pot should be provided, under which the animals may hide. The moisture of the ground (moss, filter paper, washed pebbles—no sand!) must be adapted to the necessities of the different species. A high index of moisture of the air is always of great importance. Since salamanders do not move about very much, the container may be relatively small. A round bowl of a diameter corresponding to about one and one-half times the length of the body of the animals is sufficient for two or three specimens. Most salamanders are not very demanding as far as feeding and maintenance are concerned and may be kept in good condition for years (or even decades).

REFERENCES

Ahl, E. (1930). *In* Pax und Arndt "Die Rohstoffe des Tierreiches," Vol. 1.

Bishop, S. C. (1962). "Handbook of Salamanders." Hafner, New York.

Brehm, A. (1914–1918). "Brehms Tierleben," 4th ed., Vol. 5. Kriechtiere und Lurche.

Dawson, A. B. (1920). *J. Morphol.* **31**, 487.

Dennert, W. (1924). *Z. Anat. Entwicklungsgeschichte* **72**, 407.

Drasch, O. (1894). *Arch. Anat. Physiol. Anat. Abt.* pp. 225–268.

Dunn, E. R. (1923). *Proc. Am. Acad. Arts Sci.* **58**, 443.

Dunn, E. R. (1926). "The Salamanders of the Family Plethodontidae." Smith College, Northampton, Massachusetts.

Francis, E. T. B. (1934). "The Anatomy of the Salamander." Oxford Univ. Press (Clarendon), London and New York (contains an extensive bibliography until 1932).

Heidenhain, M. (1893). *Sitzber. Physik. Med. Ges. Wurzburg* **4**, 52.

Leydig, F. (1876). *Morphol. Jahrb.* **2**, 165, 287.

Nirenstein, E. (1908). *Arch. Mikroskop. Anat. Entwicklungsmech.* **72**, 47.

Noble, G. K. (1954). "The Biology of the Amphibia." Dover, New York.

Pfitzner, W. (1880). *Morphol. Jahrb.* **6**, 469.

Schultz, P. (1889). *Arch. Mikroskop. Anat. Entwicklungsmech.* **34**, 11.

Wilder, J. (1925). "The Morphology of Amphibian Metamorphosis." Smith College, Northampton, Massachusetts.

Chapter 42

Toxicology, Pharmacology, Chemistry, and Biochemistry of Salamander Venom

GERHARD HABERMEHL

INSTITUT FÜR ORGANISCHE CHEMIE DER TECHNISCHEN HOCHSCHULE, DARMSTADT, GERMANY

I. INTRODUCTION

The black and yellow spotted fire salamander *(Salamandra maculosa maculosa, Salamandra maculosa taeniata)* has always been considered venomous. In ancient times Nicander of Kolophon and the younger Plinius reported the symptoms of poisoning from salamander venom and possible antidotes; but, obviously, there was nothing known about the poison itself. In the Middle Ages, the writings of the ancient physicians and philosophers were accepted without question; no attempt was made to investigate the toxicity of the venom and one operated only with hypotheses and speculations. Curiously, aside from beliefs such as: the sight of the animal causes sickness or that ingestion of the ashes of salamanders causes death, there are rather exact descriptions of the symptoms of poisoning. The first serious book concerning the symptoms and the possible antidotes, *De venenis animalibus,* was written by Senertius in 1676.

The first experiments on the toxicology of the salamander were made by Laurentius in 1768, and he discovered the skin gland secretion to be the only source of the venom of the salamander. This work was neglected until the middle of the nineteenth century since prior to this time the most curious descriptions are found in the literature that indicate that the writers never directly observed a salamander.

II. TOXICOLOGY AND PHARMACOLOGY OF SALAMANDER VENOM

The investigations on salamander venom since 1860 have been primarily concerned with toxicology. Detailed experimentation was first made by Zalesky (1866). He found that the skin gland venom of the salamander has the character of an alkaloid, that the venom was soluble in alcohol and dilute acids, that it could be precipitated from the acid solution by ammonia, and that it formed precipitates with phosphomolybdic acid, platinic chloride and mercuric chloride. The amorphous base he obtained he named samandarine. We now know that his product was a mixture of alkaloids.

The symptoms of the poisoning appear in the following manner. At first the poisoned animals become restless, then epileptic-like convulsions appear and the pupils become grossly dilated. The reflexes become very faint and finally disappear entirely. Respiration is weak and heart action becomes irregular. Convulsions occur for a few minutes, followed by a period of quiescence until the next generally stronger attack follows. In the terminal stage there are symptoms of paralysis, especially of the hind legs. Usually the animals die within a few hours.

The dissection of the poisoned animals shows a hemorrhage in the lungs, the heart, brain, liver veins are distended. Zalesky's work showed that salamander venom is toxic for all animals, for fishes as well as for amphibians, birds, and mammals. The salamander itself can be affected by its own venom if the venom enters its circulatory system.

Recent examinations on the toxicology and pharmacology of the salamander venom were made by O. Gessner. From his experiments it is known that samandarine affects the central nervous system and the main point of attack is in the spinal cord. The vasomotor center is also affected by an increase in blood pressure. Death occurs primarily by respiratory paralysis without damage to the heart. Furthermore, Gessner (1926) found in the skin gland secretion hemolytically active peptides.

Samandarine is not used therapeutically. It differs from digitalis and toad venom in that the stoppage of the heart in diastole cannot be compensated for by atropine (Gessner and Esser, 1935; Gessner and Urban, 1937). The strong local anesthetic effect of samandarine (Gessner and Möllenhoff, 1932) is also worthy of note. The lethal dose of samandarine for the frog is 0.019 gm, for the mouse 0.0034 gm, and for the rabbit 0.001 gm per kilogram of weight; thus samandarine is one third as toxic as strychnine.

III. CHEMISTRY OF SALAMANDER VENOM

A. Isolation of the Crude Secretion and Preparation of the Pure Alkaloids

The best method of obtaining salamander venom is as follows: The animals are narcotized with gaseous carbon dioxide. Then the skin glands are emptied by means of a glass tube connected to a water pump. To avoid self-contamination, the animals are thoroughly washed. Soon after, they are fully recovered (Schöpf and Braun, 1934). The crude secretion, which has a pH of 8, is a milky glutinous liquid. It promptly stiffens to a gumlike mass; after standing for a few days in the open air, it becomes solid and brittle. The quantity of crude secretion is about 12 mg/gm animal.

In the early investigations, the crude secretion was poured into a solution of pepsin in dilute hydrochloric acid. To digest the proteins of the secretion, the mixture was kept at 37°C for 1 week. Then after having alkalized the solution, the alkaloids were extracted with ether and chloroform. Since esters are saponified under those conditions, it is preferable to grind the secretion with sand and to extract with 80 % ethanol (Habermehl, 1964b). On evaporation of the alcohol *in vacuo*, a mixture of the alkaloids and peptides is obtained. This mixture is dissolved in diluted acetic acid, alkalized with concentrated ammonia, and extracted with ether. One minor alkaloid, cycloneosamandione is only sparingly soluble in ether and can only be extracted by chloroform or methylenechloride from the alkaline solution.

The alkaloids can be separated by crystallization of salts or derivatives, thus the main alkaloid *samandarine* is isolated from the mixture as the hydrochloride (Schöpf and Braun, 1934), the corresponding ketone *samandarone* forms an ether-insoluble semicarbazone (Schöpf and Braun, 1934; Schöpf and Koch, 1942a), *samandaridine* is isolated as the sulfate (Schöpf and Koch, 1942a), *cycloneosamandione* as the hydroiodide (Schöpf and Müller, 1960), and *O-acetylsamandarine* as the hydrochloride (Habermehl, 1964a). A further separation of the alkaloids can be achieved by Craig-distribution between chloroform and diluted acetic acid (Schöpf and Müller, 1960), as well as by preparative layer chromatography on silica gel with cyclohexane/diethylamine/methanol (95:5:15) as solvent (Habermehl and Vogel, 1964). An outline of the alkaloids is given in Table I.

Recently it was found (Habermehl, 1964a) that there is a difference in the composition of the venom of the two subspecies *S. maculosa maculosa* and *S. maculosa taeniata*. While *S.m. taeniata* contains all alkaloids mentioned in Table I, and the main alkaloid is samandarine, in *S.m. maculosa* samandarine is missing and the main alkaloid is samandarone.

TABLE I

ALKALOIDS OBTAINED IN SALAMANDER VENOMS

Name	Formula	MP	$R_F{}^a$	Functional groups		
Alkaloids with an oxazolidine System						
Samandarine	$C_{19}H_{31}NO_2$	188	0.42	—NH—	—O—	=CHOH
Samandarone	$C_{19}H_{29}NO_2$	190	0.52	—NH—	—O—	=CO
Samandaridine	$C_{21}H_{31}NO_3$	290	0.20	—NH—	—O—	γ-lactone
O-Acetylsamandarine	$C_{21}H_{33}NO_3$	159	0.55	—NH—	—O—	O—Acetyl
Samandenone	$C_{22}H_{31}NO_2$	191	0.35	—NH—	—O—	C=CH—CO
Samandinine	$C_{24}H_{39}NO_3$	170	0.68	—NH—	—O—	O—Acetyl —CH(CH₃)₂
Alkaloids with carbinolamine system						
Cycloneosamandione	$C_{19}H_{29}NO_2$	119	0.08	N—C—OH		=CO
Cycloneosamandaridine	$C_{21}H_{31}NO_3$	282	0.75	N—C—OH		γ-lactone
Alkaloids without oxazolidine and without carbinolamine system						
Samanine	$C_{19}H_{33}NO$	197		—NH—		=CHOH

^{a} Paper: SiO_2-paper, Schleicher & Schüll, No. 289. Solvent: Cyclohexane/Diethylamine 9:1.

B. The Structure Elucidation of the Alkaloids

1. Samandarine

The main alkaloid samandarine, $C_{19}H_{31}NO_2$, is a saturated secondary amine with a secondary hydroxy group which can be oxidized with chromic acid to yield the ketone samandarone, which is also found in the venom (Schöpf and Braun, 1934; Schöpf and Koch, 1942a). The second oxygen atom in both alkaloids belongs to an ether bridge. The determination of C-methyl groups proved the presence of more than one, probably two methyl groups. From the empirical formula and the functional groups mentioned, it follows that samandarine contains three carbocyclic rings. The first insight into the structure of the framework of these alkaloids came from the Hofmann degradation of samandarine. When samandarine is shaken with methyliodide and sodium carbonate, the quarternary ammonium salt is formed. The des-base (II) obtainable from this contains all carbon atoms of the N-methyl methoiodide. The nitrogen atom therefore belongs to a ring. The double bond of the des-base (II) can be proved by catalytic hydrogenation to the dihydro-des-base (III) (Scheme 1). The des-base II is stable toward alkali but on being warmed with diluted sulfuric acid it adds one mole of water, forming the oxy-dihydro-des-base (IV). By this conversion the double bond and the ether-oxygen atom disappear, while a carbonyl group and a secondary hydroxy group are formed. Therewith IV shows the properties of an enol ether. The oxy-dihydro-des-base, however, is able to react tautomerically as an inner half-acetal; thus on oxidation with chromic acid it forms a lactone (V), samandesone; this proves that the carbonyl function in IV formed by the addition of water is an aldehyde.

The formation of an enol ether upon Hofmann degradation reveals that the nitrogen atom is attached to a carbon atom which is bonded to the oxygen of the ether or which is α to the carbon carrying the oxygen atom. In both cases a des-base with the properties of an enol ether might be formed. The decision between the two cases was determined by the following experiment: On warming the oxy-dihydro-des-base (IV) with acetic anhydride, not only was the des-base formed, but also a minor quantity of the quaternary starting material (VII). Obviously the tertiary nitrogen atom of the des-base IV is alkylated by the first-formed acetate VI in the same manner as trimethylamine is alkylated to tetramethylammonium acetate by methyl acetate. With this it is proved that in samandarine the nitrogen and the ether oxygen atoms are attached to the same carbon atom.

SCHEME 1

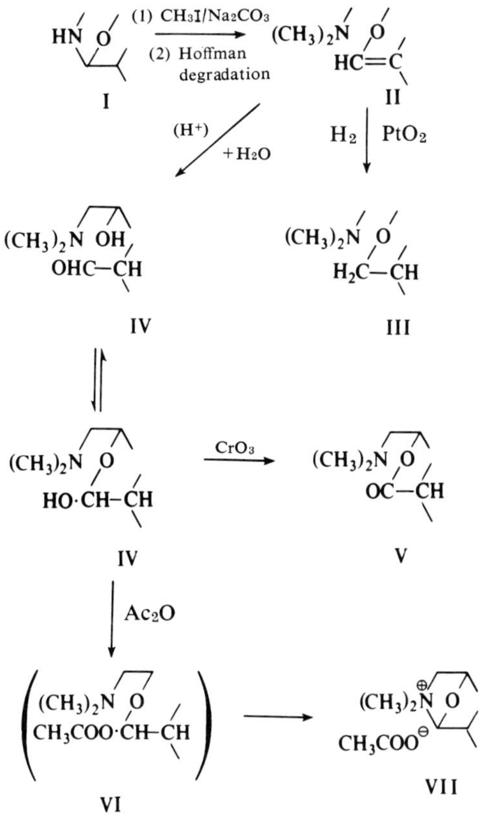

The completion of partial formula I to formula VIII results from the following reactions. Samandarine reacts with Grignard reagents to form methyl- and phenylsamandiole in a manner similar to the formation of samandiole with lithium alanate (Schopf et al., 1950). Later it was found that this reaction is characteristic for oxazolidines. The samandioles therefore contain a second secondary hydroxy group originating from the ether oxygen of the oxazolidine system, besides the original one, for they are oxidized by chromic acid to diketones which were characterized by the dioximes.

While samandarine is stable toward lead tetraacetate, the samandioles take up one mole of the reagent with the formation of one mole of formaldehyde; therefore the new hydroxy group must belong to a grouping which is attackable by lead tetraacetate. One could think of a grouping like HO—CH_2—CH—OH or HO—CH_2—CH—NH—. But as mentioned above, samandiole does not possess a primary hydroxy group and the only group remaining for an attack by lead tetraacetate is —NH—CH_2—CH—OH—. In this reaction the Schiff base (IX) is formed first and after elimination of one mole of formaldehyde is closed to (X). By combination of these results, partial formula XI for samandiole (R=H) and for methyl- and phenyl-samandiole, respectively (R=CH_3, C_6H_5) and partial formula XII for samandarine were elaborated.

$$
\begin{array}{ccc}
\begin{array}{c} CH_2\!-\!CH \\ HN \quad\;\; OH \\ CH_2\!-\!CH \end{array} &
\begin{array}{c} CH_2\!-\!OHC \\ N \\ CH_2\!-\!CH \end{array} &
 \\
XI & IX & X
\end{array}
$$

$$
\begin{array}{ccc}
HN\;O &
\begin{array}{c} CH_2\!-\!CH \\ HN \quad\;\; OH \\ CH\!-\!CH \\ R \end{array} &
\begin{array}{c} CH_2\!-\!CH \\ HN \quad O \\ CH\!-\!CH \end{array} \\
VIII & XI & XII
\end{array}
$$

These formulas do not yet contain the three carbocyclic rings mentioned above. The structure of these carbocyclic rings resulted from the following experiments.

A view of the skeleton of samandarine was given by dehydration of samandiole at 320–340°C with selenium (Schöpf and Klein, 1954). An oily mixture of hydrocarbons was formed, from which a crystalline trinitroben-zolate (m.p. 133°C) was obtained. With trinitrofluorenone an addition compound melting at 146°C was formed. By separating them on an alumina column in both cases, a mixture of two hydrocarbons was isolated, one of which ($C_{15}H_{16}$) was obtained in crystalline form. The ultraviolet spectrum revealed that the compound was a substituted naphthalene. From the comparison of the ultraviolet spectrum with those of many other substituted naphthalenes, it was deduced that the crystalline hydrocarbon must be a 1,2,5,6-tetrasubstituted naphthalene.

If the naphthalene is substituted in positions 1 and 2 each by a methyl group, there are three carbon and six hydrogen atoms remaining for a five-membered ring which then should be attached to the positions 5 and 6 equivalent to positions 1 and 2. The hydrocarbon therefore could be, 1,2-dimethyl-5,6-cyclopentanonaphthalene (XIII).

The hydrocarbon XIII was synthesized and found to be identical in melting point and infrared and ultraviolet spectra with the compound $C_{15}H_{16}$ obtained by the degradation of samandiole.

The ultraviolet spectrum of the other hydrocarbon mentioned above suggested a 1,2,5,6-tetraalkyl-substituted naphthalene, also. From groupings XII and XIII it was obvious to assume for samandarine a skeleton related to the steroids with ring A enlarged by an NH group. If this is so, there are three formulas possible for samandarine (XIV, XV, XVI).

The infrared spectrum of the oxidation product of samandarine, samandarone, shows the band for the carbonyl group at 1740/cm, which is characteristic for a ketone in a five-membered ring. The final decision between the formulas XIV, XV, and XVI was made on the basis of the X-ray analysis of a single crystal of samandarine hydrobromide—grown from methanol—by the heavy atom method (Wölfel et al., 1961). Figure 1 shows the Fourier projection along the short axis from which it is evident that for samandarine formula XVII (corresponding to XVI) is correct and the relative configuration is as depicted in XVII.

O ⟶ Z   C/2

X

FIG. 1. Fourier projection of samandarine hydrobromide.

XVII   XVIII

This structure was furthermore confirmed by a three-dimensional X-ray structure analysis (Fig. 2). From this it was seen that the skeleton of samandarine possesses the same steric arrangement as the cholic acids. The ring A of the steroid skeleton is enlarged by an NH group between the carbon atoms 2 and 3 which form an oxazolidine ring together with the carbon atom 1 and 4 and the oxygen atom linked to C–1 and C–4. So samandarine and the minor alkaloids (which will be discussed later) form a new type of steroidal alkaloids.

2. Samandarone

Among the minor alkaloids of *Salamandra maculosa taeniata,* samandarone (XIX) is predominant. In *Salamandra maculosa maculosa* it is the main alkaloid. As mentioned above, it is the ketone corresponding to samandarine. It can easily be obtained by oxidation of samandarine with chromic acid; and in turn samandarone is reduced stereospecifically by sodium and alcohol to samandarine (Schöpf and Koch, 1942a). The absolute configuration of samandarone was determined from the optical rotatory dispersion curve and from this, that of samandarine has also been determined. It is in accord with the absolute configuration of the other natural steroids, for it shows like other natural 16-keto steroids a strong, negative Cotton effect (Wölfel *et al.,* 1961). Samandarone has been synthesized by Hara and Oka (1966).

3. O-Acetylsamandarine

Another minor alkaloid is *O*-acetylsamandarine (XVIII) (Habermehl, 1964b). Its structure could be elucidated by the infrared bands at 830/840 cm^{-1} characteristic for the oxazolidine system (Habermehl, 1963c) and 1240/cm and 1725/cm which are characteristic for an ester of acetic acid. Upon saponification, samandarine as the alcohol component was isolated. The definite evidence for the correctness of structure XVIII was obtained by synthesizing *O*-acetylsamandarine. A comparison between the natural and the synthetic product showed the two substances to be identical in all respects.

4. Samandaridine

The fourth alkaloid of the series with an oxazolidine ring is samandaridine XXIII (Schöpf and Koch, 1942a). Like samandarine, it is a secondary amine. From chemical investigations together with the infrared spectrum, it is seen to possess a five-membered lactone ring. The structure of this minor alkaloid has also been elucidated by X-ray analysis of the hydrobromide (Habermehl, 1963a).

The absolute configuration has been determined by partial synthesis

Fɪɢ. 2. Result of the three-dimensional Fourier synthesis of samandarine hydrobromide.

Scheme 2

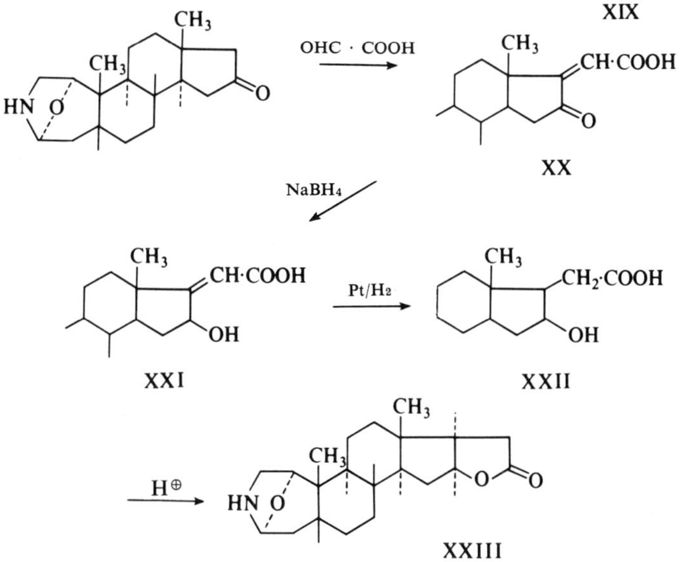

XIX

OHC · COOH

CH₃ CH·COOH

XX

CH₃ CH·COOH

O

NaBH₄

CH₃ CH·COOH OH

XXI

Pt/H₂

CH₃ CH₂·COOH OH

XXII

H⊕

HN O CH₃ CH₃ O O

XXIII

(Habermehl, 1963b) according to Scheme 2, by condensation of samandarone (XIX) with glyoxylic acid and reduction of the resulting carboxyclic acid XX with sodium borohydride. This reduction leads stereospecifically to the β-hydroxy compound XXI. The following catalytic hydrogenation yielded samandaridic acid (XXII) which on being boiled with diluted hydrochloric acid formed samandaridine (XXIII).

5. Cycloneosamandione

A further minor alkaloid is cycloneosamandione (Schöpf and Müller, 1960) (XXIV), an isomer of samandarone but without the oxazolidine ring which is characteristic for the four alkaloids discussed so far. From its infrared spectrum it is evident that one oxygen atom belongs to a keto group in a five-membered ring. The second one belongs to an aldehyde-ammonia grouping which can react in the two tautomeric forms, >N—C—OH and >NH OC<. The product of acetylation of cycloneosamandione is a derivative of the second tautomeric form since a neutral N-acetyl compound is obtained which is revealed by the infrared spectrum that shows, besides the band for the ketone in a five-membered ring at 1732/cm, bands at 1620/cm corresponding to an N-acetyl group and at 1714/cm for a carbonyl group in an unstrained system. That this band arises from an aldehyde group is proved by two further bands: the CH-valence frequency of the aldehyde-CH at 2705/ cm and the combination band of the aldehyde group at 4350/cm (Habermehl and Göttlicher, 1965). The same conclusion follows from the discussion of the NMR spectra of cycloneosamandione and its N-acetyl derivative (Habermehl and Göttlicher, 1965).

XXIV

Also in this case, the structure of the alkaloid was finally elucidated by X-ray analysis. For this the cycloneosamandione hydroiodide was examined. Three-dimensional electron density calculations proved the correct formula of cycloneosamandione to be XXIV (Habermehl and Göttlicher, 1970).

Subsequently, cycloneosamandione has the same basic structure as samandarone with the one essential difference of having an aldehyde group instead of a methyl group attached to C-10. The arrangement of the rings C and D as well as the absolute configuration were elucidated by the optical rotatory dispersion curve of cycloneosamandione. This curve shows, like the curve of samandarone, a strong, negative Cotton effect which indicates the same absolute configuration for both alkaloids. Cycloneosamandione has also been synthesized according to Scheme 3 (Habermehl and Haaf, 1968b).

XXV

XXVI

XXVII

XXVIII

6. Minor Alkaloids

Four minor alkaloids were isolated in very small quantities: *cycloneosamandaridine* (XXV) (Habermehl and Haaf, 1965), *samandenone* (XXVI) (Habermehl, 1966), *samandinine* (XXVII) (Habermehl, unpublished), and *samanine* (XXVIII) (Habermehl and Haaf, 1969). The structures of those four alkaloids have been established by means of UV-, IR-, NMR-, and mass spectral data (Habermehl and Spiteller, 1967). Furthermore *samanine* has been synthesized according to Scheme 4 (Habermehl and Haaf, 1969).

SCHEME 3

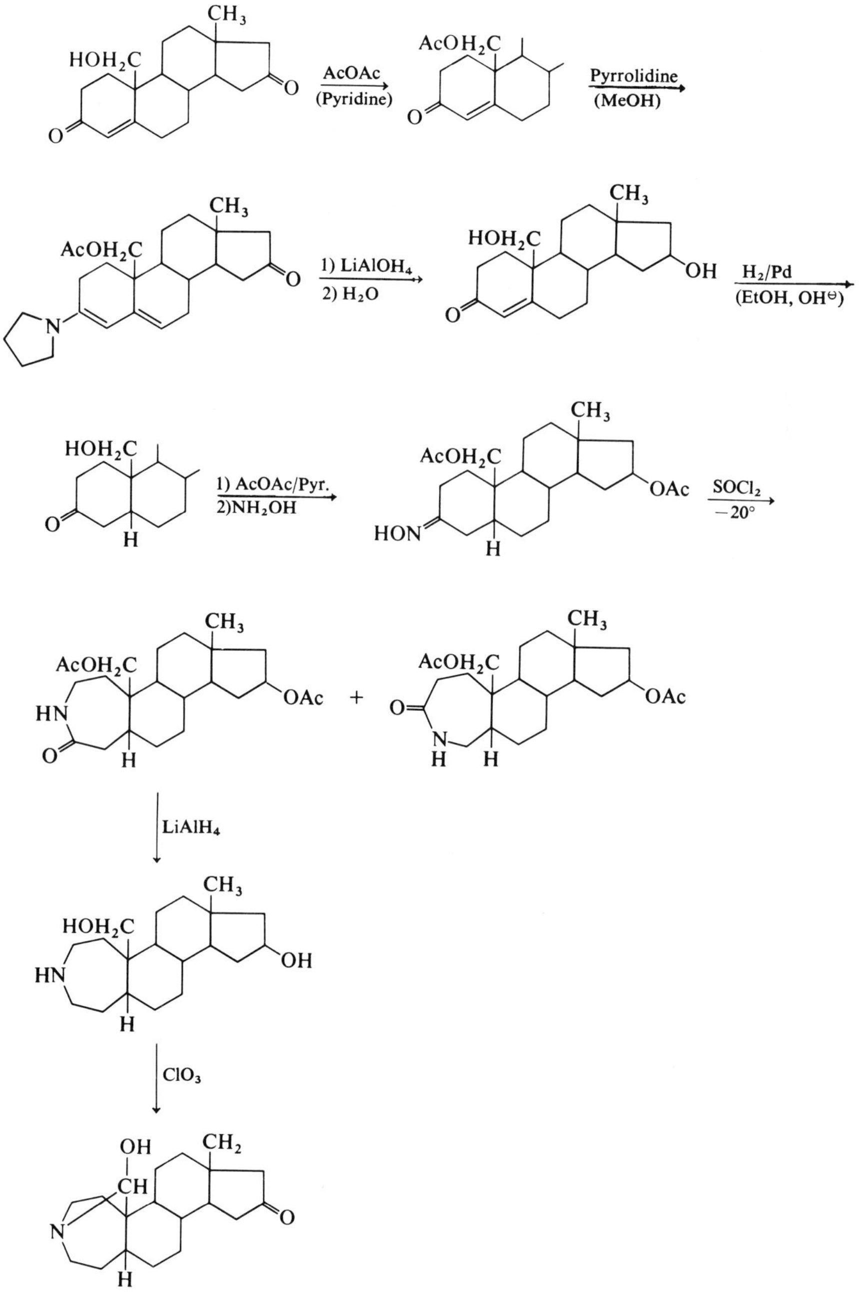

XXIV

SCHEME 4

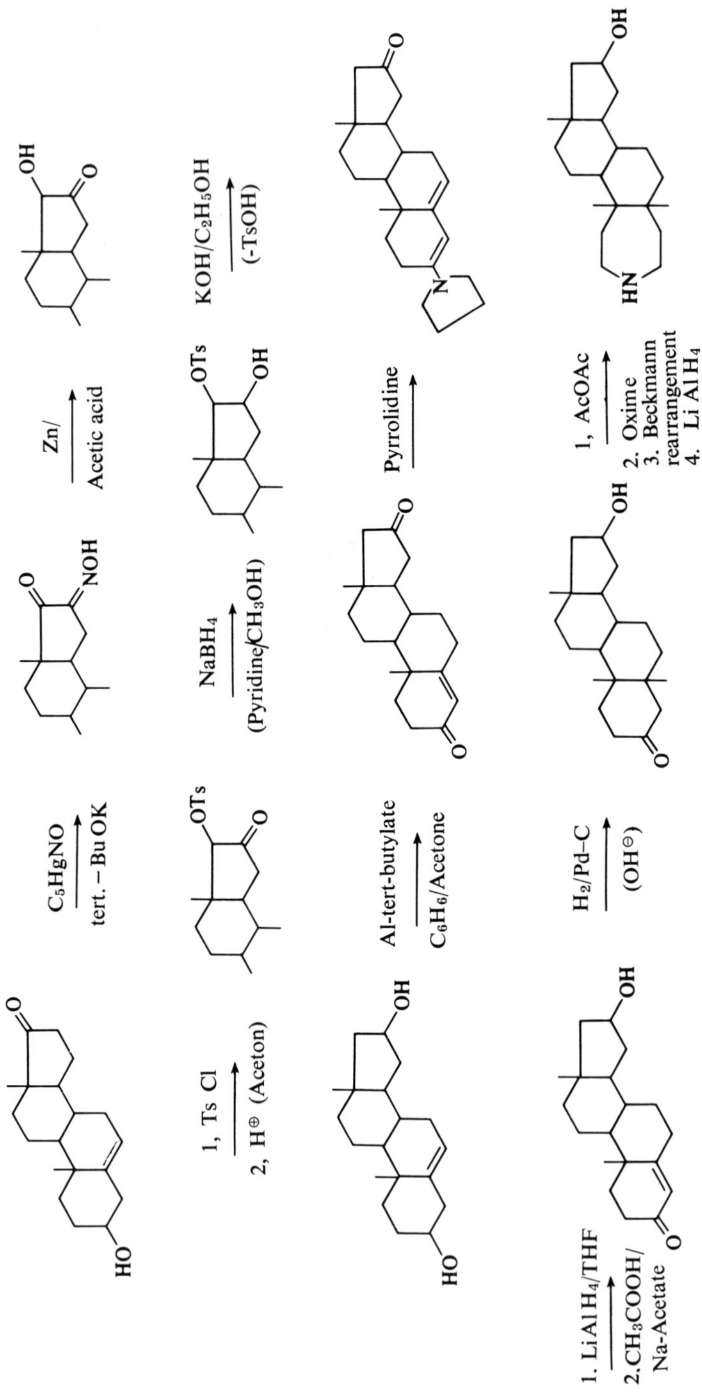

C₅HgNO
tert. – BuOK

Zn/
Acetic acid

1, Ts Cl
2, H⊕ (Aceton)

OTs

NaBH₄
(Pyridine/CH₃OH)

OTs
OH

KOH/C₂H₅OH
(-TsOH)

Al-tert-butylate
C₆H₆/Acetone

Pyrrolidine

1. LiAlH₄/THF
2. CH₃COOH/
Na-Acetate

H₂/Pd–C
(OH⊖)

1, AcOAc
2. Oxime
3. Beckmann
   rearrangement
4. LiAlH₄

IV. NEUTRAL SUBSTANCES IN THE SKIN GLAND SECRETION

Besides the alkaloids and the above-mentioned hemolytical active peptides, there could be isolated a fraction of neutral, lipophile substances (Schöpf and Braun, 1934). Six components were isolated by column chromatography on silica gel from this fraction (Habermehl, 1964a). The structures of four of them were elucidated. The main part consisted of cholesterol; the other compounds were esters of cholesterol, namely, cholesteryl stearate, cholesteryl palmitate, and cholesteryl oleate.

V. BIOSYNTHESIS OF SALAMANDER ALKALOIDS

As cholesterol is the key substance for steroid metabolism it seemed reasonable that cholesterol, found with its esters in the skin gland secretion, should be the precursor for the salamander alkaloids.

In recent investigations concerning biosynthesis, involving *in vitro* and *in vivo* experiments, we found that the salamander alkaloids are formed like other steroids from acetate via cholesterol (Habermehl and Haaf, 1968a). The enlargement of ring A may proceed after a fission between carbon atoms 2 and 3. The nitrogen inserted comes from glutamine.

The alkaloids with side chain at C-17 (samandaridine, samandenone, samandinine) might be considered as intermediates on the way from cholesterol to the alkaloids without side chains. By degradation of the cholesterol side chain (XXIX) via a cholic acid intermediate (XXX) a compound (XXXI) with the isopropyl group of samandenone and samandinine can be obtained. The formation of samandaridine (XXIII) should proceed via the malonic acid intermediate (XXXII) followed by decarboxylation (XXXIII) and subsequent ring closure.

Whether samandarine (XVI) is formed via XXXI→XVI or via XXXI→XXXII→XXXIII→XVI is not yet clear.

REFERENCES

Gessner, O. (1926). *Ber. Ges. Bef. Ges. Naturwiss. Marburg* **61**, 138.
Gessner, O., and Esser, W. (1935). *Arch. Exptl. Pathol. Pharmakol.* **179**, 639.
Gessner, O., and Möllenhoff, P. (1932). *Arch. Exptl. Pathol. Pharmakol.* **167**, 638.
Gessner, O., and Urban, G. (1937). *Arch. Exptl. Pathol. Pharmakol.* **187**, 378.
Habermehl, G. (1963a). *Chem. Ber.* **96**, 143.
Habermehl, G. (1963b). *Chem. Ber.* **96**, 840.
Habermehl, G. (1963c). *Chem. Ber.* **96**, 2029.
Habermehl, G. (1964a). *Ann. Chem.* **679**, 164.
Habermehl, G. (1964b). *Ann. Chem.* **680**, 104.
Habermehl, G., and Vogel, G. (1964). *Kolben (Sendai)* [11] **11**, 106.
Habermehl, G. (1966). *Chem. Ber.* **99**, 1439.
Habermehl, G., and Göttlicher, S. (1963). *Angew. Chem.* **75**, 247; *Angew. Chem. Intern. Ed. Engl.* **2**, 157 (1963).
Habermehl, G., and Göttlicher, S. (1965). *Chem. Ber.* **98**, 1.
Habermehl, G., and Göttlicher, S. (1970). (In press).
Habermehl, G., and Haaf, G. (1965). *Chem. Ber.* **98**, 3001.
Habermehl, G., and Haaf, A. (1968a). *Chem. Ber.* **101**, 198.
Habermehl, G., and Haaf, A. (1968b). *Z. Naturforsch.* **23b**, 1551–1552.
Habermehl, G., and Haaf, A. (1969). *Ann. Chem.* **722**, 155–161.
Habermehl, G., and Spiteller, G. (1967). *Ann. Chem.* **706**, 213.
Hara, S., and Oka, K. (1966). *J. Amer. Chem. Soc.* **89**, 1041.
Schöpf, C., and Braun, W. (1934). *Ann. Chem.* **514**, 69.
Schöpf, C., and Klein, D. (1954). *Chem. Ber.* **87**, 1638.
Schöpf, C., and Koch, K. (1942a). *Ann. Chem.* **552**, 37.
Schöpf, C., and Koch, K. (1942b). *Ann. Chem.* **552**, 62.
Schöpf, C., and Müller, O. W. (1960). *Ann. Chem.* **633**, 127.
Schöpf, C., Blödorn, H. K., Klein, D., and Seitz, G. (1950). *Chem. Ber.* **83**, 372.
Wölfel, E., Schöpf, C., Weitz, G., and Habermehl, G. (1961). *Chem. Ber.* **94**, 2361.
Zalesky, S. (1866). *Med. Chem. Untersuch. Hoppe-Seyler* **1**, 85.

VENOMOUS FISHES

Chapter 43

Venomous Fishes

BRUCE W. HALSTEAD

WORLD LIFE RESEARCH INSTITUTE, COLTON, CALIFORNIA

Zootoxins are present in a great variety of marine vertebrate animals. Among the venom-producing marine vertebrates, which include fishes and sea snakes, the fishes are the most diversified and widely represented group. A comprehensive treatise on venomous or ichthyoacanthotoxic fishes would fill a large volume. Numerous species of fishes possess venom organs, but probably less than 5% of them have been studied to date. Without a single exception, the chemical nature of these piscine venoms is unknown, and only recently have the physiological effects of three fish venoms been investigated to any extent, namely, scorpionfish (including *Synanceja*), weeverfish, and stingray venoms. Thus precise knowledge regarding the phylogenetic distribution of fish venoms and their chemical and pharmacological properties can be dealt with only in the most cursory fashion.

Although venomous fishes are worldwide in their distribution, they occur in greatest concentration and diversification within the torrid zone, and particularly within the Indo-Pacific region. As one progresses toward the extreme southern or extreme northern latitudes, venomous fishes decrease rapidly in numbers of species until they are represented only by a few elasmo-branch species, chimaeroids, the weeverfishes, and several of the more hardy members of scorpionfishes.

Space does not permit a complete discussion of all the known species of venomous fishes, so only a few representative groups and species will be discussed.

I. CLASSIFICATION OF VENOMOUS FISHES

The following classification lists the various groups of fishes which are known to contain venomous species. In some instances only one or two species within a large group are found to be venomous. This classification provides some insight as to the phylogenetic distribution of venomous fishes within the superclass Pisces. Only those phylogenetic categories are included which contain one or more venomous species. For the sake of brevity all nonvenomous groups have been eliminated. The classification provided is modified from the general classification of fishes as given by Lagler *et al.* (1962). Family names have been listed in alphabetic sequence within the order.

A List of Phylogenetic Groups Containing Venomous Fishes

Class Chondrichthyes
  Subclass Elasmobranchii (Selachii)
    Order Squaliformes (Pleurotremata)—sharks
      Family Heterodontidae—horn sharks
    Order Rajiformes (Hypotremata or Batoidei)—rays
      Family Dasyatidae—stingrays
      Family Gymnuridae—butterfly rays
      Family Mobulidae—mantas
      Family Myliobatidae—eagle rays
      Family Potamotrygonidae—freshwater stingrays
      Family Rhinopteridae—cow-nosed rays
      Family Urolophidae—round stingrays
  Subclass Holocephali
    Order Chimaeriformes (Chimaerae)
      Family Chimaeridae—chimaeras
Class Osteichthyes
  Subclass Actinopterygii
    Order Clupeiformes (Isospondyli)
      Suborder Stomiatoidei
        Family Stomiatidae—deep-sea scaly dragonfishes

Order Cypriniformes (Ostariophysi)
  Suborder Siluroidei (Nematognathi)[1]
    Family Ariidae—sea catfishes
    Family Bagridae—bagrid catfishes
    Family Ictaluridae—North American freshwater catfishes
    Family Plotosidae—plotosid sea catfishes
    Family Siluridae—Eurasian catfishes
Order Anguilliformes (Apodes)
    Family Muraenidae—morays (no venomous members known)[2]
Order Beryciformes (Berycomorphi)
    Family Holocentridae—Squirrelfishes, soldierfishes
Order Perciformes (Percomorphi)
  Suborder Percoidei
    Family Carangidae—jacks, scads, pompanos
    Family Chaetodontidae—butterfly fishes
    Family Enoplosidae—old wives
    Family Gerridae—mojarras
    Family Histiopteridae—boar fishes, bugler fishes
    Family Lutjanidae—snappers
    Family Monodactylidae—fingerfishes
    Family Percidae—perches, walleyes, darters
    Family Scatophagidae—scats
    Family Serranidae—seabasses
    Family Trachinidae—weevers
    Family Uranoscopidae—stargazers
  Suborder Siganoidei (Amphacanthini)
    Family Teuthidae—rabbitfishes
  Suborder Acanthuroidei (Teuthidoidea)
    Family Acanthuridae—surgeonfishes
  Suborder Trichiuroidei
    Family Gempylidae—snake mackerels
  Suborder Cottoidei (Cataphracti; Scleroparei; Loricati)
    Family Cottidae—sculpins

---

[1] It is believed that several Siluroid families not listed here also contain venomous catfish species, but at present there is no information regarding the nature of their venom organs, if such are present.

[2] *Muraena helena* (Linnaeus), and other species of morays, have been repeatedly reported in the older scientific literature as having venomous teeth. However, modern anatomical research fails to support these earlier claims. There is some evidence that the palatine mucosa of some morays may produce toxic substances, but there is no evidence that a true venom apparatus is present.

Family Scorpaenidae—scorpionfishes, stonefish
Family Triglidae—sea robins
Suborder Callionymoidei
Family Callionymidae—dragonets
Order Batrachoidiformes (Haplodoci)
Family Batrachoididae—toadfishes
Order Lophiiformes (Pediculati)
Family Lophiidae—goosefishes, anglerfishes

Only the groups of venomous fishes considered to be of major clinical importance will be discussed in this presentation.

II. VENOMOUS ELASMOBRANCHS

A. Horn Sharks

Venomous sharks are limited to those species that possess dorsal fin spines, namely, certain members of the families Heterodontidae, Squalidae, and possibly the Dalatiidae. Within these families there are at least 11 genera known to possess dorsal spines, but only two species, *Heterodontus francisci* and *Squalus acanthias,* are definitely known to have venomous spines. The remaining species are suspect as having venomous spines but have not been studied by the venomologist.

A List of Representative Venomous Sharks

HETERODONTIDAE—horn sharks
Heterodontus francisci (Girard)
California coast from Point Conception to Lower California and into the Gulf of California

SQUALIDAE—spiny dogfish
Squalus acanthias (Linnaeus) (Fig. 1)
Both sides of the North Atlantic and North Pacific

FIG. 1. *Squalus acanthias* (Linnaeus). Spiny dogfish. (M. Shirao.)

I. Mechanism of Envenomation

Wounds are inflicted by the two dorsal stings which are located adjacent to the anterior margins of each of the two dorsal fins.

2. Venom Apparatus

The venom apparatus of horn sharks is comprised of the dorsal fin spines and the associated glandular tissue. The dorsal stings are situated adjacent to the anterior margins of each of the two spines in most of the horn sharks. *Squalus acanthias,* which can be considered as representative of the group, has a sharp strong spine immediately anterior to each dorsal fin. The spine in cross section is roughly triangular in shape, with the apex directed anteriorly. The spine is grooved only in its exposed portion, the groove becoming more shallow toward the tip. Details of the structure can be seen in Fig. 2.

Microscopic examination of cross sections of the sting reveals it to be trigonal in shape and comprised of three principal layers: an outer layer of integument, a thick wall of hard vasodentine, and an inner core of cartilage. The glandular cells are situated in the epithelial portion of the integumentary layer in the area of the anterolateral glandular grooves and in the interdentate depression. The glandular cells are sparsely scattered in the anteroglandular groove, but are heavily concentrated in the interdentate depression. The venom cells, when stained with hematoxylin and triosin, appear as oval-shaped and contain homogeneous brown-stained material with accumulations of finely granular material. Venom production is of the holocrine type of secretion. For further details regarding the anatomy of the venom apparatus of *S. acanthias,* the reader is referred to the works of Evans (1921, 1923, 1943).

Chemistry of Venom: Unknown.

Pharmacology of Shark Venom: Unknown.

3. Clinical Characteristics

Symptoms consist of immediate, intense, stabbing pain, which may continue for several hours. Swelling and redness of the affected parts are usually present. The lesions are typically of the puncture-wound variety. Fatalities from dogfish wounds have been reported.

Treatment: See Section IX.

B. Stingrays

Clinically, stingrays constitute the most important single group of venomous fishes, since they cause the largest number of serious venomous fish stings. With the exception of the Potamotrygonidae, which are confined to the rivers of South America, most stingrays are marine, inhabiting shallow coastal waters, bays, brackish water lagoons, and river mouths.

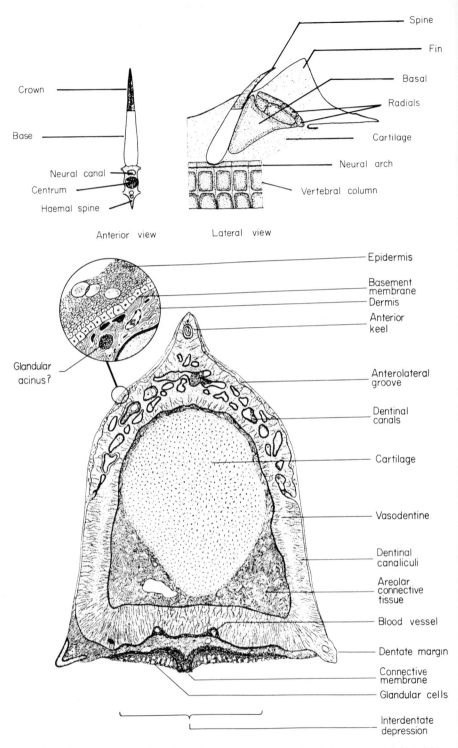

FIG. 2. Venom apparatus of *Squalus acanthias*. *Top:* Lateral view of second dorsal fin spine. *Bottom:* Cross section of anterior sting. (From Halstead and Mitchell.)

List of Representative Venomous Stingrays

DASYATIDAE—stingrays

Dasyatis dipterurus (Jordan and Gilbert) (Fig. 3)
British Colombia to Central America
Dasyatis pastinaca (Linnaeus)
Northeastern Atlantic Ocean, Mediterranean Sea, and Indian Ocean
Dasyatis say (Lesueur)
Western Atlantic from New Jersey to southern Brazil

GYMNURIDAE—butterfly rays

Gymnura marmorata (Cooper) (Fig. 4)
Point Conception, California, south to Mazatlan, Mexico

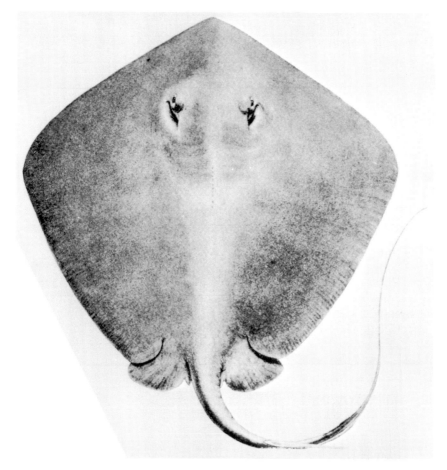

FIG. 3. *Dasyatis dipterurus* (Jordan and Gilbert). Diamond stingray. (From Walford.)

FIG. 4. *Gymnura marmorata* (Cooper). Butterfly ray. (From Hiyama.)

Gymnura micrura (Bloch and Schneider)
 Western Atlantic from Brazil to Chesapeake Bay; Gulf of Mexico

MOBULIDAE—mantas
 Mobula mobular (Bonnaterre)
  Eastern Atlantic. Ireland to West Africa; Mediterranean Sea

MYLIOBATIDAE—eagle rays
 Aëtobatus narinari (Euphrasen) (Fig. 5)
  Tropical and warm-temperate belts of the Atlantic, Red Sea, and Indo-Pacific
 Myliobatis aquila (Linnaeus)
  Eastern Atlantic, Mediterranean Sea
 Myliobatis californicus (Gill)
  Oregon to Magdalena Bay, Lower California

POTAMOTRYGONIDAE—freshwater stingrays
 Potamotrygon hystrix (Müller and Henle)
  La Plata, Argentina; Amazon, Brazil; Guianas
 Potamotrygon motoro (Müller and Henle)
  Paraguay; Amazon River; Rio de Janeiro, Brazil

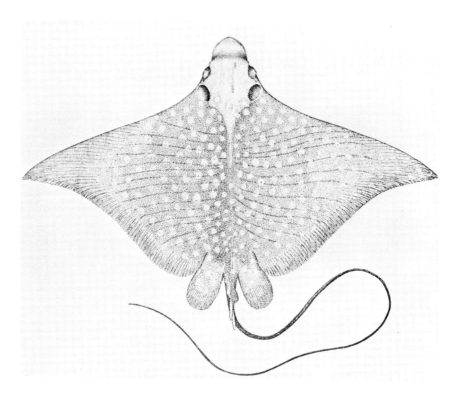

FIG. 5. *Aëtobatus narinari* (Euphrasen). Spotted eagle ray. (From Jordan and Evermann.)

RHINOPTERIDAE—cow-nosed rays
   Rhinoptera bonasus (Mitchell)
      Coastal western Atlantic from southern New England to the middle of Brazil
   Rhinoptera marginata (Geoffroy Saint-Hillaire)
      Mediterranean Sea

UROLOPHIDAE—round stingrays
   Urolophus halleri (Cooper) (Fig. 6)
      Point Conception, California to Panama Bay
   Urolophus jamaicensis (Cuvier)
      Western tropical Atlantic from southern Caribbean to Florida
   Urolophus testaceus (Müller and Henle)
      Queensland south to Victoria, Australia

FIG. 6. *Urolophus halleri* (Cooper). Round stingray. (M. Shirao.)

I. Mechanism of Envenomation

Venom is secreted and introduced into the body of the victim by the sting or venom apparatus located on the tail of the ray. Experimental studies in *Urolophus halleri* have shown that stinging is usually directed to the side by a bending of the tail toward the evoking stimulus. It has been noted that for a quick stinging response the stimulus has to be applied to the posterior third of the ray's body. The stinging response in the series of rays tested was found to be consistent, but varied with the type of stimulus used. When a stick was applied to the posterior third of the back, the ray immediately arched the tail through a vertical plane over its back and struck with precision at the probe. There were no horizontal movements or general lashing of the caudal appendage, but a very precise thrust. When a more diffuse stimulus was applied to the posterior third of the ray's body, both vertical and horizontal movements were observed.

2. Venom Apparatus

The venom apparatus of stingrays is comprised of the dorsal tail sting, its associated glandular tissue, and the caudal appendage. The venom organs of rays have been divided into four anatomical types by Halstead and Bunker (1953) on the basis of their adaptability as a defense organ.

a. Gymnurid Type (Fig. 7). This is the most weakly developed type of stingray venom apparatus. The caudal appendages in gymnurid rays are cylindrical, tapering, and greatly reduced in size. The sting is small, seldom exceeding 2.5 cm in length, and usually situated in the middle or proximal third of the tail. The striking ability of the organ is relatively feeble. Cuneiform area is poorly to moderately developed. Members of the families Gymnuridae and Mobulidae possess the gymnurid type of venom apparatus.

b. Myliobatid Type (Fig. 7). In this type the caudal appendage is cylindrical and tapers out to a long whiplike tail. The sting is generally situated on the proximal portion of the basal third of the tail and is moderate to large in size, ranging from about 5 to 12 cm or more in length. The myliobatid type of venom organs is better adapted as a striking organ than the gymnurid type. Cuneiform area is only moderately developed. Members of the families Myliobatidae and Rhinopteridae possess this type of venom apparatus.

c. Dasyatid Type (Fig. 7). The caudal appendage is depressed proximal to the sting, becoming cylindrical and tapering in cross section distally, and finally terminating in a long whiplike tail. The dorsal fin is absent. The sting is moderate to large, and may attain a length of 37 cm or longer in some species. It is usually located in the distal portion of the basal or middle third of the tail. Cuneiform area is usually poorly developed. Members of the families Dasyatidae and the Potamotrygonidae possess this type of venom apparatus.

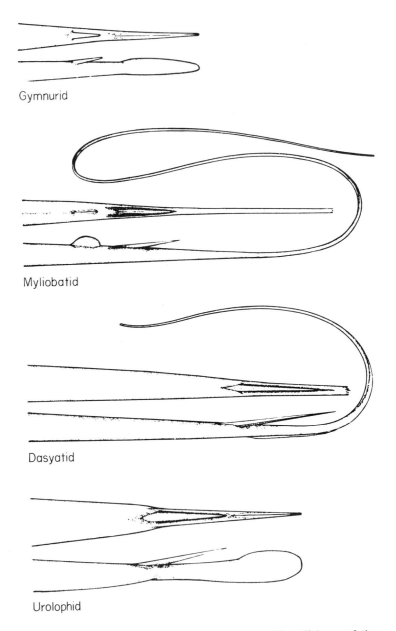

Gymnurid

Myliobatid

Dasyatid

Urolophid

FIG. 7. Anatomical types of stingray venom organs. The efficiency of the venom apparatus varies according to the placement of the sting and the muscular development of the caudal appendage. (From Halstead and Bunker.)

d. Urolophid Type (Fig. 7). The caudal appendage is relatively short, muscular, depressed proximal to sting, not produced as a whiplike structure, becomes compressed distal to the sting, and usually forms a more or less distinct caudal fin. The sting is usually located in the middle or distal third of the tail and is moderate in size, seldom exceeding 5 cm in length. The powerful muscular structure of the tail and the distal location of the sting make this a highly efficient defensive weapon. Cuneiform area is well developed. Members of the family Urolophidae possess this type of venom apparatus.

The anatomy of the venom apparatus of stingrays varies somewhat from one group and one species to the next. However, there is a basic pattern that is quite typical for all the species that have been examined to date. A description of this general pattern will suffice for the purpose of the present discussion.

The venom apparatus of stingrays consists of a bilaterally retroserrate spine, its associated glandular cells, and the enveloping integumentary sheath (Fig. 8A, B, C). The spine is an elongate tapering structure that ends in an acute sagitate tip. The spine is composed of an inner core of vasodentine which is covered by a thin layer of enamel. It is firmly anchored in a dense collagenous network of the dermis on the dorsum of the caudal appendage. The dorsal surface of the spine is marked by a number of shallow longitudinal furrows. These furrows are usually more pronounced on the basal portion of the spine and disappear distally. The serrate edges of the spine are termed the dentate margins. Medial to each dentate margin, on the ventral side, is a longitudinal groove, the ventrolateral-glandular groove. The grooves are separated from each other by the median ventral ridge of the spine. Contained within the grooves of an "unsheathed" or traumatized sting is a strip of gray tissue. The tissue lying within the ventrolateral-glandular grooves consists of glandular epithelium and blood vessels. This is the primary venom-producing area of the sting. In most stingray species there is a thickened wedge-shaped portion of the integument on the dorsum of the caudal appendage ventral to the sting, which is known as the cuneiform area. Studies have shown that the glandular cells of the cuneiform area also produce a limited amount of venom, which comes in contact with the spine when the ray is at rest. Thus, the sting is bathed in mucus and venom originating from the ventrolateral-glandular groove tissue and the cuneiform area.

Microscopic examination of histological cross sections of an intact sting reveals that it is roughly diamond shaped, and consists of a broad T-shaped dentinal structure completely enveloped by a layer of integument. The integument is comprised of two layers. The inner layer, the dermis, consists of areolar connective tissue and vascular channels. The outer layer, the epidermis, is composed of modified squamous epithelium containing many glandular cells. A cross section of the ventrolateral-glandular groove has been termed

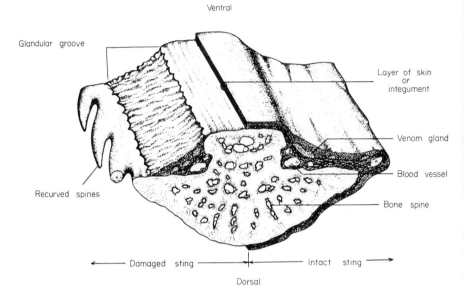

FIG. 8A.

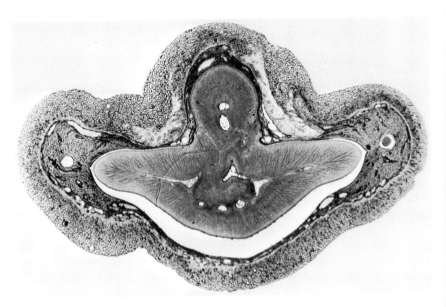

FIG. 8B.

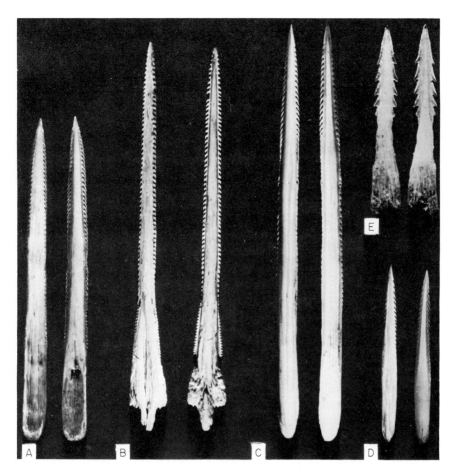

FIG. 8C.

FIG. 8. A. Cross section of the venom apparatus of a stingray showing both damaged and intact stings. Semidiagrammatic. (From Halstead.) B. Representative histological cross section of the sting from *Myliobatis californicus*. Note arrow pointing to the venom-producing glandular triangle region. × 80. C. Caudal stings from various species of stingrays. (A) *Myliobatis californicus*, (B) *Aëtobatus narinari*, (C) *Dasyatis dipterurus*, (D) *Urolophus halleri*, (E) *Gymnura marmorata*. (From Halstead and Bunker.)

the glandular triangle. There is no histological evidence to indicate that the venom is secreted through a duct.

Chemistry of Stingray Venom: Unknown.

3. Pharmacology of Stingray Venom

The only published account on some of the pharmacological properties of stingray venom is that by Russell and Van Harreveld (1954) who worked on the cardiovascular effects of the venom of the round stingray, *Urobatis halleri*. They found that in the smallest active concentrations the venom gives rise to vasodilatation followed by vasoconstriction, or vasoconstriction without preliminary dilatation. In larger and massive amounts, the toxin evokes only vasoconstriction. These observations were noted in the large arteries and veins as well as in the smaller peripheral components of the vascular system. The venom has a direct effect on the heart. It provokes both an auricular and ventricular standstill which in turn are followed by contractions of lesser amplitude and slower rate. A new rhythm is evoked from a focus outside the sinoauricular node.

Pathology: Unknown.

4. Clinical Characteristics

Pain is the predominant symptom, and it usually develops immediately after, or within a period of 10 minutes of the attack. The onset of pain usually embraces an area of approximately 10 cm in diameter about the wound. The pain generally becomes more severe during the first 30 minutes, radiating to include the entire extremity, and attaining its maximum intensity in less than 90 minutes. Although gradually diminishing, the pain may continue for 6 to 48 hours. The pain is variously described as sharp, shooting, spasmodic, or throbbing in character. Potamotrygonids or freshwater stingrays are reputed to cause extremely painful wounds. More generalized symptoms of fall in blood pressure, vomiting, diarrhea, sweating, arrhythmia, muscular paralysis, and death have been reported. Russell (1956) states that he has never observed either a spastic or flaccid paralysis in a stinging victim, and suggests that the muscular spasms may be due to contractures initiated as flexion reflexes stimulated by the intense pain. His observations are further substantiated by the fact that medications directed at alleviating the pain also relieve the muscular spasms. The nausea, faintness, vertigo, and bradycardia so often experienced by the victim within the first few minutes after the accident are believed to be attributable to the primary shock caused by the extreme pain. The fall in systemic arterial pressure due to peripheral vasodilatation, which occurs when small amounts of stingray toxin are experimentally administered intravenously, and mechanisms related to this fall may accentuate or even precipitate a transient cerebral anoxia and shock.

Stingray wounds are either of the laceration or puncture type. Penetration of the skin and underlying structure is usually accomplished without serious damage to the surrounding tissues. However, withdrawal of the sting may result in extensive tissue damage due to the retrorse serrations. The lacerated edges of a stingray wound bleed freely, but not abnormally so. Swelling in the vicinity of the wound is a constant finding. The edema may persist despite elevation of the extremity, which suggests pathological changes in the local affected tissues. Much of the edema has been attributed to lymphatic obstruc-tion, precipitated by inflammation, and damage to the lymphatics and sup-porting structures. These changes are said to be caused by the direct action of the venom. The area about the wound at first has an ashen appearance, later becomes cyanotic, and then reddened. Tissue necrosis in the vicinity of the wound frequently occurs, which suggests proteolytic properties of the venom. Although stingray injuries occur most frequently about the ankle joint and foot as a result of stepping on the ray, instances have been reported in which the wounds occurred in the chest about the heart (Halstead and Bunker, 1953; Russell, 1953b). It is estimated that about 1500 cases of stingray attacks occur each year (Russell and co-workers, 1958).

Treatment: See Section IX.

5. Prevention

It should be kept in mind that stingrays commonly lie almost completely buried in the upper layer of a sandy or muddy bottom. Stingrays are, therefore, a hazard to anyone wading in water inhabited by them. The chief danger is in stepping on one that is buried. This danger can largely be eliminated by shuffling one's feet along the bottom. Usually the body of the ray is pinned down by the weight of the victim, thereby permitting the beast to make a successful strike. Pushing one's feet along the bottom eliminates this danger, and at the same time, routs the stingray from its lair. It is also recommended that a stick be used to probe along the bottom in order to rid the area of hidden rays.

III. VENOMOUS CHIMAERAS

Chimaeras, elephantfish, or ratfish, as they are sometimes called, are a group of cartilaginous fishes having a single external gill opening on either side, covered over by an opercular skin fold with cartilaginous supports leading to a common branchial chamber into which the true gill clefts open. Externally, chimaeroids are more or less compressed laterally, tapering posteriorly to a slender tail. The snout is rounded or conical, extended as a long, pointed beak, or bearing a curious hoe-shaped proboscis. There are two dorsal fins. The first fin is triangular, usually higher than the second, and is edged anteriorly by a strong sharp-pointed, bony spine that serves as a venom apparatus.

List of Representative Venomous Chimaeras

Chimaera monstrosa (Linnaeus) European chimaera
 North Atlantic from Norway and Iceland to Cuba, the Azores, Morocco, and the Mediterranean Sea, South Africa

FIG. 9. *Hydrolagus colliei* (Lay and Bennett). Pacific ratfish. (M. Shirao.)

Hydrolagus colliei (Lay and Bennett) (Fig. 9) Pacific ratfish
  Pacific coast of North America

Chimaeroids have a preference for cooler waters and have a depth range from the surface down to 1400 fathoms. They are weak swimmers and die soon after being removed from the water. They have well-developed dental plates and can inflict a nasty bite.

1. Mechanism of Envenomation

Wounds are inflicted by the single dorsal sting located on the anterior margin of the first dorsal fin.

2. Venom Apparatus

The venom apparatus of chimaeroids consists of the dorsal spine, the glandular epithelium of the spine and connecting membrane, and the enveloping integumentary sheath. The mature spine is elongate, tapers to an acute point, and is composed of a cartilaginous core, covered by a sheath of vasodentine. In cross section, the spine is roughly trigonal in outline with a pronounced anterior keel and a shallow posterior depression, termed the interdentate depression. Lying within this depression is a strip of soft, grayish, sparsely pigmented tissue, the glandular or venom-producing tissue. The spine is covered externally by a thin layer of integument that consists of two layers—an outer epidermis and an inner dermis. The epidermis is avascular, comprised of squamous epithelium, and rests on an acellular basement membrane. Scattered throughout this epithelium are large glandular cells, in addition to mucous cells, which possess secretory activity, as evidenced by the increased vacuolization of the cytoplasmic area, and phantom nuclei.

Chemistry of Toxins: Unknown.

3. Pharmacology of Chimaeroid Venom

Tissue extracts prepared from the glandular tissue of the spines of *Hydrolagus colliei* and injected intraperitoneally into laboratory mice revealed in one series of tests mild agitation, paralysis of the hind limbs, inactivity, erection of the hair, jerking motions of the body, and death. Similar extracts prepared from the skin taken from the dorsum of the same fish failed to produce any symptoms. There is no other information available regarding the pharmacological or chemical nature of the venom (Halstead and Bunker, 1952).

Pathology: Unknown.

4. Clinical Characteristics

There are no reliable data regarding the clinical characteristics of a chimaeroid sting. Evans (1943) states that *Hydrolagus affinis* is reputed to produce a serious wound with the dorsal spine, but does not present any clinical details.

5. Prevention

Care should be taken in removing chimaeroids from nets or hooks to avoid contact with the dorsal spine.

Treatment: See Section IX.

IV. VENOMOUS CATFISHES

The suborder Siluroidei includes a group of fishes having a wide variety of sizes and shapes. Their body shape may vary from short to greatly elongate, or even eellike. The head is extremely variable, sometimes very large, wide or depressed, again very small. The mouth is not protractile, but the lips are sometimes greatly developed, usually with long barbels, generally with at least one pair from rudimentary maxillaries, often one or more pairs about the chin, and sometimes one from each pair of nostrils. The skin of these fishes is thick and slimy, or consists of bony plates. No true scales are ever present. About 1,000 species are included within this group, most of which are found in the freshwater streams of the tropics. A few species are marine.

List of Representative Venomous Catfishes

Galeichthys felis (Linnaeus)—Mexican catfish
    Cape Cod to the Gulf of Mexico
Clarias batrachus (Linnaeus)—catfish
    Netherlands Indies to India, Philippines

Heteropneustes fossilis (Bloch)—Indian catfish
    Vietnam, Ceylon, and India
Plotosus lineatus (Thunberg)—Oriental catfish (Fig. 10)
    Indo-Pacific area

The catfishes listed above may be encountered in marine or brackish waters, and are commonly taken from shallow water around entrances to rivers.

1. Mechanism of Envenomation

Venomous catfishes have a single, sharp, stout spine immediately in front of the soft-rayed portion of the dorsal and pectoral fins. The venom glands are attached to these spines. The spines of some species are also equipped with a series of sharp retrorse teeth, which are capable of severely lacerating the victim's flesh, thus facilitating absorption of the venom and subsequent secondary infection. Wounds are usually contracted as a result of attempting to grasp the fish in the region of the pectoral and dorsal fins.

2. Venom Apparatus

The venom apparatus of catfishes consists of the dorsal and pectoral stings, and the axillary venom glands. The dorsal and pectoral spines are comprised of modified or coalescent soft rays that have become ossified, and so constructed that they can be locked in the extended position at the will of the fish. The mature dorsal spine is a stoutly elongate, compressed, tapered, slightly arched, osseous structure bearing a series of retrorse dentations along the anterior and posterior surfaces, and having an acute sagitate tip. The spine is enveloped by a thin layer of sparsely pigmented skin, the integumentary sheath, which is continuous with that of the soft-rayed portion of the fin. There is no external evidence of a venom gland. The shaft of the pectoral spine is similar to the dorsal spine in its general morphology. Microscopic

Fig. 10. *Plotosus lineatus* (Thunberg). Oriental catfish. (M. Shirao.)

examination reveals that at the level of the middle third, the sting may be divided into three distinct zones; a peripheral integumentary sheath, an intermediate osseous portion, and a central canal.

The integumentary sheath is comprised of a relatively thick outer layer of epidermis and a thin layer of dermis. The glandular cells that comprise the venom gland are most concentrated at the anterolateral and the posterolateral margins of the sting where they are sometimes clumped two or three cells deep within the epidermal layer. The venom glands of most of the catfish species that have been studied appear as a cellular sheet wedged between the pigment layer and the stratified squamous epithelium of the epidermis. The microscopic anatomy of the dorsal and pectoral stings are similar in appearance. The axillary pore, which is the outlet of the axillary gland, is located just below the vertical center of the posthumeral process of the cleithrum. The gland is enclosed within a capsule of fibrous connective tissue, and is divided into three or four lobes, which are further subdivided into a variable number of lobules. The lobules are composed of large secretory cells.

Chemistry of Venom: Unknown.

3. Pharmacology of Catfish Venom

The only catfish venoms that have been studied to any extent are those of *Plotosus lineatus* and *Heteropneustes fossilis*. Subcutaneous, intraperitoneal, and intravenous injections of crude extracts of the venom result in immediate symptoms of muscular spasm, respiratory distress, and death. *Plotosus* venom is said to have both neurotoxic and hemotoxic properties (Toyoshima 1918; Bhimachar, 1944).

Pathology: Unknown.

4. Clinical Characteristics

The severity of the symptoms, incident to an attack, varies with the species of catfish and the amount of venom received. The pain is generally described as an instantaneous stinging, throbbing, or scalding sensation which may be localized or may radiate up the affected limb. Some of the tropical species, such as *Plotosus,* are capable of producing violent pain, which may last for 48 hours or more. The area about the wound becomes ischemic immediately after being stung. The pallor about the wound is soon followed by a cyanotic appearance, and then by redness and swelling. In extreme cases, there may be a massive edema involving the entire limb, accompanied by lymphadeno-pathy, numbness, and gangrene of the area about the wound. Primary shock may be present and manifested by such symptoms as faintness, weakness, nausea, loss of consciousness, cold clammy skin, rapid weak pulse, low blood pressure, and respiratory distress. Improperly treated cases frequently result in secondary bacterial infections of the wound. Wounds inflicted by some species of catfishes may take weeks to heal, but in most instances the wounds are of minor consequence. Deaths have been reported from the stings of some of the tropical catfishes.

Treatment: Symptomatic. See Section IX.

5. Prevention

Catfish stings are generally encountered when attempting to remove the fish from a net or hook. Avoid contact with the dorsal and pectoral spines.

V. VENOMOUS WEEVERFISHES

The weeverfishes of the family Trachinidae are among the more venomous marine fishes of the temperate zone. Weevers are small marine fishes, all of which attain a size of less than 18 inches in total length.

List of Representative Species of Venomous Weeverfishes

Trachinus draco (Linnaeus)—great weever (Fig. 11)
Norway, British Isles, southward to the Mediterranean Sea, and along the coast of North Africa.
Trachinus vipera Cuvier—lesser weever
North Sea, southward along the coast of Europe, and the Mediterranean Sea

Weeverfishes are primarily dwellers on flat, muddy, or sandy bays. They tend to bury themselves in the soft sand or mud, with only the head partially exposed, and to dart out rapidly at their prey. Despite their sedentary habits, weevers can move swiftly and are said to be able to strike an object with great accuracy. When trachinids are at rest, their dorsal fin is lowered. When provoked, the dorsal fin is instantly erected and the opercula expanded. The slightest contact with the body of the fish is said to lead to a quick strike.

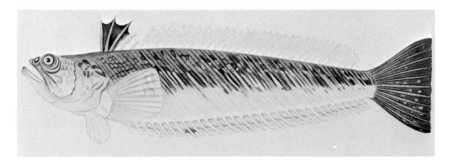

Fig. 11. *Trachinus draco* (Linnaeus). Great weever. (M. Shirao.)

I. Mechanism of Envenomation

Wounds are inflicted by the opercular and dorsal stings.

2. Venom Apparatus

The venom apparatus of the weeverfish consists of the dorsal and opercular stings, and their associated glandular tissue. The dorsal spines of weevers vary from five to seven in number. Each of the spines is enclosed within a thin-walled integumentary sheath from which protrudes a needle-sharp tip. Removal of the integumentary sheaths shows a thin, elongate, fusiform strip of whitish spongy tissue, lying within the distal portion of each of the anterolateral-glandular grooves of the spines. The tissue within these grooves is the glandular epithelium or venom-producing portion of the sting. The term "sting" refers to the spine in combination with the glandular tissue. Removal of the thin, integumentary sheath of the operculum shows a broad, compressed, "daggerlike" opercular spine ending in an acute tip (Fig. 12A, B). Attached to the superior and inferior margins of the spine, one on each margin, are two flattened, pyriform, glistening white, soft spongy masses—the venom-producing tissue. Microscopic examination of the glandular tissue of the dorsal stings reveals masses of large, polygonal, distended, light pink cells of variable size, filled with finely granular secretion. The glandular cells of the operculum are similar in morphology to those of the dorsal stings.

Chemistry of Weever Venom: Unknown.

3. Pharmacology of Weever Venom

The physiological effects of weeverfish venom are only poorly known. In frogs, Pohl (1893) and De Marco (1936b, 1937, 1938) saw paralysis and death result from injection of weever venom. However, Pohl reported no increased reflex excitability, whereas De Marco found that convulsions and hyperactive reflex activity preceded prostration and death. He suggested that the venom may directly excite the sensory-motor cortex and even result in an epileptiform crisis. The venom may cause adverse cardiovascular effects. Pohl (1893) found in frogs a reduced cardiac excitability terminating in diastolic stasis. Atropine, helleborine, caffeine, and hydrastinine were without effect. Evans (1907), on the other hand, reported a biphasic drop in blood pressure accompanied initially by an increased force of contraction and terminating in a marked hypotensive death. Briot (1902) attributed death in rabbits injected with weever venom to respiratory paralysis. It was his opinion that the effect was due primarily to paralysis of the respiratory muscles.

Briot (1902) and Evans (1907) found weeverfish venom to have hemolytic properties. Briot reported that both active and passive immunity to the venom could be produced by sublethal injections. In addition, Evan's work indicates

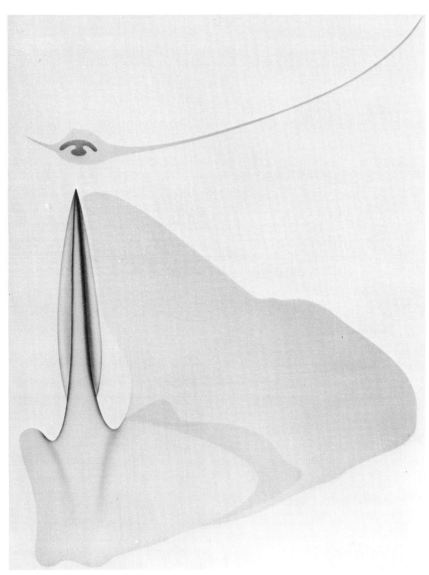

Fig. 12A.

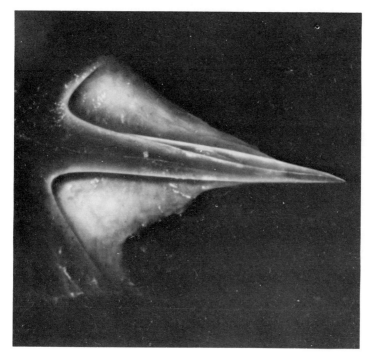

FIG. 12B.

FIG. 12. Opercular venom apparatus of the weeverfish, *Trachinus*. A. Opercular spine of the weever. (From Halstead and Modglin.) B. Opercular venom apparatus of *Trachinus draco*. Opercular skin has been removed to reveal the mass of venomous glandular tissue bordering the opercular spine.

that the toxic properties may be separable from the hemolytic and antigenic factors by filtration.

Pathology: Unknown.

4. Clinical Characteristics

Weever wounds usually produce instant pain, described as a burning, stabbing, or "crushing" sensation, initially confined to the immediate area of the wound, then gradually spreading through the affected limb. The pain progressively increases to reach an excruciating peak, usually within 30 minutes. The severity is such that the victim may scream, thrash wildly, and faint. In most of the cases reported, morphine fails to give relief. Untreated, the pain commonly subsides within 2 to 24 hours. Tingling, followed by numbness, finally develops about the wound. Injection of the venom at first produces an ischemic zone around the wound, followed soon by redness, heat, and swelling. Edema is progressive; within a half hour or more, the entire member is involved. Movement of the affected part is much restricted. The swelling may endure to 10 days or longer. These initial symptoms of pain and inflammation may be associated with, or followed by, headache, fever, chills, delirium, nausea, vomiting, syncope, sweating, cyanosis, joint aches, ankylosis, aphonia, torpidity, cardiac palpitation, bradycardia, pronounced psychic depression, respiratory distress, clonic or tonic convulsions, and death. Secondary infections occur commonly in cases improperly treated. Lymphangitis, lymphadenitis, pyogenesis, necrosis, and sloughing of the tissues around the wound are frequent. Gangrene may develop, necessitating amputation. Primary shock is a common complication to be guarded against. The recovery period is greatly variable depending on the patient, the dose of the venom, species of weever, season of the year, and possibly the locality, but the period varies usually from a few days to several months. Muscular atrophy, peripheral neuritis, and ankylosis have been reported as residual complications.

Treatment: See Section IX.

5. Prevention

Weeverfish stings are most commonly encountered while wading or swimming along sandy coastal areas of the eastern Atlantic or Mediterranean seas. The habit of these fishes to lie partially buried in the sand or mud poses a practical problem to workers in regions inhabited by weeverfish. Persons wading in waters where trachinids abound should wear adequate footwear. Skin divers should attempt to avoid antagonizing these fishes since they are easily provoked into stinging.

VI. STARGAZERS

Stargazers are bottom-dwelling marine fishes, members of the family Uranoscopidae, having a cuboid head, an almost vertical mouth with fringed lips, and eyes on the flat upper surface of the head. Uranoscopids spend a large part of their time buried in the mud or sand with only their eyes and a portion of the mouth protruding. *Uranoscopus scaber* (Linnaeus) (the European stargazer), which is found in the eastern Atlantic and Mediterranean Sea (Fig. 13), is representative of the group. Other species are found in the Indo-Pacific area, Japan, Korea, China, and the Philippines.

1. Mechanism of Envenomation

Intoxication results from contact with the shoulder spines.

2. Venom Apparatus

The only published description of the venom apparatus of *Uranoscopus* appears to be that by Bottard (1889). The venom consists of two shoulder

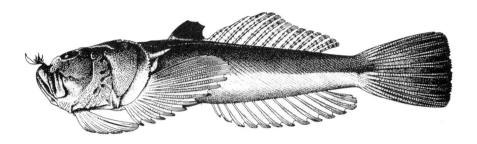

FIG. 13. *Uranoscopus scaber* (Linnaeus). European stargazer. (Courtesy of the Smithsonian Institution.)

spines, bilaterally situated, each of which protrudes through an integumentary sheath. The spine is said to have a double groove through which the venom flows. The venom gland of each spine is described as having a superior and inferior portion. There are no data available regarding the microscopic anatomy of the venom gland, nor the nature of the venom. Stings from *U. scaber* in some instances have been known to be fatal. For treatment, see page 625 on the management of fish stings.

Chemistry of Venom: Unknown.

Pharmacology of Venom: Unknown.

Pathology: Unknown.

Clinical Characteristics: Unknown.

Prevention: Avoid contact with the shoulder spines.

VII. VENOMOUS SCORPIONFISHES

The family Scorpaenidae is an important group of venomous fishes because of their wide distribution throughout all tropical and temperate seas. Some of the members of this group are extremely venomous. Morphologically, the venom organs of scorpionfishes appear to fall into three distinct types as exemplified by the genera (1) *Pterois,* the zebrafish; (2) *Scorpaena,* the scorpionfish proper; and (3) *Synanceja,* the stonefish.

Pterois is among the most beautiful and ornate of coral reef fishes. Zebrafish are generally found in shallow water, hovering about in a crevice or at times swimming unconcernedly in the open. They are also called turkeyfish because of their interesting habit of slowly swimming about, spreading their fanlike pectorals and lacy dorsal fins, like a turkey gobbler displaying its plumes. They are frequently observed swimming in pairs, and apparently are fearless in their movements. Acceptance of the invitation to reach out and grab one of these fish results in an extremely painful experience, because, hidden beneath the "lace," are the needle-sharp fin stings. The fearlessness of the zebrafish makes it a particular menace to anyone working in the shallow-water coral reef areas it inhabits.

Members of the genus *Scorpaena* are for the most part shallow-water bottom-dwellers, found in bays, along sandy beaches, rocky coastlines, or coral reefs, from the intertidal zone to depths of 50 fathoms or more. Their habit of concealing themselves in crevices, among debris, under rocks, or in seaweed, together with their protective coloration, which blends them almost perfectly into their surrounding environment, makes them difficult to see. When they are removed from the water, they have the defensive habit of erecting their spinous dorsal fin and flaring out their armed gill covers, pectoral, pelvic, and anal fins. The pectoral fins, although dangerous in appearance, are unarmed.

Stonefishes are largely shallow-water dwellers, commonly found in tide pools and shoal reef areas. *Synanceja* has the habit of lying motionless in coral crevices, under rocks, in holes, or buried in sand and mud. They appear to be fearless and completely disinterested in the careless intruder, swimming sluggishly about at infrequent intervals. Because of a thick coating of slime, irregular wartylike texture of the skin, and the habit of burying itself in the sand, stonefishes frequently become coated with bits of coral debris, mud, and algae. The phenomenal ability of the stonefish to camouflage itself makes it a real menace to bathers and those wading about in areas inhabited by this fish.

List of Representative Venomous Scorpionfishes

Scorpionfishes having a venom apparatus, as exemplified by *Pterois*
  Dendrochirus zebra (Quoy and Gaimard)—lionfish, zebrafish, etc.
    Tropical Indo-Pacific, from East Africa to the East Indies, Philippines, Micronesia, Polynesia
  Pterois volitans (Linnaeus) (Fig. 14)—zebrafish, lionfish, turkeyfish, etc.
    Red Sea, Indian Ocean, China, Japan, Australia, Melanesia, Micronesia, Polynesia
Scorpionfishes having a venom apparatus, as exemplified by *Scorpaena*
  Apistus carinatus (Bloch and Schneider)—bullrout, sulky, waspfish
    Coasts of India, East Indies, Philippines, China, Japan, Australia
  Centropogon australis (White)—waspfish, fortescue
    New South Wales and Queensland, Australia

FIG. 14. *Pterois volitans* (Linnaeus). Zebrafish. (W. Braun, courtesy of Skin Diver Magazine.)

Notesthes robusta (Günther) (Fig. 15)—bullrout
  New South Wales and Queensland, Australia
Scorpaena guttata (Girard)—scorpionfish, sculpin
  Central California south into the Gulf of California
Scorpaena plumieri Bloch—scorpionfish, sculpin
  Atlantic coast from Massachusetts to the West Indies and Brazil
Scorpaena porcus Linnaeus—scorpionfish, rascasse, sea pig, etc.
  Atlantic coast of Europe from the English Channel to the Canary Islands,
  French Morocco, Mediterranean Sea, Black Sea
Scorpaena scrofa Linnaeus—scorpionfish, rascasse, sea pig, etc.
  West coast of France south to Cabo Blanco, northwest Africa, and
  Mediterranean Sea
Scorpaenopsis diabolus (Cuvier) (Fig. 16)—scorpionfish
  East Indies, Australia, Melanesia, Polynesia
Snyderina yamanokami Jordan and Starks—yama-hime
  Japan

FIG. 15. *Notesthes robusta* (Günther). Bullrout. (After Bleeker.)

FIG. 16. *Scorpaenopsis diabolus* (Cuvier). Scorpionfish. (From Hiyama.)

Scorpionfishes having a venom apparatus, as exemplified by *Synanceja*
   Choridactylus multibarbis Richardson
     India, China, Philippines, Polynesia
   Inimicus japonicus (Cuvier) (Fig. 17)
     Japan
   Minous monodactylus (Bloch and Schneider)—hime-okoze
     South Pacific, China, Japan
   Synanceja horrida (Linnaeus) (Fig. 18)—stonefish
     India, East Indies, China, Philippines, Australia

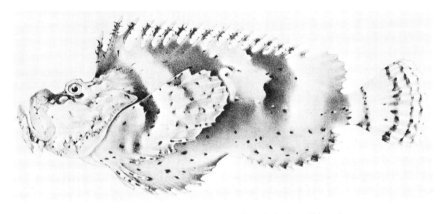

FIG. 17. *Inimicus japonicus* (Cuvier). (M. Shirao.)

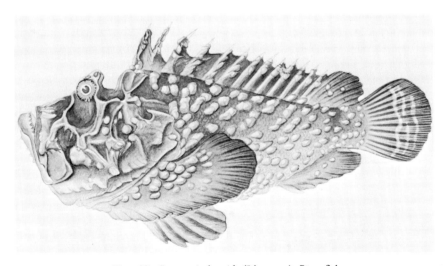

FIG. 18. *Synanceja horrida* (Linnaeus). Stonefish.

1. Mechanism of Envenomation

Wounds are inflicted by scorpionfishes with the use of their dorsal, anal, and pelvic stings. The opercular spines of some scorpaenids are also thought to be venomous and should be handled with caution.

2. Venom Apparatus

The venom organs of scorpionfishes vary markedly from one genus to the next. On the basis of their gross morphology, their venom organs can be divided into three groups (Table I) (Fig. 19). The venom apparatus of *Pterois volitans,* the zebrafish; *Scorpaena guttata,* the scorpionfish; and *Synanceja verrucosa,* the stonefish, are representative of these three groups.

TABLE I

A COMPARISON OF THE VENOM ORGANS OF *Pterois, Scorpaena,* AND *Synanceja*[a]

Structure	*Pterois*	*Scorpaena*	*Synanceja*
Fin spines	Elongate, slender	Moderately long	Short, stout
Integumentary sheath	Thin	Moderately thick	Very thick
Venom glands	Small-sized, well developed	Moderate-sized, well developed	Very large, highly developed
Venom duct	Not evident, or poorly developed	Not evident, or poorly developed	Well developed

[a] From Halstead (1959). See also Fig. 19.

The venom apparatus of *Pterois volitans,* the zebrafish, includes 13 dorsal spines, 3 anal spines, 2 pelvic spines, their associated venom glands, and their thin, enveloping, integumentary sheaths. The spines are for the most part long, straight, and slender. Located on the anterior aspect of the spine, one on either side, are the anterolateral-glandular grooves which appear as deep channels extending the entire length of the shaft. Situated within each of the glandular grooves is a slender, elongate, fusiform strand of gray or pinkish tissue—the venom glands. Microscopic examination of cross sections of the venom glands shows that located in the dermal layer within the anterolateral-glandular grooves is a cluster of large polygonal glandular cells with pinkish-gray, finely granular cytoplasm. The large venom-producing cells have a pinnate, heart-shaped arrangement, and vary greatly in size and morphology.

The venom apparatus of *Scorpaena guttata,* the scorpionfish, includes 12 dorsal spines, 3 anal spines, 2 pelvic spines, their associated venom glands, and their enveloping integumentary sheaths. If the integumentary sheath is removed, a slender, elongate, fusiform strand of gray or pinkish tissue can be observed lying within the glandular grooves on either side of the spine. The

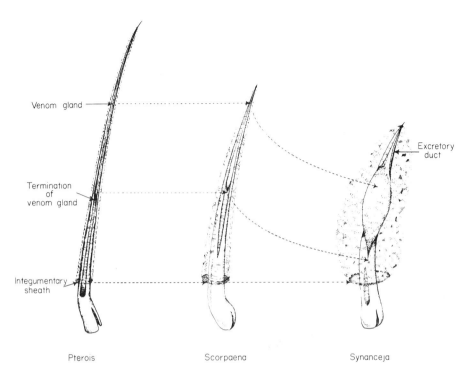

Venom gland

Termination of venom gland

Integumentary sheath

Excretory duct

Pterois   Scorpaena   Synanceja

FIG. 19. A comparison of the three types of scorpionfish dorsal stings. Diagrammatic. (From Halstead.)

spines are moderately long and heavy in comparison with those of *P. volitans.* The microscopic anatomy of the venom glands is similar to that of *P. volitans.*

The venom apparatus of *Synanceja horrida,* the stonefish, includes 13 dorsal spines, 3 anal spines, 2 pelvic spines, their associated venom glands, and their thick, warty, enveloping, integumentary sheath (Fig. 20). The fin spines are relatively short, stout, and straight. Extending throughout the entire length of the spine, one on either side, are two anterolateral-glandular grooves. The venom glands appear as two large masses, bilaterally situated, attached by tough strands of connective tissue to the middle or distal thirds of the spine. The distal ends of the glands terminate in ductlike structures lying within the glandular grooves. These ducts extend to the distal tips of the spine.

Chemistry of Scorpionfish Venom: Unknown.

3. Pharmacology of Scorpionfish Venom

Pohl (1893) was the first to work with the venom of a scorpionfish, *Scorpaena porcus*. He believed that the venom acted as a cardioinhibitor when

FIG. 20. Integumentary sheath removed to show one of the dorsal stings of the stone-fish, *Synanceja horrida*. Note the large venom glands, one on either side of the spine, and the ducts extending from the superior portion of the glands to the tip of the spine.

tested on frogs. Tests of *S. porcus* venom on a variety of laboratory animals have produced motor paralysis, respiratory distress, convulsions, and death (Dunbar-Brunton, 1896; Briot, 1904, 1905; Phisalix, 1931; Lumière and Meyer, 1938). *Synanceja horrida* venom is believed to have both neurotoxic and hemotoxic properties. The venom produces hemolysis of guinea pig, sheep, and human red blood cells (Bottard, 1889; Duhig and Jones, 1928; Duhig, 1929; Endean, 1961).

Pathology: Unknown.

4. Clinical Characteristics

The symptoms produced by the various species of scorpionfishes is essentially the same, varying in degree rather than quality. The pain is usually described as immediate, intense, sharp, shooting, or throbbing, and radiates from the affected part. The area about the wound becomes ischemic and then cyanotic. The pain produced by most scorpionfishes generally endures for only a few hours, but wounds produced by *Synanceja* may be extremely painful and continue for a number of days. Pain caused by *Synanceja* is sometimes so severe as to cause the victim to thrash about wildly, scream, and finally lose consciousness. The area in the immediate vicinity of the wound gradually becomes cyanotic, surrounded by a zone of redness, swelling, and heat. Subsequent sloughing of the tissues about the wound site may occur. In the case of *Synanceja* stings, the wound becomes numb, and the skin some distance from the site of injury becomes painful to touch. In some instances complete paralysis of the limb may ensue. Swelling of the entire member that is affected may take place, frequently to such an extent that movement of the part is impaired. Cardiac failure, delirium, convulsions, various nervous disturbances, nausea, vomiting, lymphadenitis, lymphangitis, joint aches, fever, respiratory distress and convulsions may be present, and death may occur. General health and vitality may be adversely affected. Complete recovery from a severe *Synanceja* sting may require many months.

Treatment: See Section IX.

5. Prevention

Zebrafish stings, *Pterois,* are usually contracted by persons who are attracted by their slow movements and lacy-appearing fins, and who attempt to pick up the fish with their hands. Most scorpionfish stings result from the individual removing them from the hook or nets and being jabbed by their venomous spines. Stonefish are especially dangerous because of the difficulty of detecting them from their surroundings. Placing one's hands in crevices or in holes inhabited by these fishes should be done with caution.

VIII. VENOMOUS TOADFISHES

The Batrachoididae, or toadfishes, are small, bottom fishes that inhabit the warmer waters of the coasts of America, Europe, Africa, and India.

Toadfishes are of little commercial value, and are not generally considered suitable for food, although they are eaten in some countries. Batrachoid fishes, with their broad, depressed heads and large mouths, are somewhat repulsive in appearance. Most toadfishes are marine, but some are estuarine or entirely fluviatile, ascending rivers for great distances. They hide in crevices and in burrows, under rocks and debris, among seaweed, or lie almost completely buried under a few centimeters of sand or mud. Toadfishes tend to migrate to deeper water during the winter months where they remain in a torpid condition. They are experts at camouflage. Their ability to change their color to lighter or darker shades at will, and their mottled patterns, make these fishes difficult to see. Most toadfishes tend to be sluggish in their movements, but when after food they can dart out with surprising rapidity. Toadfishes are said to be quite vicious and will snap at almost anything upon the slightest provocation. When they are disturbed or touched, they immediately erect their dorsal spines and flare out their opercular spines in defiance.

List of Representative Venomous Toadfishes

Barchatus cirrhosus (Klunzinger) (Fig. 21)—toadfish
　Red Sea
Batrachoides didactylus (Bloch)—toadfish
　Mediterranean Sea and nearby Atlantic coasts
Batrachoides grunniens (Linnaeus)—toadfish, munda
　Coast of Ceylon, India, Burma, and Malaya

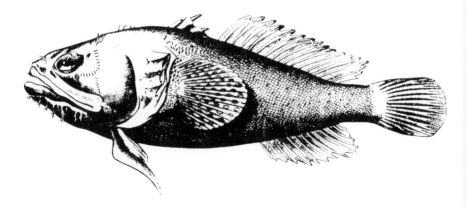

FIG. 21. *Barchatus cirrhosus* (Klunzinger). Red Sea toadfish. (From Klunzinger.)

Opsanus tau (Linnaeus)—toadfish, oysterfish
    Atlantic coast of United States, Massachusetts to West Indies
Thalassophryne dowi (Jordan and Gilbert)—toadfish, bagre sapo, sapo
    Pacific coast of Central America, Costa Rica to Panama
Thalassophryne reticulata Gunther—toadfish, bagre sapo, sapo
    Pacific coast of Central America

1. Mechanism of Envenomation

Wounds are generally contracted as a result of stepping on toadfishes which are partially buried in sand or mud. In some instances, wounds have been received from the opercular spines because of careless handling.

2. Venom Apparatus

The venom apparatus of toadfishes consists of two dorsal fin spines, two opercular spines, and their associated venom glands (Fig. 22). In the case of *Thalassophryne dowi,* which can be considered as typical of the group, there are two dorsal spines, which are enclosed together within a single integumentary sheath. The dorsal spines are slender and hollow, slightly curved, and terminate in acute tips. At the base and tip of each spine is an opening through which the venom passes. The base of each dorsal spine is surrounded by a glandular mass from which the venom is produced. Each gland empties into the base of its respective spine. The operculum is also highly specialized as a defensive organ for the introduction of venom. The bone is angular in shape. The horizontal limb of the operculum is a slender hollow bone which curves slightly, and terminates in an acute tip. Openings are present at each end of the spine for the passage of venom. With the exception of the extreme distal tip, the entire opercular spine is encased within a glistening, whitish, pyriform mass. The broad, rounded portion of this mass is situated at the base of the spine, and tapers rapidly as the tip of the spine is approached. The pyriform mass consists of a tough, saclike outer covering of connective tissue in which is contained a soft, granular, gelatinelike substance having the appearance of fine tapioca. This mass is the venom gland. The gland empties into the base of the hollow opercular spine, which serves as a duct. Microscopic examination of the venom glands shows strands of areolar connective tissue, large distended polygonal cells filled with finely granular secretion, and vascular channels. In some instances the polygonal cells will appear to have undergone complete lysis and there remain only areas of amorphous secretion. The microscopic anatomy of the dorsal and opercular venom glands is essentially the same.

Chemistry of Toadfish Venoms: Unknown.

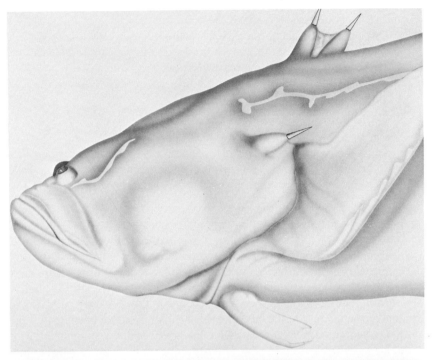

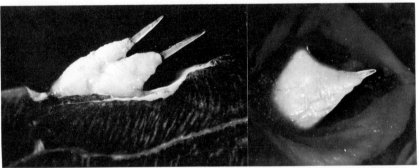

FIG. 22. *Top:* Head of toadfish, *Thalassophryne,* showing the location of the dorsal and opercular stings. *Bottom:* Dorsal and opercular stings of *Thalassophryne dowi.* Integumentary sheath has been removed to reveal venom glands. (R. Kreuzinger.)

3. Pharmacology of Toadfish Venom

The only published reports on toadfish venom are those by Froes (1933a,b) on *Thalassophryne.* Injections of the venom into guinea pigs and chicks resulted in mydriasis, ascites, paralyses, necrosis about the injection site, convulsions, and death. The author concluded that the venom of *Thalassophryne* has both proteolytic and neurotoxic properties.

Pathology: Unknown.

4. Clinical Characteristics

The pain from toadfish wounds develops rapidly, is radiating and intense. Some have described the pain as being similar to that of a scorpion sting. The pain is soon followed by swelling, redness, and heat. No fatalities have been recorded in the literature. Little else is known about the effects of toadfish venom.

Treatment: See Section IX.

5. Prevention

Persons wading in waters inhabited by toadfishes should take the precaution to shuffle their feet through the mud in order to avoid stepping on them. Removal of toadfishes from a hook or nets should be done with care.

IX. TREATMENT OF STINGS FROM VENOMOUS FISHES

Efforts in treating venomous fish stings should be directed toward achieving three objectives: (1) alleviating pain, (2) combating effects of the venom, (3) preventing secondary infection. The pain results from the effects of the trauma produced by the fish spine, venom, and the introduction of slime and other irritating foreign substances into the wound. In the case of stingray and catfish stings, the retrorse barbs of the spine may produce severe lacerations with considerable trauma to the soft tissues. Wounds of this type should be promptly irrigated with cold salt water or sterile saline, if such is available. Fish stings of the puncture-wound variety are usually small in size, and removal of the poison is more difficult. It may be necessary to make a small incision across the wound, and then apply immediate suction, and possibly irrigation. At any rate, the wound should be sucked promptly, in order to remove as much of the venom as possible. However, it should be kept in mind that fishes do not inject their venom in the manner employed by venomous snakes, so at best, results from suction will not be too satisfactory. There is a division of opinion as to the advisability and efficacy of using a ligature in the treatment of fish stings. If used, the ligature should be placed at once between the site of the sting and the body, but as near the wound as possible. The ligature should be released every few minutes in order to maintain adequate circulation. Most workers recommend soaking the injured member in hot water for 30 minutes to 1 hour. The water should be maintained at as high a temperature as the patient can tolerate without injury, and the treatment should be instituted as soon as possible. If the wound is on the face or body, hot moist compresses should be employed. The heat may have an attenuating effect on the venom, since boiling readily destroys stingray venom *in vitro*. The addition of magnesium sulfate or Epsom salts to the water is believed to be useful (Russell and Lewis, 1956; Halstead, 1959).

ACKNOWLEDGMENT

The author is indebted to the Cornell Maritime Press for the use of some of the illustrations presented in this chapter.

REFERENCES

Bhimachar, B. S. (1944). *Proc. Indian Acad. Sci.* **19**, 65–70.
Bottard, L. A. (1889). "Les poissons venimeux." Octave Doin, Paris.
Broit, A. (1902). *Compt. Rend. Séanc. Soc. Biol.* **54**, 1169–1171.
Broit, A. (1904). *Compt. Rend. Séanc. Soc. Biol.* **56**, 1113–1114.
Briot, A. (1905). *Compt. Rend. Assoc. Franc. Adv. Sci*, **33**, 904.
De Marco, R. (1936a). *Riv. Patol.* **47**, 204–208.
De Marco, R. (1936b). *Boll. Soc. Ital. Biol. Sper.* **11**, 767–768.
De Marco, R. (1937). *Arch. Fisiol.* **37**, 398–404.
De Marco, R. (1938). *Riv. Biol.* **25**, 225–234.
Duhig, J. V. (1929). *Z. Immunforsch. Exp. Ther.* **62**, 185–189.
Duhig, J. V., and Jones, G. (1928). *Mem. Queensland Mus.* **9** (2), 136–150.
Dunbar-Brunton, J. (1896). *Lancet* **2**, (3809), 600–602.
Endean, R. (1961). *Australian J. Marine Freshwater Res.* **12**, 177–190.
Evans, H. M. (1907). *Brit. Med. J.* **1**, 73–76.
Evans, H. M. (1921). *Brit. Med. J.* **2**, (3174), 690–692.
Evans, H. M. (1923). *Philos. Trans. Roy. Soc. London* **212B**, 1–33.
Evans, H. M. (1943). "Sting-fish and Seafarer." Faber & Faber, London.
Froes, H. P. (1933a). *Bahia Med.* **4**, 69–71.
Froes, H. P. (1933b). *J. Trop. Med. Hyg.* **36**, 134–135.
Halstead, B. W. (1959). "Dangerous Marine Animals." Cornell Maritime Press, Maryland.
Halstead, B. W. (1967). *Mem. Inst. Butantan, Simp. Internac.* **1**, 1–26.
Halstead, B. W., and Bunker, N. C. (1952). *Copeia No.* 3, 128–138.
Halstead, B. W., and Bunker, N. C. (1953). *Am. J. Trop. Med. Hyg.* **2**, 115–128.
Lagler, K. F., Bardach, J. E., and Miller, R. R. (1962). "Ichthyology." Wiley, New York.
Lumière, A., and Meyer, P. (1938). *Compt. Rend. Soc. Biol.* **127**, 328–330.
Phisalix, M. (1931). *Notes Stn. Océanogr, Salammbô, Tunis* **22**, 3–15.
Pohl, J. (1893). *Prager Med. Wochschr.* **18**, 31–33.
Russell, F. E. (1953a). *Eng. Sci.* **17**, 15–18.
Russell, F. E. (1953b). *Am. J. Med. Sci.* **226**, 611–622.
Russell, F. E. (1956). Unpublished data.
Russell, F. E., and Lewis, R. D. (1956). *Venoms. Publ. Am. Assoc. Adv. Sci. No.* **44**, 43–53.
Russell, F. E., and Van Harreveld, A. (1954). *Arch. Intern. Physiol.* **62**, 322–333.
Russell, F. E., Panos, T. C., Kang, L. W., Warner, A. M., and Colket, T. C. (1958). *Am. J. Med. Sci.* **235**, 566–584.
Toyoshima, T, (1918). *J. Japan. Prot. Soc.* **6**, 45–270

Author Index

Subject Index